2nd International Conference on Oleo and Petrochemical Engineering (ICOOPChE 2017)

IOP Conference Series: Materials Science and Engineering
Volume 345

Pekanbaru, Indonesia
29 - 30 November 2017

Editor:

Evelyn

ISBN: 978-1-5108-6161-9
ISSN: 1757-8981

Printed from e-media with permission by:

Curran Associates, Inc.
57 Morehouse Lane
Red Hook, NY 12571

Some format issues inherent in the e-media version may also appear in this print version.

Printed by Curran Associates, Inc. (2018)

For permission requests, please contact the Institute of Physics
at the address below.

Institute of Physics
Dirac House, Temple Back
Bristol BS1 6BE UK

Phone: 44 1 17 929 7481
Fax: 44 1 17 920 0979

techtracking@iop.org

Additional copies of this publication are available from:

Curran Associates, Inc.
57 Morehouse Lane
Red Hook, NY 12571 USA
Phone: 845-758-0400
Fax: 845-758-2633
Email: curran@proceedings.com
Web: www.proceedings.com

Table of contents

Volume 345

2nd International Conference on Oleo and Petrochemical Engineering (ICOOPChE 2017)

29–30 November 2017, Pekanbaru-Riau, Indonesia

Accepted papers received: 23 March 2018
Published online: 12 April 2018

Preface

Papers

IOP Conf. Series: Materials Science and Engineering **345** (2018) 011001 doi:10.1088/1757-899X/345/1/011001

Proceeding of the 2nd International Conference on Oleo and Petrochemical Engineering (ICOOPChE 2017)

Preface

These Conference Proceedings contain the papers accepted for presentation at the 2nd International Conference on Oleo and Petrochemical Engineering (ICOOPChE) Universitas Riau, which was held on the 29-30 November 2017 in Pekanbaru-Riau Indonesia. The conference was organized by LPPM and Department of Chemical Engineering Universitas Riau, in accordance to its mission to create creative, innovative, and competitive academics through research and devotion to community, based on science and technology 2035.

From historical point of view, Chemical Engineering has successfully played a key role in the development of traditional chemical and petrochemical industries. This conference aimed at disseminating, exchanging, and accumulating the knowledge on the field to coordinate the co-operation between researchers from academic and industrial parts in ASEAN and beyond. Active discussions hence strongly welcomed for our sustainable future, as highlighted in the conference's theme "Chemical Engineering for Sustainable Development".

The 2017 edition presents four outstanding keynotes that are a promise of an unforgettable conference. In addition, there were 49 presenters presenting their papers which have passed through peer-review process. The authors came from various countries, such as Australia, New Zealand, Thailand, Malaysia and Indonesia. Oleo and petrochemical process technology, bioenergy technology, food and bioprocess, renewable energy, chemical technology and chemical industry, chemical reaction engineering and catalysis, polymer technology, materials and membranes, and sustainable chemical processes are some of the subjects presented during the conference.

We wish to thank all participants of the Conference for their fruitful work and personal contribution to the development of these Conference Proceedings.

Editor in Chief

Evelyn, PhD.

IOP Conf. Series: Materials Science and Engineering **345** (2018) 011002 doi:10.1088/1757-899X/345/1/011002

The 2nd International Conference on Oleo and Petrochemical Engineering (ICOOPChE 2017)

Advisory Board

Prof. Dr. Ir. Aras Mulyadi, DEA.	(Rector of Universitas Riau)
Prof. Dr. Almasdi Syahza, SE. MP.	(Director of LPPM Universitas Riau)
Prof. Dr. Syaiful Bahri, Msi.	(Secretary of LPPM Universitas Riau)
Dr. Ir. Ari Sandyavitri, MSc.	(Dean of Faculty of Engineering Universitas Riau)
Ahmad Fadli, ST. MT. PhD.	(Head of Department of Chemical Engineering Universitas Riau)

Keynote Speaker

Prof. Shaobin Wang	(Department of Chemical Engineering Curtin University Australia)
Prof. Mohd. Roslee Othman	(Department of Chemical Engineering Universiti Sains Malaysia, Malaysia)
Prof. Misri Gozan	(Department of Chemical Engineering Universitas Indonesia)
Prof. Titania Nugroho	(Department of Chemistry Universitas Riau)

Conference Chair

Prof. Edy Saputra, ST. MT. PhD.	(Department of Chemical Engineering Universitas Riau)

Main Committee

Prof. Zuchra Helwani, ST. MT. PhD.	(Department of Chemical Engineering Universitas Riau)
Evelyn, ST. MSc. MEng. PhD.	(Department of Chemical Engineering Universitas Riau)
Komalasari, ST. MT.	(Department of Chemical Engineering Universitas Riau)
Syelvia Putri Utami, ST. MEng.	(Department of Chemical Engineering Universitas Riau)

Reviewer

Prof. Shaobin Wang	(Department of Chemical Engineering Curtin University Australia)
Prof. Mohd. Roslee Othman	(Department of Chemical Engineering Universiti Sains Malaysia, Malaysia)
Prof. Misri Gozan	(Department of Chemical Engineering Universitas Indonesia)
Assoc. Prof. Yogi Budhi, ST. MT.	(Department of Chemical Engineering Bandung Institute of Technology)
Dr. Nasrul Arahman, ST. MT.	(Department of Chemical Engineering Syiah Kuala University)
Juwari, ST. MEng. PhD.	(Department of Chemical Engineering Sepuluh Nopember Institute of Technology)
Prof. Arief Budiman, MS. Deng.	(Department of Chemical Engineering Gadjah Mada University)
Dr. Tuty Emilia Agustina, ST. MT.	(Department of Chemical Engineering Sriwijaya University)

IOP Conf. Series: Materials Science and Engineering **345** (2018) 011002 doi:10.1088/1757-899X/345/1/011002

Prof. Dr. Syaiful Bahri, MSi.	(Department of Chemical Engineering Universitas Riau)
Prof. Edy Saputra, ST. MT. PhD.	(Department of Chemical Engineering Universitas Riau)
Prof. Amun Amri, ST. MT. PhD.	(Department of Chemical Engineering Universitas Riau)
Evelyn, ST. MSc. MEng. PhD.	(Department of Chemical Engineering Universitas Riau)
Dr. Yelmida, MSi.	(Department of Chemical Engineering Universitas Riau)
Zulfansyah, ST. MT.	(Department of Chemical Engineering Universitas Riau)
Panca Setia Utama, ST. MT.	(Department of Chemical Engineering Universitas Riau)
Muhammad Iwan Fermi, ST. MT.	(Department of Chemical Engineering Universitas Riau)

ICOOPCHE 2017

LPPM UR

Meneliti, Mengabdi, Berkarya

UNIVERSITAS RIAU

Peer review statement

All papers published in this volume of *IOP Conference Series: Materials Science and Engineering* have been peer reviewed through processes administered by the proceedings Editors. Reviews were conducted by expert referees to the professional and scientific standards expected of a proceedings journal published by IOP Publishing.

IOP Conf. Series: Materials Science and Engineering **345** (2018) 012001 doi:10.1088/1757-899X/345/1/012001

Reactivity improvement of $Ca(OH)_2$ sorbent using diatomaceous earth (DE) from Aceh Province

M Mariana[1,2], M Mahidin[1,2], F Mulana[1,4], T Agam[3] and F Hafdiansyah[3]
[1] Department of Chemical Engineering, Syiah Kuala University, Banda Aceh – Darussalam 23111, Indonesia
[2] Environmental and Natural Resource Research Center, Syiah Kuala University, Banda Aceh – Darussalam 23111, Indonesia
[3] Undergraduate Program of Chemical Engineering, Syiah Kuala University, Banda Aceh – Darussalam 23111, Indonesia
[4] Faculty of Chemical Engineering, University Teknologi MARA (UiTM), Shah Alam 40450, Malaysia

E-mail: mariana_hasyim@yahoo.com

Abstract. In this study, the diatomaceous earth (DE) from Aceh Province was used to increase the reactivity of $Ca(OH)_2$sorbent. The high silica (SiO_2) content of about 97% in the diatomaceous earth allows the increasing reactivity of $Ca(OH)_2$sorbent by forming calcium silicate hydrate (CSH). The CSH improved the porosity characteristic of the sorbent. The improvement process was performed by mixing $Ca(OH)_2$sorbent, diatomaceous earth and water in a beaker glass at the $Ca(OH)_2$/DE weight ratio of 1:10 for 2 hand then dried at 120 °C for 24 h. The dried sorbent was calcined at 500 °C and 800 °C for 2 h. The activated sorbent was characterized using Scanning Electron Microscopy (SEM) for the morphological properties; X-Ray Diffraction (XRD) for the materials characteristics. The adsorption capacity of thesorbent was tested by methylene blue adsorption. The results showed that the $Ca(OH)_2$/DEsorbent had a higher porosity than the $Ca(OH)_2$ adsorbent.The results also showed that $Ca(OH)_2$/DE which was calcined at higher temperature of 800 °C had a higher adsorption capacity compared to $Ca(OH)_2$/DE which was calcined at lower temperature of 500 °C.

1. Introduction

There are various SO_2 removal process technologies in the gas purification technology and the most commonly used process is dry process and wet process (Iswandi, 1996). The use of dry process has several advantages compared to wet process, among others such as cheap, easy handling and non producing liquid waste. One of the dry method that has an effective, simple, handy and inexpensive SO_2 separation processes is to use bag filter reactor by using $Ca(OH)_2$sorbent. However, this process also has several weaknesses among others such as the low conversion of sorbent and the low efficiency of gas separation [1].

To improve the efficiency of gas separation and increase the reactivity of sorbents (conversion of calcium), various additives can be added to the sorbent. Sorbents obtained by mixing calcium hydroxide with a silica-containing substance can increase the conversion of calcium compared to the calcium conversion obtained without additive mixture [2]. Diatomaceous earth (DE) is a pozzolonic material containing SiO_2, Al_2O_3 and CaO [3].

2. Method

Experiment was conducted as follows: (1) Equipment of Material; and (2) Sorbent Preparation.

2.1. Equipment and material

Equipment used in this study were mixer, water bath, hot plate, magnetic stirrer, oven, erlenmeyer, balance, ball mill, sieve shaker, thermometer, SEM and XRD. Further, materials used in this study were $Ca(OH)_2$, diatomaceous earth (DE), methylene blue and aquadest.

2.2. Sorbent preparation

Sorbent was prepared by mixing $Ca(OH)_2$ and DE together with 100 ml of water in a beaker glass with ratio of $Ca(OH)_2$/DE was 1:10. The pH and density values of the sorbent were previously also analysed. The beaker was then closed and inserted into the water bath, then stirred at various temperatures for 2 hours. After stirring, the pH value and the density of slurry was measured again and then dried at temperature of 120 °C for 24 h. The dried slurry was analysed the SiO_2content that bound in $Ca(OH)_2$ sorbent. The solids were then calcined at 500 °C and 800°C for 2 hours. The characterization of the samples was performed using SEM and XRD.

3. Result and discussion

3.1. Chemical content

To see the component of $CaO.SiO_2.2H_2O$ was formed or not as a result of the reaction, it can be seen from Analysis of X-ray Diffraction Patterns. XRD analysis was conducted at Indonesian Institute of Sciences, Center for Physical Research of PUSPIPTEK Serpong Complex. Figure 1 to Figure 3 show the results of XRD analysis for $Ca(OH)_2$/DE = 10: 1 sorbents with particle sizes of 200 mesh in various reaction temperature. XRD analysis showed that the reaction between $Ca(OH)_2$ and SiO_2 produced CSH ($CaO.SiO_2.2H_2O$). This is in accordance with the results of research that has been done by Mariana, et al [4,5].

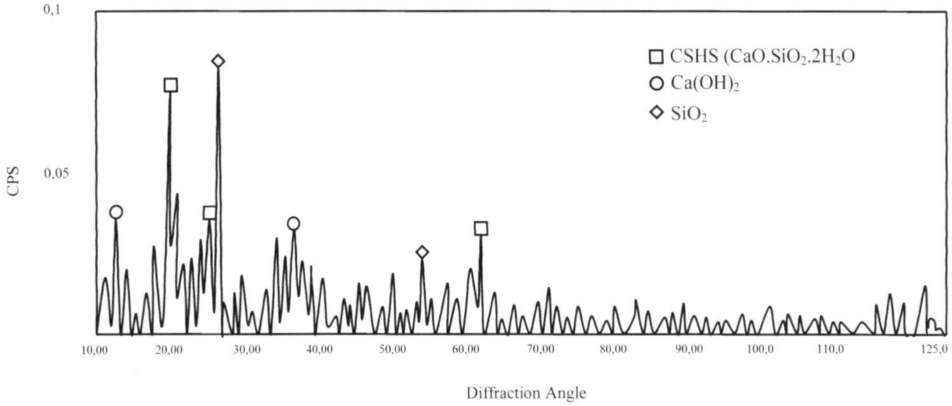

Figure 1. XRD profile of $Ca(OH)_2$/DE = 10: 1 sorbent with a particle size of 200 mesh at a reaction temperature of 65 °C

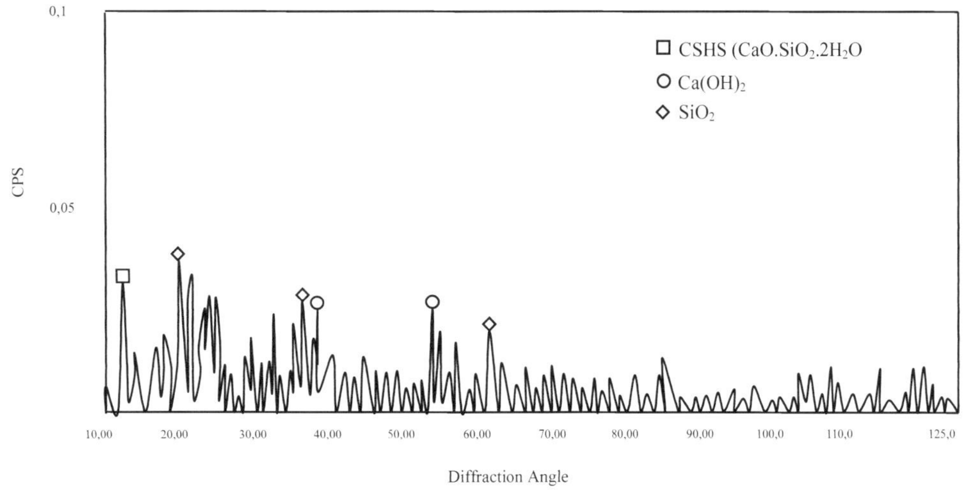

Figure 2. XRD profile of $Ca(OH)_2$/DE = 10: 1 sorbent with a particle size of 200 mesh at a reaction temperature of 70 °C

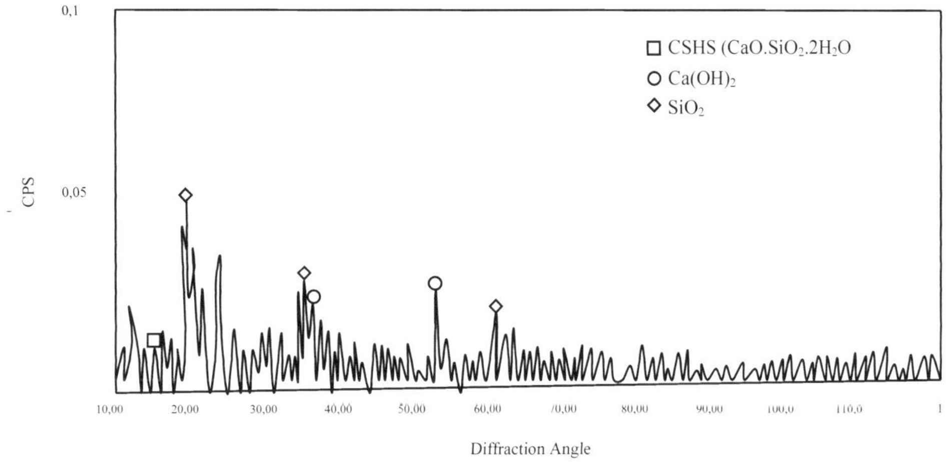

Figure 3. XRD profile of $Ca(OH)_2$/DE = 10: 1 sorbent with a particle size of 200 mesh at a reaction temperature of 75 °C

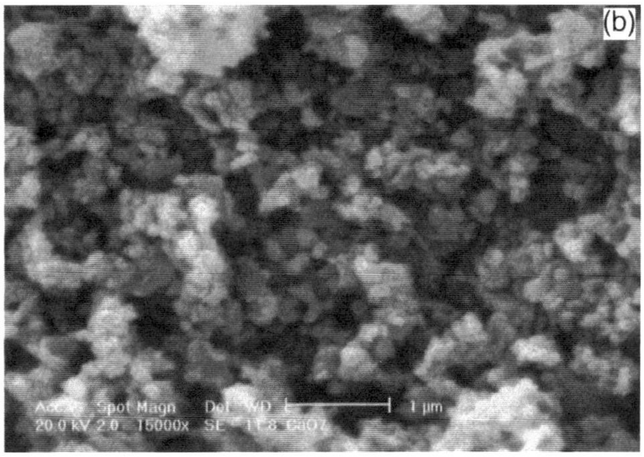

Figure 4. SEM profile of $Ca(OH)_2$.

Figure 5. SEM profile of $Ca(OH)_2$/DE = 10: 1 sorbent before reaction on a particle size of 200 mesh.

Figure 6. SEM profile of $Ca(OH)_2$/DE = 10: 1 sorbent after reaction on a particle size of 200 mesh

3.2 Morphology

Figure 4 to figure 6 showed some differences between sorbent of $Ca(OH)_2$:DE before and after the reaction where the apparent difference is that the pore of particle of sorbent after the reaction appears to be more open than the one before the reaction. As a result of this open pore would affect on adsorption capacity where adsorption capacity would be increased. The porosity of $Ca(OH)_2$ sorbent is smaller than the Ca(OH)2:DE sorbent as shown in Figure 4.

3.3 Sorbent activity test

In this study, the sorbent activity was measured by conducting adsorption process of methylene blue on the sorbent. The results of adsorption test were shown in Table 1 and Table 2. Table 1 and Table 2 show that the adsorption of methylene blue on $Ca(OH)_2$/DE sorbent increases with increasing of adsorption time. Table 1 and table 2 also show that the calcination temperature of the sorbent affects the sorbent adsorption capacity.

Table 1. Adsorption capacity of methylene blue solution on activated $Ca(OH)_2$/DE sorbent with particle size of 200 mesh and calcined at temperature of 500 °C

Adsorption time (hr)	Absorbance	Concentration of methylene blue after adsorption (ppm)	Adsorption capacity (mg Met. Blue/g DE)
1	0.219	1.522	7.454
3	0.135	1.072	7.467
5	0.0876	0.809	7.475
7	0.0424	0.563	7.483
9	0.0107	0.3908	7.488

Table 2. Adsorption capacity of methylene blue solution on activated $Ca(OH)_2$/DE sorbent with particle size of 200 mesh and calcined at temperature of 800 °C

Adsorption time (hr)	Absorbance	Concentration of methylene blue after adsorption (ppm)	Adsorption capacity (mg Met.Blue/g DE)
1	0.0395	0.547	7.4835
3	0.0380	0.539	7.4838
5	0.0441	0.573	7.4828
7	0.0203	0.443	7.4866
9	0.0168	0.424	7.4872

The sorbent adsorption capacity would increase with increasing calcination temperature. At the calcined temperature of 800 °C, the adsorption of methylene blue into sorbent reached a maximum value within a relatively short adsorption time of 1 hour. As for sorbent at calcination temperature of 500 °C, the adsorption time need longer time and reach a maximum adsorption capacity at 9 hours. The results also showed that the adsorption capacity of methylene blue on sorbent of $Ca(OH)_2$/DE was higher when compared with adsorption capacity by only using $Ca(OH)_2$ sorbent. This is due to the formation of a porous calcium silicate hydrate (CSH) in the sorbent of $Ca(OH)_2$/DE as shown in Figure 7. This is in consistent with a study by Mariana (2003) which states that sorbent of $Ca(OH)_2$/DE has a larger porous than $Ca(OH)_2$ sorbent alone as shown in Figure 8. The larger porous or larger surface area of the sorbent caused the higher of adsorption capability [6].

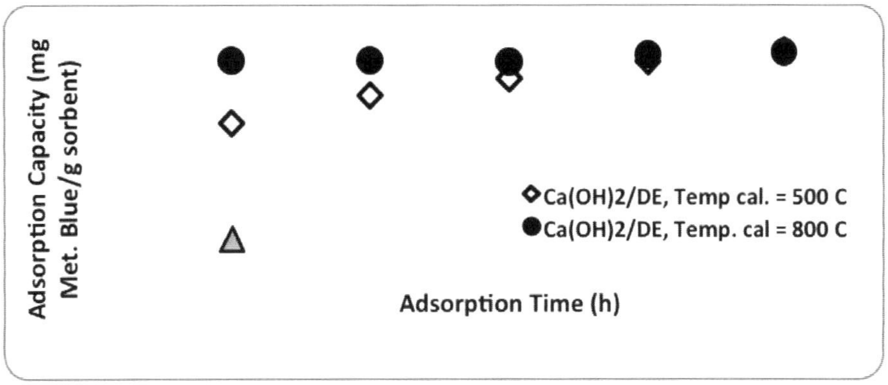

Figure 7. Effect of adsorption time on adsorption capacity of Ca $(OH)_2$/DE sorbent.

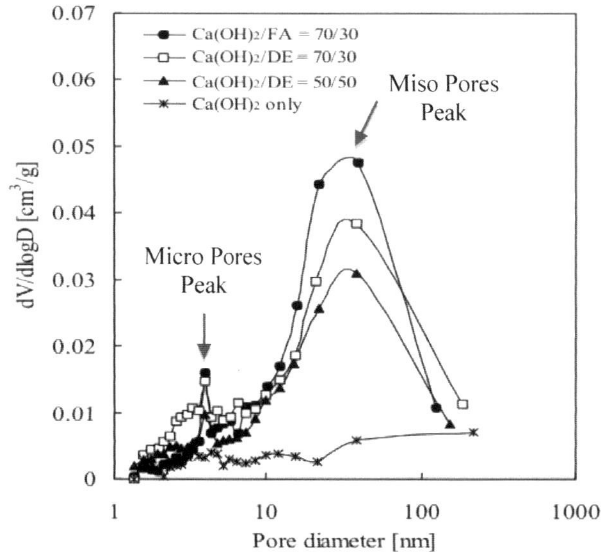

Figure 8. Effect of sorbent type on pore diameter [4].

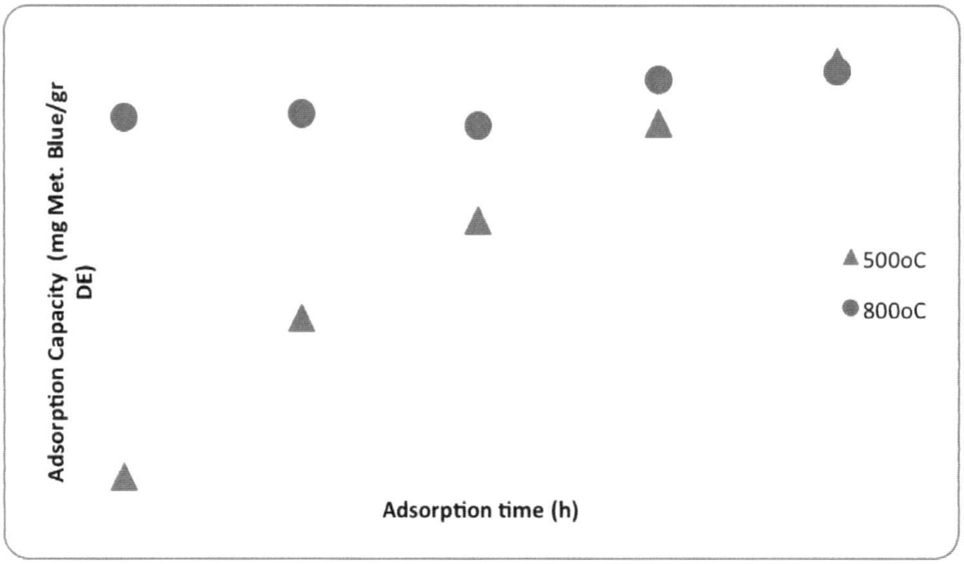

Figure 9. The effect of adsorption time toward the adsorption capacity of methylene blue solution on $Ca(OH)_2$/DE with particle size of 200 mesh after calcination at temperature of 500 °C and 800 °C.

Figure 9 shows that the adsorption capacity of methylene blue solution on $Ca(OH)_2$/DE sorbent both after calcination temperature of 500 °C and 800 °C has the same tendency, that is tend to increase with the increase of adsorption time. This shows that the adsorption capacity of of methylene blue solution on $Ca(OH)_2$/DE increase at time of 1 to 9 hours, because the surface of $Ca(OH)_2$/DE sorbent is still active to perform adsorption process. However, at the adsorption time of 5 hours of $Ca(OH)_2$/DE sorbent that activated at calcination temperature of 800 °C,the adsorption capacity decreases, this is probably due to the possibility of the surface of $Ca(OH)_2$/DE becomes less active, since desorption may also occur. At the adsorption time of 5 hour there is an adsorption equilibrium in which the amount of adsorbed agent is almost proportional to the amount of the desorbed substance [7].

4. Conclusions

From the result of this research, it can be concluded that: the reaction temperature and reaction time affect the reaction of product formation and the amount of product formed. The presence of diatomaceous earth (DE) containing of much silica in $Ca(OH)_2$ sorbents may increase the reactivity of the sorbent. The adsorption capability of $Ca(OH)_2$/DE sorbent toward methylene blue is greater than that of $Ca(OH)_2$ sorbent itself; and $Ca(OH)_2$/DE which was activated at temperature of 800 °C has the highest adsorption capacity compared to $Ca(OH)_2$/DE which is activated at temperature of 500 °C. This is due to the surface and pores of $Ca(OH)_2$/DE sorbent which are activated at temperature of 800 °C was not covered by water and other impurities.

References

[1] Garea A, Viguri J R and Irabien A 1996 *Chem. Eng. J.* **66** 171-179
[2] Ishizuka, T, Tsuchiai H, Murayama T, Tanaka T and Hattori H 2000 *Ind. Eng. Chem. Res.* **39** 1390-1396
[3] Jozewicz W, Chang J C S, Sedman C B, and T G Brna 1988 *J. Air Pol. Cont. Ass.* **38** 1027-1034
[4] Mariana, Chen C, Tsujimura Y, Maezawa A and Uchida S 2003 *J. Chin. Inst. Chem. Engrs.* **34** 2 2011-209
[5] Mariana, Maulana F and Satriyo P 2013 *Jurnal Reaktor* **12** 3 2018-224
[6] Rizki A P and Sanjaya A S 2015 *J. Nas.* **1** 2
[7] Al Quinabait M H 2005 *J. Coloidal Interface Sci.* **283** 316-321

IOP Conf. Series: Materials Science and Engineering **345** (2018) 012002 doi:10.1088/1757-899X/345/1/012002

Synthesis of biodiesel using local natural zeolite as heterogeneous anion exchange catalyst

R. Hartono[1,2], A Wijanarko[1] and H Hermansyah[1]
[1]Department of Chemical Engineering, Faculty of Engineering, Universitas Indonesia, Depok 16424, Indonesia
[2]Departemen of Chemical Engineering, University of Sultan Ageng Tirtayasa, Banten, Indonesia

E-mail: rudi.hartono@untirta.ac.id

Abstract. Production of biodiesel using homogen catalyst: alkaline catalysts, acid catalysts, biocatalysts, and supercritical methanol are very inefficient, because these catalysts have a very high cost production of biodiesel and non-ecofriendly. The heterogeneous catalyst is then used to avoid adverse reaction of biodiesel production. The heterogeneous catalysts used is ion exchanger using natural zeolit catalists bayah banten ($ZABB_{rht}$) and macroporous lewatit that can be used to produce biodiesel in the solid phase so that the separation is easier and can be used repeatedly. The results of biodiesel reach its optimum in engineering ion exchange catalyst natural zeolit bayah and macroporous lewatit which has been impregnated and calcinated at temperature 60 ^{0}C at reaction time 2 hours, are 94.8% and 95.24%, using 100 gr.KOH/100 mL Aquadest.

1. Introduction

Fuel consumption in Indonesia is covering the fields of transport, industry and power plants. Currently the number of user equipment and diesel vehicles has been increased. In line with this increasing number, the need for fuel for diesel engines, namely diesel also increased. Another problem that arises from the use of diesel fuel is environmental pollution. Alternative energy source that produce a more environmentally friendly diesel fuel emissions and does not add to the accumulation of CO_2 in the atmosphere is required, thus reducing the effects of global warming.

Biodiesel is produced through the transesterification of vegetable oils, animal fats or waste oil with methanol using a catalyst to produce fatty acid methyl ester (FAME). Biodiesel can be used as additives in petroleum fuels, mainly for the petrodiesel (PD), known as solar, generally used by 20% biodiesel in diesel mixture known as (B20). The advantage of Biodiesel compared to petrodiesel (PD) in reducing exhaust emissions, could be described, as a high flash point, lubrication large and renewable sources. Biodiesel has a higher oxygen content than petrodiesel (PD), and when it is used in diesel engines showed a reduction in particulate emissions, carbon monoxide (CO), sulfur, polyaromatic compound, hydrocarbons (HC), smoke and noise (1).

Biodiesel in the future will be produced from raw materials that can be refurbished as oils derived from plants and animal fat so that the price of biodiesel will be competitive with petroleum.

The process of making biodiesel using homogeneous catalyst on vegetable oil have limitations and problems of both catalysts alkali, acid, lipase and supercritical including the removal of free fatty acid content, the water content in the reaction mixture, the need for downstream the processing and higher operating costs of biodiesel production.

These problems can be minimized by the use of heterogeneous catalysts in the transesterification process. The use of heterogeneous catalysts is more economical and has several advantages such as non-corrosive, environmentally friendly, easily separated from the melting product, activity and selectivity (2). Heterogeneous catalysts show high potential, strong and durable, able to free fatty acids up to 40% (3).

The use of heterogeneous catalysts is an alternative substitute of the use of homogeneous catalysts, and it is easier to separate and reuse. (4) (5) produces continuous biodiesel with ion-exchange cation acid and ion-exchange cations base by conversion is 71% with molar ratio1:1 between jatropha oil and methanol. (6) Transesterification of biodiesel from palm oil by using KOH / zeolite as a heterogeneous catalyst at various concentrations of NaOH and reaction time to obtain maximum yield of biodiesel. The maximum yield of biodiesel produced is 95.09% at impregnation of 100 grams KOH in 100 ml of aquadest within 2 hours at 60 ^{0}C and the ratio of palm oil and methanol (1: 7), the amount of catalyst used 3% of palm oil. (7) The catalyst used was made by impregnating alumina with KI 35% for 3 hours then dried at 393 K for 16 hours. (KI / Al2O3) solids were calcined at 773 K for 3 hours, the weight of the catalyst used ranged from 1- 5%) weight. Biodiesel yield at optimum is 95,2%. (3) used a heterogeneous catalyst to make biodiesel from rice bran oil with free fatty acid content (FFA) (20 - 50%). The catalyst used was Asamchlorosulfonate-Circonia (HClSO3-ZrO2). The conversion result is 92% at 120 ^{0}C, molar ratio of methanol / oil (12; 1), time 12 hours and 6% by weight of catalyst. (8) Al_2O_3 was modified with MgZnO as a catalyst (MgZnAlO) and activated to produce biodiesel. Effect of parameters such as molar ratio of methanol/oil, 3.32 weight catalyst, reaction time of 6 hours and operating temperature of 182 ^{0}C. Catalysts can be reused for five cycles without reactivation. The resulting biodiesel conversion is 98%. (9) Transesterification of biodiesel from palm oil by using KOH/modernit as a heterogeneous catalyst at various concentrations of NaOH and reaction time to obtain maximum yield of biodiesel. The maximum yield of biodiesel produced is 96.7%. Those researches have successfully improve the surface area of contact and the acidity increased leading to activation as a catalyst of solute is increased.

This research focused on the modification process of anion exchanger using a natural zeolit bayah banten with KOH impregnation method that can be used as a catalyst in the manufacture of biodiesel by heterogeneous catalyst transesterification using vegetable oil. As comparison, Lewatite mocroporous also with the same way. Effect of base concentration (KOH) and the reaction conditions in the acquisition of biodiesel yield is observed on both heterogeneous catalysts.

2. Materials and method

2.1 Materials

Zeolite obtained from natural Bayah Banten that has been crushed by using crusher with a particle size <50 mesh is prior to use. Lewatit was purchased from Sigma-Aldrich Chemical Co., Lewatit MR64 Form Chloride Anion Exchange resin cross linked, Polystyrene matrix, free base, medium base macroporous. Vegetable oil obtained from restaurants at Cilegon after three-time usage. KOH Made in Germany with index-No.019-002-00-8, Brand KGA, 6427 / Darmstadt Germany, Tel +49 (6) 615172-2440 Millipore EMD Corporation. Methanol for analysis made in German.

2.2. Method

The research methodology focused on ion exchange on as heterogeneous catalysts for biodiesel synthesis to obtain optimal results with variation of time, the catalyst and temperature by first activating anion exchange, the Natural zeolit bayah Banten (ZABBrht) and Lewatit mesoporos. The research is conducted at the Laboratory of Bioprocess Engineering Department of Chemical Engineering, Faculty of Engineering, University of Indonesia, Campus UI Depok 16424.

3. Catalyst preparation anion exchanger (zeolite and mesoporous Lewatit)

3.1. Zeolit

Preparation of Zeolit is done to make the zeolite has the ability as a base catalyst, while the steps are as follows, taken from the zeolite used natural zeolite bayah banten ($ZABB_{rht}$) offerings through the stages of destruction, then screening to obtain a desired zeolite size in accordance with the size of its diameter. Activation of natural zeolite bayah banten ($ZABB_{rht}$) is done to modify the structure of the framework or the non-zeolite framework with the aim to enhance its power. Zeolite activation is done by physically and chemically. Physically aim is to evaporate the water caught in the pores of the Zeolite's structure, and the process is conducted at a temperature 110 ^{0}C so that the surface area of the pores of the crystals increases for approximately 24 hours. Chemical activation is done by using an alkaline solution KOH aims to clean the surface of the pores, removing impurities compounds and rearrange the location of atoms exchange. The impregnation process is carried out by using a solution of KOH / 100mL of distilled water to the zeolite at a temperature of 60 ^{0}C, for 2 hours and then the mixture is placed in a preheated oven at 60 ^{0}C for 24 hours and separated by a vacuum pump filter with filter paper, dry it in an oven temperature of 110 ^{0}C for 24 hours and Zeolite calcination to keep the catalyst obtained to be relatively stable at high temperatures. This process is done by heated the zeolite at a temperature of 450^{0}C. Calcination time is calculated starting from temperature 450 ^{0}C for 4 hours. The catalyst is then cooled and ready to use.

3.2. Lewatit Mesoporous

Preparation Lewatit is made to make Lewatit have the ability as a basic catalyst. The steps are as follows, the catalyst is kept at a temperature of 110 ^{0}C for approximately 24 hours. Then, the process of impregnation is carried out by using a solution of KOH / 100mL of distilled water to Lewatit at a temperature of 60 ^{0}C, for 2 hours and then the mixture is placed in a preheated oven at 60 ^{0}C for 24 hours and separated by a vacuum pump filter with filter paper, dry it in an oven temperature of 110 ^{0}C for 24 hours. The catalyst is then cooled and ready to use.

4. Results and discussion

4.1. The effect of various grams of KOH / 100mL towards percent yield of biodiesel

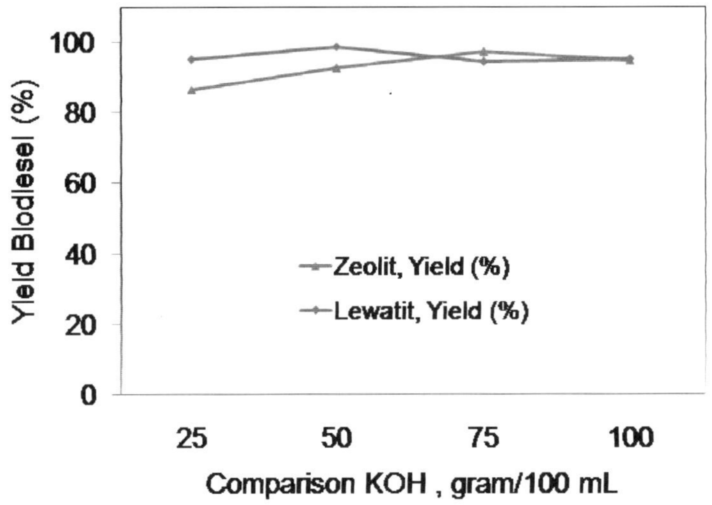

Figure 1. The Effect of various grams of KOH / 100 mL towards percent yield of biodiesel.

Experiments were performed to determine the effect of various gram KOH / 100ml zeolite to the acquisition percent yield biodiesel using used cooking oil. The optimum conditions of biodiesel is reached when using the ratio of reactants to methanol (1: 7) (Kusuma at al, 2013) to move to the right reaction and collide wider.

Zeolite Optimum condition was achieved in 75 grams of KOH / 100 mL with a percent yield 92.28%, is caused by many methoxide trapped in zeolites. Lewatit optimum condition was achieved in

50 grams of KOH / 100 ml of distilled water with 98.88% percent biodiesel yield caused by the amount of metal oxides that react.

4.2. The effect of various Gram KOH / 100mL towards the product density

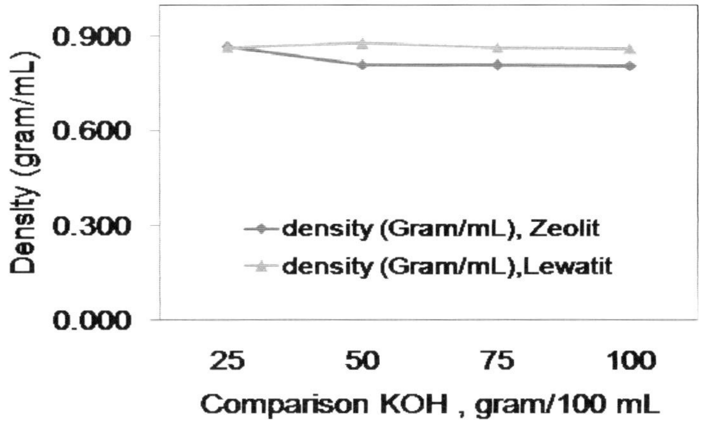

Figure 2. The effect of various Gram KOH / 100 mL towards the product density.

Figure 2 in zeolite chart showed that the higher use of gram KOH / 100mL resulted the lower the density produced. The density will affect the quality of biodiesel produced, the smaller the value the greater density caloric value.

On the use of catalysts Lewatit optimum conditions achieved with the use of a solution of 100 g KOH / 100mL distilled water. All the conditions included in the density of standard biodiesel.

4.3. The effect of various Gram KOH / 100mL towards biodiesel viscosity

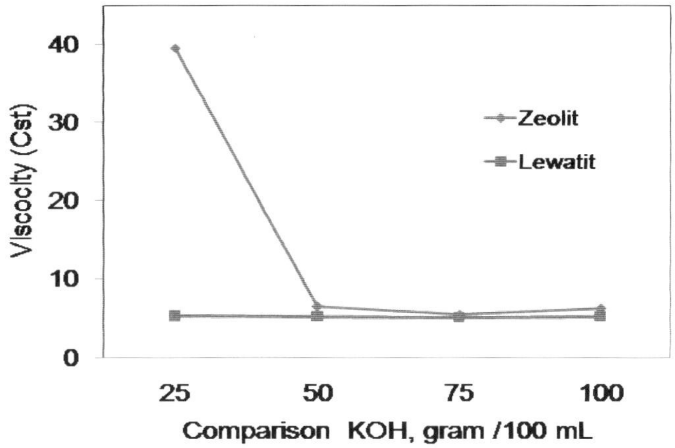

Figure 3. The effect of various gram KOH / 100mL towards biodiesel viscosity.

The more catalysts used the lower the viscosity of the resulting, it indicates that the more sodium methoxide is reacting with cooking oil so that the process of cracking the larger chains of fat is successful and results to the lower viscosity grades. Low viscosity grades of biodiesel will be more easily flow in order to improve the performance of the engine combustion system.

5. Conclusions

Based on the research that has been done, it can be concluded that biodiesel production using natural zeolite bayah banten (ZABBrht) as heterogeneous catalysts can be carried out with specific preparation and treatment and the biodiesel produced meets the Indonesia National Standard (SNI).

References

[1] Kattimani V R, Venkatesha B M and Ananda S 2014 *Adv. Chem. Engineer. Sci.* **4** (03) 361
[2] Om Tapanes N C, Gomes Aranda D A, de Mesquita Carneiro J W and Ceva Antunes O A 2008 *Fuel* **87** (10–11) 2286-2295
[3] Zhang Y, Wong W T and Yung K F 2013 *Bioresour. Technol.* **147** 59-64.
[4] Kitakawa N S, Hiromori K, Ihara T, Nakashima K and Yonemoto T 2015 *Fuel* **139** 11-17
[5] Kitakaw N S, Kanagawa K, Keiichi, Nakashima K and Yonemoto T 2013 *Bioresour. Technol.* **142** 732-736
[6] Kusuma R I, Hadinoto, Johan P, Ayucitra, Aning, Soetaredjo, Felycia E, Ismadji S 2013 *App. Clay Sci.* **74** 121-126
[7] Evangelista J P C, Chellappa T, Coriolano A C F, Fernandes Jr, Valter J, Souza L D, Araujo, A S 2012 *Fuel Proc. Technol.* **104** 90-95
[8] Intarapong P, Iangthanarat S, Phanthong P, Luengnaruemitchai A and Jai-In S 2013 *J. Energ. Chem.* **22** (5) 690-700
[9] Olutoye M A and Hameed B H 2013 *Bioresour. Technol.* **132** 103-108

IOP Conf. Series: Materials Science and Engineering **345** (2018) 012003 doi:10.1088/1757-899X/345/1/012003

Enhanced photocatalytic hydrogen production from water-ethanol solution by Ruthenium doped La-$NaTaO_3$

H Husin[1], P N Alam[1], M Zaki[1], Sofyana[1], Jakfar[1], Husaini[2] and F Hasfita[3]
[1]Department of Chemical Engineering, Syiah Kuala University, Darussalam, 23111, Banda Aceh, Indonesia
[2]Department of Mechanical Engineering, Syiah Kuala University, Darussalam, 23111, Banda Aceh, Indonesia
[3]Department of Chemical Engineering, Malikussaleh University, Lhoekseumawe, 24300, Aceh Utara, Indonesia

E-mail: husni_husin@che.unsyiah.ac.id

Abstract. The photocatalytic hydrogen production from ethanol aqueous solution, with the use ruthenium doped La-$NaTaO_3$ has been investigated. Ruthenium doped La-$NaTaO_3$ catalysts are prepared by impregnation method. The catalysts are by X-ray diffraction (XRD), scanning electron microscopy (SEM) and transmission electron microscopy (TEM). The Ru co-catalyst demonstrated from the TEM image shows a good dispersion on the surface of La-C-$NaTaO_3$ with an average particle size between 4-5 nm. The photocatalytic reaction is carried out in a closed reactor with a gas circulation system. The catalytic activity of La-$NaTaO_3$ improved markedly (6.6 times from pure water) when Ru is loaded onto its surface. The hydrogen production is notably enhanced in the presence of ethanol as electron donors. This result is much higher when compared with the rate of hydrogen production in the sample without co-catalysts about 9.4 times higher after Ru deposition from ethanol aqueous solution. Increasing the production of hydrogen on the Ru/La-$NaTaO_3$ closely related to the decrease in recombination between electron-hole pairs.

1. Introduction

Hydrogen one of the main challenges facing humanity is the generation of alternative clean and renewable fuels that allowing meeting the future energy needs. Environmental pollution and energy consumption are two of the most serious issues affecting the planet and the human race at present. Moreover, the contribution of greenhouse gasses to global warming, in particular, carbon dioxide emissions is driving the urgent development of clean, sustainable energy sources. About 85% of our energy consumption is provided by fossil fuels, which will not be able to keep up with the increasing energy demand [1-3].

One promising approach to alleviate the energy crisis is the use of hydrogen from water and light, due to it is environmentally friendly energy sources [4, 5]. Photocatalytic water splitting represents one of the promising technologies for clean and environmentally friendly hydrogen production and provides a way to use sunlight for generating hydrogen as a renewable green fuel. The photocatalytic process is initiated by the absorption of a photon with energy equal to or larger than the band gap of the semiconductor. The photoexcited carriers separate and migrate to the surface without recombination, and adsorbed species are reduced and oxidized by the photogenerated electrons and holes to produce H_2 and O_2 [6-8].

However, photogenerated electrons are also very easy to recombine of photo-generated electron/hole pairs in the semiconductor and the fast backward reaction of hydrogen and oxygen to form water. Thus, RuO_2 as co-catalyst can help push electron transfers and enhance photocatalytic activity [9, 10]. Although the loading with a co-catalyst can reduce the charge carrier recombination to some extent, hydrogen production from pure water splitting is difficult to achieve since the recombination of the electron-hole pairs cannot be completely eliminated. Some works reported in the last decades have shown that it was possible to increase the hydrogen production reaction rate using water–methanol systems. Therefore, electron donors as well as other mediators are usually required to avoid this problem. Regarding the photocatalytic of H_2 production, different substances e.g., methanol, EDTA, sulphide, acting as electron donors, have been generally used in many studies [11-13]. Ethanol is regarded as a credible and important source for H_2 production because of its can act as electron donor for the photocatalytic H_2 production on semiconductor, in comparison, its rarely being employed and a renewable nature.

Based on the above background, the aim of the present work is to study the influence of Ru as co-catalyst and ethanol as sacrificial reagents on photocatalytic water splitting reactions to produce of hydrogen under ultraviolet (UV) irradiation. An attempt was made to synthesis of RuO_2 deposited on La-$NaTaO_3$ at ambient temperature by inception wetness method. The obtained RuO_2 deposited on La-$NaTaO_3$ shown excellent properties photocatalysts.

2. Materials and methods

2.1. Chemical reagents

In this work, tantalum chloride ($TaCl_5$, 99.9%): Acros organics), sodium hydroxide (NaOH, 99,9%: Acros organics), $La(NO_3)_2.2H_2O$ (Acros Organics; 99,9%), anhydrous citric acid ($C_6H_8O_7$, 99,5%: Merck), ammonia (NH_4OH, 35%: Fisher Scientific, ethanol (C_2H_5OH, 99,5% Merck) was used as precursor for the synthesis of La-$NaTaO_3$. Triruthenium dodecarbonyl ($Ru_3(CO)_{12}$, Aldrich, 99%) were used for the synthesis of RuO_2. Deionized water was used in whole experiments. Argon gas (Ar:99,99%), and nitrogen gas (N_2: 99,99%) were used in calcination process. All gases were purchased from Serikat Company Banda Aceh, Indonesia.

2.2. Synthesis of RuO_2/La-$NaTaO_3$ nanocrystalline

All the reagents with analytical grade were used without further purification. La-$NaTaO_3$ was synthesized by a sol-gel method as previously reported by our group elsewhere [14]. In a typical preparation of $TaCl_5$ was dissolved in excess ethanol solution under vigorous stirring. A required amount of NaOH was slowly added into the $TaCl_5$ solution followed by vigorous stirring for 1 h. $La(NO_3)_3.6H_2O$ was dissolved separately in 15 ml deionized water and the solution containing $La(NO_3)_3$ added dropwise into the above solution. Citric acid was dissolved in 50 ml of deionized water and then dropwise into the above solution by stirring for 2 h to produce a transparent sol. The pH of the solution was adjusted to 4 by adding an ammonia solution to ionize the citric acid. The transparent solution was then heated at temperature of 90 °C, followed by stirring to obtain white gels. The obtained gels were poured into crucible porcelain and dried in oven at 110 °C during 7 h. The resulting samples was crushed and then pre-heated at 450 °C for 4 h in muffle furnace. The powder was continued calcination at temperature of 800 °C in air for 8 h. Finally, the resulting powder was crush by mortar and characterized by several methods.

Loading ruthenium oxide as active phase onto the obtained La-$NaTaO_3$ was performed by an impregnation method with a $Ru_3(CO)_{12}$ (Aldrich, 99%). Typically, 1.0 g of La-$NaTaO_3$ was first filled in beaker glass. Triruthenium dodecarbonyl ($Ru_3(CO)_{12}$) was dissolved in distilled water and then mixed with La-$NaTaO_3$ powder. The samples were mixed by stirring at atmospheric pressure and room temperature for 6 h. Thereafter, the slurry was dried at 110 °C overnight. The obtained sample was further oxidized in flowing air (100 mL/min at 300 °C for 3 h) to form RuO_2.

2.3. Characterization of nanocrystalline

The morphology of the powders was inspected by a scanning electron microscope (JEOL 5600 SEM). HRTEM images were obtained by a JEOL Model JEM 2010 EX instrument at an accelerating voltage of 200 kV. X-ray diffraction patterns (XRD) of tantalates were obtained with a a Shimadzu 7500 X-ray advanced diffractometer with a Cu-Kα radiation (λ = 1.5406 Å). The accelerating voltage and the applied current were 40 kV and 40 mA, respectively. The phases present in the samples were identified with the help of JCPDS (Joint Committee of the Powder Diffraction Standard) database files.

2.4. Photocatalytic hydrogen generation

The photocatalytic reaction of hydrogen production from water-ethanol solution was carried out in in a Pyrex closed gas circulation reactor. In a typical experiment, 0.3 g of the photocatalyst (Ru/La-$NaTaO_3$) was suspended in 350 ml ethanol water mixture (of 10 vol.% aqueous ethanol solution). A 400 W Mercury Lamp was used as a light source irradiated inside of the reactor. Before irradiation, the quartz flask was then connected to the experimental set-up and Ar gas was bubbled through the reaction mixture for 30 min to remove dissolved oxygen. The reactor temperature was maintained at 25 °C by means of a cooling fan. The hydrogen evolution was collected by collector gas. The hydrogen evolution rate was measured at room temperature under atmospheric pressure. Gas samples were analyzed using a thermal conductivity detector (TCD) gas chromatograph (Shimadzu, GC-8A).

3. Results and discussion

3.1. Surface morphology

The morphology of the synthesized catalysts was analyzed by scanning electron microscopy (SEM) and high resolutions (HR-SEM). Figure 1 shows SEM images of the RuO_2/La-$NaTaO_3$ catalysts. It can be seen in Figure 1, the RuO_2/La-$NaTaO_3$ catalyst are quite homogeneous nano-sized particles.

Figure 1. SEM image of RuO_2/La-$NaTaO_3$ photocatalyst.

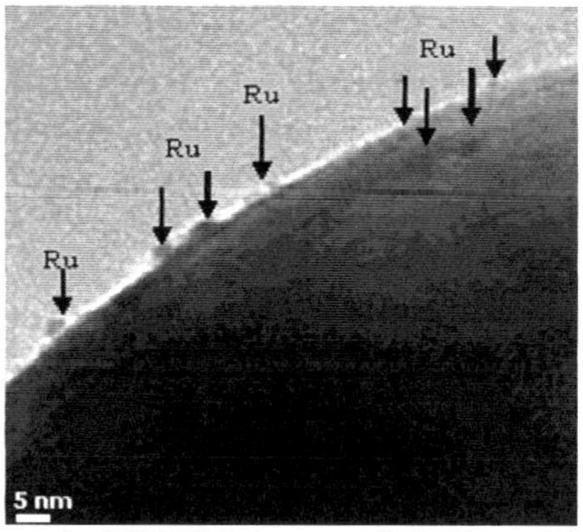

Figure 2. HRTEM image of RuO_2 oxide on La-$NaTaO_3$ surface.

All La-$NaTaO_3$ particles have similar morphologies of orthorhombic shape $NaTaO_3$. There is no serious aggregation of nanocrystalline. As demonstrated in Figure 1, all of the photocatalyst particles are very small with sizes of 50 - 150 nm, which is much smaller than the particle sizes synthesized by traditional solid state reaction method, indicating that sol-gel method can effectively inhibit the agglomeration of the $NaTaO_3$ nanoparticles. This morphology is typically observed for well-prepared nanocrystalline reported in the literature [15]. High crystallinity of the particles, for La-$NaTaO_3$, was confirmed by TEM analyses, showing smooth surface of the particles (Figure 1).

To further study the effect of the ruthenium content on the surface catalyst, several catalysts were prepared by deposition of Ru via inception wetness. The morphology and dispersion of Ru deposited

IOP Conf. Series: Materials Science and Engineering **345** (2018) 012003 doi:10.1088/1757-899X/345/1/012003

particles on La-$NaTaO_3$ were also confirmed by high resolution transmission electron microscopy. A representative TEM image of 0.6 wt.% RuO_2 is illustrated in Figure 2. It is clearly seen that Ru oxide particles in the range of 4 - 5 nm were well dispersed on the surface of La-$NaTaO_3$ with no aggregation. The size distribution of ruthenium particles is fairly uniform.

3.2. Crystal structure and composition

The crystal structure of RuO_2/La-$NaTaO_3$ was analysed by X-ray diffractometer (XRD). Figure 3 shows XRD patterns for the RuO_2/La-$NaTaO_3$ catalysts. The XRD spectra of RuO_2/La-$NaTaO_3$ in Figure 3 reveal that the RuO_2/La-$NaTaO_3$ photocatalysts had a crystal structure that was identical to that of the un-doped $NaTaO_3$. The peaks at $2\theta = 22.73, 32.54, 52.72^o$ can be assigned to the diffraction pattern of pure $NaTaO_3$ of orthorhombic structure. As depicted in Figure 3, high crystallinity of the particles was confirmed by XRD analyses, showing narrow peaks, which could effectively inhibit the recombination of photogenerated carriers [16]. Loading a small amount of ruthenium and doping of lanthanum confirming that no crystalline phase involving ruthenium oxides and lanthanum oxides was recorded.

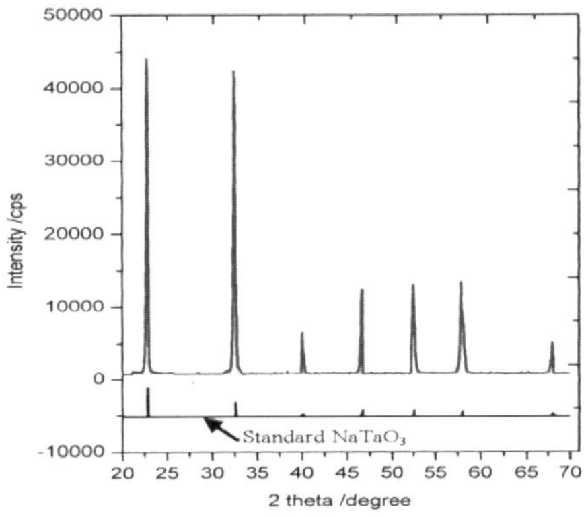

Figure 3. X-ray diffraction patterns of RuO_2/La-$NaTaO_3$ photocatalyst.

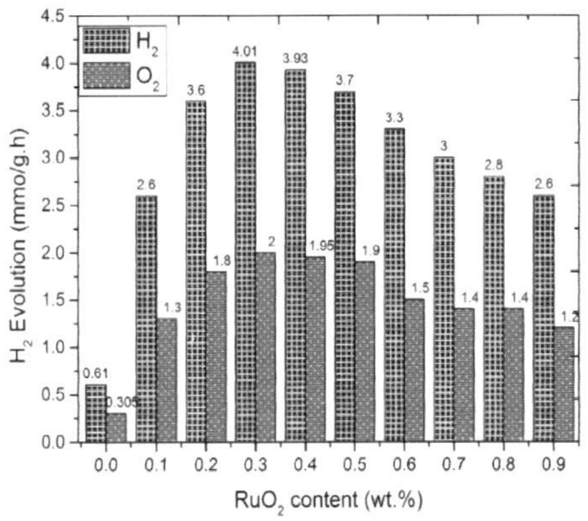

Figure 4. Formation rate of H_2 and O_2 as a function of RuO_2 content on La-$NaTaO_3$ from pure water (350 ml solution and 3 g catalyst).

3.3. Photocatalytic activities

The photocatalytic hydrogen evolution over the RuO_2/La-$NaTaO_3$ catalysts from water splitting as a function of ruthenium loading is displayed in Figure 4. The experiments were conducted using reactant solution of 350 ml of pure water. In a typical reaction, the catalyst powder (0.3 g, containing Ru: 0.0 - 0.9 wt.%) was suspended into the inner irradiation cell made of quartz under UV-light (400W high-pressure Hg lamp), and operated in the atmospheric and constant temperature of 25 °C. It can be seen that the bare La-$NaTaO_3$ exhibit very low hydrogen evolution (0.61 $mmole.h^{-1}.g^{-1}$). The amount of O_2 generated was approximately half as that of H_2. A significant increase in the hydrogen evolution is observed when the addition amounts of Ru as H_2 evolution site. The rate of hydrogen formation on RuO_2/La-$NaTaO_3$ samples is 6.6 times much higher than that obtained with the pure La-$NaTaO_3$. As can be seen in Figure 4, the photoactivity of RuO_2/La-$NaTaO_3$ increased with enhancing the RuO_2 loading from 0.1 wt%, and reaches a maximum H_2 evolution rate (4.01 $mole.g^{-1}.\ h^{-1}$) at 0.3 wt.%. While it exceeds 0.3 wt.%, the H_2 evolution rate begins to decrease, in line with recent literature results [17]. The benefit of RuO_2 deposition is due to the capability of these metals to trap electrons. It can be considered that the enhanced photocatalytic activity by co-catalysts may be possible from the interaction between RuO_2 and La-$NaTaO_3$ to increase the transfer of photogenerated carriers [18] [19] On the other hand, the excessive Ru loading will mask the La-$NaTaO_3$ surface, reducing the photoexcitation capacity of the La-$NaTaO_3$ photocatalyst [18,19].

The same trend also demonstrated in Figure 5 that the amount of hydrogen evolution increases significantly when introducing the ethanol (10 vol.%) in the solution as sacrificial reagent. In these cases, the rate of hydrogen production was found to be much higher, compared to that obtained from photocatalytic of pure water. In fact, ethanol, are not only additional proton sources, but also efficient hole scavengers which undergo fast and irreversible oxidation, thus making photo-promoted electrons more readily available for proton reduction to hydrogen [20].

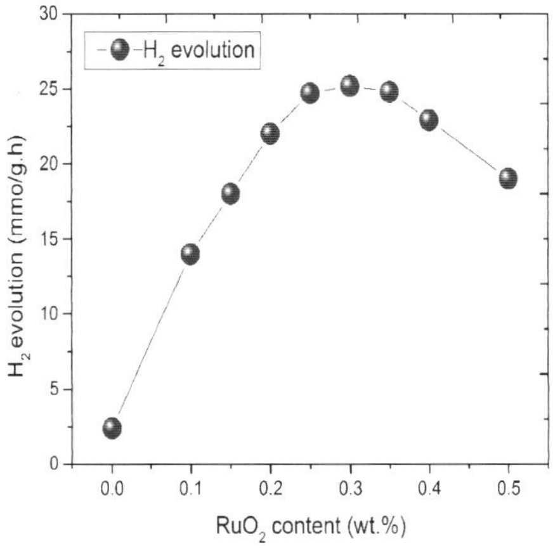

Figure 5. H_2 evolution as a function of RuO_2 content on La-$NaTaO_3$ from aqueous ethanol solution (350 ml, containing 10 vol.% ethanol).

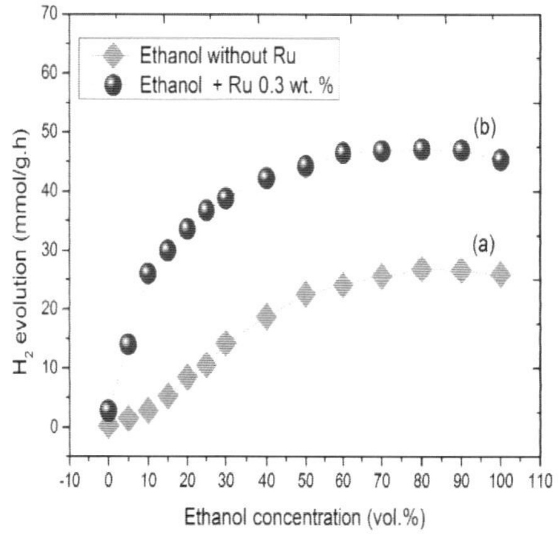

Figure 6. H_2 evolution from different amount of ethanol on La-$NaTaO_3$ (a) without RuO_2 (b) with RuO_2.

This fact can be explained if the ethanol acts as sacrificial electron donor and consumes photogenerated holes and/or oxygen to produce protons (H^+) in the valence band, while Ru traps the electron from the surface to suppress the electron-hole recombination on the semiconductor's surface [4]. Similar report have studied hydrogen generation with Rh deposited on carbon and $RuO_2/NaTaO_3$ and it application in hydrogen production and petrochemical, proven that oxidation of ruthenium oxide occurred during reaction induced high hydrogen generation activity [21, 22].

Figure 6 shows the changes in the photocatalytic activity for H_2 production with the addition amount of ethanol over La-$NaTaO_3$ and RuO_2/La-$NaTaO_3$ suspension. The amount of H_2 production on La-$NaTaO_3$ at ethanol of 10% was 2.8 mmole.h^{-1}.g^{-1} (see Figure 6.a), and reached 26.4 mmole.h^{-1}.g^{-1} when RuO_2 introducing on La-$NaTaO_3$ (Figure 6.b). The increase of hydrogen production rate was about 9.4 times higher than that of in the absence of RuO_2 on the surface of La-$NaTaO_3$. It is observed that, combination of La:$NaTaO_3$ with RuO_2 gives the highest hydrogen production rate.

The effect of ethanol concentration on the photocatalytic hydrogen production at 0.3 wt.% ruthenium loading was also observed. It is demonstrated that the rate of hydrogen evolution over La-$NaTaO_3$ initially was 2.79 mmole.$h^{-1}g^{-1}$, then increase of 13.5 mmole.h^{-1}.g^{-1} when introducing the ethanol of 5% , and goes through a maximum at ethanol of 80%, which reaches a value of about 47.1 mmole.h^{-1}.g^{-1}. When the ethanol concentration exceeds 80%, the hydrogen evolution decreased and then tented to saturation. In this regard, the co-catalyst and ethanol plays a very important role for hydrogen evolution rate, because RuO_2 is able to capture electrons and decrease the over potential of H^+/H, and the effect of added ethanol, which acts as sacrificial electron donor, consumes photogenerated holes and/or oxygen, leading to the decrease of the electron-hole recombination. These phenomenons are in agreement with reported by other group, who reported the photocatalytic of hydrogen rate from methanol and ethanol aqueous solution [23-25].

4. Conclusions

The photocatalytic hydrogen production over RuO_2/La-$NaTaO_3$ nanoparticles has been studied using ethanol as a sacrificial reagent. The deposition with RuO_2 particles was found to greatly enhance the photocatalytic activity of La-$NaTaO_3$ toward hydrogen production from aqueous ethanol solution. The addition of an appropriate amount of ethanol to the reactant solution was shown to enhance H_2 evolution, which is very important to practical application. RuO_2 is able to capture electrons and decrease the over potential of H^+/H, and ethanol acting as an electron donor to scavenge photoinduced holes quickly so that the recombination of photoinduced electrons and holes at the RuO_2/La-$NaTaO_3$ surface are inhibited.

Acknowledgment

We gratefully acknowledge the financial support from the Syiah Kuala University Banda Aceh, Republic of Indonesia.

References

[1] Luna A L, Novoseltceva E, Louarn E, Beaunier P, Kowalska E, Ohtani B, et al. 2016 Synergetic effect of Ni and Au nanoparticles synthesized on titania particles for efficient photocatalytic hydrogen production *Appl. Catal. B: Environ.* **191** 18-28

[2] Carrasco-Jaim O A, Ceballos-Sanchez O, Torres-Martínez L M, Moctezuma E and Gómez-Solís C 2017 Synthesis and characterization of PbS/ZnO thin film for photocatalytic hydrogen production *J. Photochem. Photobiol. A: Chem.* **347** 98-104

[3] Husin H, Chen H M, Su W N, Pan C J, Chuang W T, Sheu H S, et al. 2011 Green fabrication of La-doped NaTaO3 via H2O2 assisted sol-gel route for photocatalytic hydrogen production *Appl. Catal. B* **102** 343-351

[4] Husin H, Nien S W, Jern P C, Yue L J, John R and Chiang Y S 2013 Pd/NiO core/shell nanoparticles on $La_{0.02}Na_{0.98}TaO_3$ catalyst for hydrogen evolution from water and aqueous methanol solutio *Intl. J. Hydrogen Energ.* **38** 13529-13540

[5] Husin H, Mahidin M, Yunardi Y and Hafita F 2015 Visible light driven photocatalytic hydrogen evolution by lanthanum and carbon-co-doped $NaTaO_3$ photocatalyst *Key Eng. Mater.* **659** 231-236

[6] Maeda K, Teramura K and Domen K 2007 Development of Cocatalysts for photocatalytic overall water splitting on $(Ga_{1-x}Zn_x)(N_{1-x}O_x)$ solid solution *Catal. Surv. Asia* **11** 145-157

[7] Husin H 2012 Produksi hidrogen secara fotokalitik dari air murni pada katalis $NaTaO_3$ *Jurnal Rekayasa Kimia dan Lingkungan* **9** 53-58

[8] Husin H, Pontas K, Sy Y, Syawaliah S and Saisa S 2014 Synthesis of nanocrystalline of lanthanum doped NaTaO3 and photocatalytic activity for hydrogen production *J. Eng. Technol. Sci.* 46 318-327

[9] Huang B S, Chang F Y and Wey M Y 2010 Photocatalytic properties of redox-treated Pt/TiO_2 photocatalysts for H_2 production from an aqueous methanol solution *Intl. J. Hydrogen Energ.* **35** 7699-7705

[10] Husin H, Su W N, Chen H M, Pan C J, Chang S H, Rick J, et al. 2011 Photocatalytichydrogen production on nickel-loaded LaxNa1-xTaO_3 prepared by hydrogen peroxide-water based process *Green Chem.* **13** 1745-1754

[11] Rosseler O, Shankar M V, Du M K L, Schmidlin L, Keller N and Keller V 2010 Solar light photocatalytic hydrogen production from water over Pt and Au/TiO_2(anatase/rutile) photocatalysts: Influence of noble metal and porogen promotion *J. Catal.* **269** 179-190

[12] Miwa T, Kaneco S, Katsumata H, Suzuki T, Ohta K, Chand Verma S, et al. 2010 Photocatalytic hydrogen production from aqueous methanol solution with CuO/Al_2O_3/TiO_2 nanocomposite *Intl. J. Hydrogen Energ.* **35** 6554-6560

[13] Xu S and Sun D D 2009 Significant improvement of photocatalytic hydrogen generation rate over TiO_2 with deposited CuO *Intl. J. Hydrogen Energ.* **34** 6096-6104

[14] Husin H, Adisalamun, Sy Y, Asnawi T M and Hasfita F 2017 Pt nanoparticle on $La_{0.02}Na_{0.98}TaO_3$ catalyst for hydrogen evolution from glycerol aqueous solution *AIP Conf. Proc.* p 030073.

[15] Husin H, Mahidin M, Zuhra Z and Hafita F 2014 H_2 evolution on lanthanum and carbon co-doped $NaTaO_3$ photocatalyst *Bull. Chem. Reaction Eng. Catal.* **9** 81-86

[16] Jiang W, Jiao X and Chen D 2013 Photocatalytic water splitting of surfactant-free abricated high surface area $NaTaO_3$ nanocrystals *Intl. J. Hydrogen Energ.* **38** 12739-12746

[17] Husin H 2012 Efficient Photocatalytic Hydrogen Generation by Pd Modified La-NaTaO3 Nanoparticle *Proc. Ann. Intl. Conf. Syiah Kuala University-Life Sciences & Engineering Chapter* p. 226-231

[18] Lu C, Chen Y, Li Y, Ma C, Guo Y and Li Y 2015 The effect of different co-catalysts (CuO, MoS_2 and Pt) on hydrogen production of Er^{3+}:$YAlO_3/NaTaO_3$ by visible-light-induced methanol splitting *Energ.* **93** 749-757

[19] Chen W T, Chan A, Sun-Waterhouse D, Moriga T, Idriss H and Waterhouse G I N 2015 Ni/TiO_2: A promising low-cost photocatalytic system for solar H_2 production from ethanol-water mixtures *J. Catal.* **326** 43-53

[20] Dozzi M V, Chiarello G L, Pedroni M, Livraghi S, Giamello E and Selli E 2017 Highphotocatalytic hydrogen production on Cu(II) pre-grafted Pt/TiO_2 *Appl. Catal. B: Environ.* **209** 417-428

[21] Gómez-Solís C, Ballesteros J C, Torres-Martínez L M and Juárez-Ramírez I 2016 RuO_2-$NaTaO_3$ heterostructure for its application in photoelectrochemical water splitting under simulated sunlight illumination *Fuel* **166** 36-41

[22] Zhang Y, Ligthart D A J M, Quek X.-Y, Gao L and Hensen E J M 2014 Influence of Rh nanoparticle size and composition on the photocatalytic water splitting performance of Rh/graphitic carbon nitride *Intl. J. Hydrogen Energ.* **39** 11537-11546

[23] Oros-Ruiz S, Zanella R, Collins S E, Hernández-Gordillo A and Gómez R 2014 Photocatalytic hydrogen production by Au–M_xO_y (MAg, Cu, Ni) catalysts supported on TiO_2 *Catal. Comm.* **47** 1-6

[24] Strataki N, Antoniadou M, Dracopoulos V and Lianos P 2010 Visible-light photocatalytic hydrogen production from ethanol-water mixtures using a Pt-CdS-TiO_2 photocatalyst *Catal. Today* **151** 53-57

[25] Melián E P, Suárez M N, Jardiel T, Rodriguez J M D, Caballero A C and Araña J 2014 Influence of nickel in the hydrogen production activity of TiO_2 *Appl. Catal. B: Environ.* **152-153** 192-201

IOP Conf. Series: Materials Science and Engineering **345** (2018) 012004 doi:10.1088/1757-899X/345/1/012004

Comparison of kinetic model for biogas production from corn cob

L M Shitophyta and Maryudi
Department of Chemical Engineering, Faculty of Industrial Technology, Universitas Ahmad Dahlan, Postal Code 55164, Yogyakarta, Indonesia

E-mail: lukhi.mulia@che.uad.ac.id

Abstract. Energy demand increases every day, while the energy source especially fossil energy depletes increasingly. One of the solutions to overcome the energy depletion is to provide renewable energies such as biogas. Biogas can be generated by corn cob and food waste. In this study, biogas production was carried out by solid-state anaerobic digestion. The steps of biogas production were the preparation of feedstock, the solid-state anaerobic digestion, and the measurement of biogas volume. This study was conducted on TS content of 20%, 22%, and 24%. The aim of this research was to compare kinetic models of biogas production from corn cob and food waste as a co-digestion using the linear, exponential equation, and first-kinetic models. The result showed that the exponential equation had a better correlation than the linear equation on the ascending graph of biogas production. On the contrary, the linear equation had a better correlation than the exponential equation on the descending graph of biogas production. The correlation values on the first-kinetic model had the smallest value compared to the linear and exponential models.

1. Introduction

The rising impact of fossil energy consumption on global climate patterns and the environment has initiated the invention of renewable energy sources. Biomass is a type of renewable energy which has zero carbon emission [1]. Biomass can be converted into biofuel such as biogas by anaerobic digestion. Biogas is a mixture composed of methane (CH_4), carbon dioxide (CO_2), hydrogen (H_2), nitrogen (N_2) and hydrogen sulfide (H_2S) [2]. Anaerobic digestion is a process that breaks down organic matter through the action of microorganisms [3]. The steps of biomass conversion into biogas consist of three stages, i.e. hydrolysis, acid formation, and methane production [4]

Corn cob can be used as a raw material for biogas production. Corn cob is one of the agricultural wastes which is widely available in Indonesia. The total production of corn in Indonesia is 19.61.435 in 2015 [5]. Corn is also considered as an important agricultural crop for over 70 million farm families worldwide [6]. In addition to corn cob, food waste can also be used as biogas feedstock. Currently, food waste only is utilized as animal feedstock or landfilling, as a result, it causes serious health and environmental problems [7]. Food waste has high organic and moisture content [8]. Several factors which affect biogas production include temperature, pH, Carbon/Nitrogen ratio, total solid, volatile solid [9]. This study concentrated on the kinetic rate of biogas production at the variation of total solid (TS) content. Hence, the objective of this research was to compare kinetic models of biogas production from corn cob.

2. Material and methods

2.1. Feedstock and inoculum

Corn cob was dried and chopped into 2-3 cm then stored at room temperature. Corn cob was used as feedstock. Food wastes were used as co-digestion. Cow's rumen was used as an inoculum.

2.2. Solid-state anaerobic digestion

Corn cob, food waste, and inoculum were mixed and loaded into 2 L digester. Water was added to adjust TS content of 20%, 22%, and 24%. This research was conducted at room temperature and neutral pH. Biogas volume was measured every 2 days by water displacement method.

3. Kinetic model for biogas production

The kinetic of biogas production was modeled by linear and exponential equations. On the linear equation, the rate of biogas production will increase along with the increase of time, then it attains a peak value and decreases linearly to a zero point. The linear equation is stated as follows [10] :

$$y = a + bt \tag{1}$$

where, y = the biogas production rate (ml/g/day); t = the biogas production time (day); a = the intercept (ml/g/day); b = the slope (ml/g/day). On the ascending graph of biogas production, b is positive, while on the descending graph, b is negative.

On the exponential equation, the rate of biogas production increases exponentially with the digestion time. After reaching a peak value, it will decrease exponentially to a zero point. The exponential equation is expressed as follows [11] :

$$y = a + b\,exp(ct) \tag{2}$$

where, y = the biogas production rate (ml/g/day); t = the biogas production time (day); a, b = constants (ml/g/day); c = constant (ml/g/day). c will be positive on the ascending graph of biogas production, while c will be negative on the descending graph.

The kinetic of biogas production can be also modeled using first-order kinetic model as performed previously by Brown et al [12]. The first-kinetic model is stated as follows

$$ln\left(\frac{y_m}{y_m - y_t}\right) = kt \tag{3}$$

where, y_m = biogas yield obtained in 40 days (ml/g/day); y_t = biogas yield obtained at the time t (ml/g/day); k = biogas rate constant (1/day); t = biogas production time (t)

4. Results and discussion

The kinetic of biogas production was performed by linear and exponential equation models. The kinetic study was divided into two rate periods, namely the ascending rate period and the descending rate period. The linear graph of biogas production is shown in Figure 1.

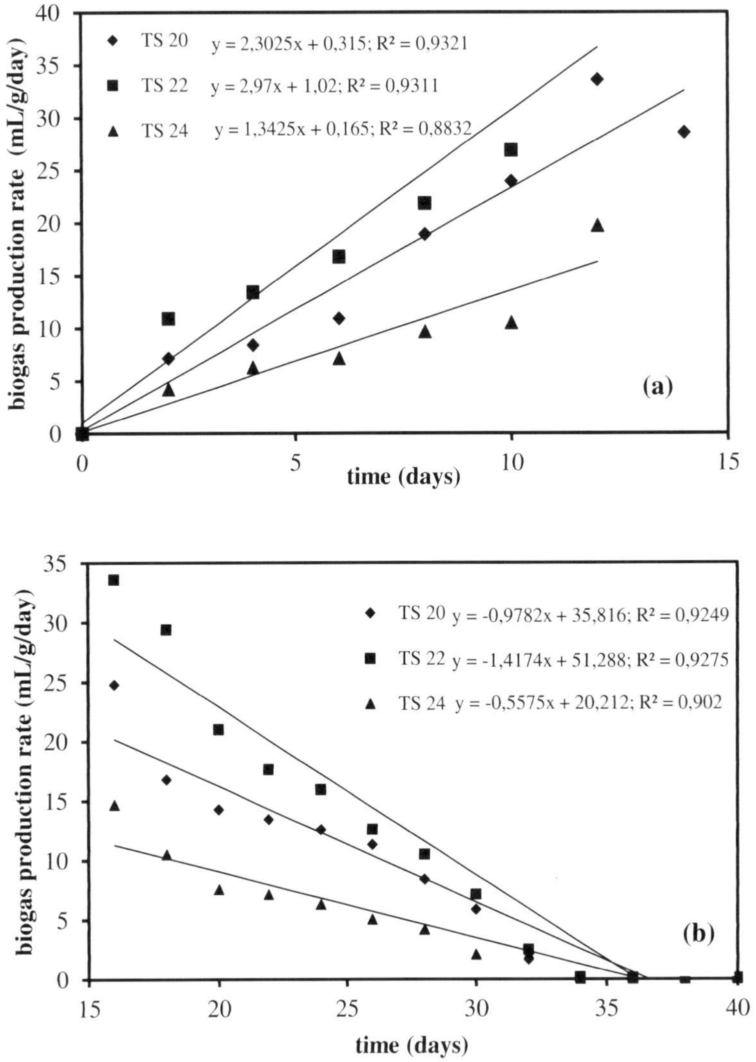

Figure 1. The linear graph of biogas production rate (a) the ascending period of biogas production rate (b) the descending period of biogas production rate.

Figure 1a and 1b shows the linear graph of biogas production rate at the TS contents of 20%, 22%, and 24%. As shown in Fig. 1(a), the regression coefficient (R^2) ranged from 0.88 to 0.93, while in Fig. 1(b) the R^2 ranged from 0.90 to 0.92.

The descending biogas production graph of the linear equation had the $R^2 > 0.9$ at the TS contents of 20%, 22%, and 24%. On the contrary, the ascending biogas production graph obtained the $R^2 > 0.9$ only at TS contents of 20% and 22%.

The exponential graph is presented in Figure 2.

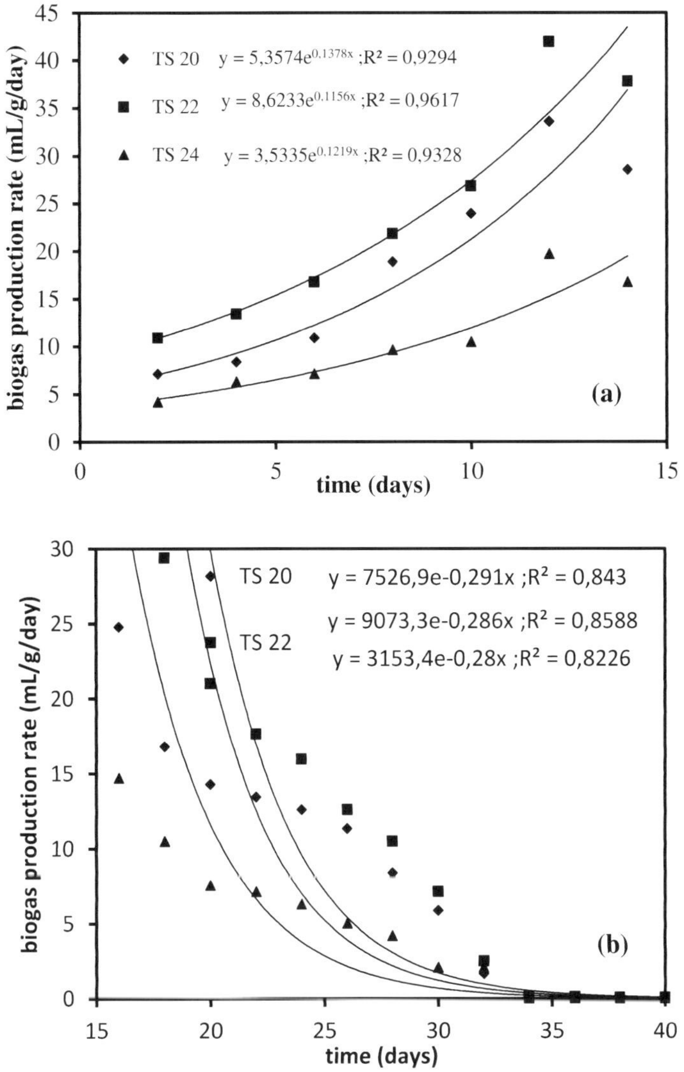

Figure 2. The exponential graph of biogas production rate (a) the ascending period of biogas production rate (b) the descending period of biogas production rate.

As presented in Figure 2, the R^2 ranged from 0.82 to 0.96. The R^2 on the ascending graph using exponential equation was greater than the R^2 on the ascending graph using linear equation. Therefore, the exponential equation had a better simulation than the linear equation on the ascending graph. The similar result was also reported by Lo et al. [11] who stated that the exponential equation model had a larger R^2 compared to the linear equation model.

On the ascending graph of the exponential equation, the $R^2 > 0.9$ was obtained at TS contents of 20%, 22%, and 24%, but on the descending graph of the exponential equation had the $R^2 < 0.9$. As seen in Figure 2b, the R^2 on descending graph was smaller than descending graph using linear equation. Therefore, the linear equation had a better simulation than the exponential equation on the descending graph of biogas production. The rate constants of biogas production by the first-kinetic model is shown in Table 1.

Table 1. The rate constants and regression coefficient of biogas production from corn cob by a first kinetic model

TS contents (%)	k (1/day)	R^2
20	0.2219	0.8351
22	0.2160	0.8477
24	0.2114	0.8461

The value of k represents the rate constant of biogas. As shown in Table 1, the greatest value of k was obtained at TS content of 20%. However, the value of R^2 had a smaller value compared to the value of R^2 using linear and exponential models. Therefore, the kinetic of biogas production from corn cob was more suitable modeled by linear or exponential models compared to the first-kinetic model

5. Conclusions

The exponential equation has a better correlation value than the linear equation during the ascending period of biogas production. While linear equation has better correlation than exponential equation on the descending period of biogas production. The R^2 obtained by first-order kinetic model has the smallest value than both linear and exponential models.

References

[1] Poudel J and Oh S C 2014 Effect of torrefaction on the properties of corn stalk to enhance solid fuel qualities *Energ.* **7** 5586-5600

[2] Eze J I and Ojike O 2012 Anaerobic production of biogas from maize wastes *Int. J. Phys. Sci.* **7** 982-987

[3] Grando R L, Valeria F, Maria A, and Antunes D S 2017 Mapping of the use of waste as raw materials for biogas production *J. Environ. Prot.* **8** 120-130

[4] Horváth I S, Tabatabaei M, Karimi K, and Kumar R 2016 Recent updates on biogas production - A review *Biofuel Res. J.* **10** 394-402

[5] BP Statistical 2016 *BP Statistical Review of World Energy*, June 2016

[6] Zhang Y, Ghaly A E and Li B 2012 Physical properties of corn residues *Am. J. Biochem Biotechnol.* **8** 44–53

[7] Mir M. A, Hussain A, and Verma C 2016 Design considerations and operational performance of anaerobic digester : A review *Cogent Eng.* **28** 1–20

[8] He M, Sun Y, Zou D, Yuan H, Zhu B, Li X, and Pang Y 2012 *Proc. Int. Conf. on Waste Management and Technology* vol 16 (Procedia Environmental Sciences) p 85–94

[9] Length F 2013 A comparative study of biogas production using plantain/ almond leaves and pig dung, and its applications *Int. J. Phys. Sci.* **8** 1291–1297

[10] Ghatak M D and Mahanta P 2014 Comparison of kinetic models for biogas production rate from saw dust *Int. J. Res. Eng. Technol.* **3** 248–254

[11] Lo H M, Kurniawan T A, Sillanpaa M E T, Pai T Y, Chiang C F, Chao K P, Liu M H, Chuang S H, Banks C J, Wang S C, Lin K C, Lin C Y, Liu W F, Cheng P H, Chen C K, Chiu H Y, and Wu H Y 2010 Modeling biogas production from organic fraction of MSW co-digested with MSWI ashes in anaerobic bioreactors *Bioresour. Technol.* **101** 6329–6335

[12] Brown D, Shi J and Li Y 2012 Comparison of solid-state to liquid anaerobic digestion of lignocellulosic feedstocks for biogas production *Bioresour. Technol.* **124** 379–386

IOP Conf. Series: Materials Science and Engineering **345** (2018) 012005 doi:10.1088/1757-899X/345/1/012005

Synthesis and catalytic activity of Birnessite-Type Manganese Oxide synthesized by solvent-free method

S S Siregar and A Awaluddin
Department of Chemistry, University of Riau, Kampus Bina Widya, Pekanbaru, 28293 Indonesia

E-mail: amirawaluddin01@gmail.com

Abstract. Redox reaction between $KMnO_4$ and glucose usingsolvent-free method produces the octahedral layer birnessite-type manganese oxide. The effects of mole ratios, temperatures, and calcinations time on the structures and crystallinity of the oxides were studied throughthe X-ray powder diffraction analysis. The mole ratio of $KMnO_4$/glucose (1:3) produces the purebirnessite with low crystallinity, whereas the mole ratio of $KMnO_4$/glucose (3:1) yields high crystalline birnessite with minor components of hausmannite-type manganese oxide.The increasing of the temperature and calcinations times (300-700 °C and 3-7 h, respectively) willimprove the crystallinity and the purity of the as-synthesized oxide. Further experiments also showed that the as-syntesized octahedral layer birnessite-type manganese oxides have catalytic activity on the degradation of methylene blue (MB) dye with H_2O_2 as oxidant. The results revealed that the effective degradation could be achieved only in the presence of both the birnessite and H_2O_2, whereas without the addition of catalyst (H_2O_2only) or addition of H_2O_2 (catalyst only), the 3.5% and 15.5% of MB removal were obtained, respectively.

1. Introduction

Synthetic dyes are widely used in textile, paper, printing, and plastic industries. The wastewater generated from this industry contains of residual dyes with low biodegradability [1]. The dye wastes become a major source of water pollution and may cause many significant environmental problems. Thus, dye wastewater must be processed to remove or reduce dyestuff content before released into water body. Methylene blue dye is one of the most commonly used synthetic dyes in a textile industry. The use of methylene blue may cause adverse effects, such as nausea, vomiting, cyanosis, irritation to the gastrointestinal tract and skin in direct contact [2]. This compound is relatively stable, therefore is difficult to oxidize using conventional method.

Removal of organic dyes from natural and industrial wastewater has drawn attention because of the highly toxic effect of these contaminations on animals, plants, and human beings. One of effective methods currently being used for dye removal is called Fenton-like process using a metal oxide as a heterogeneous catalyst and hydrogen peroxide (H_2O_2) as an oxidant. Octahedral layer manganese oxides having similar structure with birnessite mineral is known to have catalytic activities on dye degradation. These oxides have been prepared using a wide range of methods, resulting various physicochemical properties. In the Fenton-like process, the oxide reacts with H_2O_2 to produce various types of active radicals as an intermediate for dye degradation [3]. Formation of •OH, $O_2•^-$, and $HO_2•$ to degrade methylene blue by β-MnO_2/H_2O_2 was reported, this β-MnO_2 were synthesized by reflux and hydrothermal methods [4-5]. Birnessite (δ-MnO_2) prepared by hydrothermal [6]and sol gel [7]method can also be used for methylene blue degradation by using H_2O_2 as an oxidant.

Different synthesis methods of manganese oxide can produce different types of polymorphs. The conditions of preparation and precursor also affect the morphology, surface area, pore volume, and crystallinity of manganese oxide [8]. These structures and surface properties give a considerable effect on the catalytic activity inmethylene blue degradation [9-10]. Herein, we introduce a simple one-step solvent-free method for preparing of manganese oxide. This synthetic method eliminates the use of the solvent, give short synthesis time, low cost, and high yield. To date, few reports have been made on preparation of manganese oxide via this method [11-12]. In this study, we synthesized manganese oxide using $KMnO_4$ and glucose as a precursor using free-solvent method.Moreover, the effect of various mole ratio between $KMnO_4$ and glucose, time and temperature of calcination on the crystallinty of the products were investigated. Further experiments revealed that the as-synthesizedbirnessite exhibited excellent catalytic efficiency for the degradation of methylene blue (MB, used as the model) by a Fenton-like reaction.

2. Experimental

2.1. Materials
All the chemical reagents were purchased from Merck (Germany), the used of analytical purity is without any further purification.

2.2. Synthesis
The manganese oxide was prepared viasolvent-free method. The appropriate amount of $KMnO_4$ and glucose areshown in Table 1. The resulting fine-powder of $KMnO_4$ and glucose were then mixed and ground in one mortar. The mixture was calcinedat the certain condition (Table 1). The final product was cooled up to room temperature, washed using HCl 0.1 M and distilled water. The black product was dried at 110 °C.

2.3. Characterization
X-ray powder diffraction (XRD) was performed on a Shimadzhu 700 MaximaX X-ray diffractometer operating at 40 kV and a current of 30 mA with Cu Kα radiation (λ = 1.54060 Å). The data were collected with scan speed 2° per minute in 2θ, from 2θ = 10° to 90°. The scanning electron microscopy (SEM) images were taken on a ZEISS EVO-50 instrument. The Brunauer-Emmett-Teller (BET) surface area measurements and Barret-Joyner-Halenda (BJH) adsorption were carried out using a Quantachrome Nova nitrogen adsorption apparatus. Prior to the adsorption measurements, the samples were degassed under vacuum for 3 h at 120 °C.

2.4. Catalytic activity
The catalytic reaction was carried out in the Beakerglass, contained 25 mL MB dye solution (100 mg/L), 60 mL deionized water, and 25 mg catalyst. The suspension was continuously stirred magnetically for 30 min to achieve adsorption/desorption equilibrium. Following adding 15 mL of 30 wt% H_2O_2 solution, the mixture was allowed to react at room temperature. At given time intervals, 2,5 mL of the mixture was pipetted into a volumetric flask, quickly diluted with distilled water to 25 mL, and placed in iced bath. For optical absorption measurements, the diluted solution was immediately centrifuged at 10000 rpm for 3 min to remove the catalyst particles. Then the centrifuged dye solution put into a quartz cell, and the changes of absorptions at 664 nm were taken to identify the concentrations of MB, using a Shimadzu UV mini 1240 UV-vis spectrophotometer.

IOP Conf. Series: Materials Science and Engineering **345** (2018) 012005 doi:10.1088/1757-899X/345/1/012005

Table 1. Variation condition of synthesis birnessite-type manganese oxide

Mole ratio $KMnO_4$:glucose	Temperature of calcination (°C)	Time of calcination (h)
1 : 3	500	5
1 : 1	500	5
3 : 1	500	5
3 : 1	300	5
3 : 1	700	5
3 : 1	700	3
3 : 1	700	7

3. Results and discussion

3.1. Characterization of manganese oxide

Octahedral layer Manganese oxide birnessite was synthesized using free-solvent method. This method was realized by mixing two different precursors ($KMnO_4$ and glucose) to form solids followed with the heating process. Prior to the mixing process, each precursor in the groundseparating with mortar, then $KMnO_4$ and glucose were mixed in a single mortar. This process is a exothermic reaction, it generates sparks when $KMnO_4$ and glucose were in theground together in one mortar. The variation of mole ratio, temperature, and time of calcination in the synthesized of manganese oxide can be seen in Table 1.

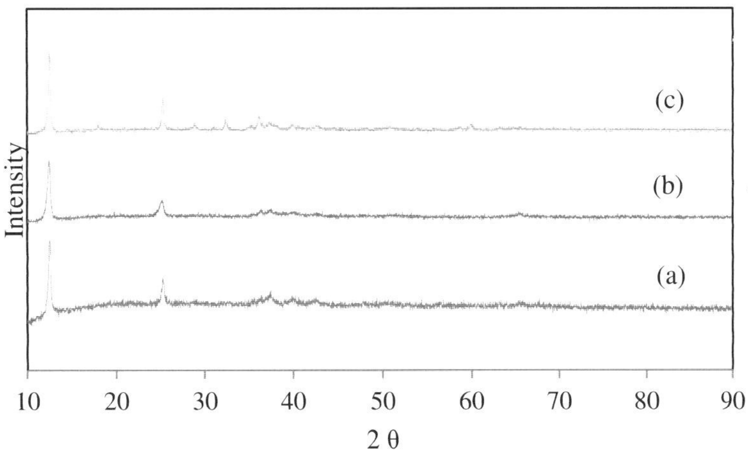

Figure 1. XRD patterns of the obtained products at different mole ratio ($KMnO_4$:glucose): (a) 1:1; (b) 1:3; (c) 3:1

X-ray diffraction was used to characterize the phase and crystallinity of the as-synthesized manganese oxide samples. Figure 1 shown that the XRD patterns of the resulting manganese oxide products synthesized under different mole ratio of $KMnO_4$and glucose. The diffraction peaks of these manganese oxide match well with the standard pattern of moniclinic phase of birnessite (JCPDS card no. 80-1098). The diffraction peaks at 2θ values of 12.59°; 25.23°; 37.40°; and 42.54° can be ascribed to the reflections of (001), (002), (-111), and (-112) planes of the birnessite. As shown in Figure 1, the peaks at 2θ values of 12.59, and 25.23°C increase in intensities of the mole ratio of $KMnO_4$/glucose. The mole of $KMnO_4$ is higher than mole of glucose (Figure 1c), it will generate birnessite-type manganese oxide with relatively good crystallinity, with small impurities due to hausmannite-type manganese oxide at 2θ values of 17.98°, 28,58°; and 32,35°. However, if mole ratio of $KMnO_4$is

1:3in Figure 1b, the impurities peaks will be disappeared and resulting more pure birnessite phase. This results indicate that the small ratio of $KMnO_4$/glucose facilitates the formation of more pure birnessite phase with lower crystallinity, whereas the high mole ratio of $KMnO_4$/glucose produce more crystalline phase of birnessite with minor impurities. The different crystal phases of manganese oxide are known to correlate with the potassium content. Ching et al., reported that the high concentration of $KMnO_4$ generates octahedral layered birnessite, and the lower concentration of $KMnO_4$ produces cryptomelane-type manganese oxide octahedral molecular sieve [13].

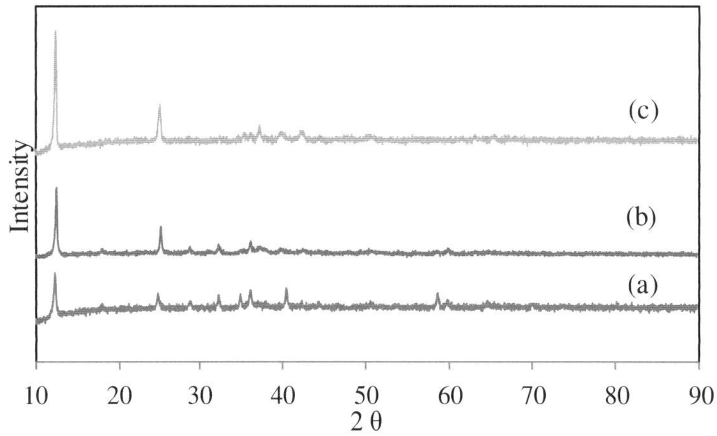

Figure 2. XRD patterns of the obtained products with mole ratio of $KMnO_4$:glucose = 3:1 at different temperature of calcination: (a) 300; (b) 500; (c) 700 °C

The effect of temperatures on the structure and crystallinity of as-synthesized manganese oxides is shown in Figure 2. The calcination temperatures have a sizably impact in the structure and crystallinity of the as-synthesized manganese oxide. The increasing of the calcination temperature increases the intensity of the main reflection peaks of birnessite (2θ values of 12.38°; 24.91°).At low temperature (300 °C), the oxide consists of mixtures of birnessite and hausmannite.The increasing of the temperature (700 °C in Figure 2), the intensities of two main peaks of birnessite also increases, reflecting more crystalline phase of birnessite and the peaks due to hausmannite phase disappear. Thus, the higher temperature produces more crystalline and pure birnessite.

Effect of the calcination time on the structure and crystallinity of the manganese oxides can be seen in the Figure 3. When the calcination time is 3 hours, the intensities at 2θ value of 12.54°; 25.26°; 35.30°; 37.33°; 39.90° and 42.66° are relatively low, indicating the low crystallinity of birnessite phase. For a 5-hour calcination, the diffraction peaksat of 2θ at 12.50°; 25.12° of birnessite phase increase substantially. Further increasing of the calcinations time (7hours) reduces the intensity at of 2θ at 12.50°, but produces sharper peaks. The calcinations times from 5 and 7 hours do not give significant effect in the crystallinity of birnessite.

IOP Conf. Series: Materials Science and Engineering **345** (2018) 012005 doi:10.1088/1757-899X/345/1/012005

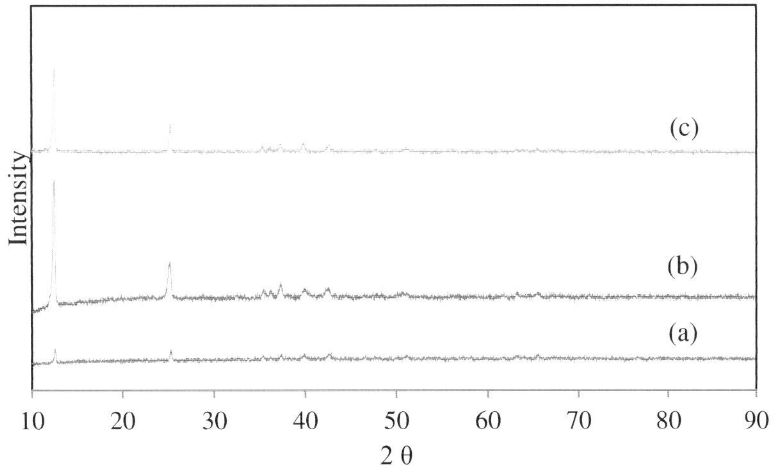

Figure 3.XRD patterns of the obtained products with mole ratio of $KMnO_4$:glucose = 3:1 at different time of calcination: (a) 3; (b) 5; (c) 7 hours

The morphologies of the as-synthesized birnessite were observed by SEM images, as shown in Figure 4(a) and 4(b), respectively. From the low-magnification SEM image (Figure 4(a)), the birnessiteconsisted of aggregate particles with irregular sizes.

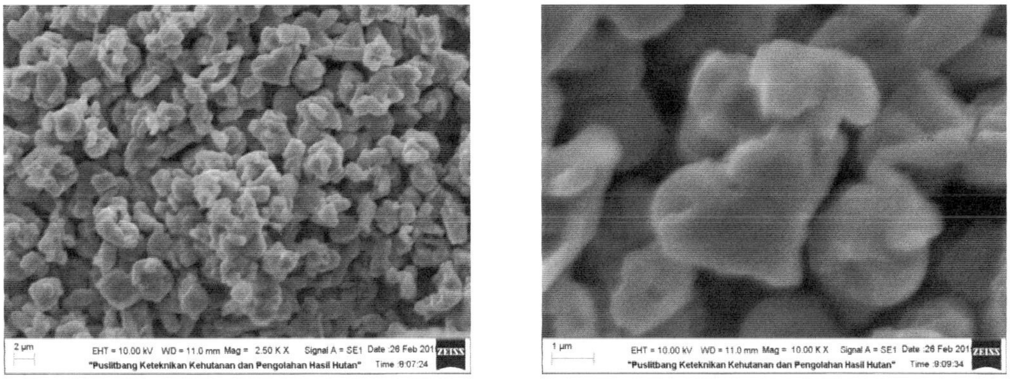

(a) **(b)**

Figure 4. SEM images of the birnessite as-synthesized at 700 °C for 7 hours, Mag.x (a) 2,500 (b) 10,000 (mole ratio of $KMnO_4$:glucose = 3:1)

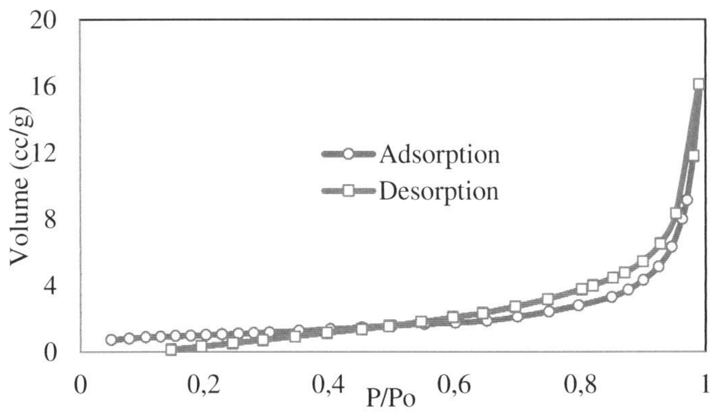

Figure 5. N_2 adsorption/desorption isothermal plot of the birnessite synthesized at 700 °C for 7 hours.

Figure 5 shown the adsorption-desorption isothermal plot of the as-synthesized birnessite prepares at 700 °C for 7 hours. The pore volume and pore size distribution were obtained using the Barrett, Joyner and Halenda (BJH) method from desorption branch of the isotherms and surface area were determined by the BET method using N_2 adsorption-desorption isotherms. The isotherm obtained is type III according to the IUPAC classification with the Brunauer-Emmett-Teller (BET) specific surface area of 23.1278 m^2/g. Figure 6 shown the mesopore size distributions calculated by BJH method. The BJH plots show a pore size distribution in the range of 20-80 Å, indicating mesoporous materials.

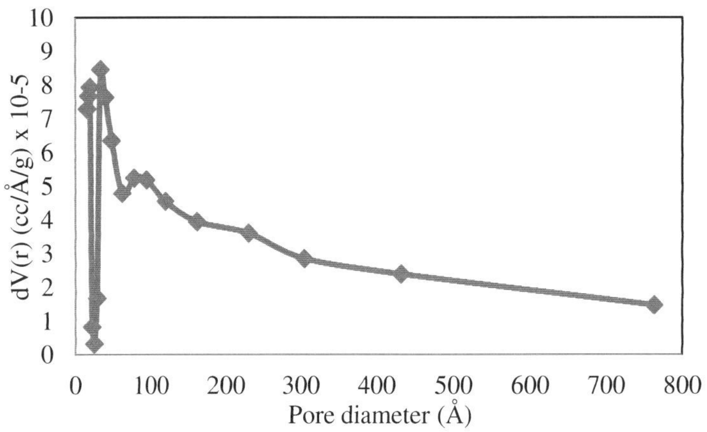

Figure 6. Mesopore size distribution plot by BJH method of the birnessite synthesized at 700 °C for 7 hours.

3.2. Catalytic degradation of methylene blue

Previous studies have reported that manganese oxides with various structures exhibited excellent catalytic properties in degradation of dyes using oxidizing agent such as H_2O_2[3-7].Several species are responsible for the degradation, included the free radical species, such as •OH, $O_2•^-$, and $HO_2•$ .The degradation of dyes using the manganese oxides proceeds by an adsorption–oxidation–desorption mechanism [3,14,15]. Herein, the as-synthesized birnessite-type manganese oxides were tested for their catalytic performance in the degradation of methylene blue as a model under controlled conditions. In the absence of a catalyst (only MB+H_2O_2), no obvious dye degradation was observed after 120 min. With the as-synthesized birnessite as catalyst but no H_2O_2 (only MB+ birnessite), the degree of decoloration reaches to 15% within 60 minute and then tends to be saturated, which can be ascribed to adsorption of the dye molecules on thebirnessite. In the contrast, the presence of the birnessite and H_2O_2, the degree of degradation reached to about 50% within 90 min (Figure 7).

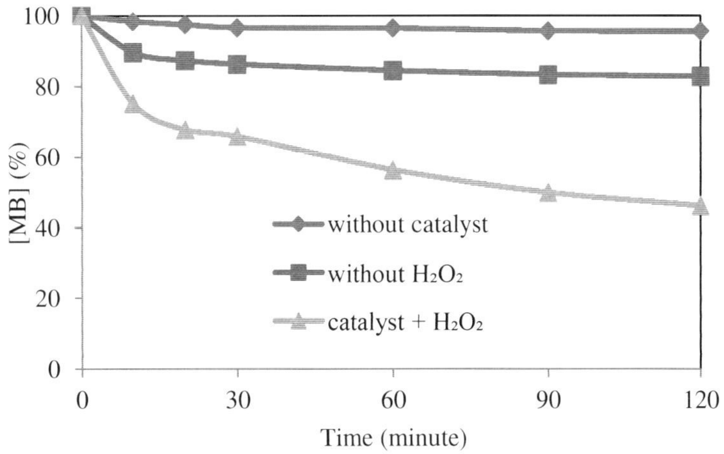

Figure 7. Time profil of MB degradation: (♦) only MB + H_2O_2(■) only MB + birnessite (▲) MB + birnessite + H_2O_2(reaction condition: $[MB]_o$=25 mg/L; [Birnessite]=0,25 g/L; [H_2O_2,30 wt%]=15 mL; at room temperature.

The UV-vis spectra of the MB solution and the dye degradation at different reaction time are showed in figure 9 and 10. In this study, the characteristic absorption peak at about 664 nm of MB was used as a monitored parameter during the catalytic degradation process. The absorption spectra at t = 0 minute is used for the starting solution of MB with a concentration of 25 mg/L (without the birnessite and H_2O_2). The four characteristic absorption peaks (245, 292, 614, and 664 nm) of the MB dye could be observed in the absorption spectrum. As soon as H_2O_2 was added to the mixture of MB and birnessite,the characteristic absorption peak of MB (664 nm) dropped (in 10 minute observation), and became very broad and weak with the time prolonged. In addition, the peaks at 614 and 663 nm are not observed in the spectrum even in the early of 10 minute reaction. The color of the reaction mixture slowly turned from deep blue to blackish, indicating that most of the MB dye was degraded with the degradation efficiency 95.67%.

Figure 10 shown thatUV-Vis spectra of MB with the birnessite (no addition H_2O_2) with different reaction times. The four characteristic absorption peaks of methylene blue decrease considerably, but remain even after 180 minutes of reaction time. This process indicates that the MB is believed to be only adsorbed on the surface of the birnessite-type manganese oxide.

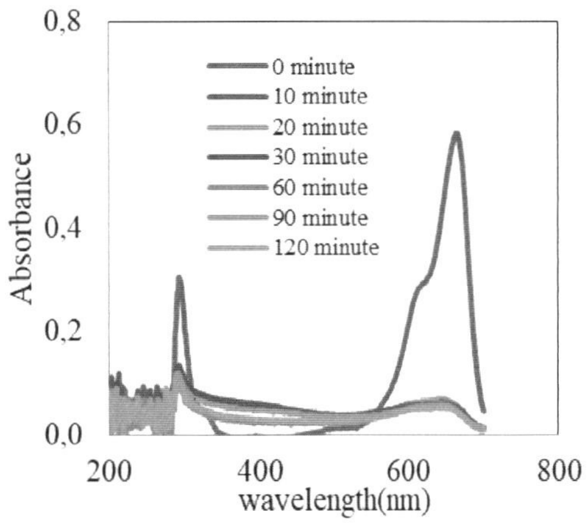

Figure 8. UV-Vis spectra of MB with addition of birnessite and H_2O_2, (reaction condition: $[MB]_o$ = 25 mg/L; [birnessite] = 2 g/L; [H_2O_2, 30 wt%] = 5 mL; at room temperature)

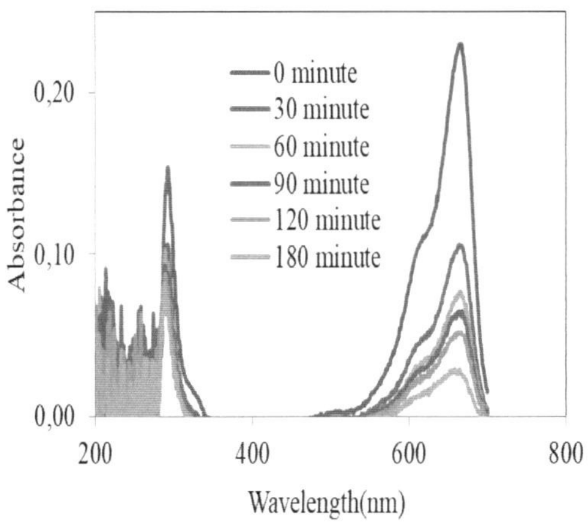

Figure 9. UV-Vis spectra of MB without addition of H_2O_2 (only birnessite); (reaction condition: $[MB]_o$ = 25 mg/L;[birnessite] = 2 g/L; at room temperature)

Dye molecules comprise of two key components, the chromophore, responsible for producing color, and the auxochrome, which is a functional group of atoms attached to the chromophores. This functional group is able to modify the ability of the chromophore to absorb light.Changes in the chemical structure of MB dye molecule are detected by decreasing color intensity (decolorization) that can be used as early indicators of dye degradation[16]. Further analysis with GC-MS is required to find out the final product and mechanism of the catalytic degradation of MB using the birnessite.

4. Conclusion

A solvent-free synthetic method of the octahedral layer birnessite-type manganese oxidehas been developed from the redox reaction between $KMnO_4$ and glucose. The Mole ratio $KMnO_4$ /glucose is critical in this process, where small mole ratio of $KMnO_4$/glucose (1:3) generates pure birnessite with low crystallinity. However, when mol ratio of $KMnO_4$/glucose is high (3:1), high crystalline birnessite is produced with minor contaminant. In addition, higher calcination temperatures lead to substantial improvements of the both crystallinity and purity of the as-synthesized oixeds. The as-synthesized octahedral layer birnessite manganese oxides have catalytic activities on the degradation of MB dye using Fenton-like reaction. To obtain the optimal performance in degradation, appropriate amount of catalyst and H_2O_2 should be applied.

References

[1] Kulandaivel S, Kaleeswari P and Mohanapriya P2014 *Int.J.Curr.Microbiol.App.Sci.* **3,** 865-882
[2] Fil B A, Özmetin C, Korkmazand M2012 *Bull. Korean Chem. Soc.* **33,** 3184-3190
[3] Yu C, Li G, Wei L, Fan Q, Shu Q andYu J C 2014 *Catal. Today* **224,** 154-162
[4] Cui H J, Huang H Z, Fu M L, Yuan B L and Pearl W2011 *Catal. Commun.* **12,** 1339-1343
[5] Cheng G, Lin T, Yang R, Sun M, Lan B, Yang L and Deng F2014 *J. Solid State Chem.* **217,** 57-63
[6] Ma Z, Wei X, Xing S and Li J 2015 *Catal. Commun. J.* **67,** 68-71
[7] Zhang L, Nie Y, Hu C and Hu X2011 *J. Hazard. Mater.* **190,** 780-785
[8] Ye Q, Lu H, Zhao J, Cheng S, Kang T, Wang D and Dai H 2014 *Appl. Surf. Sci* .**317** 892-901
[9] Kim E J,Oh D, Lee C S, Gong J, Kim J and Chang Y S 2016 *Catal. Today* **282,** 71-76
[10] Cao G, Su L, Zhang X and Li H2010 *Mater. Res. Bull.* **45** 425-428
[11] Ding Y S, Shen X F, Sithambaram S, Gomez S, Kumar R, Crisostomo V M B, Suib S L and

Aindow M 2005 *Chem. Mater.* **17** 5382-5389
[12] Rocha R P, Soares O S G P, Órfão J M Pereira M F Rand Figueiredo J L 2013 **99** 95411
[13] Ching S, Roark J L, Duan N and Suib S L 1997 *Chem. Mater.* **4756,** 750-754
[14] Yu T T, Li K L, Guo X L, Li F, Huang J M and Zhang Y X 2015 *J. Phys. Chem. Solids* **87** 196-202
[15] Zhang W, Yang Z, Wang X, Zhang Y, Wen X and Yang S 2006 *Catal. Commun.* **7** 408-412
[16] Martani E, Margino S and Nurnawati E 2011 *J. Mns. dan Lingkung.* **18** 127-136

IOP Conf. Series: Materials Science and Engineering **345** (2018) 012006 doi:10.1088/1757-899X/345/1/012006

Experimental study of Pb (II) solution sorption behavior onto Coffee Husk Bioactivated Carbon

Z Fona and U Habibah
Department of Chemical Engineering, State Polytechnic of Lhokseumawe, Jl. B. Aceh-Medan, km.280.5 Buketrata, Lhokseumawe, Aceh, Indonesia

E-mail: zahrafona@yahoo.com

Abstract. Coffee husk which is abundantly produced in the coffee plantations is potential to be a challenging adsorbent. The fate of Pb (II) solution in the sorption mechanism onto the adsorbent has been investigated. This paper aimed to study the efficiency of Pb (II) aqueous solution removal using activated carbon from coffee husk (CAC). The sorption characteristics were using two isotherm models, Langmuir and Freundlich, were also reported. The coffee husk from local plantations in Middle Aceh was carbonized and sieved to 120/140 mesh. The charcoal was activated using hydrochloric acid before contacted with the different initial concentrations of Pb (II) solution. The remaining concentrations of the metal in the specified contact times were determined using Atomic Adsorption Spectrophotometer at 283.3 wavelength. The result showed that the equilibrium concentrations were obtained in about 30 minutes which depended on the initial concentration. The sorption mechanism followed Freundlich isotherm model where the adsorption constant and capacity were accordingly 1.353 and 1.195 mgg^{-1}. The iodine sorption was up to 1,053 mgg^{-1}. Based on the ash and moisture content, as well as iodine sorption, the activated carbon met the national standard.

1. Introduction

The natural water resources quality nowadays is worrying. Therefore, using natural water directly is not recommended anymore due to the lack of quality, high possibility of contaminant from industrial process, including pharmaceutical and cosmetic, agricultural activity, and household wastewater, where go to and contaminate water body. Additionally, natural water resources might contain contaminants come from natural organic matters which also have some negative impact to human health. However, the artificial contaminants such as toxic organic compounds: heavy metals, dyes, pesticides, and pharmaceuticals and even the wastes of water treatment [1]. These, might cause to enhance the toxic compounds on the natural occurring pollutants in the water body [2].

Heavy metals can occur as solution in water, such as lead, and they are non-degradable. Therefore, it is very important to find out the effective, efficient and inexpensive materials to eliminate the compounds. Many research has been conducted to eliminate the heavy metal solution in the water. Among the other methods, adsorption using activated carbon found to be the most effective and low-cost process. Hence, it is widely used in water purification process, and industrial systems as well as home applications [3-4].

Activated carbon, also named activated charcoal is carbon which tremendously high surface area, where the adsorption is taking place. The surface area can be from 500 to 1500 m^2. It can adsorb a wide diversity of organic and inorganic pollutants in aqueous solution [5-7]. However, the adsorption efficiency might depend on the ion type in the solution as well as the origin of carbon itself. Every biomass carbon source has a specific characteristic in the carbon active form, beside the chemical used for activation or the activated method performed.

The activated carbon can be produced from natural resources such as biomass waste. Huge amount of biomass waste disposed improperly all over the world [8]. Using them as carbon active is one of the competitive way offered. Some organic waste from agricultural process or crop residues such as tea seed shells [9], rice husk [5], coconut shell [6], doom palm shell [10], and many other sources. In current research, coffee husk collecting from community plantation in Middle Aceh converted to be activated carbon (CAC). Biomass activated carbon, nevertheless, become the most promising adsorbent due to it enormously high adsorption capacity, very cheap, easy to be produced and used.

In advanced method of production of activated carbon, nitrogen is used as a treating gas to produce highly porous carbon. However, this process is costly and complicated to perform. In this research, the carbon produced from coffee husk (coffee husk activated carbon/CAC) using a simple method, carbonization and activation using dilute hydrochloric acid solution. The fate of heavy metal pollutant, Pb (II) were investigate, the equilibrium concentrations and isotherm characteristics using Langmuir and Freundlich models.

2. Materials and method

Coffee husk biomass waste was collected from the coffee plantation in Middle Aceh. Hydrocloric acid and $Pb(NO_3)_2$ from Merck, Germany.

2.1 Preparation of CAC

Coffee husk was washed and dried under sunshine. The carbonization was performed using traditional process in closed vessel for 2 hours. The charcoal was sieved to 40, 60, 80, 100, and 120 mesh and activated using 0,1 N hydrochloric acid for 48 hours. Finally, the activated carbon (CAC) was filtered, neutralized using aquadest, and dried in oven of 105 ^{o}Ctemperature for 3 hours.

2.2 Batch experiment of sorption process

Activated carbon was weighed 1 g, put into 5 different Erlenmeyer flasks, each poured with lead solution of 10, 20, 30, 40, and 50 ppm. Therefore, they were stirred using the electric shaker of 200 rpm, and taken after 10, 20, 30, 40, and 50 minutes to analyze the concentration using Atomic Absorption Spectrophotometer at 283.3 mm wavelength.

2.3 Characteristics of CAC determination

Water content of the activated carbon were analyzed using Moisture Analyzer Mx 50 by heating sample to 105°C until reach constant weight, while the ash content determined according to SNI 1995, by heating sample to 650 C for 3 hours. Iodine sorption capacity of the CAC was determined using national standard (SNI 1995) by titration method. Iodine solution of 0.1 N was poured into the flask containing 0.25 g CAC and stirred for 15 minutes, the adsorbed iodine measured as iodine adsorption capacity.

3. Results and discussion

3.1 Equilibrium concentration

The lead solution concentrations after different contact time adsorption onto CAC through different initial concentrations have been measured. Therefore, the equilibrium concentrations obtained by plotted the concentration each time per initial concentration (C/Co) to time, as shown in Figure 1. The equilibrium stage was one of the specific characteristic depends on natural of the adsorbate and adsorbent used.

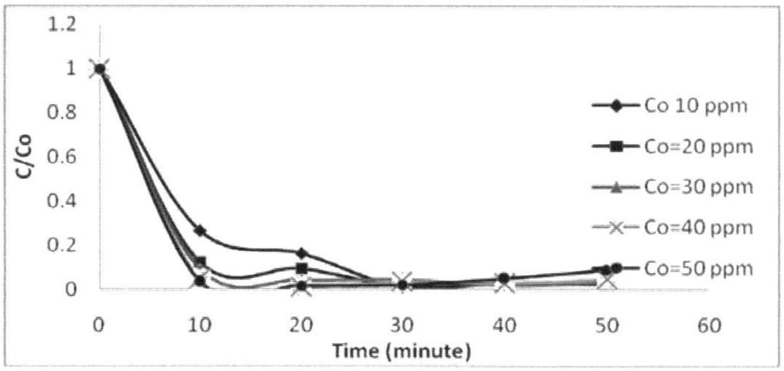

Figure 1. Equilibrium stage of lead solution adsorption under different initial concentration.

The equilibrium obtained approximately in 30 minutes varies along with the initial adsorbate concentrations. Higher concentration yielded shorter time to meet the balance between the ion in the solution and the ion adsorb onto sorbent, and vice versa.

3.2 Removal of lead solution

The removal percentage of lead solution onto CAC depended on contact time and initial lead concentrations. Figure 2 shows the removal percentage in each initial concentration and contact time, where 73.31-98.77% of lead removed.

At high initial concentration, the lead ions existing for the sorption onto the CAC surface are very high. According to El-Ashtoukhy [11], at low concentration the adsoption occurred in specific site, while in higher concentrations the sites are saturated and load the exchange place. Moreover, M. Karnib, et al., 2014 found that the removal percentage of activated carbon decreased by increasing the metal ion concentrations.

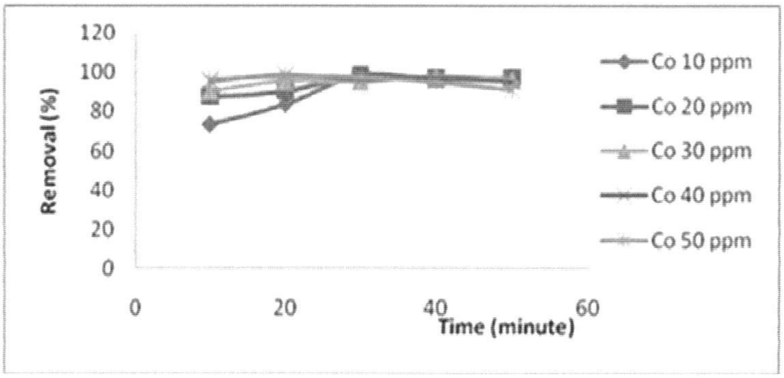

Figure 2. Lead solution removal under different initial concentration.

3.3 Adsorption isotherm

Adsorption isotherm shows the natural characteristics of the sorbent in accordance to solute-sorbent interface interaction. The sorbate concentration and deposit onto sorbent at constant temperature was also indicated in the isotherm. The equilibrium data from the experiments were fitted to the Langmuir and Freundlich isotherm models.

Langmuir isotherm sorption model suppose the adsorption of the sorbate onto the sorbent occurs only in monolayer. It has the uniform energies over the surface. There is no sorbate migration in the surface plane [12-13]. Langmuir model was explained by Equation 1.

$$\frac{Ce}{Qe} = 1 / q_0 b + C_e / q_0 \quad (1)$$

IOP Conf. Series: Materials Science and Engineering **345** (2018) 012006 doi:10.1088/1757-899X/345/1/012006

where *Ce* is the equilibrium concentration (mgL^{-1}), *qe* is the amount of adsorbed concentration at equilibrium (mgg^{-1}) and q_0 and bare Langmuir parameters related to maximum adsorption capacity (mgg^{-1}) and bonding energy of adsorption (Lmg^{-1}) respectively. Plot of Ce/Qe to Ce is represented in Figure 3. According to Langmuir model, the linear equation y = 0.143x + 0.583. The maximum adsorption capacity (q_0) and bonding energy adsorption (b) are 6.993 mgg^{-1} and 0.245 Lmg^{-1} respectively (See Table 1). Unfortunately, the linearization parameter R^2 for the isotherm was very low, 0.332. It pointed out that the adsorption of lead solution onto CAC did not fit to the model.

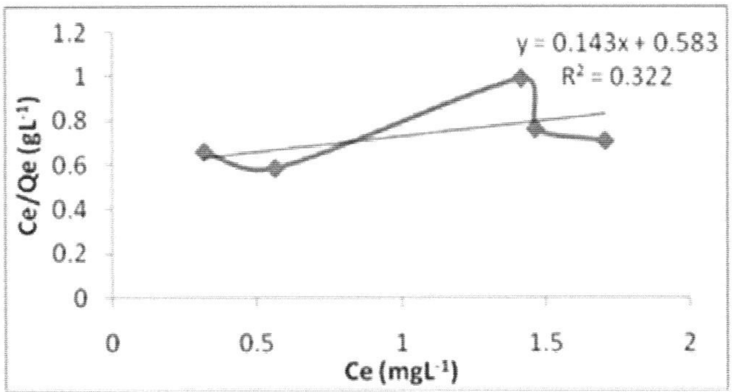

Figure 3. Langmuir isotherm model

The Langmuir model has a dimensionless constant separation factor called equilibrium parameter, R_L. It indicates the characteristics of isotherm: $R_L = 1$ means linear, $R_L>1$ unfavorable, $R_L = 0$ irreversible, and $0< R_L<1$ favorable. The parameter is determined [14] by Equation 2.

$$R_L = 1/(1+b.Co) \quad (2)$$

Figure 4 shows the correlation between R_L to initial concentration of adsorbate. The R_L values ranged from 0.050 to 0.300, which specify that the adsorption was favorable. Furthermore, lower R_L value (closer to 0) more favorable the adsorption process would be. Increasing initial adsorbate concentration produced lower RL value. This result was according to Gao, et al [10], beside it was found that the R_L value more favorable at lower temperature. Freundlich isotherm model is suitable to the sorption of relatively low sorbate concentration and heterogeneous surface of adsorbent [12]. The model was expressed as:

$$Qe = Kf. Ce^{1/n} \quad (3)$$

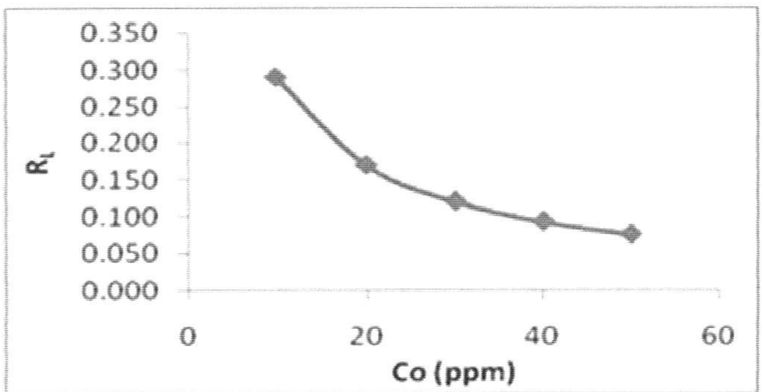

Figure 4. R_L values under different initial lead solution concentration.

where Qe is the amount of sorbate adsorb per unit adsorbent (mgg^{-1}), Ce is the equilibrium concentration, 1/n and Kf are Freundlich constants related to the favorability of adsorption process and capacity of sorbate respectively. The value of $1/n < 1$ means the adsorption is favorable. Figure 5 represented the plot of log Qe to log Ce of Freundlich isotherm model.

The n parameter shows the adsorption intensity, which is the energies of metal ions and adsorbent. Additionally, the Kf value is the constant of multilayer adsorption, which shows the bond strength. The calculated values of each parameter are shown in Table 2.

In Table 2, the constant of 1/n from the adsorption of lead solution onto CAC was 0.837. It indicated that the process was favorable. The n value describes the adsorption intensity of the adsorbent, where 1.195 for CAC. Furthermore, the bond strength (Kf) was 1.353 mgg^{-1}.

The values of correlation coefficient (R^2) were important to find out the best-fitted linear equation. From the result, R^2 of Freundlich model was 0.937, while the Langmuir correlation coefficient was 0.332. According to the values, the adsorption of lead solution onto CAC was the best fitted to Freundlich model, which imply that the process appeared in heterogeneous surface. This result was in accordance to Karnib, et al., [7] where the adsorption of lead solution onto activated carbon was in shape to Freundlich isotherm model.

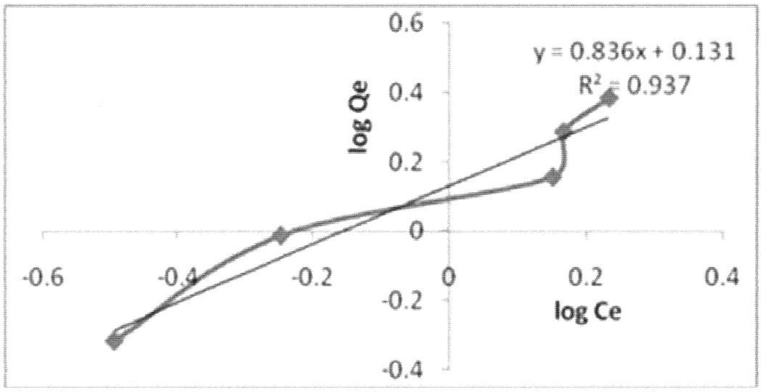

Figure 5. Freundlich isotherm model.

Table 1. Isotherm model parameters and correlation coefficients for adsorption of lead solution onto CAC

Langmuir isotherm model			Freundlich isotherm model			
Q_0 (mgg^{-1})	b (Lmg^{-1})	R^2	Kf (mgg^{-1})	1/n	n	R^2
6.993	0.245	0.322	1.353	0.837	1.195	0.939

3.4 Characteristics of CAC

The characteristics including moisture and ash content, as well as iodine adsoption capacity have been investigated. The characteristics obtained were compared to national standard. Tabel 2 reported the current results compared to Indonesian National Standard (SNI).

Table 2. CAC characteristics

Parameter	National Standard SNI 06-3730-1995	Current work
Water content	Max 15%	5.36 %
Ash content	Max 10%	3.00 %
Iodine adsorption capacity	Min750 mgg^{-1}	723.3-1053.3 mgg^{-1}

Moisture in AC affects significantly the adsorptive quality. High water content of AC decreases the ability of adsorbent to adsorb ions, the surface area occupied mostly by the water molecules. Water content in CAC was 5.36%, less than the standard where max 15%. Ash content indicates the metal oxide and other impurities contained in the activated carbon. The residue obtained after heating the sample to 650 °C for 3 hours was weighed as ash. The ash content in the CAC activated by 0.1 N HCl was 3%. This value was under the ash content of national standard for activated carbon. In another research, Huda, et al., [15] reported that the ash content of activated carbon originated from coffee waste was 9.2%.

4. Conclusion

The removal of lead ions up to 98.77% at 10 ppm initial concentration for 30 minutes adsorption. Lead ions adsorption onto CAC best fitted to Freundlich isotherm model with the adsorption intensity and relative capacity were accordingly 1.195 and 1.353 mgg^{-1}. Activated carbon from coffee husk was favorable for elimination of lead solution. In accordance to ash, water content and iodine adsorption capacity, CAC produced met the national Indonesian standard.

Acknowledgement

We are grateful for financial support of "DIPA PNL 2017" of Indonesia and Wirda Putri for the undefined hard work and help during performing this research.

References

[1] Schwarzenbach RP, Escher BI, Fenner K, Hufstetter TB, Johnson CA, Von Gunten U and Wherli B 2006 *Sci.* **313** 1072
[2] Gavrilescu M, Demnerova K, Aamand J, Agathes S and Fava F 2015 *New Biotechnol.* **32** 147-156
[3] Moreno-Castilla C 2004 *Mater. Carbon.* **42** 83-94
[4] Sweetman M J, May S, Mebberson N, Pendleton P, Vasilev K, Plush S E and Hayball J D 2017 *J. of Carbon Research.* **3** (18) 3-29.
[5] Ahiduzzaman and Islam S 2016 *Springerplus* **5** (1248) 2-14
[6] Lim A P and Aris A Z 2014 *Rev. Env. Sci. Biotechnol.* **13** 163-181
[7] Karnib M, Kabbani A, Holail H, and Olama Z 2014 *Energ. Procedia.* **50** 113-120
[8] Domingues R R, Trugilho P F, Silva C A, de Melo ICNA, Melo L C A, Magriotis Z M and Sanchez-Monedero M A 2017 *Plos One* 1-19
[9] Gao J, Qin Y, Zhou T, Cao D, Xu P, Hochstetter D and Wang Y 2013 *J. Biomed. Biotechnol.* **14** (7) 650-658
[10] Gaya UI, Otene E and Abdullah A H 2015 *Springer Plus* **4** (458) 1-18
[11] El-Ashtoukhy, Amin N and Abdelwahab O *Desalination.* **223** 162-173.
[12] Baccar R, Bouzid J, Feki M and Montiel A 2009 *J. Hazard. Mat.* **162** (2-3) 1522-1529
[13] Coulson J M and Richardson J F 2002 *Chemical Engineering* Vol 2, 5th ed. Butterworth-Heinemann
[14] Hall K, Eagleton L, Acrivos A and Vermeulen T 1966 *Ind. Eng. Chem. Fundam.* **5** (2) 212-223.
[15] Huda H, Ardi Z and Johansyah, A A 2015 *Jurnal IPTEK* **19** (2) 45-53

IOP Conf. Series: Materials Science and Engineering **345** (2018) 012007 doi:10.1088/1757-899X/345/1/012007

Effect of Lanthanum-Natural Zeolite, La/NZA catalyst on biodiesel production from crude palm oil

A Setianingsih, Wisrayetti, Khairat and S Bahri
Department of Chemical Engineering, University of Riau, Kampus Bina Widya Jl.HR. Soebrantas KM 12.5 Pekanbaru

E-mail: syaifulbahri@eng.unri.ac.id

Abstract. Biodiesel can be produced from vegetable oils through the trans-esterification process. In this study, potential vegetable oil of Crude Palm Oil (CPO) was used as sample. The purposes of this research were to produce biodiesel from CPO as an alternative fuel, having study the ratio of impregnation of Lanthanum on NZA, and its catalyst weight to the biodiesel yield. The La/NZA catalyst is made as followed, first the natural zeolite size was reduced using grinding, then activated using HCl 6 N and NH_4Cl 1 N, followed with the drying process. La is impregnated into NZA as solution having variations of 1 and 3% (w/w) of NZA, then it was followed with dried in an oven, calcination, oxidation and reduction. Production of biodiesel is carried out through two stages of esterification and transesterification processes. In the trans-esterification process conducted with the various variation of catalyst weight i.e. 1, 2 and 3% of La/NZA (w/w) for a total weight of 80 grams of CPO sample, having the ratio of oil : methanol 1 : 9. Reaction was lasted for 60 minutes at 60°C having 400 rpm stirring speed. From the result, the conversion of 85.37% is given by the run on using 3% La/NZA catalyst having catalyst weight 1%.

1. Introduction

Energy is decisive factor for the economic development of a country [1]. Nowadays, we know the production of the petrol oil were decreasing, and therefore increasing the energy crisis. So we need to develop production of renewable energy from vegetable oil or biomass. Biodiesel is one of the renewable alternative fuels, it has been used to combine with the petrodiesel, it called biosolar.

Biodiesel is made from vegetable oil derived from renewable natural resources. Crude palm oil is one of big source for production biodiesel because the palm oil plantation is very large. The development of biodiesel from CPO has more advantages than biodiesel from other vegetable oils, such as better physical and chemical properties.

Biodiesel, which is also known as fatty acid methyl ester (FAME), is a mixture of methyl esters resulting from the transesterification reaction of edible or non-edible vegetable oil with aliphatic alcohols with the aid of an acid or base catalyst [2]. Transesterification can be conducted homogeneously or heterogeneously. A heterogeneous catalytic process is more advantageous than a homogeneous. A heterogeneous catalyst has higher activity, easier to separate and reuse, and release less pollution to the environment than a homogeneous catalyst, therefore, the former reduces the overall production costs [3].

IOP Conf. Series: Materials Science and Engineering **345** (2018) 012007 doi:10.1088/1757-899X/345/1/012007

One of the heterogeneous catalysts is zeolite. Availability of zeolite is much enough and has not been used optimally, so zeolite has potentially to developed as a catalyst [4]. In an effort to improve the catalyst performance it is necessary to add metal to the zeolite. One of the right metals used is Lanthanum (La) which serves to increase the activity, stability and selectivity of the catalyst.

The use of La/NZA heterogeneous catalyst in the biodiesel manufacturing process is expected to increase yield and make the biodiesel production process more efficient. The aims of this research are to create and to determine the characterization of La/NZA catalyst for, producing biodiesel from CPO as an environmentally friendly for then can be used as alternative fuel.

2. Crude palm oil (CPO)

The palm oil plant (*Elaeis guineensis Jacq.*) is a plant originating from West Africa. Currently, there are 17 types of palm oil is globally traded feeding oil. The palm oil plant has the optimum growth in the area around the tropical and wet equator with temperatures ranging from 24 to 32°C in a year, abundant sunshine, high rainfall (~ 2,000 mm). Currently, Indonesia is the world's major producer of palm oil that produces more than 44% of the palm oil.

The development of CPO production in Indonesia has increased every year. This can increase the potential for producing biodiesel. Until now, palm oil is still the mostly exported in the form of CPO, while in the country about 80% of palm oil is processed into food products, especially cooking oil [5].

Palm oil produces two types of oil, namely palm kernel oil or palm kernel oil obtained from palm kernel and crude palm oil (CPO) obtained from pulp containing 50% oil [6]. CPO is the result of processed palm fruit flesh (mesocarp) through sterilization process of fresh fruit bunch, threshing, pressing and screening [7]. CPO is a bit thick yellow orange reddish.

3. Catalyst

The catalyst is a chemical that can increase the rate of reaction by lowering the activation energy and directing the reaction to achieve equilibrium without getting involved in the reaction permanently [8]. Catalysts are widely used in nature, laboratory and industry. The catalysts that are in the same phase (liquid) with the reactants are referred to as homogeneous catalysts. While the catalysts that are in different phases with the reactants (can be solids, non-mixing liquids or gases) are referred to as heterogeneous catalysts [9].

3.1 Heterogeneous catalysts

Heterogeneous catalysts consist of two types of acidic and heterogeneous catalysts. Several heterogeneous catalysts have been synthesized in both acidic and alkaline properties. The most commonly used heterogeneous base catalysts are metal oxide compounds such as alkali metals, alkaline earths as vegetable oil transesterification catalysts. The alkaline earth metal oxides (MgO, CaO, SrO, and BaO) are known as single metal oxides [10].

The use of heterogeneous solid catalysts reduces the problems posed by homogeneous catalysts. Heterogeneous catalysts can be recycled and used multiple times with better and product separation. In addition, the use of this catalyst also minimizes raw material costs and production costs. This process is environmentally friendly and can be applied to batch or continuous processes without requiring a purification stage [10]. Heterogeneous catalysts are not as homogeneous catalysts because they can reduce the cost and time of water washing and reduce the neutralization stage to separate and catalyze recovery [11].

3.2 Natural zeolite catalyst

Zeolite is a mineral, it is soft and easy to dry. The color of the zeolite is white-gray, white-greenish, or yellowish white. The size of zeolite crystals is mostly no more than 10 - 15 microns [4]. This mineral has a specificity to loses water very quickly. Zeolite consisted of a chemical compound of alumino - silicate dihydrate with sodium, potassium and barium cations.

Zeolites have a three-dimensional structure, formed from tetrahedral alumina and silica with inner cavities containing metal ions, usually alkaline freely moving water molecules. This makes the zeolite particles have a pore that can be distinguished to macropore (>50 nm), micropore (<2 nm) and between them mesopores. The basic framework of the zeolite structure consisted of tetrahedral units of AlO_4 and SiO_4 which are interconnected through the O atom.

Natural zeolites generally have low catalytic activity, low crystallinity and non-uniform pore size. Therefore, it needs to be activated before use as a catalyst. In addition to eliminating impurities found in natural zeolites, the zeolite activation process is also intended to modify the properties of zeolites, such as surface area and acidity. The increased surface area and acidity will cause the catalytic activity of the zeolite become increase. One of the advantages of zeolite is have surface area and acidity that is easily modified.

3.3 Lanthanum okside (La_2O_3)

Lanthanum (La) is a chemical element of rare-earth metal of class III on the periodic table, which is a prototype of the lanthanide element series. Lanthanum belongs to a rare-earth group of ductile, soft, silvery white metal that is soft enough to cut with a knife. Lanthanum was discovered in 1893 by a Swedish chemist named Carl Mosander. Lanthanum is the second most reactive of rare-earth metal after europium.

Lanthanum is oxidized in air at room temperature to form La_2O_3. Lanthanum is quite electropositive, slowly lanthanum reacts with water and quickly dissolves in dilute acids, except hydrofluoric acid (HF) due to the formation of a protective fluoride layer (LaF_3) on the metal surface. These metals react directly with carbon, nitrogen, boron, selenium, silicon, phosphorus, sulfur and halogens.

The lanthanum metal is paramagnetic from 6 K (-267 °C, or -449 °F) to the melting point at 1,191 K (918°C, or 1,684°F) with specific gravity of 6.1545 g/cm^3. Lanthanum becomes superconducting at atmospheric pressure below 6.0 K (-267.2 °C, or -448.9 °F) in cubic β - f face - centered or 5.1 K (-268.1 °C, or -450.5 °F) in the close-packed hexagonal α-phase.

4. Production of biodiesel

4.1. Esterification

Esterification is the reaction between a fatty acid and an alcohol using an acid catalyst. Esterification with acidic catalyst converts free fatty acid (FFA) to alkyl esters. Esterification generally uses a homogeneous acid catalyst such as sulfuric acid (H_2SO_4) and hydrochloric acid (HCl). The esterification stage is usually followed by the transesterification stage. The reaction of fatty acids to methyl ester as follows:

$$\underset{\text{Fatty acid}}{RCOOH} + \underset{\text{Methanol}}{CH_3OH} \underset{\text{catalyst}}{\leftrightarrow} \underset{\text{Methyl Ester}}{RCOOH_3} + \underset{\text{Water}}{H_2O}$$

The esterification reaction is effected by several factors including free fatty acid and the amount of methanol reactant, reaction time, temperature, catalyst concentration and water content in oil. The higher amount of methanol used and the content of free fatty acids in the oil, the higher yield of methyl esters and the smaller the free fatty acid content at the end of the reaction. The longer reaction time then the yield of methyl ester obtained large. The temperature of 60 °C already provides adequate methyl ester yield. However, the excessive amount of catalyst does not increase with the actual recovery of methyl esters [12].

4.2. Transesterification

Transesterification is the reaction between triglycerides and alcohols to form alkyl esters (biodiesel) and glycerol. The reactions that occur are as follows in Figure 1:

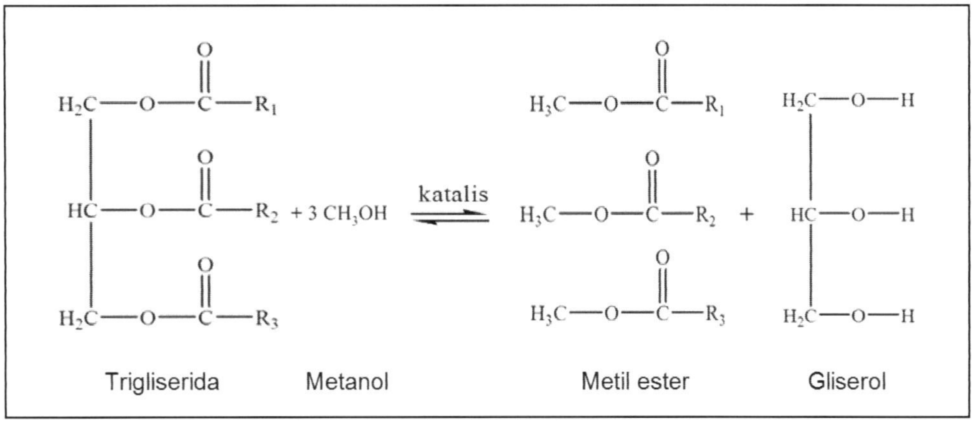

Figure 1. Reaction of transesterification [12].

Transesterification reactions are very sensitive to FFA levels that present in vegetable oils. The permissible FFA levels in raw materials of vegetable oil are 1 - 2.5%. High fatty acid content can lead to side reactions between the basic catalyst and fatty acid that will form the soap or known as saponification reaction. The soap in the transesterification reaction can complicate the separation process of the product (alkyl ester) with the catalyst because the soap will diffuse the mixture during leaching. Meanwhile, the formed water may react with the alkyl ester through the hydrolysis reaction to form fatty acids. This reaction actually reduces the product [12].

5. Biodiesel

Biodiesel is chemically defined as a methyl ester derived from natural oils/fats, such as vegetable oils, animal fats or used cooking oil. Biodiesel belongs to renewable fuels because its raw materials come from natural materials that can be renewed. Biodiesel is biodegradable, non-toxic and essentially sulfur-free.

Methyl ester produced as a substitute for conventional petroleum fuels, so it must fulfill biodiesel standards. The characteristics of biodiesel generally include density, kinematic viscosity, cetane number, combustion calor, pour point, incandescent point, and cloud point. Biodiesel quality standards are not based on the basic materials used in producing biodiesel but based on the performance of biodiesel [13].

6. Methods

6.1. Materials

Materials used in this research are CPO, natural zeolite, N_2 gas, O_2 gas, and H_2 gas, HCl 6 N, $AgNO_3$ 0,1 N, NH_4Cl 1 N, aquadest, methanol (pa), ethanol 96%, La_2O_3, H_2SO_4 1,2 N, NaOH 0,1 N, and KOH 0,1 N.

Tools that used in is mortal porcelain, sieve -100 and +200 mesh, reactor, magnetic stirrer, heating mantel, filter paper, becker glass, oven, stirrer, hot plate, furnace tube, neraca analytic, tube and regulator gases N_2, O_2, H_2, condensor, thermometer, picnometer, viscometer Oswald, flask, volatile cup, buret, erlenmeyer, pipette, gas chromatography-mass spectroscopy (GC-MS) and SEM-EDX.

6.2. Making catalyst La/NZA

The activation of natural zeolite was carried out by immersing the finely ground zeolite (-200 + 100 mesh) in 500 ml of HCl 6 N solution for 30 min at 50 °C, and refluxing while stirring with speed 200 rpm. Then it was filtered and washed repeatedly until ion Cl^- was not detected by $AgNO_3$ solution, dried at 130 °C for 3 hours in the oven. The zeolite was soaked in a 500 ml NH_4Cl 1 N solution at 90 °C for 1 week while stirring, then filtered, washed, dried.

The solid La_2O_3 dissolved in 200 ml of aquades, then the 50 gram NZA mixed into La_2O_3 solution by stirring using a magnetic stirrer while heated at 60 °C for 3 hours, then filtered and washed repeatedly, dried at 120 ºC on oven for 3 hours. The amount of metal variation that is 0, 1 and 3% w/w of NZA.

The result of impregnation is La/NZA catalyst must be calcinated at 500 ºC in while a 400 ml of nitrogen gas was delivered, followed by oxidation at 400 ºC using 400 ml / min of oxygen gas for 2 h and reduction at 400 ºC using 400 hydrogen gas ml / min for 2 hours. The catalyst was characterized using SEM-EDX.

6.3. Biodiesel production

The CPO has a free fatty acid content above 2%, therefore the CPO should be esterified first using catalyst 1% H_2SO_4 1,2 N. Transesterification process at temperature 60 ºC, mole ratio oil : methanol 1 : 9, reaction time 60 minutes, with variation of catalyst weight 1, 2 and 3%. Then the product must be purified to obtain biodiesel and calculated yield.

7. Result and discussion

7.1. Characterization La/NZA catalyst

The dispersion of metals into zeolites can be determined by using SEM-EDX analysis. The result of SEM analysis can be seen in Figures 2, 3 and 4, EDX analysis can be seen in Table 1.

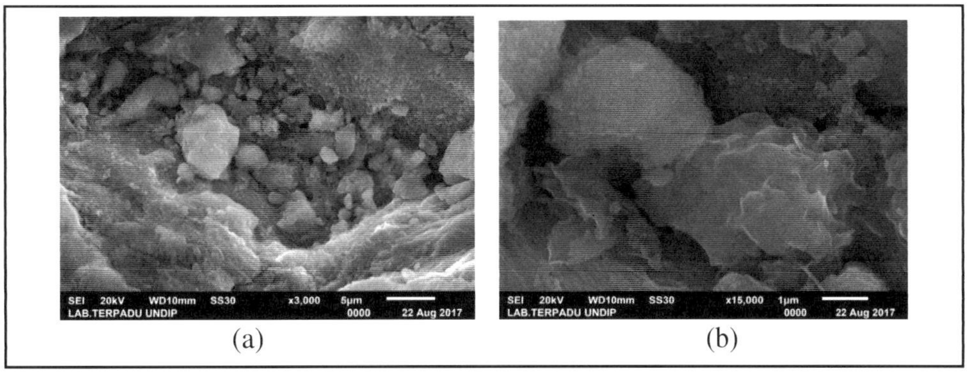

Figure 2. NZA catalyst surface morphology, (a) magnification 3,000 times and (b) magnification 15,000 times.

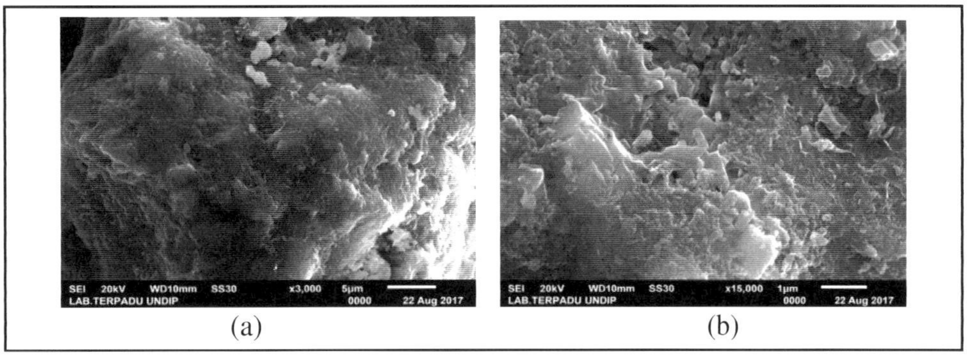

Figure 3. 1% La/NZA catalyst surface morphology, (a) magnification 3,000 times and (b) magnification 15,000 times.

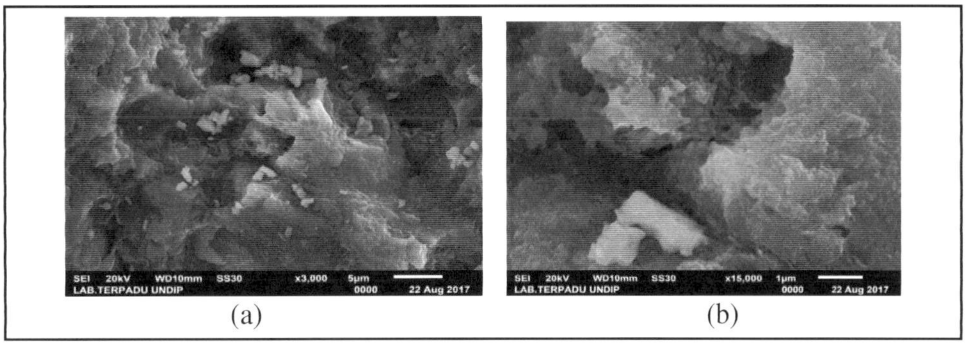

Figure 4. 3% La/NZA catalyst surface morphology, (a) magnification 3,000 times and (b) magnification 15,000 times

Figures 2, 3 and 4 shown the morphology surface of the catalyst having a characteristic crystal structure consisting of layered lamellar and there are many pores with different sizes. The crystal structure is distinctive and has many pores causing the catalyst surface area to increase. In Figures 3 and 4, there is a La metal attached to a zeolite surface with an irregular round or oval shape. This causes a morphological change in the catalyst to become more rough and there are many pores. The porous catalyst surface structure is probably due to the calcination process at high temperatures. Calcination causes the structure on the catalyst surface to change due to the evaporation of some chemical and water compounds.

Table 1. EDX analysis of catalysts

No.	Sample	NZA	1% La/NZA	3% La/NZA
	Componen	Composition (%weight)	Composition (%weight)	Composition (%weight)
1.	C	15,94	16,35	13,25
2.	Na	-	0,83	-
3.	Al	8,16	7,60	3,60
4.	Si	74,99	69,48	79,33
5.	K	0,91	2,13	0,59
6.	Fe	-	0,86	1,46
7.	Cu	-	1,02	-
8.	La	-	1,73	1,77
9.	Si/Al Ratio	9,19	9,14	22,04

The results of the EDX analysis seen in Table 1 shown the amount of composition of each compound or component present in the catalyst. In 1% La/NZA catalysts, impregnated La metals contained 1.73%, excess La is likely to be caused by La metal in the form of lanthanum oxide which has not been reduced by hydrogen. In the case of 3% La/NZA catalysts of impregnated La/NZA as much as 1.77%, the decrease of La content in the catalyst is probably due to the less homogeneous mixing of the impregnation process so that only a few of the La metals are impregnated into NZA and the rest soluble with aquades and when separation of the catalyst in the form of solids with liquid (aquades).

The activity and selectivity of the catalyst can be seen from the Si/Al ratio contained in the catalyst. The Si/Al ratio is very influential because if fewer Al atoms means less negative charge on the zeolite, so there are fewer balancing cations available. Zeolite with high Si compositions have high electronegativity and affinity for hydrocarbons, more stable in strong acid environments, where acidity of catalysts increases. As the Si/Al ratio increases, the cracked methyl esters become short-chain hydrocarbons into more.

7.2. Biodiesel characterization

7.2.1. The physical analysis of biodiesel

The physical properties analysis of biodiesel can be seen in Table 2. Table 2 shown the analysis of all biodiesel products from this research and biodiesel physical properties based on SNI 04-7128-2015.

Table 2. Physical characterization of biodiesel

Catalyst	Weight	Density at 40°C (kg/m^3)	Viscosity at 40°C (cSt)	Acid number (mg-KOH/g)	Flash point (°C)	Yield (%)
NZA	1%	873	5,210	0,736	125	82,87
	2%	881	5,121	0,771	127	83,87
	3%	865	4,894	0,701	124	85,25
1%La/NZA	1%	875	5,413	0,701	124	84,13
	2%	868	5,120	0,771	128	84,25
	3%	874	5,082	0,701	125	84,50
3%La/NZA	1%	857	5,391	0,718	125	85,37
	2%	868	5,049	0,800	128	84,63
	3%	867	4,979	0,736	126	82,00
SNI 04-7128-2015		850 – 890	2,3 – 6,0	Maks. 0,8	Min. 100	

Table 2. shown that the difference in yield of biodiesel is not significant, this is caused by Si/Al ratio of catalysts not much different. The biodiesel density of the research results is in the range of values based on SNI. The biodiesel density value is effected by the molecular weight of the components contained in the biodiesel. Increasing the weight of components make increasing the density. Smaller density is better to use as alternative biofuel.

The biodiesel viscosity obtained from this research is within the range of 2006 biodiesel SNI. High viscosity causes difficult biodiesel to flow, so will increasing the cost of piping installation to distribute biodiesel. The high viscosity of biodiesel is effected by the raw materials and the transesterification process temperature does not constant.

The acid number of biodiesel has a value above its maximum standard of 0.5. The acid number expresses the free fatty acid content contained in the biodiesel, the higher the acid number the more free fatty acid content. Acid numbers determine the quality of biodiesel is good or not. High acid numbers may be caused by CPO feedstock having a high FFA that is 5.472% and possibly because there is a few amount of water in biodiesel.

Flash point indicates high low volatility and the ability to burn from a fuel. The biodiesel flash point of this research has met the quality standard based on SNI that has a value above 100°C. It states that biodiesel is at a safe limit to fire hazards during storage and during distribution. The highest flash point is 128°C and lowest is 124°C.

7.2.2 The chemical analysis of biodiesel

Analysis of the chemical properties of biodiesel was tested using GC - MS. Biodiesel analyzed is biodiesel with highest yield that is on the use of 3% catalyst La/NZA 1%. GC-MS analysis aims to determine the chemical compounds present in biodiesel. The result of GC-MS biodiesel analysis can be seen in Figure 5.

IOP Conf. Series: Materials Science and Engineering **345** (2018) 012007 doi:10.1088/1757-899X/345/1/012007

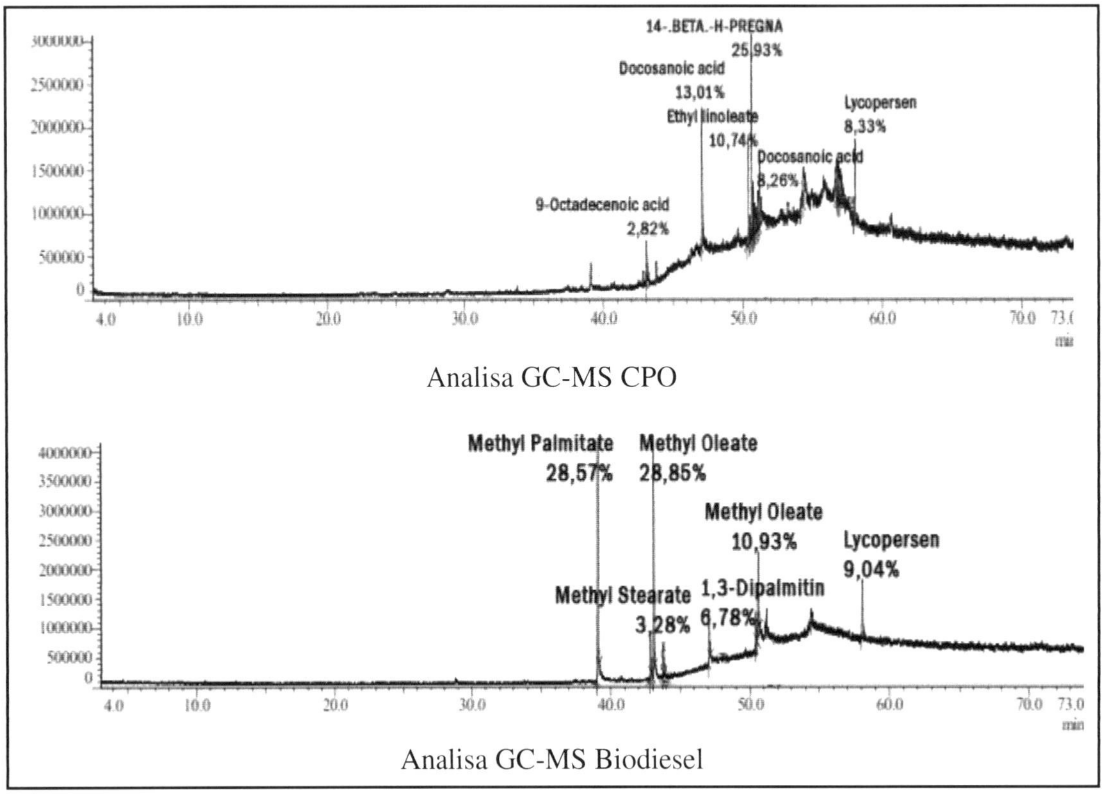

Figure 5. Comparison of GC-MS CPO analysis results with biodiesel.

CPO chromatogram and biodiesel on 3% La/NZA catalysts showed a significant difference. Where top five peaks in biodiesel are, first there is methyl palmitate with area of 28,57%, second methyl oleate 28,85%, 1,3-dipalmitin 6,7%, second methyl oleate 10,93% and lycopersen 9,04 %. From the chromatogram, the highest peak compounds are not all methyl esters, there are 1,3-dipalmitin glycerin and lycopersen which is a dye (carotenoid) that gives the orange to red color of the CPO. From these results it can be stated that the biodiesel obtained still needs to be purified to separate glycerin and lycopersen.

The ability of the catalyst to produce the desired product (selectivity) is affected by pressure, reaction temperature, reactant composition, pore size distribution and surface area. Glycerin and lycopersen formed may be caused by the temperature of the reaction is not constant, because when the research temperature reaction is measured only with a thermometer, sometimes up and down 1 – 3 °C. The second is probably caused by the poor pore size distribution of the catalyst, so the impregnated metals do not spread throughout the catalyst surface. Based on the discussion it can be concluded that the manufacture of biodiesel from CPO using La/NZA catalyst has produced good biodiesel.

8. Conclusions

The conclusions of this research is biodiesel from CPO can be produced using La/NZA catalyst. The best catalyst was 3% La/NZA with 1% catalyst, the highest yield 85.37%. The result is biodiesel has met the quality standard of biodiesel based on SNI 04-1782-2015.

References

[1] Vieira S S, Magriotis Z M, Santos N A V, Saczk A A, Hori C E and Arroyo P A 2013 Biodiesel production by free Fatty acid esterification using Lanthanum (La^{3+}) and HZSM-5 based catalysts *Bioresour. Technol.* **133** 248-255

[2] Kurle Y M, Islam M R and Benson T J 2013 Process development and simulation of glycerol-free biofuel from canola oil and dimethyl carbonate *Fuel Proc. Technol.* **114** 49-57

[3] Birla A, Singh B, Upadhyay S N and Sharma Y C 2012 Kinetics studies of synthesis of biodiesel from waste frying oil using a heterogeneous catalyst derived from snail shell *Bioresour. Technol.* **106** 95-100

[4] Ulfayana S, and Helwani Z 2014 Natural zeolite for transesterification step catalysts in biodiesel production from palm off grade *Proc. Regional Conf. Chem. Eng.* 2-3 December 2014, Yogyakarta

[5] Hambali E, Mujdalipah S, Tambunan A H, Pattiwiri A W and Hendroko R 2008 *Teknologi Bioenergi* 2nd ed. (Jakarta: Agromedia Pustaka)

[6] Formo M W, Jungermann E, Norris F A and Sonntag N O V 1979 *Bailey's Industrial Oil and Fats Product*, 4th ed., Vol. 1, ed D Swern (New York: Wiley Interscience Publication)

[7] Far R D and Farr W E 2000 *Introduction to Fats and Oils Technology* 2nd ed., ed R D O'Brien (Champaign, Illinois: AOCS Press)

[8] Hagen J, 2006 *Industrial Catalysis* (Germany: Wiley)

[9] Helwani Z, Othman M R, Aziz N, Kim J, and Fernando W J N 2009 *Appl. Catal. A-Gen* **363** 1-10

[10] Endalew A K, Kiros Y, and Zanzi R 2011 *Energ.* **36** 2693-2700

[11] Yan S, Maggio C D, Mohan S, Kim M, Salley S O, and Ng K Y S 2010 *Top Catal.* **53** 721-736.

[12] Knothe G, Krahl J, and Gerpen J V 2005 *The Biodiesel Handbook* (US: AOCS Press)

[13] Legowo E H, Gafar Q, Sijabat O, Pupung P I and Arifin Z 2001 International biodiesel workshop *Prosiding Enhancing Biodiesel Development And Use* 2-4 Oktober 2001, Indonesia

IOP Conf. Series: Materials Science and Engineering **345** (2018) 012008 doi:10.1088/1757-899X/345/1/012008

Removal of Methylene Blue from aqueous solution using spent bleaching earth

E Saputra, R Saputra, M W Nugraha, R S Irianty and P S Utama
Chemical Engineering Department, Engineering Faculty, Riau University, Pekanbaru, Riau, Indonesia

E-mail: edysaputra@unri.ac.id

Abstract. The waste from industrial textile waste is one of the environmental problems, it is required effective and efficient processing. In this study spent bleaching earth was used as absorbent. It was found that the absorbent was effective to remove methylene blue from aqueous solution with removal efficiency 99.97 % in 120 min. Several parameters such as pH, amount of absorbent loading, stirring speed are found as key factor influencing removal of methylene blue. The mechanism of adsorption was also studied, and it was found that Langmuir isotherm fitted to data of experiment with adsorption capacity 0.5 mg/g.

1. Introduction

Industrial wastes are increasing with the rapid development of industry, i.e. volume and type. One of the most disturbing problems of industrial waste is thedye content [1]. The textile industry is one of the industries that produces colored liquid waste, the dye in the textile industry is one of the main raw materials, about 10-15% of the dye has been used, unfortunatelly it cannot be reused and must be disposed [2]. In the textile industry, methylene blue is one of the most commonly used thiazine dyes, because it is economical and also easy to obtain. The use of methylene blue may cause some effects, such as gastrointestinal irritation if swallowed, causing cyanosis if inhaled, and irritation of the skin if touched by the skin [3].

Many dye removal processing methods have been tried to deal with this problem, but only the adsorption method is superior to other methods [4]. Because adsorption is a simple, easy to operate, flexible, affordable method of separation, and does not produce toxic products [5]. The first step to get an effective adsorption process is by selecting adsorbents with high selectivity, capacity and recyclable. Spent bleaching earth (SBE) is using of adsorbent generated from CPO refining process.

In this research, spent bleaching earth (SBE) will be used as adsorbent to adsorb the methylene blue dye with variation of pH 6, 8, and 10, variation of adsorbent mass 1, 1.5 and 2 g and 60 minute stirring time, 90 minutes and 120 minutes.

2. Experimental

The activation was done to process of spent bleaching earth was done by heating using furnace at 500 °C for 7 hours, then sieved using 200 mesh sieves. The preparation of the liquor was carried out by dissolving 1 gram of methylene blue into a 1 litter flask, and added aquadest to the threshold. Standard methylene blue solution with 1 ppm concentration of 10 mL was measured using UV-VIS spectrophotometer to obtain the maximum wavelength. Methylene blue solution with a concentration of 20 ppm with variation of pH 6, 8, 10 and 12 taken 200 mL and fed into a beaker. Then added adsorbents with the variation as much as 1, 1.5, and 2 grams. For then, stirred with a magnetic stirrer with a speed of 205 rpm varied for 60, 90, and 120 minutes. Subsequently the solution was filtered using filter paper, deposited on the sample bottle for analysis.

From the analysis with the adsorption method, the colour testing of the sample before and after added adsorbent spent bleaching earth is by calculating the efficiency of decreasing concentration. The adsorption capacity is calculated from the Langmuir equation or Freundlich equation [6].

3. Result and discussions

Effect of pH variation on reduced methylene blue substance concentration is presented in Figure 1. Base on the figure, it is seen that the efficiency of decreasing concentration of the highest methylene blue dye at pH 8. It was possible that the isoelectronic point of methylene blue lies at pH 8 and the functional group present in the adsorbent could bind the dye to its maximum. At lower pH, H^+ in the solution will further disrupt the bonding between the adsorbent and methylene blue, as more carboxyl and siloxane groups in the adsorbent bind the H^+ and become positively charged, resulting in the adsorbents becoming more difficult to bind to the methylene blue dye ions. At pH 10 there was a decrease, because of OH^- in the solution could not be captured by the dye, so there werefree OH^-in the solution. The carboxyl group present in the adsorbent was a positive partially charged active group, so that at the time of addition of the colour base tends to be partially negative. This led to competition between the dye and the free OH-ion to occupy the surface of the adsorbent which will decrease the dye adsorption power by adsorbent

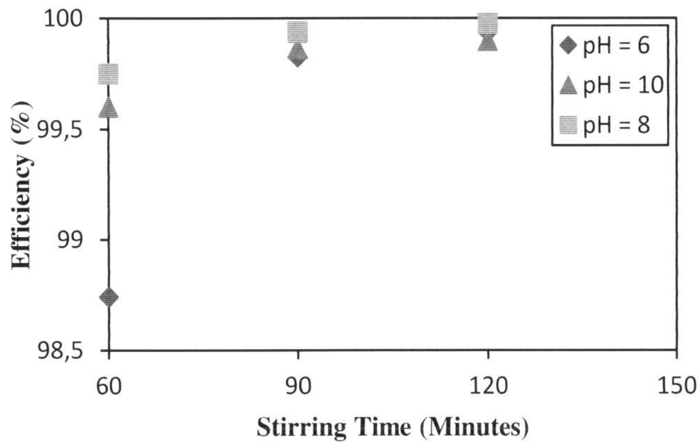

Figure 1. Effect of pH variation on decreasing methylene blue colour substance concentrations.

The use of an appropriate amount of adsorbent mass will affect to the efficiency of decreasing of the concentration of the methylene blue dye. It can be seen in the following picture that there is an increase in stress on the methylene blue dye in each adsorbent mass increase. 2 gr of adsorbent mass always has the highest decreasing efficiency. This is directly proportional to the theory put forward by Wambu [7] that the more adsorbent mass used the higher the efficiency level. The results of this study indicate that the more mass of the adsorbent undergoing 200 mL of methylene blue dye solution, the more the dye adsorbed by the adsorbent.

The agitation time used has an effect on the efficiency of decreasing concentration of methylene blue dye. It can be seen in Figure 1 and Figure 2 that there is always an increase in the efficiency of decreasing the concentration of the methylene blue dye in each increase in aggregation time. The stirring time of 120 minutes always has the highest decreasing efficiency for each pH and the adsorbent mass has been varied. The results obtained in accordance with the theory proposed [8] that the longer time of stirring it will get the results of high concentration reduction efficiency. The small particle size has a greater molecular interaction power so that its absorption becomes better.

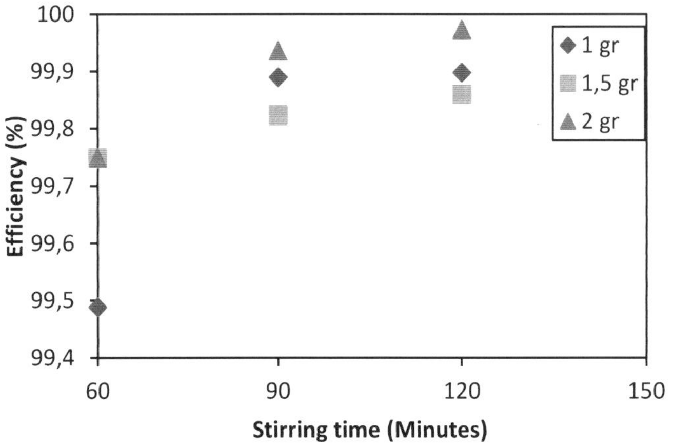

Figure 2. Effect of adsorbent mass variation on decreasing concentration methylene blue substance.

To know the adsorption capacity of pH variation, adsorbent mass and stirring speed at adsorption of methylene blue dye by spent bleaching earth then used two isotherm models namely Langmuir and Freundlich isotherms. The determination of isotherm signifies a relationship with the adsorption capacity, therefore the Ce Versus Ce / qe curve was made according to the Langmuir isotherm model in Figure (a) and the Ce versus log qe log curve according to the Freundlich isotherm model in Figure (b) By comparing the coefficient of determination (r2), an appropriate adsorption isotherm model will be selected.

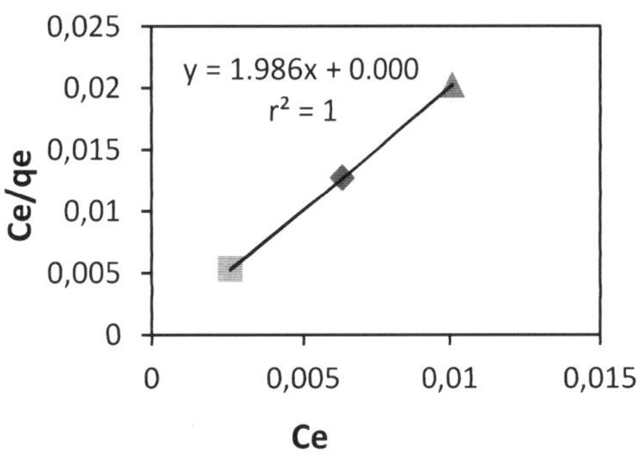

Figure 3. Freundlich isoterm from methylene blue adsorption by spent bleaching earth on a variation of pH.

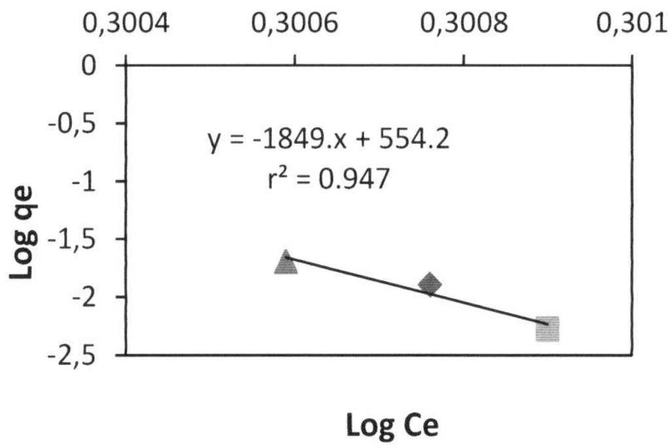

Figure 4. Langmuirfrom methylene blue adsorption by spent bleaching earth on a variation of pH.

From Figure 4 (a) and Figure (b) it can be seen that the isotherms suitable for methylene blue adsorption by spent bleachiang earth on variarian pH is Langmuir isotherm because it has a higher coefficient of determination (r^2) that is 1 whereas for isotherms Frenudlich coefficient of determination is only 0.947, the isotherm used is Langmuir isotherms with adsorption capacity (Qo) of 0.5035 mg / g. This result is in line with what was stated by [9], that the adsorption of methylene blue dye follows Langmuir isotherms. These results suggest that the possibility of the adsorption process occurring involves the formation of a strong enough bond (chemisorpsi) greater than in fisisorpsi so that it is difficult to be released again.

IOP Conf. Series: Materials Science and Engineering **345** (2018) 012008 doi:10.1088/1757-899X/345/1/012008

4. Conclusions

Efficiency of decreasing concentration in methylene blue dye was highest when pH 8, 2 gram adsorbent mass and stirring time 120 minutes with efficiency decreasing concentration of 99.973%. Adsorption of methylene blue dye by spent bleaching earth on pH variation follows Langmuir isotherm with determination coefficient (r^2) is 1, adsorption capacity of 0.5035 mg / g. Adsorption of methylene blue dye by spent bleaching earth on variation of adsorbent mass following Langmuir isotherm with determination coefficient (r^2) is 0.843, adsorption capacity of 0.3816 mg / g. Adsorption of methylene blue dye by spent bleaching earth on variation of stirring time following Langmuir isotherm with determination coefficient (r^2) is 1, the adsorption capacity is 0.5012 mg / g.

References

[1] Pratiwi Y 2010 Penentuan tingkat pencemaran industri tekstil berdasarkan nutrition value coeficient biondikator *Jurnal Teknologi* **3** (2) 129-137

[2] Dhaneshvar N, Ayazloo M, Khatae A R and Pourhassan M 2007 Biological decolourization of dye solution containing malachite green by microalgae *Cosmarium sp. Bioresour. Technol.* **29** 1-7

[3] Hamdaoui O and Chiha M 2006 Removal of methylene blue from aqueous solutions by wheat bran *Acta Chim.* **54** 407-418

[4] Cui L 2008 Performance and mechanism of methylene blue biosorption on orange peel *Environ. Technol.* **29** 1021-1030

[5] Gupta V K and Suhas 2009 Application of low-cost adsorbents for dye removal - A review *J. Environ Manage.* 2313-2342

[6] Kheang L S, *et al.* 2006 A study of residual oils recovered from spent bleaching earth: Their characteristics and applications *Am. J. Appl. Sci.* **3** 2063-067

[7] Wambu E W et al. 2009 Kinetics of copper desorption from regenerated spent bleaching earth *American-Eurasian J. Sci. Res.* **4** (4) 317-323

[8] Yenti S R and Zultiniar 2011 *Penerapan zat warna metilena biru dengan memanfaatkan bagas tebu* (Pekanbaru: Jurusan Teknik Kimia Universitas Riau)

[9] Tanasale M F J D P, Killay A and Saily M 2012 Kitosan dari kulit kepiting rajungan (portunus sanginolentus L.) sebagai adsorben zat arna biru metilena *Jurnal Natur Indonesia* **14** (2) 165-171

Effective utilizations of palm oil mill fly ash for synthetic amorphous silica and carbon zeolite composite synthesis

P S Utama[1,2], E Saputra[2], Khairat[2]
[1]Chemical Engineering Department Prince of Songkla University, Hatyai, Thailand
[2]Chemical Engineering Department University of Riau, Pekanbaru, Indonesia

E-mail: psutama@yahoo.com

Abstract. Palm Oil Mill Fly Ash (POMFA) the solid waste of palm oil industry was used as a raw material for synthetic amorphous silica and carbon zeolite composite synthesis in order to minimize the wastes of palm oil industry. The alkaline extraction combine with the sol-gel precipitation and mechanical fragmentation was applied to produce synthetic amorphous silica. The byproduct, extracted POMFA was rich in carbon and silica content in a significant amount. The microwave heated hydrothermal process used to synthesize carbon zeolite composite from the byproduct. The obtained silica had chemical composition, specific surface area and the micrograph similar to commercial precipitated silica for rubber filler. The microwave heated hydrothermal process has a great potential for synthesizing carbon zeolite composite. The process only needs one-step and shorter time compare to conventional hydrothermal process.

1. Introduction

Riau province is the largest Crude Palm Oil (CPO) producer and has the greatest palm oil plantation area in Indonesia. The crude palm oil production in 2015 was 7.8 million ton/year and the plantation area was 2.4 million ha [1]. According Kramanandita et al. (2014) production of one ton CPO will produce empty fruit bunch of 1.41 ton, fibers 0.9 ton and shells 0.3 ton [2]. Around 85% of fibers and 15% of empty fruit bunch and shells are used as fuel in palm oil industry to produce steam. It is estimated 5% of the fuel unburned and produce ash [3]. Therefore, it can be calculated in 2015 palm oil industry in Riau generated ash of 0.4 million ton. In order to minimize the waste of palm oil industry, the utilization of POMFA as a raw material for precipitated silica production has been done in the previous research. The alkaline extraction combine with the sol-gel precipitation and mechanical fragmentation was applied to produce precipitated silica, the result was 60% of silica from the ash can be extracted. The silica content in the POMFA was 39% so from 1 ton of POMFA only 234 kg silica can be extracted [4]. The process above still produces waste of 766 kg. However, if the waste can be processed further and produces valuable product the objective to minimize the waste can be fulfilled.

The main composition of POMFA is carbon of 30% and silica of 39% [4]. The extraction process will decreased the content of silica, make the POMFA richer in carbon and still has significant silica content. Moreover, the extraction process makes the carbon in the POMFA more porous because of part of silica leave the POMFA particle. Even though it can be predicted that the POMFA after extracted is a good material for adsorption but the content of carbon is too low and the content of ash is too high to be processed further as an activated carbon. One of the possible utilization of extracted POMFA is used as a raw material for synthesizing zeolite carbon composites. The zeolite carbon composite is a new material that has the properties of zeolite and carbon. The surface properties of carbon are usually hydrophobic and the nature of the zeolite surface is usually hydrophilic so that the zeolite carbon composite surface has both properties. This composite is particularly suitable for the adsorption of both organic and metal ions in the liquid and gas phases [5]. In the previous study, zeolite carbon composite was successfully synthesized from fly ash coal [6] and bottom ash coal [7] by fusion process using NaOH at 750 °C followed by hydrothermal process at temperatures below 100 °C for more than 10 hours.

The hydrothermal process at temperatures of 80 to 110 °C and time of 2 to 8 hours was applied to synthesize the NaA zeolite carbon composite from rice husk ash [8]. This is possible because the silica derived from crop material is in amorphous form, which is more reactive than silica in the coal ash. According to Pathak et al. (2014) the use of conventional hydrothermal processes has the disadvantages of low yield, long process times and heterogeneous structures. This weakness can be fixed using the microwave as a heater. In microwave heating process, heat is radiated directly into the molecule present in the reactant without being affected by the conductivity of the wall of the reactor or the reactants themselves. The result is a rapid local heating of the molecules and makes the reactivity of the molecule increases [9].

In this research, firstly the POMFA was used as a raw material to produce synthetic amorphous silica that has similar properties to commercial precipitated silica for shoe sole and mechanical rubber good filler by alkaline extraction combine with sol-gel precipitation using H_2SO_4 and mechanical fragmentation. Secondly, the preliminary study of synthesizing carbon zeolite composite from the extracted POMFA as the raw material by microwave heating hydrothermal process.

2. Methods

The raw material used in this experiment was POMFA from Sawee Industrial Palm Oil ltd. Chumporn, Thailand. The sodium hydroxide was NaOH laboratory grade with minimum content of 97% (dry basis). The sulfuric acid was commercial grade with minimum total acidity of 98% H_2SO_4 by mass. Aluminium oxide p.a. which has purity > 99.5% and average particle diameter < 10 µm (Sigma Aldrich, USA) was used as the aluminium source The equipment and the optimum condition from the previous research was used to produce the synthetic amorphous silica. The extraction process was done in 2500 cm^3 glass extractor. The extraction conditions were mass of POMFA 468.2 gram; the POMFA mass to NaOH volume ratio 0.2341 g/cm3; the concentration of NaOH 1.4 N and the stirring speed of 1065 RPM and time 60 min. The extract silica and the extracted POMFA were separated using vacuum filter. The sol-gel precipitation using H_2SO_4 10% (v/v) and mechanical fragmentation process were applied simultaneously in 1000 cm^3 glass precipitator. The conditions of sol-gel precipitation and mechanical fragmentation was stirring speed of 1160 RPM, pH of 8.75 and precipitation time of 100 min [4n this experiment, sol-gel precipitation temperature of 90 °C was used instead of 30 °C as in the previous research in order to obtain specific surface area that suitable for rubber filler. The synthetic amorphous silica wad characterized by a scanning electron microscope (SEM, JEOL JSM-5800LV, Japan), Energy-dispersive X-ray scanning electron microscope (SEM-EDX , JEOL JSM-5800LV, Japan) for element analysis and the spesific surface area (SSA) were obtained by the Brunauer-Emmett-Teller (BET) method using a surface area analyzer (Quantachrome Nova 2000 e, USA). After extraction process, the extracted POMFA was washed and dried. The chemical composition of the extracted POMFA was analyzed using C, H, N and O analyzer and XRF (X-ray Flourescence Spectrometer, PW 2400, Philips). In this preliminary study, synthesizing carbon zeolite composite the Si/Al ratio used were 5, 8 and 11 [10]; time 30 min and the input power of the microwave 800 W. The POMFA of 3 g were mixed with Al_2O_3 of 0.221, 0.128 and 0.085 g respectively to obtain the predetermined Si/Al ratio above. The mixture of fly ash and Al_2O_3 and 20 cm^3 sodium hydroxide solution of 1 N were added into 100 cm^3 polytetrafluoroethylene (PTFE) reactor. After zeolitization was performed the solid washed and dried. The dried solids obtained were characterized using XRD.

3. Results and discussions

3.1 Synthetic amorphous silica

The chemical composition of POMFA and extracted POMFA analyzed using C, H, N, O analyzer and XRF are listed in Table 1.

Table 1. Chemical composition of POMFA and extracted POMFA

Oxide	POMFA [a] (% dry weight)	Extracted POMFA (% dry weight)
SiO_2	39.02	24.50
Al_2O_3	1.19	0.86
Fe_2O_3	1.71	1.11
CaO	9.39	7.64
MgO	4.11	6.84
K_2O	6.20	2.52
MnO	0.20	0.18
P_2O_5	5.16	4.27
SO_3	0.90	0.22
CuO	0.06	0.05
Rb_2O	0.03	-
SrO	0.04	0.03
Na_2O	-	2.41
C	30.44	49.15
Total	99.63	99.79

[a] (Utama et al., 2013 [4])

In the extraction process, part of silica in the POMFA was dissolved into sodium hydroxide solvent resulting the silica content in the extracted POMFA decreased. The decreasing silica content increasing the carbon content because of the carbon is insoluble in sodium hydroxide. If it is assumed that the carbon is fixed it can be calculated that of 60.11% silica can be extracted from the POMFA. This is in accordance with Utama et al., (2016) that reported at the similar conditions around 60% of silica can be extracted from POMFA [4].

The SEM micrograph of the synthetic amorphous silica obtained and the commercial precipitated silica depicted in Figure 1 and Figure 2 respectively. It can be seen that the microstructure of the synthetic amorphous silica and commercial precipitated silica is similar. The primary particle of both silica is quite uniform and posses spherical particle shape. The primary particle sizes are in the nano-scale, however the primary particle size of the commercial precipitate silica is smaller than the synthetic amorphous silica obtained. According Quarch et al. (2010) the primary particle size of the synthetic amorphous silica depend on the rate of gelation process [11]. There are many factors influenced the rate of gelation such as temperature, pH and ionic strength of the solution. The primary particle size can be tailored by varying the temperature and pH of the sol-gel precipitation process.

Figure 1. SEM micrograph of synthetic amorphous silica obtained.

Figure 2. SEM micrograph of the commercial precipitated silica.

The SEM EDX was used to determine the elements in the synthetic amorphous silica obtained. The energy dispersive X-ray spectrometry is depicted in Figure 3. It shown that the most element impurities were sodium and carbon. The carbon was found in the POMFA as a carbonate and as a char. In the extraction process, the char is insoluble but the carbonate soluble in the sodium hydroxide solvent form sodium carbonate. The sodium and the carbon as carbonate might be trap in the silica matrix and did not completely washed. The minor element K and Al were from the POMFA and S was from sulphuric acid was used to destabilize the extract silica in the sol-gel precipitation process were might be carried over into the synthetic amorphous silica obtained.

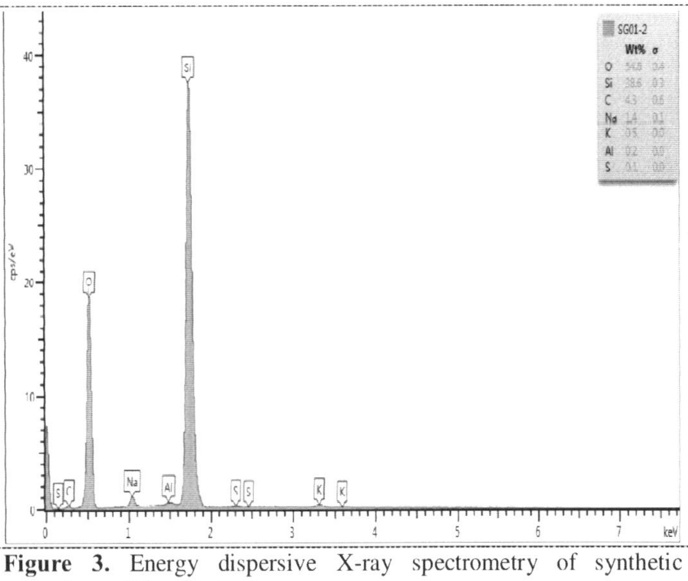

Figure 3. Energy dispersive X-ray spectrometry of synthetic amorphous silica obtained.

Figure 4 shows nitrogen adsorption–desorption isotherms of amorphous silica obtained. The type of physisorption isotherm is similar to the type II reversible isotherm in IUPAC classification that indicates unrestricted monolayer-multilayer absorption of non-porous or macro-porous adsorbent. The hysteresis loop of the isotherm is similar to H3 hysteresis loop. The H3 loop represent of as an assemblage of particles that are loosely coherent and have slit shapes pore. This fact is in accordance with the SEM micrograph that shows the primary particles of synthetic amorphous silica are form aggregates that are loosely coherent [12].

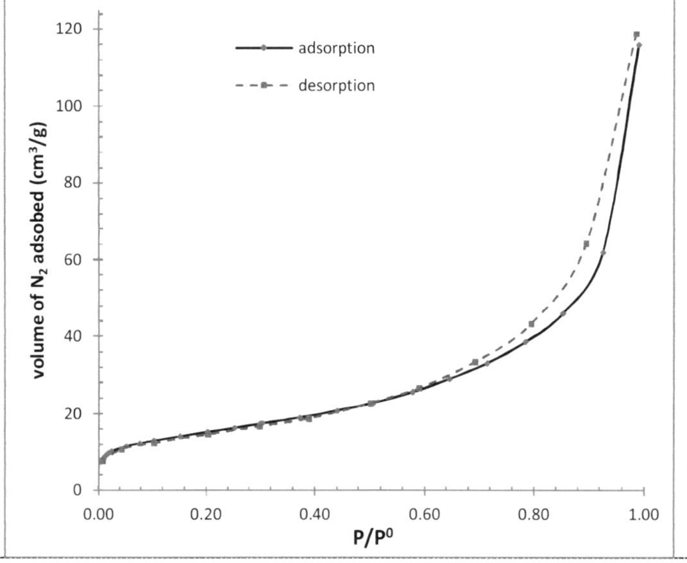

Figure 4. Nitrogen adsorption–desorption isotherms of amorphous silica obtained.

The BET-SSA analysis of the synthetic amorphous silica obtained gives SSA of 53.590m^2/g. The SSA of the silica is similar to SSA of Ultrasil* 360 of 55 m^2/g which suitable as a rubber filler for shoe sole and mechanical rubber goods [13].

3.2 Carbon zeolite composite.

The slurry of extracted POMFA, alumina and sodium hydroxide solution in PTFE reactor was heated by microwave for zeolitation process. The solid obtained after zeolitation process was analyzed using XRD, the result is depicted in Figure 5. It can be seen that for the Si/Al ratio of 5 (F1) the peaks of the XRD diffraction indicate there were silica as quartz and alumina as corundum. The peaks of XRD diffraction for the Si/Al ratio of 11 (F3) indicate there were silica as quartz, alumina as corundum and carbon as graphite. The peaks of XRD diffraction for the Si/Al ratio of 8 (F2) indicate there were alumina as corundum and aluminium silicate as kyanite. Even though the zeolitation process by microwave heated hydrothermal has not formed zeolite yet, however for the Si/Al ratio of 8 in a relative short time and only using one-step can made the silica and alumina react to form aluminium silicate. In this experiment, the sodium hydroxide concentration used was 1 N and the liquid volume to solid ratio used was 6.66 (cm^3/g). The sodium hydroxide concentration and the liquid volume to solid ratio might be too low and made the zeolite cannot be formed. The carbon zeolite synthesis can be explored further by using higher sodium hydroxide concentration and volume liquid to solid ratio.

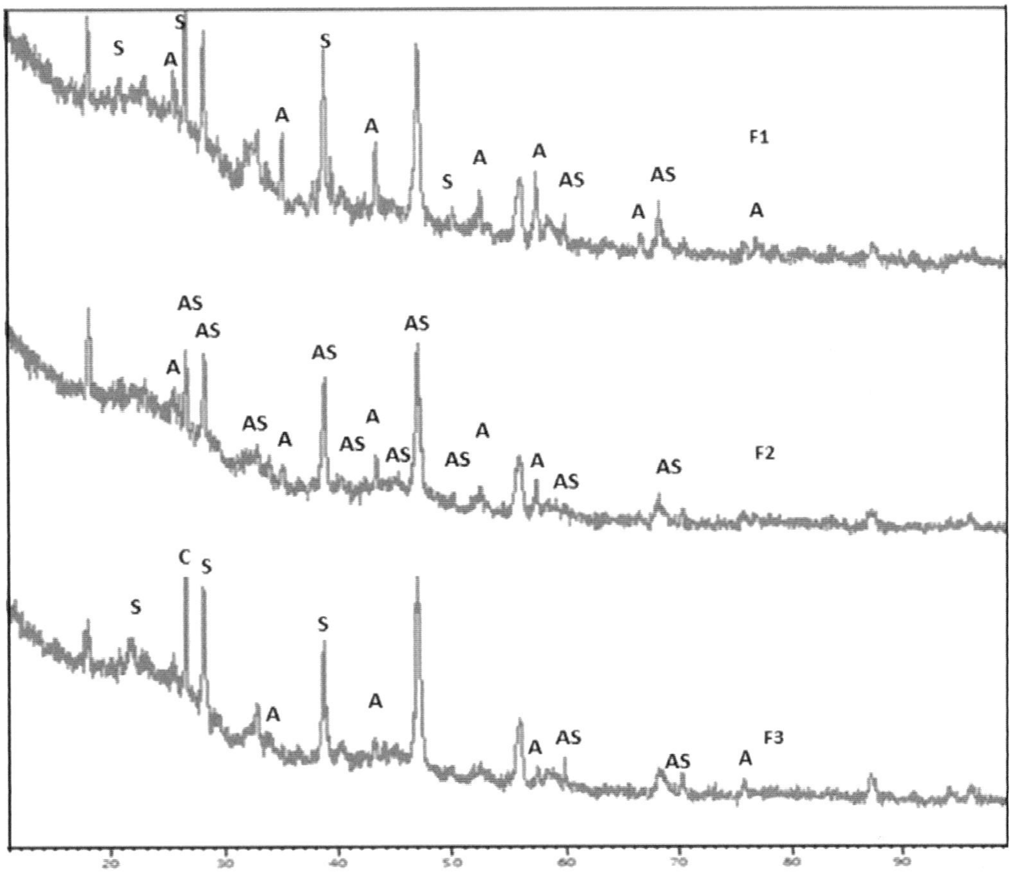

Figure 5. The XRD diffraction of solid obtained with variation of Si/Al ratio (A= alumina as corundum; C = carbon as graphite; S = silica as quartz and AS = aluminium silicate as kyanite).

4. Conclusions

The sodium hydroxide extraction combine with sol-gel precipitation using sulphuric acid and mechanical fragmentation have been successfully applied for producing synthetic amorphous silica from the POMFA. The synthetic amorphous silica obtained has chemical composition, microstructure and specific surface area similar to commercial precipitated silica that is suitable for mechanical rubber goods and shoe sole filler. The microwave heated hydrothermal process shows great potential to be applied for synthesizing carbon zeolite composite from extracted POMFA. Even though the zeolitation process by microwave heated hydrothermal has not formed zeolite yet, however the ratio of Si/Al ratio of 8 in a relative short time and only using one-step can made the silica and alumina react to form aluminium silicate.

Acknowledgments

This work was supported by DPRM, Ministry of Higher Education and Research Technology Indonesia through Penelitian Produk Terapan grant.

References

[1] Badan Pusat Statistik Propinsi Riau 2017 *Riau dalam angka 2015* (Pekanbaru: Badan Pusat Statistik Propinsi Riau)
[2] Kramanandita R, Bantacut T, Romli M and Makmoen M 2014 *Chem. Mater. Res.* **6 (8)** 46-53
[3] Tay H and Show K 1995 *Cement Concrete Comp.* **13** 27-36
[4] Utama P S, Yamsaengsung R and Sangwichien C 2016 *Key Eng. Mater.* **673** 183-192
[5] Babic B, Kokunesoski M, Gulicovski J, Prekajski M, Pantic J, Radosavljevic-Mihajlovic A and Matovic B 2011 *Process Appl. Ceram.* **5** (2) 91-96
[6] Miyake M, Kumar Jha V, Kimura Y and Matsuda M 2008 *Trans. Ecol. Environ.* **109** 203-212
[7] Widiastuti N, Zhely M, Hidayah N, Praseytoko D and Fansuri H 2014 *Adv. Mater. Lett.* **5** (8) 453-458
[8] Sirisoonthon S, Jiemsirilers S, Nilpairach S, Wasanapianpong T, Sujaridworakun P and Chuankrekkul N 2014 *Key Eng. Mater.* **608** 241-246
[9] Pathak C Y, Debananda R and Sumanta D 2014 *World J. Civil Eng. Constr. Technol.* **1** (1) 2-11
[10] Cardoso A M, Paprocki A, Ferret L S, Azevedo C M N and Pires M 2015 *Fuel* **139** 59-67
[11] Quarch K, Durand E, Schilde C, Kwade A and Klind M 2010 *Chem. Eng. Res. Des.* **88** 1639-1647
[12] Sing K S W, Everett D H, Haul R A W, Moscou L, Pierotti R A, Rouquerol J and Siemieniewska T 1985 *Pure Appl. Chem.* **57** (4) 603-619
[13] Evonik Resource Effiency GmbH 2017 *Product information ULTRASIL* 360* (Wesseling: Evonik)

Crystal-growth kinetics of magnetite (Fe_3O_4) nanoparticles with Ostwald Ripening Model approach

S P Utami, A Fadli, E O Sari and A S Addabsi
Biomaterial and Corrosion Laboratory
Department of Chemical Engineering Faculty Riau University

E-mail: fadliunri@yahoo.com

Abstract. Magnetite (Fe_3O_4) nanoparticles is a magnetic nanomaterial that have potential properties to be applied as drug delivery The purpose of this study was to determine the influence of time and temperature synthesis of magnetie characteristics and determine its crystal growth kinetics model with Ostwald ripening model approach. Magnetite nanoparticles synthesized from $FeCl_3$, citrate, urea and polyethylene glycol with hydrothermal method at 180, 200 and 220 °C for 1,3,5,7,9 and 12 hours. Characterization by *X-ray Diffraction* (XRD) indicated that magnetite formed at temperatures of 200 and 220 °C. Magnetite crystallite diameter obtained was 10-29 nm. Characterization by *Transmission Electron Mycroscope* (TEM) shows that magnetite nanoparticles have uniform size and non-agglomerated. Core-shell shaped particles formed at 200 °C and 220 °C for 3 hours. Irregular shape obtained at 220 °C for 12 hour synthesis with particle diameter about 120 nm. Characterization using *Vibrating Sample Magnetometer* (VSM) shown that magnetite has super paramagnetism behaviour with the highest saturation magnetization (Ms) was 70.27 emu/g. magnetite crystal growth data at temperature of 220 °C can be fitted by Ostwald ripening growth model with growth controlled by the dissolution of surface reaction ($n \approx 4$) with the percent error of 2.53%.

1. Introduction

Recently, nanoparticles technology have been used in various biological applications such as drug delivery or nano drug delivery. Nano drug delivery is a direct transfer method to the target (the pain part). The crucial issues showed that the importance of development nano drug delivery technology, it is because a cancer is a second of death cause in the world. WHO said if we don't controll it, they predict that 26 million people will get cancer and 17 million will die because of cancer in 2030. The chemotherapy method is less effective because not only kill cancer cells but also make the quality of patientlife will be down. Nano drug delivery is a method that will provide a good solution of the negative effects of chemotherapy.

Nanomagnetite (Fe_3O_4) is a nanomaterial which recently got more attention in various studied. In last decade, nanomagetite synthesis which has a superparamagnetism is always did for many various applications especially in biological application such as MRI contrast agent and cancer therapy via hyperthermia [1] and recently develop as drug delivery [2]. This is because nanomagnetite has the appropriate properties for medical applications that are superparamagnetic and have high biocompatibility [3].

Nanomagnetite was synthesized as a drug delivery from $FeCl_3$, citrate, urea and polyethylene glycol (PEG) with hydrothermal method at 200 °C [4]. The resulting magnetite particles are shaped *hollow* and *core shel l*has the behavior of super paramagnetism, having high water solubility which is claimed to be

IOP Conf. Series: Materials Science and Engineering **345** (2018) 012010 doi:10.1088/1757-899X/345/1/012010

the most ideal for drug delivery application. With this advantages, it would be better if particles size and morphology could be controlled because it is an important factor in drug delivery system [5] [6].

This study was to determine the magnetite (Fe_3O_4) crystal growth kinetics model with Ostwald Ripening model approach followed with the magnetite synthesis by Cheng et al. [4]. In this study, results of nanoparticle biocompatibility also improved by using polyethylene glycol (PEG) as capping agent. The study of crystalline growth kinetics modeling is needed to investigate factors that affected to the crystal growth more systematically and as the basis of planning in the synthesis process. This modeling is also required in the tailoring process.

2. Experimental

In this study, all chemicals used was Ferric (III) Chloride ,Aquadest, Urea, Sodium Citrate ($C_6H_5O_7Na_3.2H_2O$) Polyethylene Glykol (PEG), dan Ethanol 99%. Synthesis process for hydrothermal method have been done with mixtures $FeCl_3$, Sodium Citrate, urea dan aquadest into *Teflon Lined Autoclave*. Then 0.3 gr PEG added and hydrothermal method was started at 180, 200 and 220 °C. The liquid samples is taken from reaction container in various time intervals such as 1, 2, 3, 5, 7, 9 and 12 hours, and liquid and solid was decanted. The solid was washed using ethanol and let it dried during the nighttime at 60 °C. Next the solid will characterization with XRD, TEM and Magnometer.

3. Result and discussion

3.1. The Effect of temperature and reaction time of hydrothermal method for yield and composition

The variation of temperature of 180, 200 and 220 °C were used in this study. This variable was selected because the response of the kinetic reaction when the temperature decreases from 20 °C from the temperature reference (200 °C). At 180 °C, the precipitate was formed after synthesized for 5 hours. The precipitate was obtained from 5 hour synthesis reddish brown colored and drawn weak by magnet. Synthesis nanomagnetite at 200 °C, precipitate was formed after 3 hours with brown colored, while the precipitate for long time synthesis have a black color and drawn by magnet. Synthesis at 220 °C, gived a precipitate in black colored since 3 hours synthesized. The color of precipitated was more dark if the synthesis time was more. If the synthesis time is longer, the precipitate is getting more (see Figure 1) that is 33.33% at 3 hour synthesis, 36.67% at 5 hour synthesis, and then increasing to 80% at 7 hours and become 100% at 9 hour and 12 hour synthesis.

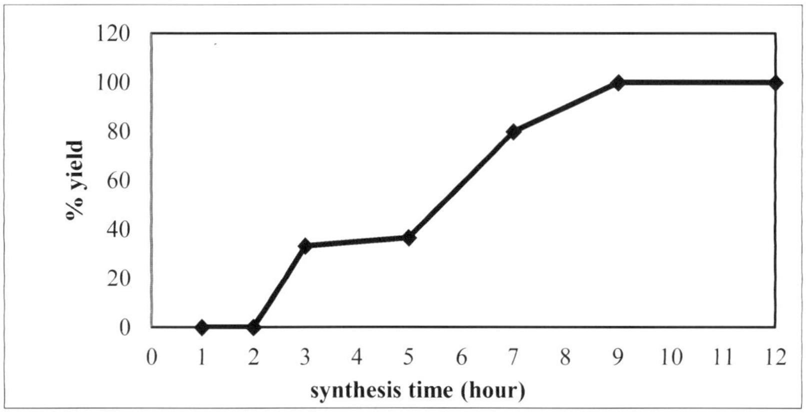

Figure 1. A relation between synthesis time and percentage of yield.

3.2 Effect of temperature and reaction time of hydrothermal method to cristal size

The size modification of the crystal particles was determined using Scherre Equation (1). At the application of Scherrer was a sistimatic error. This error can be reduced with modification of Scherrer equation into logarithmic Equation [7].

$$\beta = \frac{K\lambda}{D.\cos\theta} = \frac{K\lambda}{D}.\frac{1}{\cos\theta} \quad (1)$$

$$ln\beta = ln\frac{K\lambda}{D.\cos\theta} = ln\frac{K\lambda}{D} + ln\frac{1}{\cos\theta} \quad (2)$$

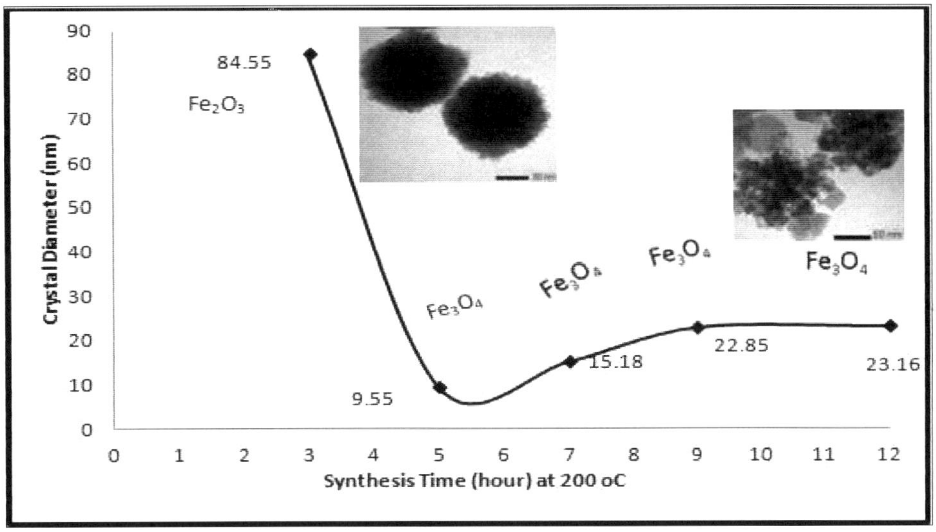

Figure 2. A Relation between synthesis time with crystal diameter at 200 °C.

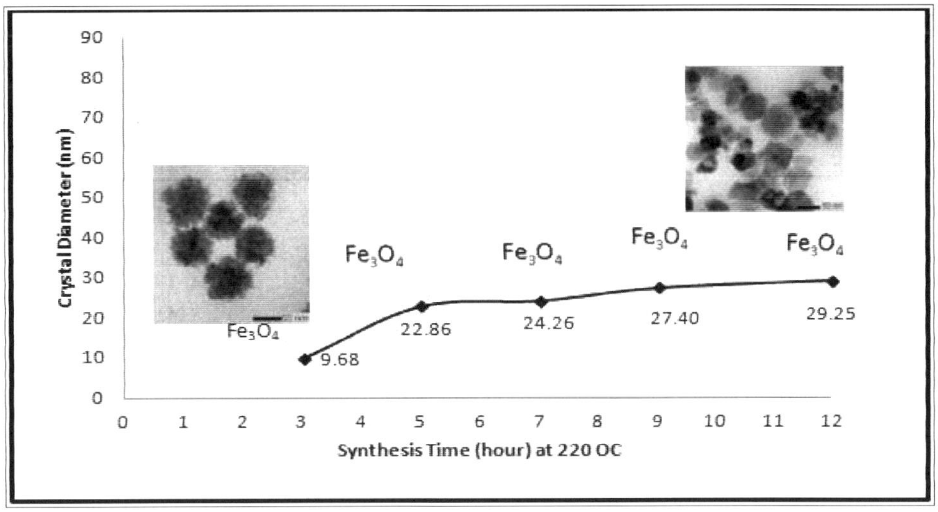

Figure 3. A Relation between synthesis time with diameter crystal at 220 °C.

The last crystal size (12 hour synthesis) at 200 °C more small like 23.16 nm if compare between at 220 °C with crystal diameter like 29.25 nm. The rate grow of crystal diameter size at the two temperature was different too. At 200 °C the rate grow of crystal diameter was more than at 220 °C was 5:2. This is would be different from mecanism growth of crystal at the two temperatures that will be prove by modeling.

3.3 Effect of adding PEG to nanoparticle morfology
PEG was selected as a *capping agent* because its prove has a high biocompatibility [8-10]. With this adding gave a hope that the result of nanoparticles size was uniform and can avoid agglomeration between particle. Nanoparticles morphology was shown by TEM at Figure 4 and Figure 5. Based on TEM in 80.000 time, in general shown that the formed particle has a nano size with size distribution was enough uniform and not agglomerated.

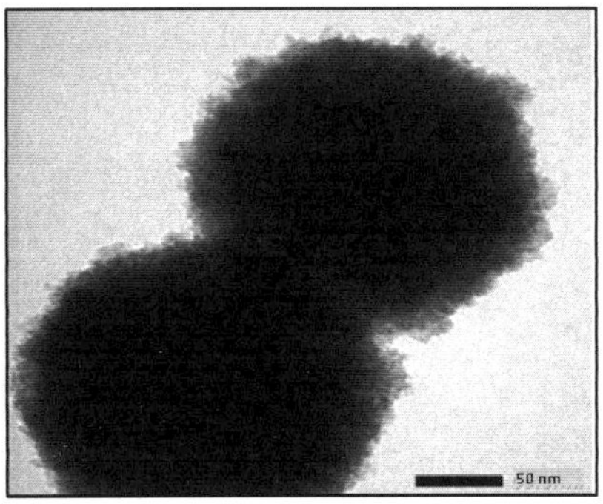

Figure 4.TEM images of sample at 200 ˚C for 3 hours.

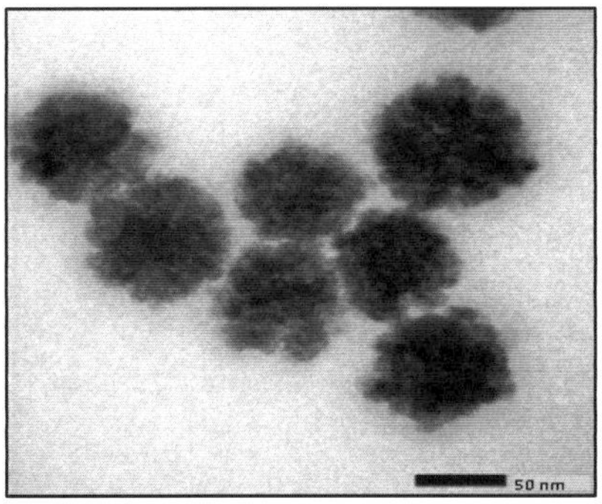

Figure 5. TEM images of sample at 220 ˚C for 3 hours.

The result at 200 ºC for 3 hour was a particle formed like a ball with 100 nm diameter and on the other side at the same time at 220 ºC the form was close to *core-shell* with *black* core and *shell* which was more light with diameter 100 nm less. This different shape and size was affected by the temperature of reaction and crystal composition.

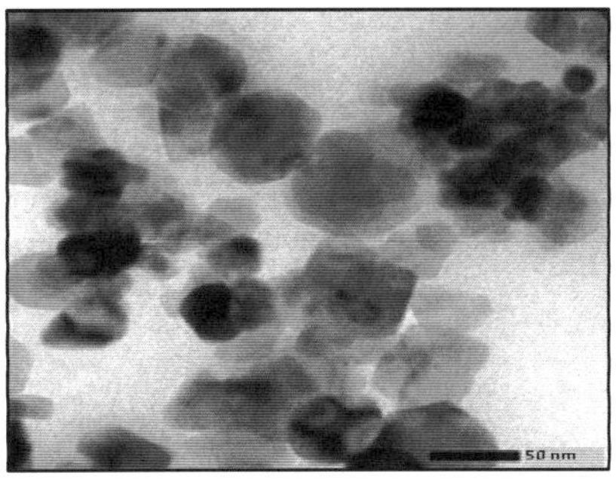

Figure 6. TEM images of sample at 200 ˚C for 12 hours.

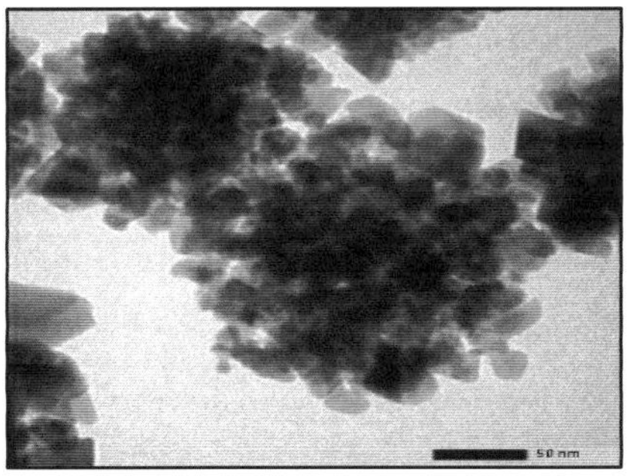

Figure 7. TEM images of sample at 220 ˚C for 12 hours.

The nanomagnetite with capping agent PEG has a morphology shape like core shell about 120 nm. So that, with this adding PEG did prevent the agglomeration and reduce nanoparticles size. The size of the nanomagnetite is suitable with drug delivery criteria that 10nm - 200 nm [11]. Core shell structure has a majority as a carrier drug delivery. The shell was role as a part of core protector from oxidation by the air and has a function as a attachment place with ligands or drug [12].

IOP Conf. Series: Materials Science and Engineering **345** (2018) 012010 doi:10.1088/1757-899X/345/1/012010

3.4 The Effect of Temperature and Synthesis Time to Magnetism Properties of Nanomagnetite

The magnetism properties was investigated using *vibrating sample magnetometer* (VSM) at 300K with magnet at range -1 T to 1 T.

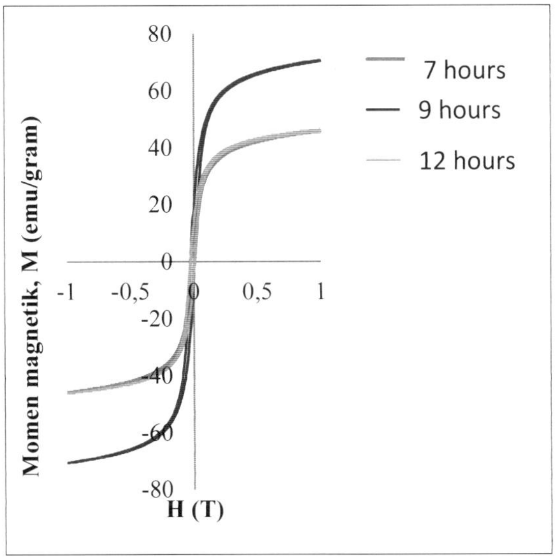

Figure 8. Hysteresis curve nanomagnetite at 200 °C for 7, 9 and 12 hours.

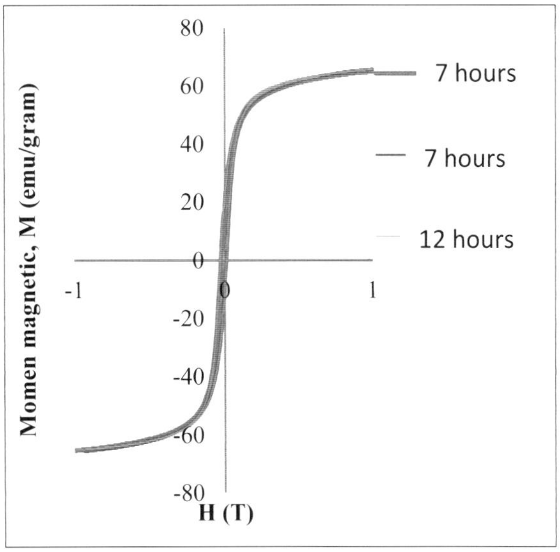

Figure 9. Hysteresis curve nanomagnetite at 220 °C for 7, 9 and 12 hours.

Generally, both of those figure shown there is loop hysteresis that so narrow with coercivity value close to zero. This shown that nanomagnetite has a superparamagnetism behavior at room temperature. Magnetite with size less than 30 nm have a super paramagnetism behavior [4]. The particle of crystal nanomagnetite in this experiment have a size about 20 nm so that it has a super paramagnetism behavior.

Table 1. The saturation of magnetization value (saturated), intensity peak, and crystal diameter from nanomagnetite sampleat temperature and different synthesis time

Temperature (°C)	Synthesis time	Ms (Emu/g)	Intensity peak XRD high (count)	Crystal diameter (nm)
	7	45.78	604.707	15.18
200	9	70.27	873.271	22.85
	12	45.78	776.103	23.16
	7	65.04	1330.228	24.26
220	9	65.29	1384.059	27.40
	12	65.41	1490.334	29.25

The saturation of Magnetization (Ms) is equal to the level crystalinity of nanoparticles. If the crystalization is high, so the saturation of magnetic is high too. This is because at high crystallization the structure is close to the structure of material bulk crystal.

3.5 Model of crystal growth of nanomagnetite

Model of crystal growth of nanomagnetite at this experiment was based on Ostwald Ripening Model approach with commonly equation $r - r_0 = Kt^{1/n}$. The suitable model is did by two approach, was (1) suitable method to determine n and k with logaritmic curve and (2) suitable curvetime versus diameter crystal in practice and theoretical.

Table 2. Result of determining n and K withlogaritmic curve method

Temperature (°C)	R^2	ln K	1/n	k	n
200	0.8229	1.76	0.37	5.80	2.72
220	0.947	1.67	0.27	5.31	3.66

Based on calculation, the value of n was obtained like 2,72 ≈ 3, on the other hand at 220 °C the value of n was 3,66 ≈ 4. n = 4 indicates that the rate of crystal growth is controlled by the rate of dilution at the interface of particle-matriks. If n=3, it shows that rate of crystal growth controlled by diffusion [13]. But, this result can not be suitable, especially for data at 200 °C because it has a low linearity and it showed by R^2value less than 0,9 which R^2=0.8229.

To increasing a confidence, approve with model curve that shows like in Figure 10 was did. Errors percentage shows at Table 3.

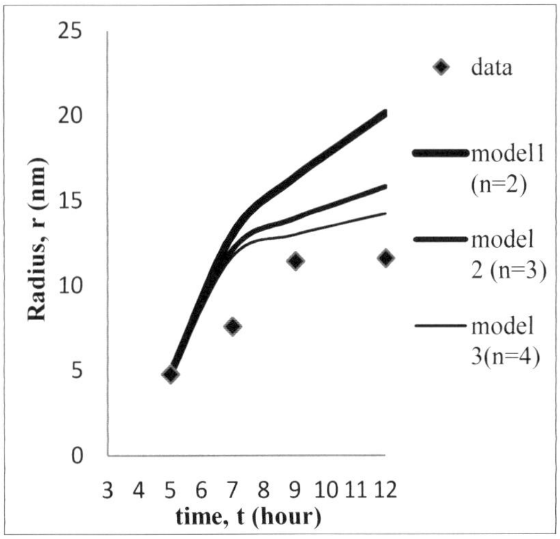

Figure 10. Curve synthesis time(t) vs crystal radius (r) at 200 °C.

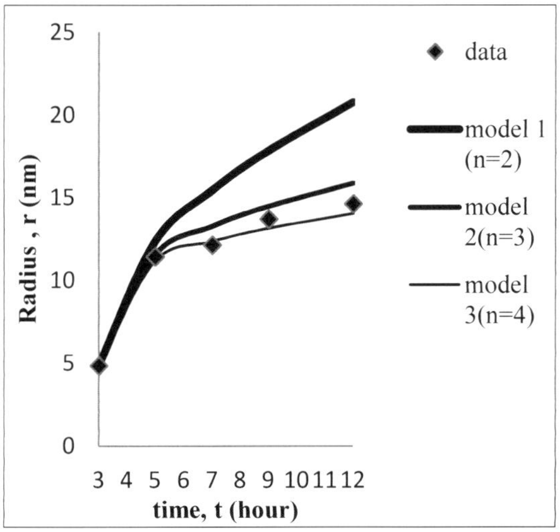

Figure 11. Curve synthesis time (t) vs crystal radius (r) at 220 °C.

Table 3. The error percentage of experiment data to model

Temperature (°C)	model	%error
200	model1 (n=2)	28.54
	model 2 (n=3)	20.45
	model 3 (n=4)	16.34
220	model1 (n=2)	20.44
	model 2 (n=3)	5.69
	model 3 (n=4)	2.53

Based on Figure 10, every model gives a ralat that too close different for the others model with one certain trend to exsperiment data. Error decline with increasing n. The large Error percentage to the three of those model (Table 3) indicates that at 200 °C the crystal growth not following the Ostwald Ripening model. In other hand at the Figure 11, model 3 with n=4 was the most close to exsperiment data with less % errorwhich 2,53% (Table 3). This is indicate that at 220 °C, the crystal grow of nanomagnetite following the Ostwald ripening model that controlled by dissolution interface of particle-matrix.

4. Conclusion

The process of nanomagnetite particles synthesis using the hydrothermal method was affected by temperature and synthesis time. If the increasing of the temperature it will made the reaction time become longer so that the reaction of magnetite formation and yield also become faster. The morphology of particles was affected by temperature and reaction time. At 200 °C 12 hour synthesis and 220 °C 3 hour synthesis gave a magnetite nanoparticle at core shell shape. Nanoparticles have a super paramagnetism behavior with coersivity value (Hc) and remenensi (Mr) hysteresist curve close to zero. Growth of nanomagnetite crystal following Oswald Ripening model at 220 °C which controlled by velocity of dissolution rate ($n \approx 4$) and K=5,3with % error 2,53%.

References

[1] Qiao R, Yang C and Gao M 2009 Superparamagnetic iron oxide nanoparticles: from preparations to in vivo MRI applications *J. Mater. Chem.* **19** 6274-6293

[2] Mahmoudi A, Simchi M, Imani U O and Hafeli J 2009 Optimal design and characterization of superparamagnetic iron oxide nanoparticles coated with polyvinyl alcohol for targeted delivery and imaging *Phys. Chem. C* **113** 8124-8131

[3] Mohapatra M and Anand S 2010 Synthesis and applications of nano-structured iron oxides/hydroxides – A review *J. Sci and Tech* **8** 127-146

[4] Cheng W, Tang K, Qi Y, Sheng J and Liu Z 2010 One-step synthesis of superparamagnetic monodisperse porous Fe3O4 hollow *J. Mater. Chem.* **2** 1799-1805

[5] Bae H Y and Park K 2011 Targeted drug delivery to tumors: Myths, reality and possibility *J. Cont. Rel.* **153** 198-205

[6] Yu M, Huang S, Yu K J and Clyne A M 2012 Dextran and polymer polyethylene glycol (PEG) coating reduce both 5 and 30 nm iron oxide nanoparticle cytotoxicity in 2D and 3D cell culture *J. Mol Sci* **13** 5554-5570

[7] Monshi A, Foroughi M R and Monshi M R 2012 Modified scherrer equation to estimate more accurately nano-crystallite size using XRD *J. Nano Sci. Eng.* **2** 154-160

[8] Cholan S, Shanmugam N, Kannadasan N, Sathishkumar K and Deivam K. 2014 Effect of poly ethylene glycol (PEG) as surfactant on cerium doped ZnS nanoparticles *J. Mater. Technol.* **3** 222–227

[9] Manson J, Kumar D, Meenan B J and Dixon D 2011 Polyethylene glycol functionalized gold nanoparticles: the influence of capping density on stability in various media *Gold. Bull.* **44** 99-105

[10] Shameli K, Ahmad M B, Jazayeri S D, Sedaghat S, Shabanzadeh P, Jahangirian H, Mahdavi M and Abdollahi Y 2012 Synthesis and characterization of polyethylene glycol mediated silver nanoparticles by the green method *Int. J. Mol. Sci.* **13** 6639-6650

[11] Winarti L 2013 Sistem penghantaran obat tertarget, macam, jenis-jenis sistem penghantaran, dan aplikasinya *Stomatognatic* **10** 75-81

[12] Luszczyk K, Kaleta J and Mech R 2014 Magnetic core-shell structures as potential carriers in drug delivery system *Int J. of Eng Sci.* **1** 1-4

[13] Huang F, Zhang H and Banfield F J 2003 Two-stage crystal-growth kinetics observed during hydhrothermal coarsening of nanocrystalline ZnS *Nano Lett* **3** 373-378

Application of Response Surface Methodology (RSM) for wastewater of hospital by using electrocoagulation

Murdani[1], Jakfar[2], D Ekawati[2], R Nadira[2] and Darmadi [2]
[1]Chemical Engineering Master Program, Syiah Kuala University, Banda Aceh, Indonesia
[2]Chemical Engineering Department, Syiah Kuala University, Banda Aceh, Indonesia

E-mail: darmadi@unsyiah.ac.id

Abstract. Hospital wastewater is a source of potential environmental contamination. Therefore, the waste water needs to be treated before it is discharged into the landfill. Various research methods have been used to treat hospital wastewater. However, some methods that have been implemented have not achieved the effluent standards for hospitals that have been set by the government. The experiment was conducted by an electrochemical method is electrolysis using aluminum electrodes with independent variable is the voltage, contact time and concentration of electrolytes. The response optimization using response surface with optimum conditions obtained by the contact time of 34.26 min, voltage 12 V, concentration electrolyte 0.38 M can decrease of COD 65.039%. The model recommended by the response surface for the three variables, namely quadratic response.

1. Introduction

Hospital is a public place that provides health services for all levels of society as well as a center for health and research education. A wide range of activities can take place in a hospital, from non-medical activities to medical activities that generate solid, liquid and gas waste. Each of these wastes will impact the soil, water and air [6]. Hospital wastewater contains high concentrations of organic compounds, as well as other chemical compounds and pathogenic microorganisms that can cause disease to the surrounding communities. the activities of the hospital will produce pathogenic microorganisms such as viruses, bacteria, including Escherichia coli, Campylobacter and Shigella, and protozoa, including G. lamblia and Cryptosporidia, and heavy metals such as arsenic (As), cadmium (Cd), chromium (Cr), lead (Pb), copper (Cu), Zinc (Zn), which are very harmful to human health that is still present in water and discharges from laboratories, pharmacies, radioactive elements and other chemicals that cannot be eliminated by simple waste treatment. Some technology has been applied in treating hospital wastewater inside it fluidized bed biofilm reactor, Contact Aeration Process, Rotating Biological Contactor, Biofilter Up Flow and Activated Sludge [7]. The use of chemicals in sewage processing produces a lot of sludge, while the microbiological processing system requires a large area of land and requires a relatively long processing time, therefore the conventional processing as described has been much forgotten. another alternative that can be tried to cover the shortage of technology that has ever been electrocoagulation. electrocoagulation is a technology that is currently developing effectively applied for treating wastewater [5] Electrocoagulation for water treatment where the anode occurs the release of active coagulant in the form of metal ions (usually aluminium or iron) into the solution, while the cathode electrolysis reaction in the form of release of hydrogen [3]. in this research, electrode used in the form of aluminum. the reaction that occurs on is as follows :

Anode : $Al^{3+} + 3H_2O \longrightarrow Al(OH)_3 + 3H^- + 3e^-$

Cathode : $2H_2O + 2e^- \longrightarrow 2OH^- + H_2$

2. Materials and method

This research used the sample of hospital wastewater from the initial storage at the Tgk Chik Ditiro Sigli Wastewater Treatment Plant. All types of liquid waste generated from all parts of the RS are collected in the tub. The initial COD content of the hospital wastewater generated is still above the standard quality standard of 107.157mg /l.

2.1 Materials and research procedure

Electrodes are prepared as much as 2 sets (300 x 150 x 2) mm, then cleaned by sandpaper with sandpaper. After that the tool is assembled for container batch system, then put into it the hospital liquid waste as much as 1,3 L to be analyzed with electric voltage which is set at rectifier as specified variable (6 V). Snippets are captured at each time interval t15; 30, and 45 minutes for COD testing. After the electrodes are sanded back at different voltages of 9 and 12 V with the same time interval.

2.2 Experimental design

Design Expert Software is a collection of statistical and mathematical techniques useful for experimental design, model fit and determines optimum operating conditions for predetermined responses. In Design Expert Software there are several methods that can be used for response optimization, but in this research the method used is Box Bhenken to optimize and see the effect of variable contact time, voltage and electrolyte to decrease COD. on the electrocoagulation of hospital waste water treatment. Based on Design Expert Software using Box Bhenken method resulted in a trial design of 17 run experiments with the repetition at the center point (center point) as much as 5 runs. Repetition is useful for calculating pure errors from sum of squares. As shown in Table 1, the factors were designed as A, B, and C. All factors were prescribed into three levels, namely +1, 0, -1 for high, intermediate and low value, respectively.

Table 1. Input variable and code factor of response surface method of Box Behnken design

Code Factor	Variable	-1	0	1
	For electrocoagulation and electro-Fenton			
A	Time (minute)	15	30	45
B	Voltage (V)	6	9	12
C	Electrolyte concentration (ppm)	0	0.5	1

Some of the factorials of bhenken box designs are calculated in ANOVA. ANOVA also results in the interaction between process variables and response variables. Where the components will be used in calculating the F-Ratio in determining the effectiveness of a model [2]. The tables available on ANOVA are usually used for the test performance of a model.

3. Result and discussion

3.1. Summary of statistical models for decrease in COD

The Statistical Model Summary provides suggestions for choosing the right model and in accordance with the experimental data. Table 2 shows the statistical model for percentage reduction of COD resulting from the Box-Behnken design.

Table 2. Summary of statistical models for decrease in COD

Response	Source	Std. Dev	R-Square	Adj-R^2	Pre-R^2	PRESS
% Decrease COD	Linear	7,04	0,8488	0,8139	0,7587	1027,83
	2FI	7,81	0,8570	0,7711	0,5774	1799,70
	Quadratic	1,82	0,9945	0,9875	0,9396	257,44 (Suggested)

From Table 2 it can be seen that the suggested model is quadratic. The quadratic model shows R^2 value closer to 1, ie 0.9945 and Adj-R^2 is 0.9875. The statistical summary model focuses on the value of R^2 obtained to see the suitability of the model. stating that the value of R^2 close to 1 indicates a high degree of correlation between observation and prediction value [10]. The value of R^2 obtained from a good model is above 0.8 and close to 1, because at that value the reaction can be well explained. R^2 is a value that explains the ratio of a ratio of variation to total variation and is a measure of the degree of suitability of the model. Similarly, the value of Adj-R^2 shown model, which is equal to 0.9875. the higher the value of Adj-R^2, indicating a more significant model. This value states that 98.75% of contact time, electrolyte and voltage variables affect the uniform response. While 0.25% influenced by other factors that are not included into the variable to be studied. Box-Behnken design also provides an actual mathematical model to predict the value of % COD decrease in wastewater processing by electrocoagulation as follows :

$$Y = -80.71725 + 3.9863833\, X_1 + 7.8601667\, X_2 + 41.9245\, X_3 - 0.045141\, (X_1)^2 - 0.274917\, (X_2)^2 - 17.887\, (X_3)^2 + 0.0082222\, X_1.X_2 - 0.033667\, X_1.X_3 - 1.94\, X_2.X_3 \quad (1)$$

The above equations are based on actual factors of each variable ie X1 for contact time, X2 for voltage and X3 for electrolyte. contact time (X1), voltage (X2) and electrolyte (X3), all of which have positive influence means the addition of the three independent variables will increase the percentage decrease of COD. From Equation 1 the predicted result of % decrease of COD of hospital effluent treated by electrocoagulation. Comparison of data of research result with result data from calculation using equation 1 can be seen in Table 3. The highest percentage of COD decrease obtained from the research result was 69.51% with 45 minutes contact time, 12 V voltage and 0.5 M electrolyte. While the percentage value of COD decrease obtained from model prediction was 70.53% at the time contact 45 minutes, 12 V voltage and 0.5 M electrolyte. The lowest percentage of COD decrease obtained from the results of the research was 18.36% with contact time 15 minutes, voltage 6 V and electrolyte concentration of 0.5 M. While the percentage value of COD decrease obtained from the model prediction is 17.34% at contact time 15 minutes, 6 V voltage and 0.5 M electrolyte.

Table 3. Experiment result data and model prediction results on% COD decrease

Run	Value of Variabel X			COD Decrease (%)	
	X_1 (Contact time)	X_2 (Voltage)	X_3 (Electrolyte)	Experiment	Prediction
1	45	12	0,5	69,51	70,53
2	45	6	0,5	58,21	56,66
3	15	9	1	24,41	24,58
4	30	6	1	47,53	48,38
5	15	12	0,5	28,18	29,73
6	30	9	0,5	57,70	56,20
7	30	12	0	56,79	55,94
8	45	9	1	63,42	64,12
9	15	6	0,5	18,36	17,34
10	30	9	0,5	54,23	56,20
11	30	9	0,5	56,55	56,20
12	45	9	0	59,23	5906
13	30	9	0,5	55,32	56,20
14	30	9	0,5	57,18	56,20
15	30	6	0	35,27	36,99
16	15	9	0	19,21	18,50
17	30	12	1	57,41	55,69

For the condition at the center point (center point) done as much as 5 runs, namely at the time of contact 30 minutes, voltage 9 V and electrolyte 0.5 M. The percentage value of COD decrease obtained from the research results are 57.7, 54.23, 56, 55, 55,32 and 57,58% while from model prediction result is 56.2% for the 5 runs. The relationship between the experimental results and the predicted results of hospital effluent treatment on the COD% decrease, illustrated in Figure 1.

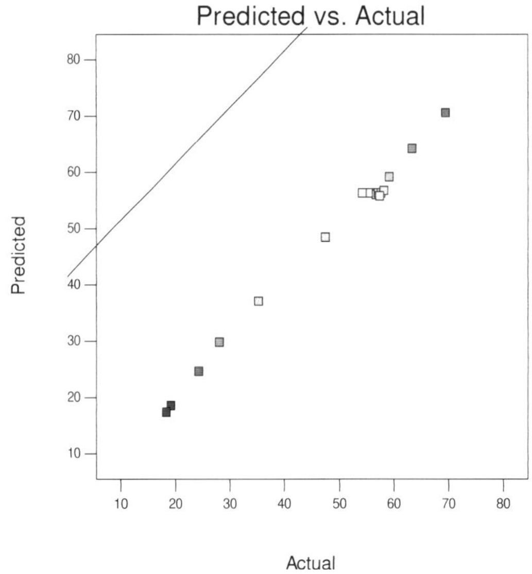

Figure 1. Relationship between research value and prediction value% decrease in COD on hospital wastewater treatment.

3.2 Analysis of variance (ANOVA) for decrease in COD

Some of the factorials of box bhenken designs are calculated in ANOVA. ANOVA also results in the interaction between process variables and response variables. Where the components will be used in calculating the F-Ratio in determining the effectiveness of a model [2]. The tables available on ANOVA are usually used for the test performance of a model.

Table 4. ANOVA results electrocoagulation of hospital effluent treatment to% COD decrease

Source	Sum of Square	df	Mean Square	F value	P-value Prob>F	Characteristics
Model	4235,81	9	470,65	141,41	<0,0001	Significant
A- Contact Time	3208,41	1	3208,41	963,97	<0,0001	Significant
B - voltage	344,79	1	344,79	103,59	<0,0001	Significant
C- electrolyte	61,99	1	61,99	18,63	0,0035	Significant
A^2	434,36	1	434,36	130,50	<0,0001	Significant
B^2	25,78	1	25,78	7,74	0,0272	Significant
C^2	84,20	1	84,20	25,30	0,0015	Not Significant
AB	0,55	1	0,55	0,16	0,6971	Not Significant
AC	0,26	1	0,26	0,077	0,7899	Not Significant
BC	33,87	1	33,87	10,18	0,0153	Significant

The first thing to do for output analysis, is to detect the suitability of the model. Based on the above table, it can be said statistically significant model. Significant models can be applied to Waste treatment to obtain optimum treatment for COD reduction. This statement can be proven by looking at

IOP Conf. Series: Materials Science and Engineering **345** (2018) 012011 doi:10.1088/1757-899X/345/1/012011

the value of P-value model (<0.0001) and F-value model (141,41). P value below 0.05 indicates that the model is significant. The calculated p value indicates a significant test for each variable and a good interaction for each of these variables. If the p value is greater than 0.05, then the model is not significant. Significant models can be applied to Waste treatment to obtain optimum treatment for COD reduction.

3.3 Response surface analysis against percentage decrease of COD by electrocoagulation.

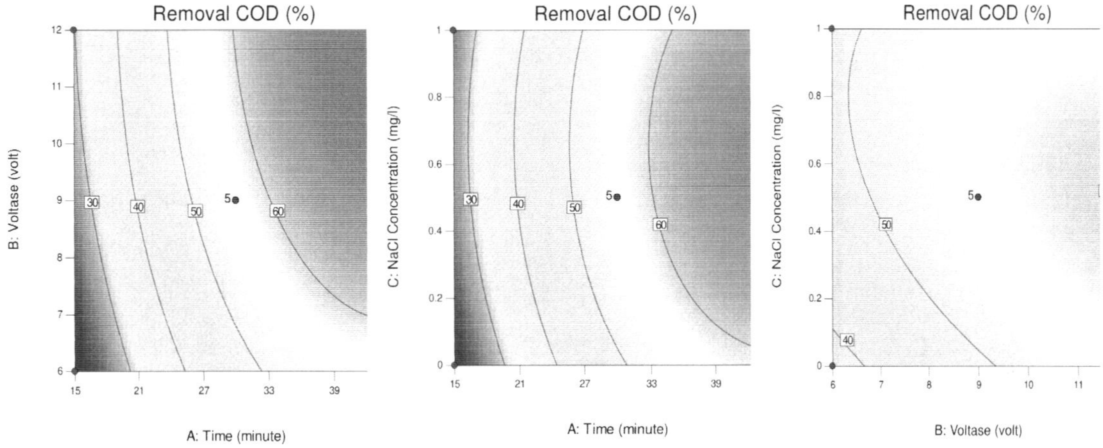

Figure 2. Two-dimensional contour plots express the effect of time, voltage and electrolyte concentration on electrocoagulation.

From the Figure presented, it appears that the three independent variables (contact time, voltage and electrolyte) significantly influence the% COD decrease. COD (Chemical Oxygen Demand) is the amount of oxygen needed to oxidize all organic matter chemically by using a strong oxidizer. In this research the voltage varied is 6 V, 9 V and 12 V. Contact time 15 min, 30 min and 45 min and electrolyte concentration ie without electrolyte 0 M, 0,5 M and 1 M. In electrochemical process with the addition of current into the process will result in oxidation reactions at the anode. Anode made of aluminum will experience an oxidation reaction forming Al3 + which will react with OH- form Al (OH) 3. Al (OH) 3. This in itself, serves as a coagulant that can bind both dissolved and suspended contaminants in solution. The higher the voltage given the coagulant to be formed will be more as well. The optimum voltage for the electrocoagulation process is 12 V [8]. In addition to being affected by the voltage, the decrease of COD in electrocoagulation is also affected by the electrolyte and the contact time. Electrolytes in the electrocoagulation process serves as a catalyst [1]. The addition of electrolytes to the electrochemical process will result in the following reactions:

$$2Cl^- \longrightarrow Cl_2 + 2e^-$$

$$Cl_2 + H_2O \longrightarrow HOCl + Cl^- + H^+$$

$$HOCl \longrightarrow OCl^- + H^+$$

The addition of electrolytes in the electrocoagulation process will decrease COD levels by up to 89% [9]. Another variable that affects COD decrease is contact time. long contact time results in longer contact between coagulants and contaminants in the waste, thereby decreasing larger COD. Electrocoagulation is generally an adsorption particle flocculation in which the coagulant at positive charged electrocoagulation will absorb negative ions in wastes such as nitrates, phosphates, nitrites and other organic compounds and form flocs that help the process of decreasing COD. This is consistent with [9] which states that the longer electrolysis time using Al electrode will increase the formation of metal hydroxide clumps that can neutralize the floc present on the surface.

4. Conclusions

This research investigates electrocoagulation performances in hospital wastewater treatment. The efficiency of COD removal is the main response of this research. The research uses response surface method (RSM) with Box-Behnken Design to get the experimental data interaction.. ANOVA provides the R2 values to verify that the processes are quadratic models. The largest percentage of COD decline was 69.51% and the lowest was 18.36% with R^2 value 0.9945.

Acknowledgments

Authors thank for the help or encouragement of colleagues, support from Chemical Engineering of Syiah Kuala University.

References

[1] Iswanto B, Silalahi M D S and Ayuningtyas U 2009. *Pengolahan Air Limbah Domestik dengan Proses Elektrokoagulasi Menggunakan Elektroda Alumunium* **5** No.1 Studi Kasus: Limbah Cair Gedung Syarif Thajeb (M), Universitas Trisakti.

[2] Hanraha G, Garza C, Garcia E and Miller K 2007 Experimental design and response surface modeling: A method development application for the determination of reduced inorganic species in environmental samples *J. Environ. Inf.* **9** (2) 71-79.

[3] Holt P K, Barton G W, Mark M and Mitchell C A 2005 The future for electrocoagulation as a localised water treatment technology *Chemosphere* **59** 335-365

[4] Garg K K and Prasad B 2015 Development of box behnken design for treatment of terephthalic acid wastewater by electrocoagulation process: Optimization of process and analysis of sludge *J. Environ. Chem. Eng.* **4** (1) 178-190

[5] Mollah M Y A, Schennach R J R 2001 Electrocoagulation (EC) science and applications. *J. Hazard. Mater.* **84** 29-41

[6] Nurdijanto S A, Purwanto and Sasongko S B 2011. *Rancang Bangun dan Rekayasa Pengolahan Limbah Cair Rumah Sakit* (Studi Kasus Rumah Sakit Kristen Tayu *Lingkungan. Universitas Dipenogoro,* Jurusan Teknik Kimia Fakultas Teknik Universitas Dipenogoro

[7] Prayitno 2011 *Teknologi Pengolahan Air Limbah Rumah Sakit* Vol. **1** No. 2 Program Doktor Kajian Lingkungan dan Pembangunan Universitas Brawijaya. Malang, Indonesia

[8] Sunardi 2007 Pengaruh Tegangan Listrik dan Kecepatan Alir Terhadap Hasil Pengolahan Limbah Cair yang Mengandung Logam Pb, Cd, dan TSS Menggunakan Alat Elektrokoagulasi *Seminar Nasional III*, Yogyakarta Indonesia.

[9] Tak B, Tak B, Kim Y, Park Y, Yoon Y and Min G 2015 Optimization of color and COD removal from livestock wastewater by electrocoagulation process: Application of Box-Behnken design (BBD) *J. Ind. Eng. Chem.* **28** 307–315

[10] Zhao Q, Kennedy J F and Wang X 2011 Optimization of ultrasonic circulating extraction of polysaccharides from asparagus officinalis using response surface methodology *Intl. J. Biol. Macromol.* **49** 181–187

IOP Conf. Series: Materials Science and Engineering **345** (2018) 012012 doi:10.1088/1757-899X/345/1/012012

Techno-economic assessment of Levulinic Acid Plant from *Sorghum Bicolor* in Indonesia

M Gozan[1], B Ryan[1] and Y Krisnandi[2]
[1]Department of Chemical Engineering, Faculty of Engineering, University of Indonesia, Depok 16424, Indonesia
[2]Department of Chemistry, Faculty of Mathematics and Natural Science, University of Indonesia, Depok 16424, Indonesia

E-mail: misrigozan@gmail.com

Abstract. The increasing need of energy is one of the main energy security issues in Indonesia. Hence, alternative energy is needed. Levulinic acid (LA) is among chemical platform used in the synthesis for avariety of high-value materials, such as fuels and commodity chemicals. It is predicted that global LA market demand to reach 3.1 tons in 2016. This study examines industrial processdesign and economic analysis for LA production in Indonesia.*Sorghum bicolor* was used as feed because of its high cellulose, low lignin contents and availability in Indonesia. The conventional economic problem from biomass based production was diminished sincethe valuable waste from pretreatment processwas sold to otherindustry.This plant was recommended to be built in an industrial estate area in *Jawa Timur* (East Java) province. Results from simulation using SuperPro Designer 9.0 was used for the techno-economic assessment.The plant assessment showed that the minimum production capacity was 7.7 ton per day to achieve an internal rate of return (IRR) and payback period (PBP) values of 19.61% and 3.93 years, respectively. Sensitivity analysis showed that product selling price was the most predominant factor for IRR, NPV, PBP and ROI. Raw material and water had low effects on those economic parameters. These values indicated that LA plant was feasible to be built in Indonesia.

1. Introduction

Currently, humankind has three major energy security problems, i.e. the increasing of fossil energy consumption, the fossil fuel depletion time and the environmental degradation caused by fossil fuel consumption. These problems are closely related to each other and create a huge challenge for the development of alternative energy source [1].

Earth has an abundant supply of biomass, especially those which located in the forest and vast ocean. The biomass availability potential in Indonesia is predicted 32,654 MW [2] or around 0.62 10^{18} J/year with load factor 0.6. The world has total biomass availability around 200 10^{18} J/year [3]. Mostly, biomass is used to feed the living creatures, especially human, but it also has another function as anenergy source. In the past decade, the researchers and industry have focused on the biomass conversion to energy source [4,5]. Accordingly, biomass utilization as an alternative energy source would provide a great benefit to overcome the aforementioned problems.

In Indonesia, the increase of energy consumption is 8% per year. This value was predicted to increase together with the growth of the economy and human population [6].Biodiesel is one of the proven alternative energy which can be used in Indonesia. The production of biodiesel in Indonesia is 2million kL/year and it is predicted to reach five million kL/year [7]. In recent years, researcher has considered LA as one of the most promising chemical platforms for producing fuel such as biodiesel [8]. Besides biodiesel, LA, or sometimes called as 4-oxopentaoic acid, also has an important role as building compound to produce another compounds, such as 5-bromolevulinic acid, valeric acid, methyltetrahydrofuran (MTHF), methyl pirolidon and many more [9,10]. Up to now, several LA plants from biomass are successfully operating around the world and more projects are underway (see Table 1).

Table 1. Existing and proposed LA plant

Company	Information	Reference
Biofine	In 1994, constructed one ton/day demonstration plant in South Glens Falls, New York (operated for 10 years)	[11]
Segetis	In 2013, started pilot plant in Golden Valley, Minnesota	[12]
GFBiochemicals	In 2015, started 10,000 MT/year commercial plant in Caserta, Italy (planned to scale-up to 8,000 MT per year by 2017)	[13]

Until this work was finalized there is no Levulinic Acid (LA) plant in Indonesia. However, there is growing global market of LA over the last decades. With a global high prices from 5 to 8 $/kg [14] it is an attractive renewable platform especially for the main consumers such as the agricultural, pharmaceutical, cosmetic and food industry. Eventhough, having a ketone and carboxylic group, it has high functionalityfor a wide range of applications like polymers, plasticizers or fuel additives[15]. Some companies such as Biofine, Segetis and GFBiochemicals, are planning to expand the global market of LA as seen in Table 1.

Production of LA and biodiesel as LA derivate may contribute to reducing Indonesian oil import.The development of LA industry was slow in 1990's because the synthetic raw material was expensive and the yield obtained from the process was small. To overcome this condition, biomass is used to replace the synthetic raw material to produce LA. Cellulose, which is contained in the biomass, can be used as main raw material. Sweet sorghum (*Sorghum bicolor*) is one of the biomass which can be used to produce LA. Sorghum is one kind of biomass which is originated from tropic and subtropic area in Southeast Pacific and Australasia regions (Australia, New Zealand, and Indonesia). The availability of *S. bicolor* is abundant in Indonesia. The cellulose composition in this species is relatively high, 31.42%, which make this species suitable to be used in LA production.

Looking at the importance of alternative energy and the availability of *Sorghum bicolor* in Indonesia, the establishment of LA plant from *Sorghum bicolor* seems to be beneficial. Since there is no LA plant in Indonesia before, a sophisticated process design is needed in which the process must be socially acceptable and economically feasible. The techno-economic assessment can be used as a tool to evaluate the uncertainty from thetechnologic and economic point of view. This assessment is a valuable asset to help the Indonesian policy makers and investors to make their decision.

Production unit design is one of the most important things to do before building a plant. There are several factors to be considered in designingthe unit processes, such as the selection of the simulator. The selected simulator should be able to perform rigorous mass and energy balance calculations of the overall process. It also needs to perform economic analysis automatically by maintaining some variables which already contained in the simulator option. The simulation result is used to determine the feasibility of the plant. SuperPro Designer 9.0 is a simulator that can fulfill those criteria. The strength of this application, such as material balance calculation of the overall process, equipment sizing, product analysis, and economic analysis made this application chosen as thesimulator in this study.

This study conducted assessment of the economic viability in LA production based on *Sorghum bicolor* in Indonesia through the economic calculation using SuperPro Designer 9.0. The variation in production capacities which simulated in SuperPro Designer 9.0 is reviewed in economic evaluation. The IRR, NPV, and PBP showed the feasibility of LA plant to be built in Indonesia. The sensitivity analysis was also performed to obtain the correlation between raw material cost, water, and selling price.

2. Material and method

2.1. Raw material

In this work, sorghum was chosen as raw material because of its suitable properties. The low amount of lignin contained in the raw material (7.57%) can make the pretreatment process more effective, and the high amount of cellulose can increase the yield of LA product during the hydrolysis process [16].

Indonesia is the ninth biggest sorghum producer after India, China,Nigeria, United States, Sudan, Argentina, Mexico, and Thailand [17]. In Indonesia, sorghum has been recognized widely, especially in Jawa, NTB, and NTT. Sorghum has many strong points, i.e. resistant to drought, high production, inexpensive production costs, and resistant to pests as well as diseases. This condition makes sorghum become one of the feasible biomass, which can be developed in Indonesia [18].

Simulation

In this work, SuperPro Designer 9.0 (then is called SuperPro) was used to design, simulate and optimize the LA production process. Firstly, Block Flow Diagram (BFD), as in Figure 2, was used to develop arrangement of unit procedures in SuperPro. Then, data from previous researches as well as literature were collected to provide operating conditions data required by the SuperPro. Economic calculations were done by SuperPro and became the basis data for setting up for production capacity.

The raw material input acts as theindependent variable while the value of IRR, PBP, ROI, NPV, and sensitivity analysis result act as the dependent variable. Through this work, the selected production capacity was expected to achieve the IRR and PBP value of more than 14% and less than five years respectively. This value refersto the MARR value for chemical industry and plant-overall lifetime, which is expected to be 15 years [19].

Prior to the whole plant simulation, preliminary economic analysis was performed.Assessment was done by simulating only the unit in charge of hydrolising lignocelllulose and converting it to levulinic acid. Conversion of lignocellulosic material to LA can be performed by acid hydrolysis. Equipment cost data used was taken from SuperPro built in cost model.

2.2. Process description and simulation

Biorefinery concept can be applied in this plant because multi products can be produce such as LA, formic acid, and furfural from *Sorghum bicolor*. LA is the main product, while formic acid and furfural are the side products. Biorefinery is a system that processed or fractionated bio-based source to obtain more than one product including bioenergy, biofuels, chemicals and high value-added compounds [20].

*Sorghum bicolor*grains were ground and given a pretreatment with dewaxing (extraction) and delignification (filtering) processes. Hydrolysis process was then carried out using homogeneous catalyst followed by neutralization. Cellulose and hemicellulose were two out of three components obtained form lignocellulosic material. These components are polymer that can be attacked through hydrolysis. The hydrolysis product of cellulose is glucose (C_6). Meanwhile, the product of hemicellulose is xylose (C_5). Decomposition of C_6 with acid as catalyst produced 5-hydroxymethyl-2-furaldehyde as an intermediate product. Further rehydration gave LA and formic acid. Decomposition of C_5gave furfural. Clarification was done to separate the remnants of lignocellulose which were not hydrolyzed. At the end of the process, multiple distillation was carried out to separate the formic acid, furfural and LA. The whole process was simulated under continuous operation mode. The process block flow diagram generated is shown in Figure 1.

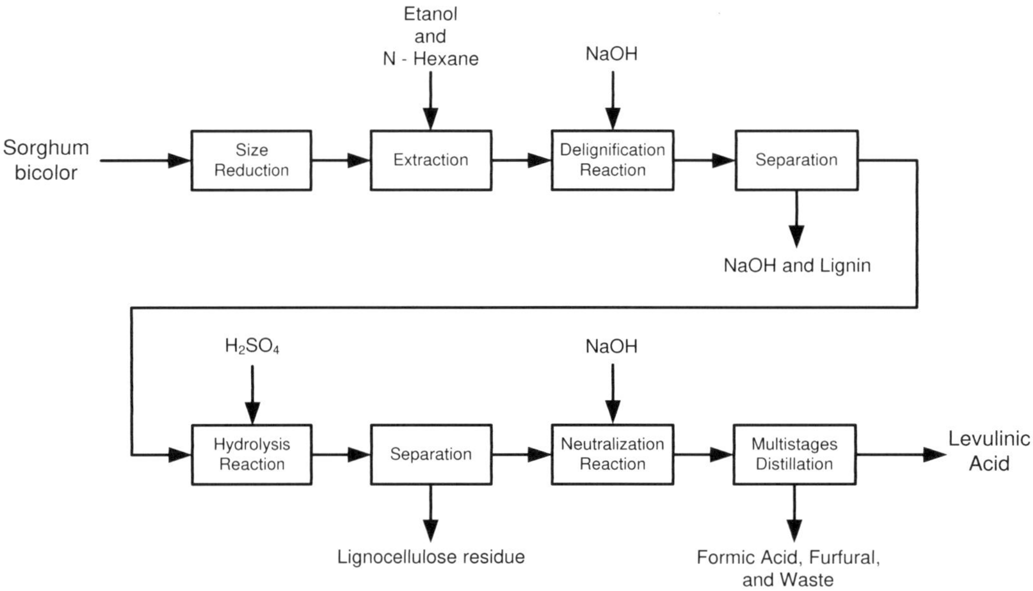

Figure 1. Block flow diagram of levulinic acid production.

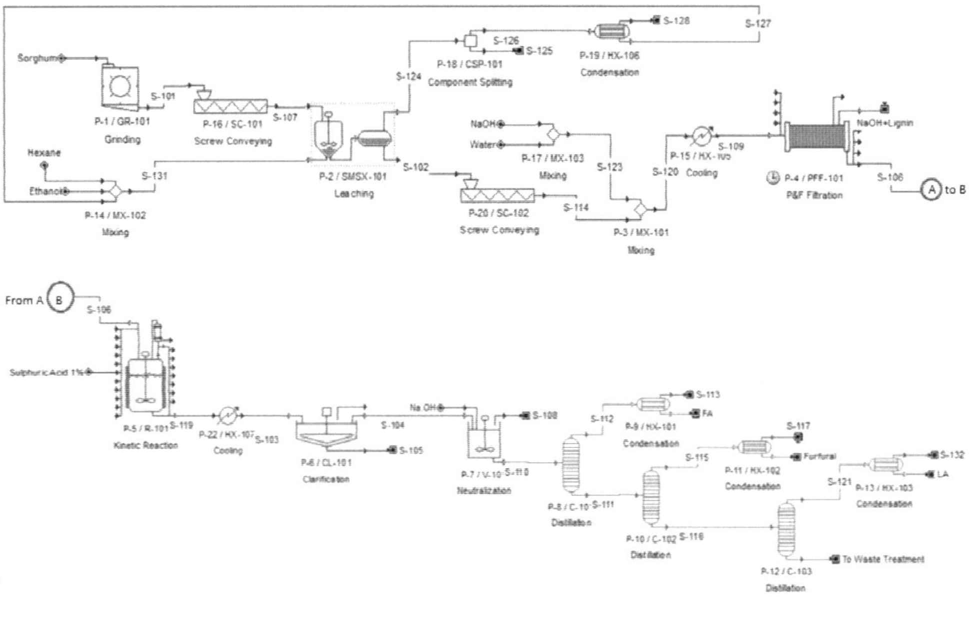

Figure 2. Simulation flowsheet of levulinic acid production.

Figure 2 shows the simulation of LA production process using SuperPro Designer 9.0. In this simulation, data of operating condition for each unit operation were obtained from open literature and industry. The data include input and output stream composition, preparation condition of each production unit, and constraint in each operating unit. The selected criteria and operating condition are shown in Table 2.

Table 2. Selected criteria and operating condition of LA process.

Process	Equipment Code	Criteria and Operating Condition	Reference
Size Reduction	GR-101	Equipment: Disk Mill Power: 1.1 kW	[21]
		Output: 315 microns	Preliminary work
Solid-Liquid Extraction (Dewaxing)	SMSX-101	Temperature: 80°C Residence Time: 6.5 hours	Preliminary work
		Solvent: Hexane & ethanol mixture (2:1)	Preliminary work
Delignification	MX-101	Solvent: NaOH	[13]
		Concentration: 10% Temperature:72.5°C	[22]
		Impeller Speed: 190 rpm	[23]
		% Component Removal • Cellulose: 20% • Hemicellulose: 41.4% • Lignin: 70%	Preliminary work
Hydrolysis	R-101	Catalyst: H_2SO_4 Concentration: 1% Residence Time: 20-25 minutes	[9,10] Preliminary work Preliminary work
		Temperature: 130°C Impeller Speed: 100 rpm	[10] [23]
Purification	C-101/ 102/103	Separation: 3 stage Distillation	Preliminary work
		Lignocellulose relative volatility	Preliminary work

Some researchers have studied the utilisation of acidic [24,25] and base pretreatments. Base pretreatment has shorter residence time (1-2 hours) compare to acidic (20-30 hours). Base pretreatment works at lower temperature (50 °C) than acidic pretreatment (120-150°C). Although we have to use higher concentration of base, the price of base is much lower .Other benefit of using base pretreatment is its broader lignin tolerance. Having these advantages, we selected base pretreatment.

For hydrolysis reaction, there are some available options [10, 26, 27], which basically fall into homogenous (H_2SO_4, HCl) and heterogenous catalyst (Mn/ZSM-5). Homogenous catalyst was selected due to its higher conversion, selectivity and LA yield. It also has lower price and modest temperature (120°-150°C). The disadvantages is its higher corosoivity. However, selection of vessel materials and concentration of acid might reduce this problem. In the hydrolysis reactor there are multiple reactions following this model,

$$k = (C_{H+})^{\alpha}.k_0.\exp(-\frac{E}{RT}) \quad (1)$$

The kinetic and Arhenius parameters of those reactions are illustrated in Figure 3.

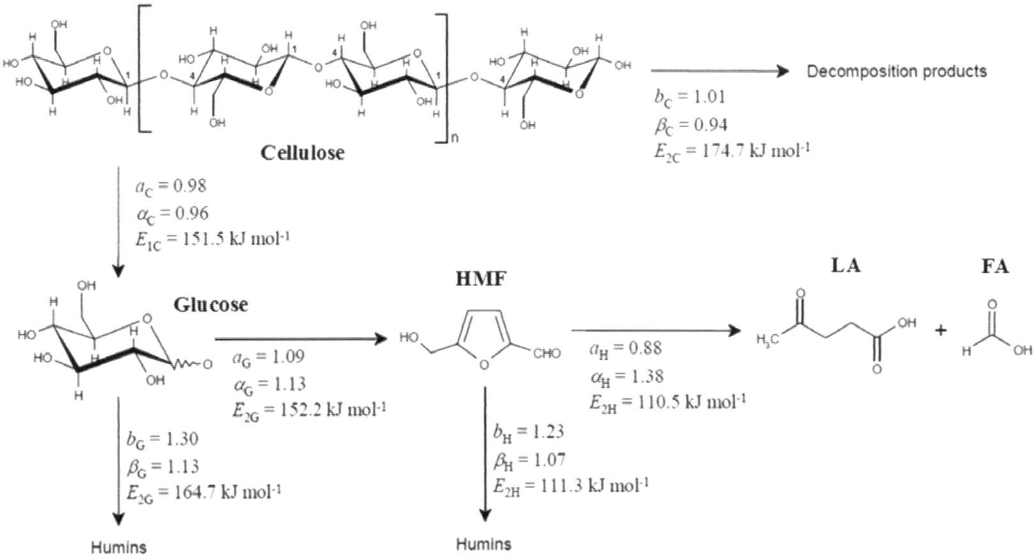

Figure 3. Kinetic parameters for hidrolysis reactions in production of Levulinic Acid and Formic Acid from cellulose, modified from [28].

3. Results and discussion

3.1. Plant location

Plant location selection is one of the main factors in determining the success of a plant. Therefore, some importants factors such as raw materials availability, utilities, market, communication access and transportation should be considered properly. Central Java and East Jawa are the two provinces with the largest sorghum producer. To determine the location of the plant, a review of some aspects of these two provinces was conducted as shown in the Table 3.

Table 3. Levulinic acid location analysis

Criteria	Central Java	East Java
The distance of proposed plant location to the nearest market target	(Reference city: Semarang)	(Reference city: Surabaya)
• Tuban (km)	210.4	103.0
• Cilacap (km)	260.5	504.1
• Gresik (km)	293.8	21.6
Total distance (km)	764.7	628.7
Raw Materials Availability (ton/years)	57,000	55,000
Raw Materials Production Area (ha)	19,000	10,000
Raw Materials Productivity (ton/ha)	3	5.5
The Largest Industrial Area (ha)	450	835

East Java is closer to the main market targets of levulinic acid, i.e. Tuban, Cilacap, and Gresik. The shorter distance to the target market renders lower distribution costs.East Java alsoposses bigger industrial area, such as Gresik Industrial Estate (140 ha), Ngoro Industrial Park Mojokerto (440 Ha), and Surabaya Industrial Estate Rungkut (835 Ha) [29]. Sorghumproduction in East Java is slightly lower than that in Central Java. However, there is a much progressive productivity and an ambitious plan in Eat Java to expand the plantation area. Therefore, East Java is selected as the location of the LA plant.

Surabaya Industrial Estate Rungkut (SIER), located in East Java, is an industrial area that was designed with the concept of an integrated industrial area, flood-free, and ready to use with the empty area offered is 401 hectares [29]. SIER also offers advantages in term of transportation. It has relatively short distance to the seaport, airport, rail station and highway gate as depicted in Table 4.

Table 4. Average mileage of SIER to transportation gate [30]

Destination	Distance (km)
Juanda International Airport	5
Tanjung Perak Seaport	19
GubengRail Station	10
Surabaya-GempolHighway Gate	5
Central Government Area of Surabaya	8

3.2. Production capacity determination

Various raw material input range from 30 to 55tonnes per day, were simulated in SuperPro influenced the values of IRR and PBP as shown in Table 5. Input of 30 tonnes per day, which produces 5.78 tonnes LA per day, results in high unit production cost for all the products (Levulinic acid, Furfural, Formic Acid and Lignin waste) which is 11.15$/kg. The SuperPro calculates Unit Production Revenue based on total revenue divided by the main stream product which was set in our simulation as LA stream. The simulation data of all products were set based on on market price, i.e. 7.62 $/kg LA; 1.2 $/kg Furfural; 0.95 $/kg Formic acid; and 0.09 $/kg Lignin waste. The calculated unit production revenue is 9.89 $/kg. Thus, the IRR and PBP were not available (N/A). Higher raw material input makes the increase in IRR and reduction of PBP.

Table 5. Variation of production capacity to IRR and PBP.

Raw Material Input(tons of *S. bicolor*/day)	Production Capacity(tons of LA/day)	Production Cost ($/kg main product)	IRR (%)	PBP (year)
30	5.78	11.15	N/A	N/A
35	6.75	9.72	3.52	9.96
40	7.75	8.66	19.61	3.93
45	8.68	7.82	31.95	2.45
50	10.15	6.64	49.14	1.52
55	10.61	6.60	52.27	1.41

Figure 4 shows the relationship between production capacity and the IRR as well as PBP. In this work the cross point, i.e. 40 tons raw material input per day, has IRR 19.61% and PBP 3.93 years. These fullfil the requirement of IRR >14% and PBP <5 years for an interesting investment. Further calculation on this work is then based on this *Sorghum bicolor* material input.

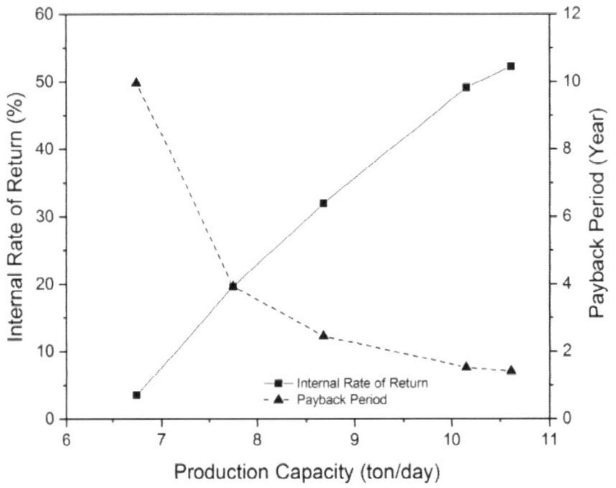

Figure 4. Production capacity versus IRR and PBP of LA production.

3.3. Economic analysis

Total capital investment consists of equipment purchase cost, bulk material cost, and other costs that can be seen from the table below:

Table 6. Total capital investment of LA plant

Component	Cost (US$)
Total Plant Direct Cost (TPDC) (*Physical Cost*)	
1. Equipment Purchase Cost	1,326,000
2. Equipment Installation	340,000
3. Process Piping	464,000
4. Instrumentation	530,000
5. Insulation	40,000
6. Electricity Installation	133,000
7. Building	597,000
8. Yard Improvement	199,000
9. Auxiliary Facilities	530,000
TPDC	4,158,000
Total Plant Indirect Cost (TPIC)	
11. Engineering	1,039,000
12. Construction	1,455,000
TPIC	2,495,000
Total Plant Cost (TPC = TPDC+TPIC)	
TPC	6,653,000
Contractor's Fee & Contingency (CFC)	
13. Contractor's Fee	333,000
14. Contingency	665,000
CFC	998,000
Direct Fixed Capital Cost (DFC = TPC+CFC)	
DFC	7,650,000
Working Capital (WC)	
WC	1,679,000
Total Capital Investment (TCI)	9,329,000

Table 7. Total operational cost of LA plant

Cost	$	%
Raw Material	9,696,000	48.53
Direct Labor Cost	434,000	2.17
Facility-Dependent	1,445,000	7.23
Laboratory/QC/QA	65,000	0.33
Waste Treatment	99,000	0.49
Utility	8,241,000	41.24
TOTAL	19,981,000	100

The raw material and utility costswere themost important component in total operating cost. The main contributors for the high raw material cost were the high amount of raw materials used in this plant. These include solvent (n-hexane and ethanol), catalyst (sulfuric acid), and *Sorghum bicolor* itself. Besides, the high amount of energy required renders high utility cost percentage in this biorefinery plant.

Table 8. Profitability analysis of LA plant.

Production Capacity		
Lignin Waste	11,501	ton /yr
Formic Acid	2,065	ton /yr
Furfural	413	ton /yr
Levulinic Acid	2,546	ton /yr
Revenue		
Lignin Waste	90	$/ton
Formic Acid	950	$/ton
Furfural	1,200	$/ton
Levulinic Acid	7,620	$/ton
Profitability Analysis		
Gross Margin	12.71	%
Return On Investment	25.46	%
Internal Rate of Return	19.61	%
Net Present Value	8,653,000	$
Payback Period	3.93	Years

Profitability analysis in LA plant has been calculated automatically by SuperPro Designer 9.0. The result is shown in Table 9. LA was not the only product obtained from the plant. The existence of furfural and formic acid as byproducts also increased the revenue. LA industrial price was within the range of US$ 5 to 8 per kilogramwhile furfural and formic acid industrial price was assumed US$ 1.2 and S$ 0.95 per kilogram respectively[14]. LA price used in this plant was US$ 7.6. To increase the plant revenue, the processable waste was also sold, such as NaOH and lignin mixture which obtained from delignification process. If we observed the plant based on its function, NaOH and lignin mixture was not used anymore. Lignin and NaOH purification process will only increase plant production cost. Because of these reasons, lignin and NaOH mixture were sold to the industry which undergo lignin purification. Based on the product selling price, total plant revenue was US$ 22,732,927. This revenue made the ROI, IRR, and PBP value of 25.46%; 19.61%; and3.93years, respectively. The value of IRR and PBP that was less than one-third of plant lifetime will attract investor to invest in this plant.

Table 9. Technology comparation.

Product (ton/ton biomassa)	Teknologies/Scenarios			
	Biofine [11]	Biofuel [31]	Biorefining [31]	Our study
Levulinic acid	0.157	NA	0.252	0.194
Furfural	0.076	0.107	0.157	0.031
Formic acid	0.062	0.100	0.100	0.002
Lignin	NA	0.180	0.180	0.154
EtilLevulinat	NA	0.313	NA	NA
AHRs	0.517	0.151	0.151	NA

The content of LA generated in this work lies between Biofine and Biorefining. The different is not only due to raw material but also the kinetic and other parameters.

3.4. Sensitivity analysis

Sensitivity analysis was performed to determine a correlation between direct-indirect cost and every component in profitability analysis, such as IRR, NPV, ROI, and PBP (Figure 5-8). In this plant, raw material cost, utility cost (water) and product selling price was chosen as the parameter in this sensitivity analysis.

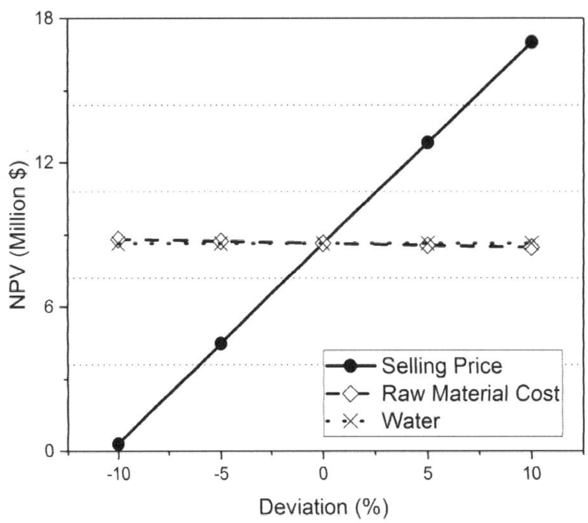

Figure 5. NPV sensitivity analysis.

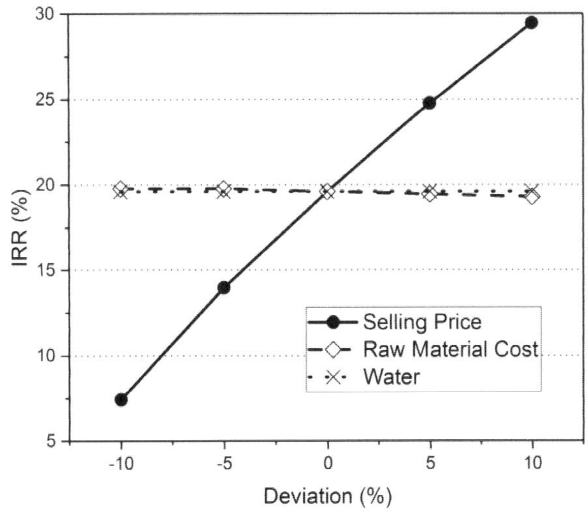

Figure 6. IRR sensitivity analysis.

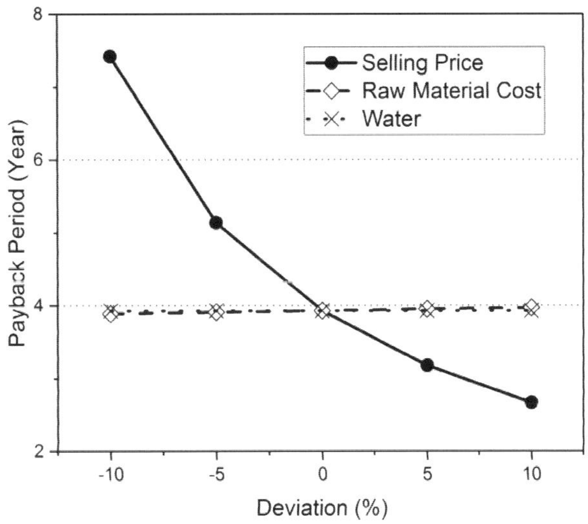

Figure 7. PBP Sensitivity Analysis

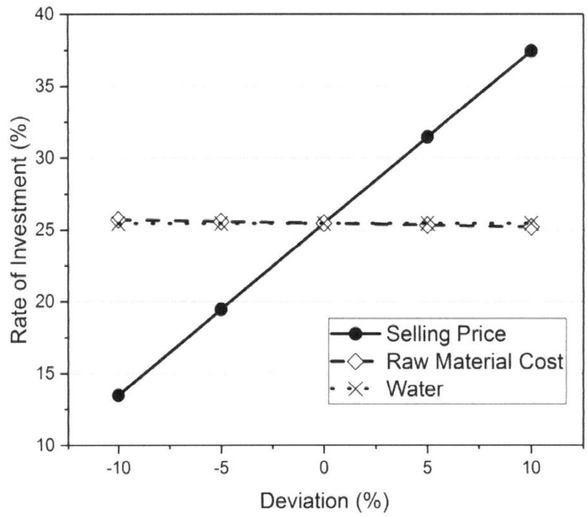

Figure 8. ROI Sensitivity Analysis

NPV and IRR value rose up with the increasing of LA price as expected. The higher the LA price, the bigger the company profit. With the increasing of the LA price, the PBP also becameshorter.The parameterwhich affects IRR, NPV, ROI, and PBP the most was product selling price itself. On the other hand, raw material cost was notsignificantly affects IRR, NPV, ROI, and PBP.Deviation of utility cost did not have significant effect on the change of NPV, IRR, PBP, and ROI. This can be seen from the stagnant graph on NPV, IRR, ROI, and PBP when utility cost increased and lowered. This may happenbecause those parameters only took asmall portion ofthe total operating cost.

4. Conclusion

In this work, LA production from *Sorghum bicolor* consisted of three main processes: pretreatment, hydrolysis, and separation process. Optimization by using SuperPro Designer 9.0 showed that the minimum production capacity was 7.71 tons of LA perday resulted from 40 tonnes *Sorghum bicolor* per day as raw materials. This meant that the LA yield was 19.4% (ton of LA resulted per ton of Sorghum bicolor in input). Besides LA, side products as furfural, and formic acid were also formed 3.1% and 0.2%, respectively. This integrated LA plant design achieved the value of IRR 19.61% and 3.93 years of PBP.

Abbreviation

CFC	Contractor's Fee & Contingency
DFC	Direct Fixed Capital Cost
IRR	Internal Rate of Return
LA	Levulinic Acid
MARR	Minimum Acceptable Rate Of Return
NPV	Net Present Value
PBP	Payback Period
POEFB	Palm Oil Empty Fruit Bunches

ROI	Return On Investment
TCI	Total Capital Investment
TPC	Total Plant Cost
TPDC	Total Plant Direct Cost
TPIC	Total Plant Indirect Cost
WC	Working Capital

Acknowledgment
The authors acknowledge financial support from Indonesia Estate Crop Fund for Palm Oil (BPDPS Research grant number Peng-01/DPKS.4/2015). Preliminary part of this research was funded by The Ministry of National Education, Republic ofIndonesia through Competitive Research Grant number 2245/H2.R/12/HKP.0500/2014. The authors also gratefully acknowledge Center of Bioindustrial Technology, Agency for Assessment and Aplication of Technology (BPPT), The Ministry of Research and Technology, Republic of Indonesiafor the technical supports.

References

[1] Andika R, Husnil Y A and Lee M 2014 Does lower energy usage mean lower carbon dioxide emissions? – A new perspective on the distillation process *Korean J. Chem. Eng.* **31** 1110 – 1114
[2] Dewan Energi Nasional Republik Indonesia 2014 Energy Outlook Indonesia *Ministry of Energy and Mining* (2014)
[3] Yokoyama S and Matsumur Y 2015 The present status and future scope of bioenergy in Japan *J. Japan Inst. Energ.* **10**
[4] Kurian J K, Nair G R, Hussain A and Raghavan G S V 2013 Feedstocks, logistics and pre-treatment processes for sustainable lignocellulosic biorefineries: A comprehensive review *Renew. Sust. Energ. Rev.* **25** 205–219
[5] Tolan J S 2006 *Iogen's Demonstration Process for Producing Ethanol from Cellulosic* Biomass, ed B Kamm, P R Gruber and M Kamm (Weinheim: Wiley-VCH) p 193-207
[6] Samsuri M, Gozan M, Hermansyah H, Prasetya B, Nasikin M and Watanabe T 2008 Ethanol production from bagasse with combination of cellulase-cellubiase in simultaneous saccharification and fermentation (ssf) using white rot fungi pre-treatment *J. Chem. Nat. Resour. Eng.* **3** 20-32.
[7] Ministry of Industry Republic of Indonesia, Indonesia produsenutama biodiesel. http://www.kemenperin.go.id/artikel/1903/Indonesia-Produsen-Utama-Biodiesel, 2013, (accessed 4th May 2014)
[8] Farone W A and Cuzens J E 2000 Method for the production of levulinic acid and its derivatives *US Patent* **6054611** A (2000).
[9] Girisuta B, Janssen L P B M and Heeres H J 2006 Green chemicals a kinetic study on the conversion of glucose to levulinic acid *Chem. Eng. Res.* **84** 339-349.
[10] Girisuta B, Janssen L P B M and Heeres H J 2006 A kinetic study on the decomposition of 5-hydroxymethylfurfural into levulinic acid *Green Chem.* **8** 701-709.
[11] Biofine, Our Company. http://biofinetechnology.com/, 2015 (accessed 18th December 2015).
[12] Segetis, Segetis announced the successful start up of their pilot plant facility in Golden Valley, Minnesota. http://www.segetis.com/news/news20.html, 2013 (accessed 18th December 2015).
[13] PCI Magazine, GFBiochemicals' levulinic acid technology ready for commercialization. http://www.pcimag.com/articles/100302-gfbiochemicals-levulinic-acid-technology-ready-for-commercialization, 2015 (accessed 18th December 2015).
[14] Grand View Research 2014 *Levulinic Acid Market Analysis and Segment Forecasts to 2020* (San Francisco: Grand View Research)

[15] Rackemann D W and Doherty W O S 2011 The conversion of lignocellulosics to levulinic acid *Bioprod. Bioref.* **5** 198-214
[16] Boonsayompoo O and Reungsang A 2013 Thermophilicbiohydrogen production from the enzymatic hydrolysate of cellulose fraction of sweet sorghum bagasse by Thermoanaerobacteriumthermosaccharolyticum KKU19: Optimization of media composition *Int. J. Hydrogen Energ.* **38** 15777-15778
[17] Sirappa M P 2003 Prospek pengembangan sorgum di Indonesia sebagai komoditas alternative untuk pangan, pakan dan industri (the development prospect of sorghum in indonesia as alternative commoditiy for food, feed and industry) *Jurnal Litbang Pertanian* **22** 133-140
[18] Suryaningsih R and Irhas 2014 Bioenergy plants in Indonesia: Sorghum for producing bioethanol as an alternative energy substitute of fossil fuels *Energ. Procedia* **47** 211-216.
[19] Seider W D, Seader J D, Lewin D R and Widagdo S 2009 *Product and Process Design Principles: Synthesis, Analysis, and Design*, 3rd Edition (Hoboken: Wiley & Sons, Inc.)
[20] Jonathan M B, Valentina A M and Carlos A 2016 Design strategies for sustainable biorefineries *Biochem. Eng. J.* **111** 122-134
[21] Chun C, Xinojian C A and Peilin C 2009 Kinetic studies on wheat straw hydrolysis to levulinic acid *Biotechnol. Bioeng.* 17 (5) 835-839.
[22] Cao C, Sun R, Liu R, Yin X and Wu 2012 Comparison of the effects of five pretreatment methods on enhancing the enzymatic digestibility and ethanol production from sweet sorghum bagasse *Bioresour. Technol.* **111** 215-221
[23] Seader J D, Henley E J, Roper D K 2010 *Separation Process Principles: Chemical and Biochemical Operations*, 3rd Edition (Hoboken: Wiley & Sons, Inc.)
[24] Xu F, Shi Y C and Wang D 2011 Process optimization for ethanol production from photoperiod-sensitive sorghum: focus on cellulose conversion *Ind. Crop. Prod.* 34 1212-1213
[25] Xu F, Shi Y C and Wang D 2013 Towards understanding structural changes of photoperiod-sensitive sorghum biomass during sulfuric acid pretreatment *Bioresour. Technol.* **1** 704-709
[26] Weingarten R, Kim Y T, Tompsett G A, Fernández A, Han K S, Hagaman E W, Conner Jr. W C, Dumesic J A and Huber G W 2013 Conversion of glucose into levulinic acid with solid metal (iv) phosphate catalysts *J. Catal.* **304** 123-124
[27] Bevilaqua D B, Rambo M, Rizzatti T M and Cardoso A 2013 Cleaner production: Levulinic acid from rice husks*J. Clean. Prod.* **47** 96-97.
[28] Dussan K, Girisuta B, Haverty D, Leahy J J and Hayes M H B 2013 Kinetics of levulinic acid and furfural production from Miscanthusgiganteus *Bioresour. Technol.* **149** 216–224.
[29] Indonesia Industrial Estate Directory, Profil kawasani ndustri daerah jawa timur (Industry profile of east java province). http://regionalinvestment.bkpm.go.id/, 2015 (accessed 18th December 2015)
[30] Badan Koordinasi Penanaman Modal (BPKM), Data kawasan industry Jawa Timur. (Industrial data region in East Java Province) http://www.bkpm.go.id/img/file/, 2015 (accessed 18th December 2015)
[31] Hayes D 2013 Report on Optimal Use of DIBANET Feedstocks and Technologies, DIBANET project, 2013.

IOP Conf. Series: Materials Science and Engineering **345** (2018) 012013 doi:10.1088/1757-899X/345/1/012013

Resin purification from Dragons Blood by using sub critical solvent extraction method

Saifuddin and Nahar
Chemical Engineering Department. Lhokseumawe of State Polytechnic, Indonesia

E-mail: odinx66@yahoo.com

Abstract. Jernang resin (dragon blood) is the world's most expensive sap. The resin obtained from jernang that grows only on the islands of Sumatra and Borneo. Jernang resin is in demand by the State of China, Hong Kong, and Singapore since they contain compounds that have the potential dracohordin as a medicinal ingredient in the biological and pharmacological activity such as antimicrobial, antiviral, antitumor and cytotoxic activity. The resin extracting process has conventionally been done by drizzly with maceration method as one way of processing jernang, which is done by people in Bireuen, Aceh. However, there are still significant obstacles, namely the quality of the yield that obtained lower than the jernang resin. The technological innovation carried out by forceful extraction process maceration by using methanol produced a yield that is higher than the extraction process maceration method carried out in Bireuen. Nevertheless, the use of methanol as a solvent would raise the production costs due to the price, which is relatively more expensive and non-environmentally friendly. To overcome the problem, this research proposed a process, which is known as subcritical solvent method. This process is cheap, and also abundant and environmentally friendly. The results show that the quality of jernang resins is better than the one that obtained by the processing group in Bireuen. The quality of the obtained jernang by maceration method is a class-A quality based on the quality specification requirements of jernang (SNI 1671: 2010) that has resin (b/b) 73%, water (w/w) of 6.8%, ash (w/b) 7%, impurity (w/w) 32%, the melting point of 88°C and red colours. While the two-stage treatment obtained a class between class-A and super quality, with the resin (b/b) 0.86%, water (w/w) of 6.5%, ash (w/w) of 2.8%, levels of impurities (w/w) of 9%, the melting point of 88 °C and dark-red colours.

1. Introduction

Dragon blood is the most expensive resin in the world and most needed in the pharmaceutical world. Since the resin has Dracohordyn compounded, which is plays an important role in the pharmaceutical world. Drachordyn is the main constituent found in dragon blood and included as anthocyanin compound and used as an effective pharmaceutical element since its biology and pharmacology activities such as an anti-microbe, anti-virus, antitumor, and cytotoxic activities [1].

Dragon blood is a resin from secretion process of rattan fruit. It adheres and covers the outside of rattan fruit. Thus, the extraction of the fruit is needed to get the resin out. The dragon blood traditionally is utilized as medicine. It is also used as a dye to paint varnish stuff.

The main chemical component of dragon blood is Esther resin and dracoresino tannol (57- 82%). Moreover, red resin also contains compounds such as dracoresene (14%), dracoalban (up to 2.5%), insoluble resin (0.3%), residue (18.4%), benzoic acid, benzoilasetat acid, dracohodin and some of pigments especially nordracorhodin and nordracorubin [2].

This research is related to the efficiency process in extracting the resin of anthocyanin derivation from a raw material of dragon blood. The combined methods between maceration and infundation are used since the engineering of water, as the solvent provides no negative effect to health, the waste gained was environmentally friendly, therefore, it is also lowly cost and has rapid extraction process, which leads to the saving of production cost. Generally, the best solvent used is methanol, however, its price and the effect given to the waste make methanol is reconsider to be used. Alternatively, the extraction using infundation method becomes the most appropriate method to be applied to the people that process the dragon blood.

The using of solvent in conventional processing by the people in Bireuen in extracting the resin were in huge amount compared to the dragon blood weight in order to extract the resin completely and this process was moderately ineffective. This research analysed resin, water, and ashes content and also the melting point of the resin after being extracted. Thus the result was adapted accordance to SNI.

2. Material and method

2.1 Material and equipment

The material used in this research was dragon blood, and Aquades as the solvent to be extracted in beaker glass 1000 ml. The extraction process was conducted in two methods; firstly by using maceration method and secondly by using simulant maceration under subcritical water in the autoclave at 900 ^{o}C and atmospheric pressure.

Figure 1. Dragon blood fruit.

2.2 Method

A 200-gram of crushed jernang, put into a beaker glass and soaked (maceration) for 1 hour, using a solvent (methanol and water) with the ratio of 1: 1, 1: 2, 1: 3, respectively. It is then fed into the autoclave for 60 minutes based on the critical point of methanol and water. Then the separation of the dregs from the solvent mixture (water or methanol) is carried out. The mixture is then distilled to obtain the resin; finally, the resin is analysed by testing the physio-chemical properties of jernang.

2.3 Extraction of critical solvent using water

The result of extraction process with maceration method was dried in autoclave with 90 °C for 30 minutes before it was filtered. The result of the extraction, which in lump form was separated from the liquid then it. It had to be dried then analysed.

2.4 Determining the yield of the resin

In determining the content of resin, the analysed sample should be sufficiently dry and usually, the used sample was taken from determined of the used sample for water content. If the sample was still wet, it is not only slowing down the extraction process but also can turn down the beaker glass so it would complicate the determining the constant weight of the beaker glass.

$$\text{Resin yield (\%)} = \frac{A}{Weight\ of\ Sample} \times 100\% \quad (1)$$

Table 1. Chemical characteristic of dragon blood resin (SNI 01-1671-1987)

No	Characteristic (Properties)	Jernang (Dragon's blood) SAD	SMJ	SNI (Indonesia National Standard) Quality I	Quality
1	Water content (%) (Moisture content)	4,4	3,2	Maks.3 (Max)	Maks.6 (Max)
2	Dirt (%) (Impurity)	16,0	5,2	Maks.14 (Max)	Maks.3 9 (Max)
3	Ash (%) (Ask content)	2,8	0,7	Maks.8 (Max)	Maks.20 (Max)
4	Melting point (°C) (Melting point)	105	80	80-120 °C	80-120 °C

2.5 Water content

Water content is the difference between material weight before and after drying/heating. Every material will come to the balance with surrounding humidity if being put in the open air. This water content is named balanced water content. Every certain relative humidity can produce certain balanced water content, too. Therefore, it can relate the balanced water content and relative humidity. Principally, it steams the water contained in the material by heating/drying, then weights the material until it reaches the constant weight, and it means that all the water has been steamed.

$$\text{Water content} = \frac{A-B}{\text{gram sample}} \times 100\% \quad (2)$$

2.6 Determining ashes content

Ashes are the rest of complete burning from organic material (unsteamed residue if burned in a certain way). Chemically, ashes can be defined as metal oxide and other unburned materials it is an indicator of a clean level where the higher level of the ashes means the higher impurity is. Naturally, there are metals in dragon blood resin. These metals are the nutrient of the plant, which plays the important role in biochemical reaction. The metals are physiologic ashes. In the preparation stage, the rattan fruit can be contaminated with sand, soil, etc. Sand is unburned silicate compound and not dissolved in acid thus it acid undissolved composer compound of ashes. For this reason, the waste of dragon blood needs to be determined to observe the impurity content that contains the waste. If the ashes content is beyond the requirement, the resulted resin is unable to be used.

$$\text{Ashes Content} = \frac{B-A}{gram\ sample}\ x\ 100\% \qquad (3)$$

2.7 Determining the melting point

The melting point is the observed temperature when the solid material (dragon blood) started to melt until the whole particles changed into liquid, where the solid temperature changes into liquid in atmospheric pressure. To determine the melting point, a one-gram sample that needs to be inserted into a tube of capillary glass. The tube contained the dragon blood was set in a determiner tool of melting point. After the tool being operated, the temperature when the sample changed from solid to liquid was noted as the melting point of dragon blood.

3. Results and discussion

In this study, the quality of the sample is determined based on SNI parameters and matched with the produced quality according to the research that has been conducted. The resin is produced by extracting the jernang with the water's sub-critical method by using water and methanol with the difference of the jernang ratio, which is carried out by heating; the purpose of heating (boiling) is to melt the resin contained in the fruit of the jernang. Based on the obtained data after heating using different solvents, the one that using water contender more than the one using a methanol solvent. This is because the solvent water is more soluble with the resin than the methanol solvent. Based on the obtained information from the SNI quality table standard, the quality class for samples using water solvents and methanol can be classified by the following tests:

3.1 Resin

The resin content is determined by heat extraction using soxhlet and diethyl ether solvent. The level testing of resin is using to notice the purity of jernang. Figure 2 illustrates the comparison of resin content to the ratio of jernang. In that figure, it can be analysed and compared that the water solvent and produced resin methanol of jernang quality requirements (SNI 1671: 2010), is B-quality ranging from 40% -41%. The produced resin content between the water solvent and the methanol solvent is almost adjacent, but the water solvent produces the highest resin content. This shows that many of the polar compounds contained in the jernang recall that only polar compounds are soluble in a polar solvent. Based on the ratio of the raw material to the solvent, obtained by more resin with a ratio of 1: 3 this can cause more solvent that penetrates the fruit cell membrane of the vein for the occurrence of diffusion. Consequently, more extracts have come out from the jernang cell membrane.

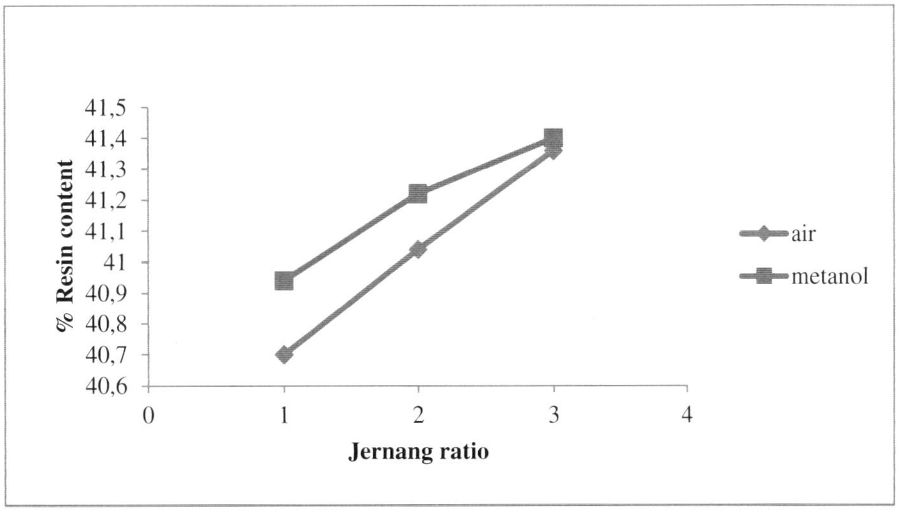

Figure 2. Comparison of resin toward jernang ratio.

3.2 Water

Water content is determined to notice the results of the product, whether it has a good quality and it also depends on the results of the analysis, by testing the quality of the jernang in accordance with the SNI. Figure 3 illustrates the comparison of water content to the ratio of jernang whereas the produced water content by the water solvent and methanol is very significantly different. However, the comparison produced based on SNI shows that the quality in the methanol solvent ranges from 8.8%, then the water levels in this test achieves the AB-quality. Furthermore, in the resulting quality water solvent the reached quality is B-quality. This indicates that the methanol solvent does not contain much water so that the water content that measured by the produced resin using the methanol solvent is significantly different from the resin obtained by using a water solvent.

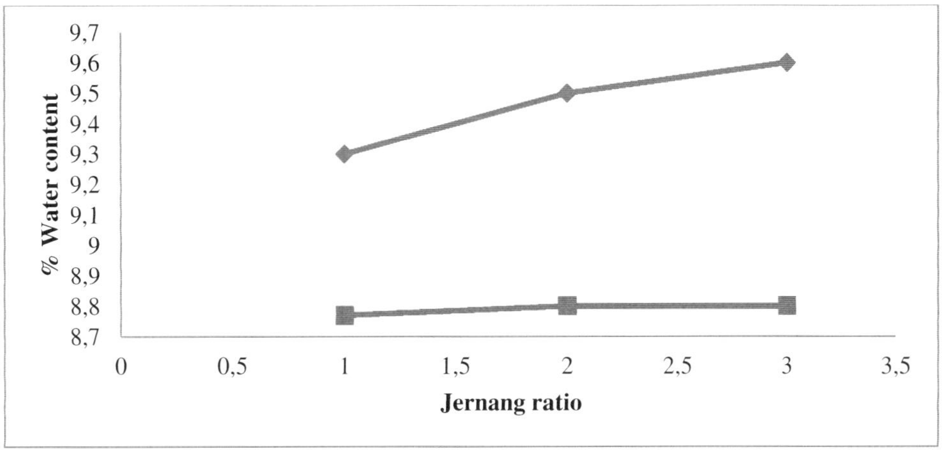

Figure 3. Comparison of water content toward jernang ratio.

3.3 Dirt (impurity)

The impurity content indicates the amount of impurity mixture present in the jernang resin. In Figure 4 we can see the ratio of the impurity content to the ratio of jernang, where the resulting ratio between the two solvents is significantly adjacent. In the water solvent, the resulting impurity content from 42-43.5% and in the water solvent the resulting impurity content from 41-42%. Requirements generated based on the quality of SNI resulting impurity reaches the B-quality.

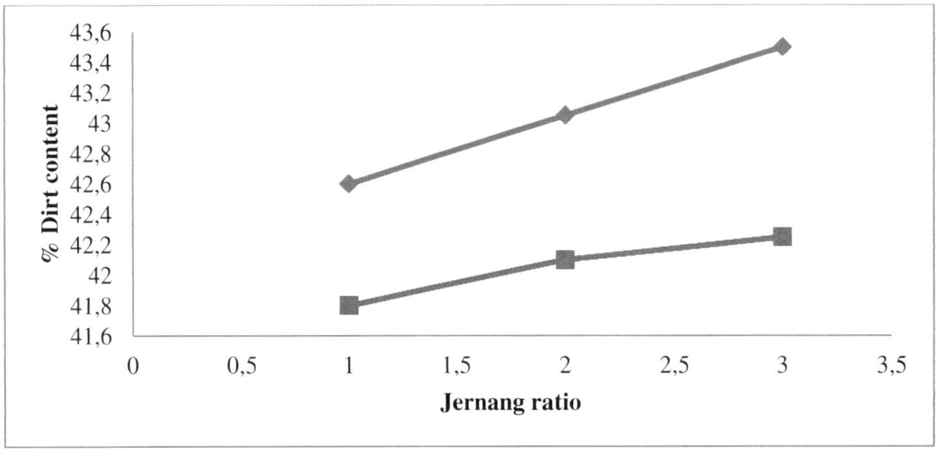

Figure 4. Comparison of dirt content toward jernang ratio.

3.4 Ashes

The determination of ash content aims to measure the level of organic material that presents in jernang. Ash content is a physical parameter that can indicate the number of other compounds that are considered as impurities contained in the jernang resin.

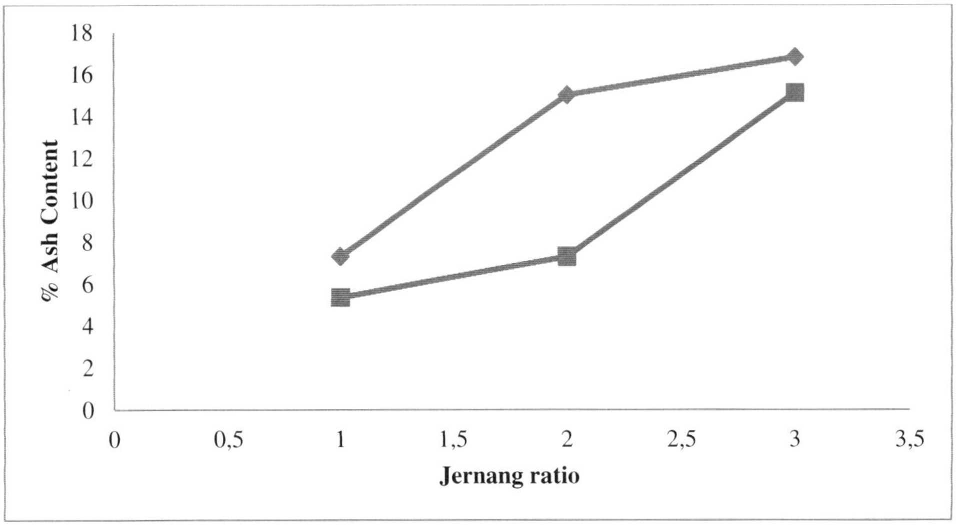

Figure 5. Comparison of ash content toward jernang ratio.

Figure 5 illustrates the highest percentage of ash obtained in water solvents; this is due to the large amount of organic material present in jernang. The percentage of ash content generated by the water solvent ranged from 7.3 to 16.8%. It means that the greater the ratio jernang, the greater it is resulting ash content. This is due to the existence of other polar compounds, which are not the compound of jernang resin that participate in dissolving into the water (polar solvent). Based on the specification of quality requirements in accordance to SNI, in the analysis of ash content obtained is A-B quality.

4. Conclusion

The quality of the produced resin by using methanol solvent is classified into B quality in accordance the quality of jernang resin in SNI 1671: 2010. Additionally, the quality of the produced resin by using water solvent is classified into B quality in accordance with the quality requirements of jernang resin in SNI 1671: 2010. However, the produced resin used more water solvents, instead of water momendipol (water engineering), the produced resin is better than the results from previous studies, which is concluded with methanol solvent that obtained more resin.

References

[1] Gupta D, Bleakley B and Gupta R K 2008 Dragon's blood: Botany, chemestry and therapeuticu ses *J. Ethnopharmaco.l* **115** (3) 361-380

[2] Risna R A 2006 Dragon's blood tumbuhan obat yang menjanjikan dari Taman Nasional Bukit Tigapuluh *Warta Kebun Raya, Pusat Konservasi Bogor–LIPI* **6** (1) 45-49

[3] Grieve M 2006 http//www.botanical.com/botanical/mgmh/d/dragon20.html. Accessed 21 January 2014

[4] Rao G S R, Gehart M A, Lee R T, Mitscher L A and Drake S 1982 Antimicrobial agents from higher plants: Dragon's blood resin *J. Nat. Prod.* **45** 646-648

[5] Anonim 2010 Getah jernang. SNI 1671:2010 (Jakarta: Badan Standardisasi Nasional)

[6] Bambang W 2010 Dragon's blood extraction at various seed maturity levels and their physico-chemical properties (RPI–5) *ITTO-Philippines-ASEAN Rattan Project* (PD 334/05 Rev. **2** (I)

[7] Coppen, J J W 1995 Gum, resins, and latexes of plant origin *Wood Forest Products Non Wood Forest Products* No. 6 (Roma: FAO)

[8] Rondao R J B L 2012 Dragon's blood *Disertasi* University of Coimbra, Coimbra

[9] Rustiami H, Setyowati F M and Kartawinata K 2004 Taxonomy and uses of Daemonorops draco (Willd.) *J. Trop. Ethnobiol.* **1** (2) 65-75

[10] Soemarna Y 2009 JPHH.**6** (1) 3-39.

[11] Sumarna Y 2005 China Butuh 400 ton jerenang rotan dari Indonesia.www.kapanlagi.com. Accessed 6 Desember 2014

[12] Shi J, Hu R, Lu Y, Sun C and Wu T 2009 Single-step purification of dracorhodin from dragon's blood resin of using high-speed counter-current chromatography combined with pH modulation *J. Sep. Sci.* **32** 4040-4047.

[13] Purwanto Y, Polosakan Y, Susiarti S, Walujo E B 2005. *Laporan Teknik Bidang Botani Puslitbang LIPI.* Bogor (ID): LIPI.

[14] Toriq U 2013. Senyawa Kimia Penciri Jernang untuk Pembaruan Parameter Standar Nasional Indonesia. Fakultas Matematika Dan Ilmu Pengetahuan Alam. Institut Pertanian Bogor. (Skripsi).

IOP Conf. Series: Materials Science and Engineering **345** (2018) 012014 doi:10.1088/1757-899X/345/1/012014

Diversifying bio-petro fuel sources for future energy sustainability and its challenges

M R Othman[1], Z Helwani[1,2] and I Idris[1]
[1]School of Chemical Engineering, Universiti Sains Malaysia, 14300 Nibong Tebal, Pulau Pinang, Malaysia
[2]Department of Chemical Engineering, Riau University, Pekanbaru, 28293, Indonesia.

E-mail: chroslee@usm.my

Abstract. Petroleum has been important in the energy industry since 19th century when the refining of paraffin from crude oil began. The industry recently appears to be in a downtown and fragile moment despite the price of oil is slowly rising. Renewable alternatives such as bio-fuels have gained increasing traction while petroleum fuel seemingly concedes to bio-fuels due to the rising public concern on the environment and stricter emission regulations. To be a strategic fuel in the energy security matrix, both fossil and bio-fuels options should be considered. However, the use of bio-fuels to achieve a degree of carbon neutrality is not without challenges. Among the challenges are land development and socio-political issue, carbon neutrality due to ILUC, high 2G bio-fuel feedstock and production cost, competing technology from electric vehicles and the impending fourth industrial revolution, NOx emissions and variation in biodiesel quality. This paper briefly reviews the potential of fuels source diversification and the challenges and how they can raise up to the challenges in order to be sustainable and attractive. In order to achieve this objective, first carbon credit through carbon trading needs to continue to stabilize the energy price. Second, 1G bio-fuel needs to forgo the use of natural, peat forest, rubber estate since these are an effective carbon sink and oxygen source. Third, advanced bio-fuels with high yield, process economics and sustainability need to be innovated. Fourth, the quality and standard bio-fuel that reduces NOx emission need to be improved. Finally and most importantly, carbon capture technology needs to be deployed immediately in fossil fuel power plants.

1. Introduction

Global bio-fuel production has increased steadily since 2005 with annual growth rate of 14.1% [1]. Biodiesel production rose by 6.5% in 2016 with Indonesia providing more than 1149 thousand tons of crude oil equivalent [1] in the world bio-diesel market. This is despite the result from a previous study reporting that the first generation (1G) biodiesel made from edible vegetable oils emit higher greenhouse gas emission and require higher cost of production than its fossil fuel counterpart [2]. The increase in bio-fuel demand is expected to continue in the future. However, with the recent European Parliament's vote to phase out the use of vegetable oil in bio-fuels by 2020 [3] and France government's plan to end the sale of petrol and diesel cars by 2040 [4] and coal by 2022 [5] in order to meet the Paris climate accord's target and carbon neutral (CN) by 2050, this trend may be short-lived. The entry of artificial intelligent (AI) to herald the fourth industrial revolution (IR-4) portends another challenge for bio-petro fuels and combustion technology. To the combustion engine, petroleum and bio-fuel manufacturers, the recent episodes may not bode well to them, and to certain extend, pose an existential threat.

For developing countries with dominant fossil fuel economy, a drastic program to phase out fossil fuels completely could lead to a catastrophe, considering the benefits of these fuels to the social, economic and political wellbeing. Beside diversifying bio-petroleum fuel sources for energy sustainability during the slow but steady transit into a completely renewable energy economy, the developing countries may be able to benefit from carbon trading through installation of carbon capture technology in their existing and future fossil fuel power plants and infrastructure [6].

This paper briefly reviews the challenges and immediate and midterm potential solutions in order for the industries to remain resilient and resourceful to ensure the sustainability of the fuels and the economic well-being generated by using the fuels.

2. Bio-fuel challenges

The first generation (1G) biodiesel made from edible feedstock reportedly exacerbated the scarcity of food [7]. Attempting to move towards inedible 2G bio-fuels such as Jatropha oil [8-9], algae (3G) bio-fuel feedstock [10-11] and genetically modified algae (4G) bio-fuels is also met with significant challenges. In this paper, few of these challenges and the potential solutions to overcome the challenges are proposed.

2.1. Illegal forest clearance

Bio-fuel is perceived as carbon neutral because the CO_2 released to the atmosphere on combustion is consumed back by the bio-fuel crops during photosynthesis. As long as bio-fuel supply can meet the population growth and demand, the fuel can provide the needed energy security without affecting the environment. But, bio-fuel use was promoted more than necessary, causing the global bio-fuel demand and production to increase. Illegal planters emerge to fill the demand and natural forest was encroached to make way for illegal oil palm cultivation. This has become a great challenge for the government and legal oil palm industry to overcome. Laws that protect natural forests need to be enforced aggressively, some land for food crops cultivation need be reserved in order to ensure food security and, finally the 1-2G biodiesel industrial estate outside of natural forest, within the existing oil palm estate, coconut and fallow area needs to be developed. Reasons for such proposals shall be discussed shortly.

2.2. Indirect land use change

Figure 1 shows the equivalent carbon dioxide (CO_2) emission from indirect land use change (ILUC) for bio-diesel and bio-ethanol production [2]. The carbon emission from ILUC for all the four food crops to produce biodiesel is the same. The difference by the direct emission is possibly due to the different source of electricity. Biodiesel produced higher CO_2 emission than fossil fuel, indicating that it is not effective in achieving carbon neutrality (CN). Bio-ethanol made from maize and sugar on the other hand, achieved 50% carbon neutrality or more, which is on the right track toward 100 percent carbon neutral. The challenge would be for the bio-diesel manufacturers to reduce the emission further so that it can be an attractive fuel of the future.

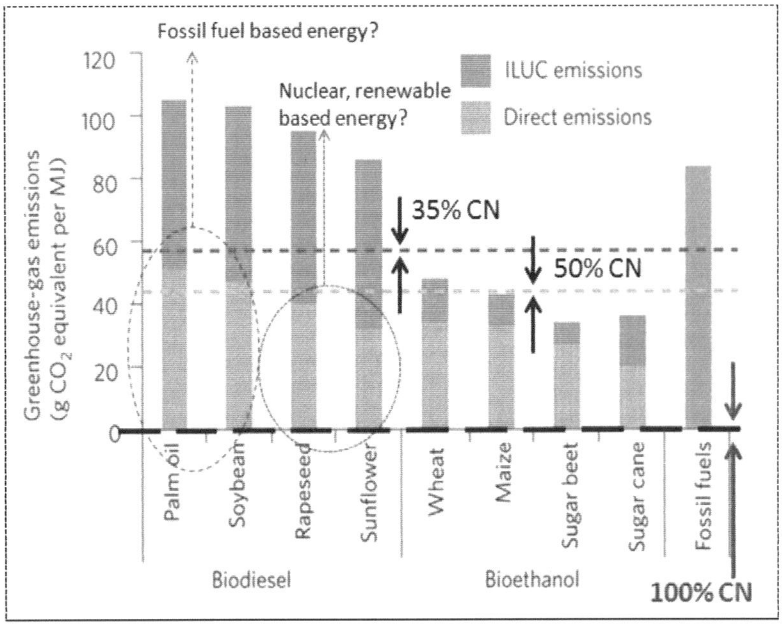

Figure 1. Comparison of CO_2 emission by bio-fuels [2].

Figure 2 shows the equivalent CO_2 emission by palm oil due to indirect land use change (ILUC) [12]. Bars on the right represent positive carbon emission (unfavorable), whereas bars on the left represent negative carbon emission (favorable). This study signals a clear message that natural forest, peat forest and a land designated for natural rubber should never be allowed for oil palm cultivation. In order to reduce the CO_2 emission, the most suitable places to cultivate oil palm would be the land designated for oil palm plantation or existing oil palm land, coconut estate and tropical fallow. These places are an effective carbon-sink according to the study.

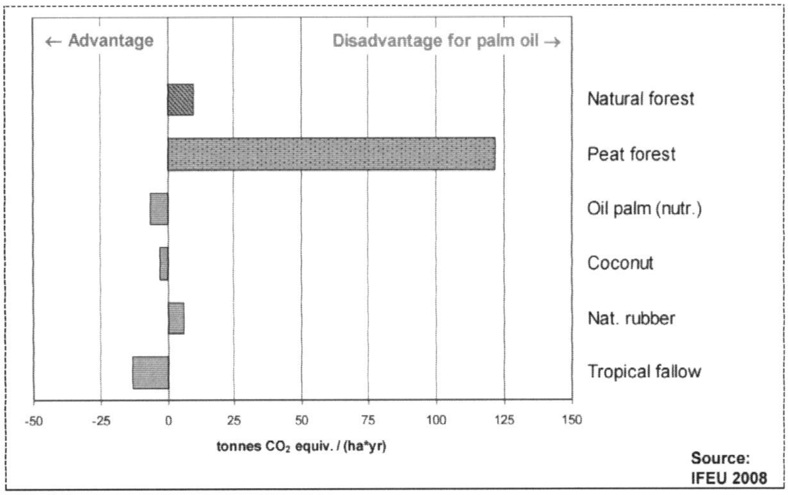

Figure 2. CO_2 emission by palm oil due to indirect land use change (ILUC) [12].

2.3. High biodiesel production cost

Production of bio-diesel from inedible (2-4G) feedstock incurs high expenses associated with acquiring the feedstock and transportation. Some inedible feedstock such as waste cooking oil that requires pretreatment and processing are significantly more costly than processing the edible oil. Companies that ventured into the biodiesel business could not get the return on investment (ROI) or cope with the expenses. Some companies had to file a bankruptcy notice [13].

Company may raise up to the challenge by becoming more resourceful in acquiring the needed feedstock alternatives and logistics. The feedstock that resembles the properties of hydrocarbon may use the existing crude refinery to produce bio-fuels. This would eliminate the huge capital upfront and make the bio-fuel production more feasible.

2.4. Competing technology

Norway, French and UK revealed their decision to ban petrol and diesel vehicles to herald a new era of electric vehicles [14]. Figure 3 shows the comparison of CO_2 emission from fossil fueled vehicles and electric vehicles [15]. Electric vehicles produce CO_2 emission but the amount is largely dependent on the power plants that generate the electricity. The higher electricity demand created by the electric vehicles is met by higher fuel consumption in the power plants. Higher fuel consumption subsequently emits more CO_2 emission. Coal power plant generates the highest CO_2 and hydro-electric plant generates the lowest CO_2. The figure suggests that a complete switch to electric vehicles will continue to produce CO_2 emission indirectly regardless of the fuel source types.

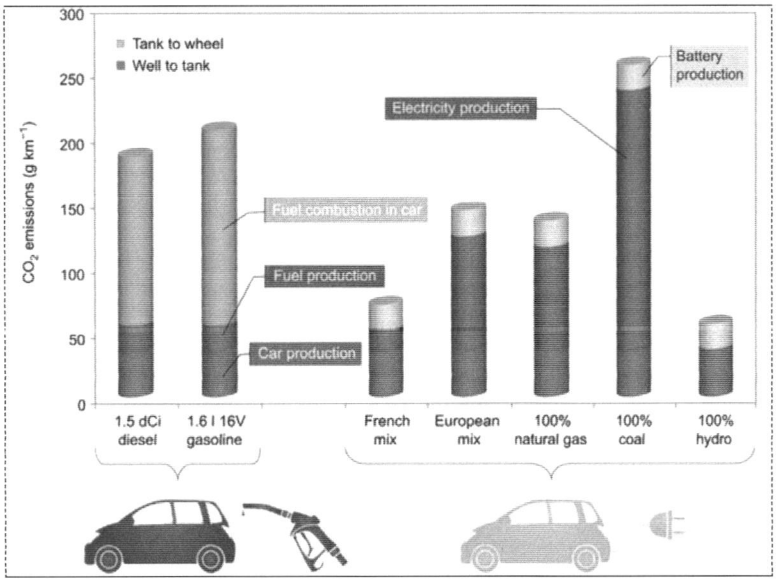

Figure 3. CO_2 emission by petrol-diesel and electric vehicles [15]

However, closed examination of the graph indicates that there is possibility for fossil-fueled vehicles to become environmentally more attractive than electric vehicles if there is significant reduction in the emission. The CO_2 emission from petrol-diesel vehicles can be reduced further by using better quality fuels and additives, blending the gasoline with bio-ethanol for gasoline engines or installing more effective catalytic converters that transform NO_x into harmless gases from diesel engines.

NO_x reduction technology is readily available [16] and so is the carbon capture to reduce CO_2 emission from large stationary sources such as power plants [17-22] that generate the electricity. These technologies can be deployed immediately to assist the industries in confronting the emerging challenges and to become more resilient.

2.5. NO_x emission

There is a report on public health impacts of excess NO_x emissions from diesel passenger vehicles in Germany [23]. The study indicates that higher public exposure to NO_x would lead to higher mortality rate. Germany reportedly has the highest mortality rate among the European countries due to the excess NO_x emission allegedly coming from the diesel emission scandal [24].

Figure 4 shows that higher biodiesel content in the blended fuel reportedly causes higher NO_x emission [24]. However, increasing biodiesel concentration in blended fuel reduces particulate matter (PM), carbon monoxide (CO) and unburned hydrocarbon (HC) emissions due to the better combustion by the former than its petroleum diesel counterpart. Although emission from PM, CO and HC are reduced markedly, NO_x emission is a major concern as it affects public health. NO_x reportedly increases premature mortality rate. A bio-diesel blend or 100% use of bio-diesel, despite has the potential to achieve the carbon neutrality (provided that there is no emission from ILUC and direct emission), it could pose a serious threat to the public health due to the NO_x exposure. This is one of the most challenging issues that bio-diesel manufacturers should overcome to render bio-diesel attractive and promising fuel for both mobile and stationary sources.

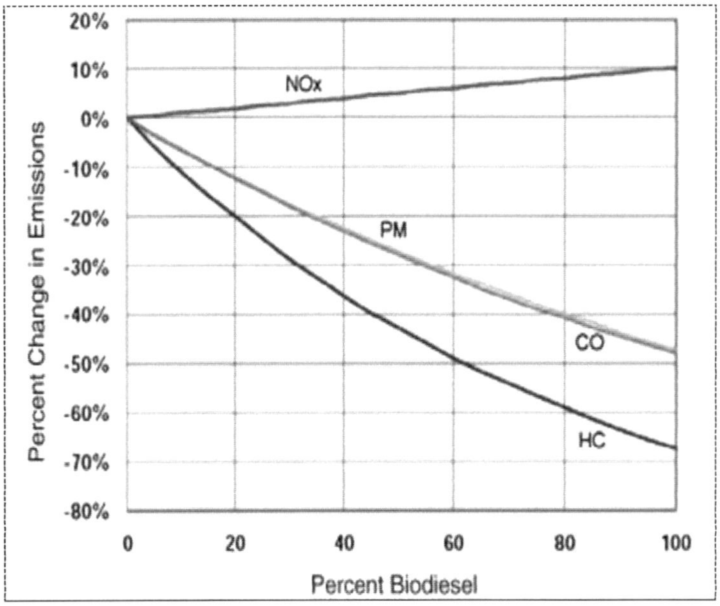

Figure 4. NO_X emission by biodiesel engines [24].

There are 7 approaches that can be exploited by the bio-diesel industry to reduce NO_x emission from diesel engines [16]. The first approach called humid air method feeds water vapour into the combustion chamber. The second method recycles the exhaust gas to the combustion chamber. The third method uses water injection and water emulsion. The fourth method uses high scavenge pressure and compression ratio in the engine. The fifth method uses selective catalytic reduction. The sixth method uses two-stage turbocharger and finally the seventh method requires engine component modification.

Alternatively, di-tertiary butyl peroxide and ethylhexylnitrate may be blended with bio-diesel since they are known to reduce NO_x without PM trade-off. Additionally, tertiary butylhydroquinone, an antioxidant, is an effective NO_x reducing agent, may also be used but at the expense of slight PM increase.

3. Conclusion

Bio-fuel is important in fossil fuel based economy in order to achieve a degree of carbon neutrality and energy security. Although bio-petro fuels exhibit the highest environmental impact and produce the most CO_2 on burning, they have the potential to perform favourably if their carbon emission is reduced significantly through carbon capture technology. This would be a logical choice for countries with dominant fossil fuel economy such as the United States of America and other developing countries to consider in order to power their countries sustainably. One of the contributing factors for the United State withdrawal from Paris climate accord [25] but continue to take part in the international climate talks [26] possibly as a result of this contemplation.

While bio-ethanol shows high potential towards carbon neutrality, bio-diesel on the other hand lags far behind from that objective. Bio-diesel quality that reduces emission significantly needs to be re-engineered. Carbon credit is important for bio-fuels to stabilize the energy price. Carbon capture technology to capture the carbon dioxide emission from large stationary sources such as power plants needs to be annexed to the existing and new plants. An immediate action plan would be to deploy carbon capture technology in fossil fuel power plants. Lands for food security and security of natural forest need to be reserved. There should also be policies to bid energy security through clean, renewable & sustainable sources. There should also be aggressive law enforcement on illegal forest clearance & open burning. Since natural, peat forest, natural rubber estates are an excellent carbon sink and oxygen source, 1G bio-fuel needs to forgo the use of these lands. Advanced bio-fuel with high yield, process economics and sustainability is needed in order to cater for the expected increase in energy demand. The designated lands for high yield 1-2G biodiesel resource cultivation/development need to be properly planned, executed and controlled. A new market for bio-petro fuels needs to be created in the future when electric vehicles have established their presence. A good example would be a liquid fuel cell technology that converts bio-petro liquid fuel directly into electricity. Finally, any dispute arising from energy issues needs to be resolved diplomatically between governments of the countries.

Acknowledgements

The authors gratefully acknowledge support from Fundamental Research Grant Scheme (FRGS) for Geran Sanjungan Penyelidikan (GSP), Ministry of Higher Education (MOHE), Malaysia

References

[1] https://www.bp.com/content/dam/bp/en/corporate/pdf/energy-economics/statistical-review-2017/bp-statistical-review-of-world-energy-2017-full-report.pdf
[2] Renssen S V 2011 A biofuel conundrum *Nat. Climate Change* **1** 389–390
[3] http://www.europarl.europa.eu/news/en/press-room/20170329IPR69057/meps-call-for-clampdown-on-imports-of-unsustainable-palm-oil-and-use-in-biofuel
[4] https://www.theguardian.com/business/2017/jul/06/france-ban-petrol-diesel-cars-2040-emmanuel-macron-volvo
[5] https://arstechnica.co.uk/tech-policy/2017/07/france-wants-to-ban-sale-of-gas-and-diesel-cars-by-2040-end-coal-by-2022/?comments=1&post=33610895
[6] Othman M R, Martunus, Zakaria R and Fernando W J N 2009 *Energ. Policy* **37** 1718-1735.
[7] Tenenbaum D J 2008 Food vs. Fuel: Diversion of crops could cause more hunger *Environ. Health Perspect.* **116** (6) A254-A257
[8] Helwani Z, Aziz N, Bakar M Z A, Mukhtar H, Kim J and Othman M R 2013 Conversion of Jatropha curcas oil into biodiesel using re-crystallized hydrotalcite *Energ. Convers. Manage.* **73** 128–134
[9] Helwani Z, Aziz N, Kim J and Othman M R 2016 Improving the yield of Jatropha curcas's FAME through sol-gel derived meso-porous hydrotalcites *Renew. Energ.* **86** 68-74

[10] Ly H V, Galiwango E, Kim S S, Kim J, Choi J H, Woo H C and Othman M R 2017 Hydrodeoxygenation of 2-furyl methyl ketone as a model compound of algal Saccharina Japonica bio-oil using iron phosphide catalyst *Chem. Eng. J.* **317** 302–308
[11] Le T A, Ly H V, Kim J, Kim S S, Choi J H, Woo H C and Othman M R 2014 Hydrodeoxygenation of 2-furyl methyl ketone as a model compound in bio-oil from pyrolysis of Saccharina Japonica Alga in fixed-bed reactor.*Chem. Eng. J.* **250** 157-163
[12] Institute for Energy and Environmental Research (IFEU), Heidelberg, Germany (2008)
[13] https://www.reuters.com/article/us-abengoa-bankruptcy-usa/abengoa-unit-files-for-u-s-bankruptcy-with-up-to-10-billion-in-debt-idUSKCN0VX2OH
[14] http://money.cnn.com/2017/07/26/autos/countries-that-are-banning-gas-cars-for-electric/index.html
[15] Larcher D and Tarascon J M 2015 Towards greener and more sustainable batteries for electrical energy storage *Nat. Chem.* **7** 19–29
[16] https://www.marineinsight.com/tech/10-technologiesmethods-for-controlling-nox-sox-emissions-from-ships/
[17] Wiheeb A D, Helwani Z, Kim J and Othman M R 2016 Pressure swing adsorption technologies for carbon dioxide capture: A review *Sep. Purif. Rev.* 45 108-121
[18] Helwani Z, Wiheeb A D, Kim J, Othman M R 2016 In-situ mineralization of carbon dioxide in a coal-fired power plant *Energ. Source Part A* **38** (4) 606-611
[19] Martunus, Helwani Z, Wiheeb A D, Kim J and Othman M R 2012 In-situ carbon dioxide capture and fixation from a hot flue gas *Intl. J. Greenhouse Gas Control* **6** 179-188.
[20] Martunus, Helwani Z, Wiheeb A D, Kim J and Othman M R 2012 Improved carbon dioxide capture using metal reinforced hydrotalcite under wet conditions *Intl. J. Greenhouse Gas Control.* **7** 127-136
[21] Martunus, Othman M R and Fernando W J N 2011 Elevated temperature carbon dioxide capture via reinforced metal hydrotalcite *Microporous Mesoporous Mater.* **138** 110–117
[22] Othman M R, Martunus, Fernando W J N and Kim J 2010 Fractal rate of adsorption and surface diffusivity of carbon dioxide across mesoporous adsorbents *Adsorption Sci. Technol.* **27** (10) 893-906
[23] Chossière G P, Malina R, Ashok A, Dedoussi I C, Eastham S D, Speth R L, Barrett S R H 2017 Public health impacts of excess NOx emissions from Volkswagen diesel passenger vehicles in Germany *Environ. Res. Lett.* **12** 034014
[24] Hoekman S K and Robbins C 2012 Review of the effects of biodiesel on NOx emissions *Fuel Proc. Technol.* 96 237-249
[25] https://www.reuters.com/article/us-un-climate-usa-paris-id USKBN1AK2FM
[26] http://www.independent.co.uk/news/world/americas/us-politics/trump-climate-change-talks-un-attend-exit-paris-agreement-a7877991.html

IOP Conf. Series: Materials Science and Engineering **345** (2018) 012015 doi:10.1088/1757-899X/345/1/012015

Thermal behaviour and microanalysis of coal subbituminus

Heriyanti, W Prendika, R Ashyar and Sutrisno
Department of Chemistry, Faculty of Science and Technology, Jambi University
Kampus Pinang Masak UNJA Mendalo, Jambi Province, Indonesia

E-mail: herasutrisno@gmail.com

Abstract. Differential scanning calorimetry (DSC) and X-ray powder diffraction (XRD) is used to study the thermal behaviour of sub-bituminous coal. The DSC experiment was performed in air atmosphere up to 125 °C at a heating rate of 25 °C min^{-1}. The DSC curve showed that the distinct transitional stages in the coal samples studied. Thermal heating temperature intervals, peak and dissociation energy of the coal samples were also determined. The XRD analysis was used to evaluate the diffraction pattern and crystal structure of the compounds in the coal sample at various temperatures (25-350 °C). The XRD analysis of various temperatures obtained compounds from the coal sample, dominated by quartz (SiO_2) and corundum (Al_2O_3). The increase in temperature of the thermal treatment showed a better crystal formation.

1. Introduction

Coal is physically heterogeneous and chemically complex consisting organic and inorganic components, which contains the types of mineral, bound in it, which are Carbon (C), Oxygen (O), Silicon (Si), Aluminium (Al), Magnesium (Mg) and Calcium (Ca). The mineral as impurities must be taken into account because of the higher the content, the lower the carbon content, causing the lower the heat value of the coal. One of the important factors in the use of coal for an industry is its mineral properties in the combustion processes. Minerals impurities are the main source of the elements that formed ash when coal is burned. The main supplement of coal-fired residue is elements in concentrations of 0.02% to over 0.5% in coal, and usually includes aluminium, calcium, iron, and silicon [1].

The need to understand the implications of coal-based properties due to their use as a fuel has led to the development of advanced techniques to characterize their heterogeneity. The use of coal as fuel demands an understanding of the implications of nature to characterize its heterogeneity. It takes a multidimensional approach to characterize the heterogeneity of coal properties. The type of applications, the ash problem, and the thermal behaviour of the burning process also affects the characterization approach [2].

2. Materials and methods

2.1. Materials

The coal was obtained from PT. Tambang Bukit Tambi, Jambi Province Indonesia. First, it was ground using a grinder FIRTSCH Pulveisette-16, followed by sieving using shaker by Changsha Kaiyuan Instruments (60 Mesh, 100 Mesh and 150 Mesh) and Electrolab *sieve shaker* EMS-8.

2.2. Coal characterization

The coal samples were characterized by using SEM-EDX and Calorimeter Bomb. The Scanning Electron Microscope (SEM) and Energy Dispersive X-ray Spectrometry were used to observe the morphology and to conduct the elemental mapping of the samples. The SEM-EDX used was software from Particle Metric PhenomProX and the electron optical magnification range 80-130.000x, a resolution higher than 10 nm, acceleration voltages about 5-15 kV. The PhenomproX with Energy Dispersive X-ray Spectroscopy (EDS) detector. The calorimeter analysis was using Calorimeter Bomb, type 5E-C5500 Automatic Calorimeter AXT.

2.3. Thermal heating treatment

The thermal treatment on the samples were conducted using an oven CKiC 5E-DHG, Muffle Furnace Changsha Kaiyuan Instruments, hotplate Boeco MSH-420 with temperature variations of 25, 150, 200, 250, 300, and 350 °C for 2 hours. The samples were then analysed by several instruments including proximate analysis, structural analysis using XRD, and determine the dissociation energy using Differential Scanning Calorimetry (DSC).

The proximate analysis procedure performed refers to the ASTM standard method, for the determination of the water content of ASTM D3173, for ASTM D3174 ash determination and for the determination of ASTM D3175 volatile matter.

To determine the crystal structure was using an X-Ray Diffraction (XRD) PAN analytical Xpert3 powder goniometer PW3050/60 type of Bragg-Brentano at the Faculty of Science and Technology at Jambi University. Data diffraction of XRD is collected using a target atom Cu (1.54 Å) at 40 kV and 30 mA at range 2θ 15°-70°, step 0.040 and measurement time of 0.5 s per step.

The dissociation energy was determined by using PerkinElmer DSC4000. The heat rate used was 50°C min^{-1} at 25-350 °C. The sample weight of 0.053 grams corresponded to the aluminium pot as the sample site and the carrier gas flow rate (N_2) of 20 ml min^{-1}.

3. Results and discussion

3.1. Coal characterization

3.1.1. The SEM-EDX analysis

The SEM-EDX analysis was used to evaluate the morphology and the components contained in the coal sample. The SEM-EDX result is presented in Figure 1. The darker part of the coal particle corresponds to a higher hydrogen content, the shiny part corresponds to a higher carbon content, where some very bright parts may also correspond to mineral grains in coal particles [2]. The image also shows clearly separate mineralized structures with irregular morphological forms. This indicated that the minerals contained in coal were in the form of crystal structures, where the crystal structure has a fixed structure because the atoms were arranged regularly in a pattern that repeats periodically over an infinite long-range.

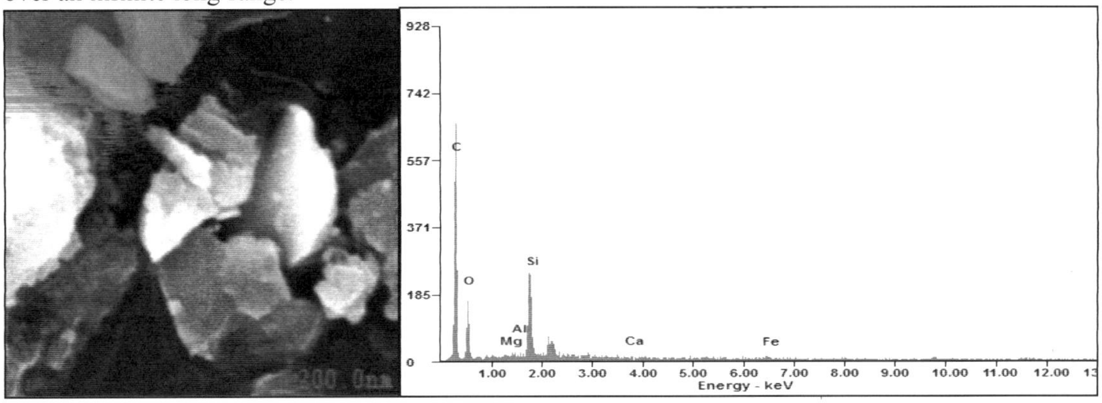

Figure 1. SEM-EDS analysis of the coal

From the EDX results, the dominating elements were C and O with the percentage of 56.13% and 35.13%. Sub-bituminous coal has a carbon value of 50-70% [1] and oxygen contained in coal with a content of 20-30% w/w derived from a variety of accumulating plant material or derived from oxygen inclusions that occur when of contact of the source layer with oxygen in open air or water at the time of sedimentation. Besides C and O elements, it also found other elements, which are Si, Fe, Al, Ca and Mg with a percentage of 10.97%, 2.21%, 0.49%, 0.7%, and 0.43%, respectively. Silicon was one of the macro elements in coal with 2.82% concentration in coal of anthracite level and 10.38% for sub-bituminous level [3]. The potential of mineral content from this coal sample was SiO_2, Al_2O_3, MgO, $CaCO_3$ and Fe_2O_3.

3.1.2. The calorimeter analysis

The result of coal calorific value analysis was 4905.6 Cal g^{-1}, this heat value indicated that the sample was categorized as sub-bituminous coal type (low-calorie coal). In the previous research, the value of calories for coal with sub-bituminous rank was 5793.2 and 5706 Cal g^{-1}, which was classified as sub-bituminous coal type. General coal in Indonesia, especially Sumatra can be categorized in the type of young coal or low-calorie that has a calorific value between 3500-6100 Cal g^{-1}. For the sub-bituminous grades of coal, the heating value obtained in this study was low, this low calorific value may be due to the oxidation process with air to the coal sample at the time of storage of the sample prior to the analysis. The low calorific value in coal was caused by the high level of ash that is impurities inside the coal, thus causing the caloric value to decrease.

3.2. The thermal heating treatment

3.2.1. Proximate analysis

Proximate analysis of coal is one of the first important variables to determine the quality and the initial identifier of the type of coal used. The proximate analysis results are presented in Table 1.

Table 1. Proximate analysis of the coal with thermal heating in various temperature

Parameter	Temperature (°C)					
	25	150	200	250	300	350
Inherent Moisture (%)	3,07	0,79	0,42	0	0	0
Volatile Matter (%)	20,7	10,8	5,51	3,81	2,2	1
Ash (%)	19,2	86,2	91,9	94,2	97,1	98,6
Fixed Carbon (%)	56,9	2,21	2,17	1,18	0,69	0,39

The moisture content in coal consists of several different forms of condensed water (surface free water), water bound to polar groups and cations (water absorbed), and water arising from the chemical decomposition of both organic and inorganic materials [4]. Based on Table 1, moisture content decreases with increasing thermal treatment temperature. The sub-bituminous coal has an inherent moisture of 0.1-1.4% [5].

As is the case with inherent moisture content of coal samples, the higher the temperature the volatile matter in coal decreases. This was because the flying substance is the organic part of the coal that evaporates when heated at a certain temperature. When coal is heated slowly, the volatile matter in coal is decreasing. Coal with bituminous and sub-bituminous grades has a volatile matter content of 15-45% [6].

The analysis of ash content at various temperatures shows that the ash content obtained was higher along with the increase in temperature. Ash is the residue of burning at high temperatures i.e. 950°C where the carbon element will be oxidized and the minerals contained in coal will bind to oxygen to form minerals such as oxides such as SiO_2, Al_2O_3, Fe_2O_3, CaO and other oxides. The higher ash content was due to the chemical changes in coal occurring in the presence of the filling process, the quantity of ash can be more than, equal to, or less than the amount of mineral material in coal, depending on the nature of the minerals and the chemical changes occurring in the proliferation process. Various changes that occur include (1) loss of water from silicate minerals, (2) loss of carbon dioxide from carbonate minerals, (3) oxidation of iron oxide pyrite [6]. The higher the ash content will be able to degrade the quality of the coal production itself. High ash levels also cause problems in handling and disposing of large amounts of ash produced during the combustion process [6]. Sub-bituminous coal has an Ash content of 4-15% [6].

Fixed carbon on coal related to combustion heat. Where the higher the temperature then the carbon content moored in coal will decrease further. This was due to the high ash content contained in the coal sample during the combustion process; the condition will be able to degrade the quality of the coal being analysed. Sub-bituminous coal has a fixed carbon content of 50-70% [6].

From the result of proximate analysis for inherent moisture, volatile matter, ash, and fixed carbon of coal from PT. Bukit Tambi, Jambi Province, Indonesia was categorized into sub-bituminous levels.

3.2.2. The XRD analysis

The result of initial coal characterization using XRD at 25°C (Figure 1 and Table 2) indicated that the XRD spectrum was dominated by Si in Quartz (SiO_2) form, which was indicated by the presence of some sharp peaks at 2θ = 20.9014 and 26.7465 with d-spacing [Å] 4.25018 and 3.3330 with trigonal crystal system (axis hexagonal) (JCPDS 00-046-1045). As for the Aluminium element identified in the spectra in the form Corundum (Al_2O_3). Corundum crystal structure has a characteristic of 2θ = 68.4323, with a d-spacing value of 1.3698 Å and an intensity of 4.36%. The characteristic of this position is almost in accordance with JCPDS 00-010-0173 standard that Corundum has 2θ position characteristics at 68.1980.

Other minerals were calcite ($CaCO_3$) with a rhombohedral crystal system found at the peak of 2θ 39.5245 with d-spacing [Å] 2.28008 (JCPDS 00-05-0586), and hematite (Fe_2O_3) minerals with rhombohedral crystals, found at the peak of 2θ 65.9097 with d-spacing [Å] 1.41605 (based on JCPDS 00-033 -0664). The minerals identified by XRD correspond to the initial characterization results using SEM, where the dominant elements present in the coal content were Si, Al, Fe and Ca elements that were possible in the form of SiO_2, Al_2O_3, Fe_2O_3, and $CaCO_3$.

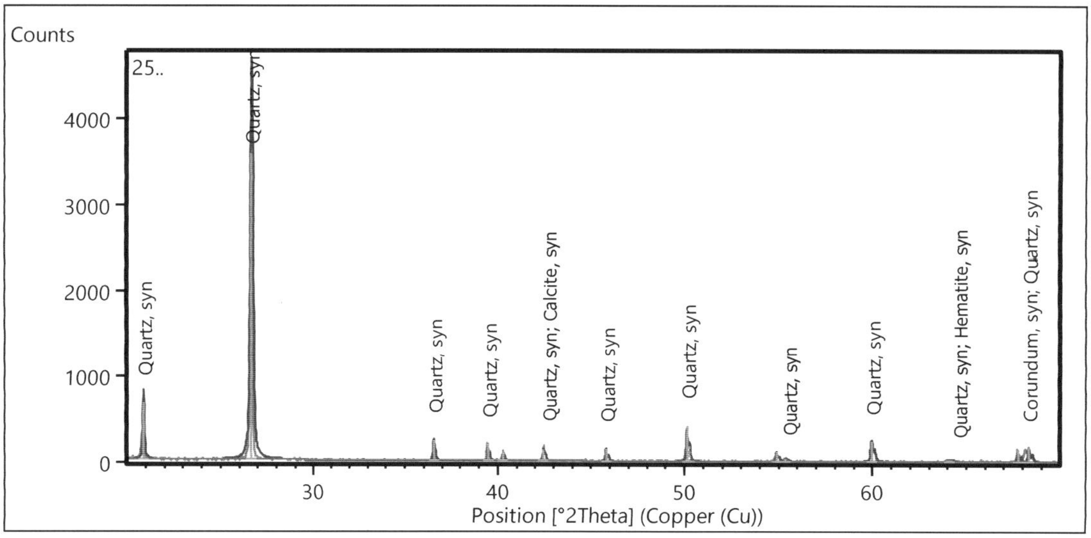

Figure 2. XRD Spectra of the coal at 25 °C

Table 2. Interpretation data of XRD of the coal at 25 °C

2θ [°]	d-spacing [Å]	Mineral	Intensity (%)
20.970	4.2363	Quartz	15.5
26.745	3.3315	Quartz	100
36.643	2.4524	Quartz	9.6
39.553	2.2784	Quartz; Calcite	5.09
40.396	2.2328	Quartz	3.57
42.527	2.1257	Quartz	5.67
45.877	1.9784	Quartz	2.96
50.240	1.8145	Quartz	11.8
54.973	1.6689	Quartz	3.96
60.054	1.5393	Quartz	8.8
64.049	1.6706	Quartz; Hematite	3.6
68.432	1.3698	Corundum	4.3

The XRD analysis of coal at various temperatures was done on temperature variations of 150-350°C with a range of 50 °C to see mineral changes or changes in crystal structure based on the position of 2θ. Changes in the structure of a crystal were strongly influenced by temperature, burning coal at high temperatures below 500 °C increased the crystalline structure of oxide minerals in coal such as SiO_2 [7]. Changes in minerals or changes in the structure of coal crystals by heat treatment are shown in Figure 3. Coal in lignite and sub-bituminous grades have high impurity minerals, the high percentage of ash and impurities is caused by the loss of most of the OH groups in silicates when the coal is isolated at high temperatures, where the XRD results state that SiO_2 minerals have the greatest percentage followed by pyrite, kaolinite, calcite and gypsum [8].

Figure 3 shows the XRD analysis with the temperature variations of 150 to 350 °C. The spectra obtained almost similar to the previous analysis conducted at a temperature of 25 °C (Figure 2), where these peaks show the dominant and consistent peaks of SiO_2 minerals at 350 °C temperature value 2θ = 26.7559 with a 100% intensity detected by JCPDS 00-046-1045 standard. Other mineral peaks also appear as $CaCO_3$, Fe_2O_3 and Al_2O_3. The shift value 2 on each increase in temperature variation is shown in Table 3.

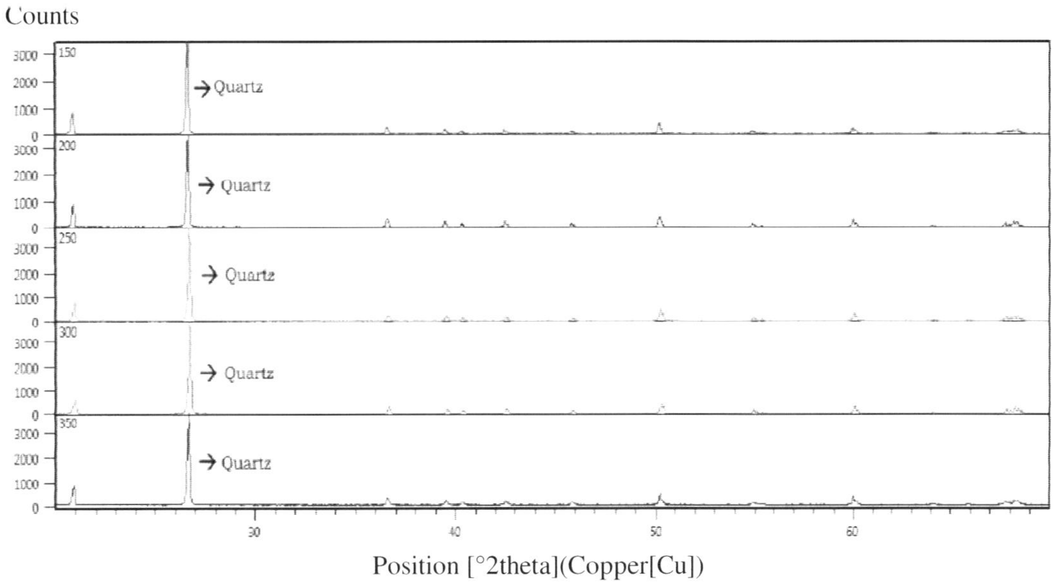

Figure 3. XRD spectra of coal with thermal heating (150-350 °C)

Table 3. Interpretation data of XRD of coal with thermal heating (150-350 °C)

150°C		200°C		250°C		300°C		350°C	
2θ [°]	Int (%)	2θ [°]	Int (%)	2θ [°]	Int (%)	2θ [°]	Int (%)	2θ [°]	Int (%)
20.89	23.2	20.91	24.5	20.95	24.8	20.99	25.2	21.00	26.1
26.66	100	26.67	100	26.73	100	26.75	100	26.76	100
36.57	8.28	36.58	8.9	36.63	6.2	36.66	8.9	36.68	8.5
39.48	6.5	39.50	5.9	39.55	5.9	39.57	4.7	39.51	4.5
40.32	3.5	40.33	3.5	40.38	3.5	40.37	3.5	40.39	3.5
42.47	4.8	42.49	4.8	42.52	4.8	42.55	4.8	42.58	4.8
50.31	5.1	50.17	5.1	50.22	5.1	50.39	5.66	50.65	5.1
54.89	3.8	54.91	3.8	54.96	3.8	54.97	3.42	55.38	3.8
59.98	6.8	60.00	6.8	60.04	6.8	60.05	7.03	60.19	5.11
60.15	4	60.17	4	60.17	4	60.17	3.29	60.23	4
64.07	1.14	64.13	1.21	64.11	1	64.12	1	64.07	1
67.75	4.7	67.77	4.8	67.81	4.31	67.83	4.92	67.98	4.98
68.15	4.82	68.17	5.72	68.21	5.9	68.24	6.3	68.21	6.5

Based on Table 3, the increasing temperature used in the thermal process caused an increase in the value of 2θ, especially for SiO_2 which was in the range of 26.6609 with intensity 100%. The XRD analysis was consistent with the results of the proximate analysis, i.e. the higher temperature used in coal combustion was resulting the higher ash content and the lower the carbon fixed of the coal. It was caused by the high levels of silica formed. The analysis of coal samples using temperature variations of 50 to 1200 °C for ± 2 hours using SEM and XRD was dominated by SiO_2 minerals [9].

The shift in the values of 2θ were due to an increase in the energy that allows the atoms to vibrate at a greater distance between atoms that occurs in a heated material. When an atom has enough energy to break its bond, there will be a diffusion process where the atoms will jump to a new position. The higher energy (temperature) given to a material causes the larger spacing between the atoms and affect the phase and structure of the material. The phase changes and mineral structure of SiO_2 can also be seen from the crystal size, which can be known from the Scherer equation (D), where the higher temperature used the larger the crystal size. The comparison of crystal size formed by temperature variation in the thermal process is shown in Table 4.

Table 4. Size of SiO_2 crystals at various temperatures

Temperature (°C)	2θ	d-spacing [Å]	D (nm)	Intensity (%)
150	26.6609	3.34366	10.37	100
200	26.6726	3.34222	10.65	100
250	26.7312	3.33502	14.63	100
300	26.7503	3.33269	14.88	100
350	26.7559	3.33206	15.79	100

Silica (SiO_2) is the simplest silicate material having three crystalline forms: quartz, cristobalite and tridymite. Based on the JCPDS 00-046-1045 standard, the SiO_2 crystalline phase was obtained in XRD analysis results in crystals of quartz. The quartz form of SiO_2 can be distinguished into 2, i.e. α-quartz and β-quartz. The difference between the two is in the crystal system and the value of its space group, where the quartz has a trigonal crystal system (a hexagonal shaft) and has a space value of P3121 and

P3221 where the quartz has a hexagonal crystalline system and has the value of space group P6222 and P6422.

In the results of the XRD analysis, the data obtained according to JCPDS 00-046-1045 standard shows that the value of space group owned for quartz mineral in this research is P3221 with trigonal crystal system (hexagonal shaft), it was identified that quartz mineral (SiO_2) obtained on the results of XRD analysis was classified in a α-quartz crystalline structure. The β-quartz was not formed in this study was due to the thermal treatment used in the low-temperature range of 150-350 °C so that SiO_2 atoms will be difficult to move for greater inter-atomic spacing positions. The crystalline-quartz system has a smaller atomic spacing than the quartz-crystalline system. The thermal treatment of SiO_2 at < 570°C low quartz temperature with a crystalline α-trigonal structure (hexagonal shaft) and for temperature 570-870 °C formed high quartz having a crystal β-hexagonal structure [10]. The SiO_2 crystal parameters were obtained from XRD analysis in this study are shown in Table 5 and the crystal structure in Figure 4.

Table 5. The parameter of SiO_2 crystal system

Parameters	Quartz
Formula	SiO_2
Crystal structure	Trigonal
Space group	P 3221
No space group	154
a (Å)	4.9134
b (Å)	4.9134
c (Å)	5.4052
Alpha (°)	90.0000
Beta (°)	90.0000
Gamma (°)	120.0000
h k l	1 0 1
d-spacing (Å)	3.34347

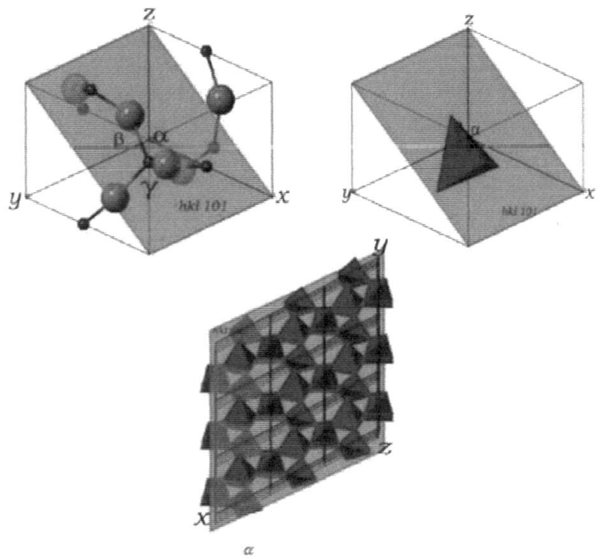

Figure 4. The crystalline α-quartz structure of the trigonal system (hexagonal shaft).

3.2.3. The DSC analysis

The DSC curve of the sample is presented in Figure 4. In this study, DSC curve shows two endothermic peaks, namely at 145 and 320 °C. The first endothermic peak at 145 °C, which was caused by the hydrolysis reaction occurring to water from coal. Within the temperature range of 100-200°C, there was a release of a chemical combination of water, followed by the release of compounds of light molecular weight that were weakly bound to the carbon group in the coal. At temperatures below 200°C coal undergo an initial decomposition stage where coal loses maximum water [11][12]. At higher temperature, the breakdown of coal chemical bonds began to occur and results in the formation of tar and light hydrocarbons. At higher temperatures, the breakdown of coal chemical bonds began to occur and results in the formation of tar and light hydrocarbons.

The percent of the sample weight lost during the DSC analysis was 1.5%. The mass loss occurs due to the release of moisture (H_2O) in coal. In the first stage, the release of a chemical combination of water or hydrolysis was the breaking process of the H_2O bond due to the heating process. This corresponded to the results of the analysis of proximate analysis on the determination of moisture content that was influenced by temperature rise. These results indicated a decrease in water content along with rising temperatures; this was due to the evaporation process. On thermal heating, low rack coal would experience shrinkage during the thermal heating process, causing moisture to escape from the pores of the coal particles at 100-110 °C, the collapse of the pore structure in the coal due to moisture loss and evaporation [4]. After the heating process, the moisture on the coal would become steam and attempt to exit the gap or cavity between coal particles into the air [13]. Thermal analysis of bituminous coal using PerkinElmer DSC7 with 50°C $\leq$ T $\leq$ 200 °C shows endothermic curve followed by the evaporative process of H_2O bonding [14]. The thermal analysis of coal by DSC and Thermogravimetric explained that in the temperature range 75-118°C is the process of loss of water molecules with endothermic peaks in regions 50 to 150 °C [15].

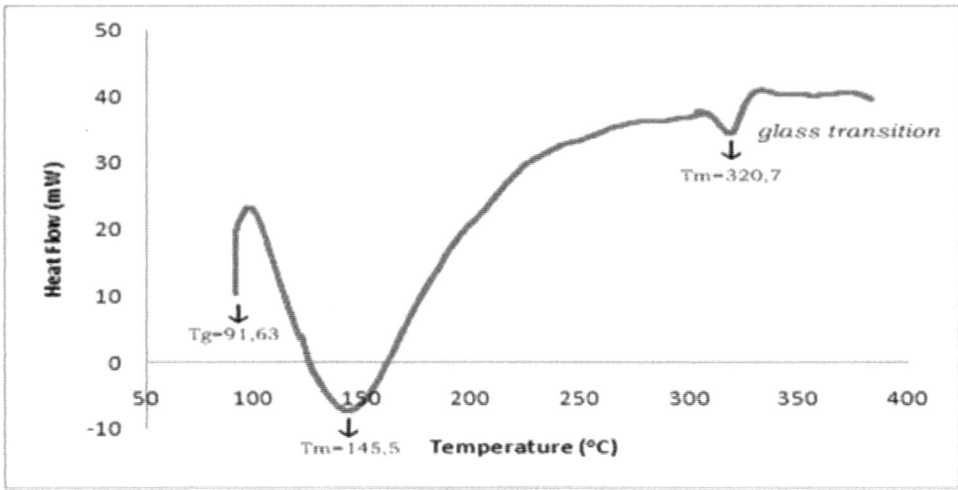

Figure 5. DSC curve of sub-bituminous coal.

The second endothermic peak was found at 320 °C, this stage is called the active thermal decomposition stage. At this stage, the decomposition temperature can be used as a determinant of the degree of quality of coal. After losing water, at 200-400 °C, coal began to lose particle mass from the decomposition of phenol structure and carbon oxidation of carboxylate and carbonyl group [12]. At temperature 200-350 °C, sixth coal as research sample started experiencing pyrolysis reaction that is decomposed phenol and carbon oxide structures of carboxylates and carbonyl groups, which are represented by a short endothermic peak on the DSC curve [16].

The DSC curve also provides information about glass transition. The glass transition is a reversible transition from solid to amorphous material. The amorphous structure also represents the outside of a combustion product. This amorphous structure usually involves a major constituent element in chemical reactions associated with burning ash. At the beginning of the active thermal decomposition shows the decomposition of compounds containing carbon, hydrogen, oxygen, and the release of volatile matter such as CO_2 (carbon dioxide) and CH_3 (methane) and will begin to form pre-plastic phase (150-400 °C) [17].

The determination of an amorphous phase of a sample was also influenced by the heat capacity (Cp) and temperature, called glass transition temperature (Tg). In the glass transition (Tg), the solid will acquire sufficient mobility to convert the solid phase to the plastic phase. It can be observed that the inflexion point of the endothermic peak indicates the glass transition temperature in each sample used, and the glass transition temperature of each sample will vary within the range of temperatures used during the analysis.

The endothermic peak indicated that the thermal analysis performed on coal at a temperature of 100 to 350 °C is an endothermic reaction. The results of the DSC curve of the coal using the heating range of 20 and 50 °C min^{-1} also found an endothermic peak that was at 140 and 330 °C, while with 100 °C min^{-1} heat flow emerging exothermic peak and in the temperature range 425-450 °C [16]. The endothermic peak indicated the glass transition temperature of the coal sample [18]. On the DSC samples of bituminous coal samples also appear of weak endothermic curves of glass transition at a temperature of 90 – 180 °C that can be associated with evaporation of water, and water binds to hydrogen [14].

The presence of sharp endothermic peaks displayed on the DSC curve can be used to determine dissociation energy. The dissociation energy is a measure of the heat changes that occur at the breaking of the bonds expressed by the value of the enthalpy value (ΔH). In the dissociation energy can also be determined energy changes that occur during the reaction called the change in free energy Gibbs (ΔG). One aspect that is associated with the energy of a reaction is entropy (ΔS) which shows the degree of irregularity in a system, some of these energy determination parameters can be calculated using known Tm (onset), Tm (peak), and heat capacity (ΔCp) of the peak slope on the DSC curve. The parameters that can be determined from the endothermic peak slope shown from the DSC analysis can be seen in Table 6.

Table 6. The DSC curve reading parameters from glass transition

Heat flow (°C min^{-1})	Tg (°C)	Tm onset (°C)	Tm peak (°C)	Enthalpy (ΔH, Jg^{-1})	Entropy (ΔS, $Jg^{-1}K^{-1}$)	Gibbs Energy (ΔG)
50	91.63	91.7	145.4	617.1	1.147	0
		305.1	320.7	55.8	0.094	0

The value of the dissociation energy is the number of enthalpy changes (ΔH) required for the breaking of the bond, where the positive value indicates that when the bond breaking process occurs the heat absorption process, called the endothermic reaction. The value of enthalpy (ΔH) at a peak value of 145.4°C is 617.1 Jg^{-1}. This high enthalpy value may explain that the energy required for the change is considerable. It can also be seen from the wide endothermic peak, seen in the temperature range 91.7-178.4°C. The release of moisture content in sub-bituminous coal is considerable, i.e., it can be seen from the amount of energy required for the process at 96.8 °C which has a wider and sharper peak than coal with anthracite and bituminous species that have shorter and smaller peaks.

The sharp endothermic peak at Tm peak 320.7 °C was the peak of active thermal decomposition that can determine the degree of quality of the coal. At this temperature occurred the release of phenol compounds and the oxidation of carboxylic acids and carbonyl groups. This corresponded to the results of the analysis of the volatile matter, where the higher the temperature will cause the level of volatile matter decreases. In this study, the active thermal decomposition peak was in the 300-340 °C temperature range, this temperature range may indicate that the coal sample used is included in the oxidized sub-bituminous coal level, the occurrence of this oxidation may be caused by coal that has reacted with oxygen in a period of time.

4. Conclusion

The characterization of coal PT. Tambang Bukit Tambi, Indonesia with temperature variation i.e. moisture content, volatile matter and fixed carbon have decreased with the increase of temperature, otherwise the higher ash content. This is due to the oxidation of carbon and other elements in coal and binds to oxygen to form oxide minerals.

The dominant mineral elements of C, O and Si with amorphous dominating structures were shown in the XRD results at various temperatures, the presence of this amorphous structure is amplified by the DSC results showing the two peaks of the endothermic curves of glass transition at 145 and 320 °C. The XRD analysis of various temperatures obtained compounds from the coal sample, dominated by quartz (SiO_2) and corundum (Al_2O_3). The existence of two endothermic curves of the glass transition in DSC analysis can determine dissociation energy value (ΔH) of coal, which is 617.1 Jg-1and 55.8 Jg-1.

References

[1] Speight J G 2006 *The Chemistry and Technology of Petroleum*, 4th Edition (London: Taylor & Francis Group)
[2] Gupta R 2007 *Energ. Fuel* **21** 451-460
[3] Twardowska I, Tinripathi P S, Singh G and Kyziol J 2003 *Kluwer Acad.* 25-26
[4] Xianchun L I, Hui S, Qi W, Chatphol M, Terry W and Jianglong Y U 2009 *J. Environ. Sci.* **21** S127-S130
[5] Donahue C J and Rais E A 2009 *J. Chem. Educ.* **86** 222-224
[6] Speight J G 2005 H*andbook of Coal Analysis* vol 166 (New Jersey: John Wiley & Sons, Inc.)
[7] Lager G A, Jorgensen J D and Rotella F J 1998 *J. Appl. Phys.* **53** 6751
[8] Maldonado-hbdar F J and Mastral-lamarca A M 1995 *Fuel* **74** 6 818-822
[9] Qiu S, Zhang S, Zhang P, Qiu G and Zhang Q 2016 *Characterization of Minerals, Metals, and Materials* (Switzerland: Springer International Publisher)
[10] Shinohara Y and Kohyama N 2004 *Ind. Health* **42** 227-285
[11] Tahmasebi A, Yu J, Li X, and Meesri C 2011 *Fuel Process. Technol.* **92** 10 1821-1829
[12] Mashingo P and Jhon G 2012 *Seco Intl. Conf. Adv. Eng. Technol.*
[13] Lester E and Kingman S 2004 *Fuel* **83** 1941-1947
[14] West R D, Markevicius G, Malhotra V M, and Hofer S 2012 *Fuel* **98** 213-217
[15] Stenberg K J and Stenberg V 1989 *Fuel* **68** 95-99
[16] Elder P 1983 *Fuel* **63** 262-267
[17] Menendez J A, Diez M A, Puente G. de la, Fuente E E, Alvarez R, Pis J J 1998 *J. Anal. Appl.* **45** 75-87.
[18] Erol M, Kucukbayrak S, and Ersoy-Mericboyu A 2007 *Fuel* **86** 706-714

Torrefaction of oil palm frond: The effect of process condition to calorific value and proximate analysis

W Susanty, Z Helwani and Zulfansyah
Department of Chemical Engineering, Faculty of Engineering, University of Riau

E-mail: susantywenny@gmail.com

Abstract. Oil palm frond can be used as alternative energy source by torrefaction process. Torrefaction is a treatment process of biomass into solid fuel by heating within temperature range of 200-300°C in an inert environment. This research aims to result solid fuel through torrefaction and to study the effect of process variable interaction. Torrefaction of oil palm frond was using fixed bed horizontal reactor with operation condition of temperature (225-275 °C), time (15-45 minutes) and nitrogen flow rate (50-150 ml/min). Responses resulted were calorific value and proximate (moisture, ash, volatile matter and fixed carbon). Analysis result was processed by using Design Expert v7.0.0. Result obtained for calorific value was 17.700-19.600 kJ/kg and for the proximate were moisture range of 3-4%; ash range of 1.5-4%; volatile matter of 45-55% and fixed carbon of 37-46%. The most affecting factor signficantly towards the responses was temperature then followed by time and nitrogen flow rate.

1. Introduction

Biomass is a primary source of renewable carbon that can be utilized as a feedstock for solid fuel production in order to achieve energy independence. As the largest producer of palm oil in Indonesia, Riau has potential to produce more lignocellulosic biomass. Riau has the widest oil palm plantation which reached about 1.7 million hectares in 2011 [1]. This extens has made Riau as the biggest contributor of CPO for national production that reached 6.2 million tons. Beside producing abundant vegetable oil, this extens also produces so many oil palm frond. Oil palm tree that can be planted in one hectare of oil palm plantation is about 140 trees [2]. Harvesting of fresh fruit bunches, 1-2 pieces of palm frond are cut in order to facilitate the next pollination and harvest. There is 6,400-7,500 palm frond in one hectare per year, about 4.5 kg per palm frond [3]. So that the potential of palm fronds produced from oil palm plantations in Riau in 2014 reached 7.7 million tons of dry palm frond.

Oil palm frond (OPF) is still being an unutilized waste maximally, that is often to be left in plantation. OPF combustion will cause bad environment. In order to give economical value-added of oil palm frond, Oil palm frond can be used as alternative energy source. OPF utilization for energy-based industry has not been done maximally, while energy potential in OPF is about 1,3 TJ/year [4].

Torrefaction is recommended as an efficient way to enhance solid fuel properties by water removal, reduction of the hygroscopic range and increased grindability. Torrefaction is a thermal pretreatment process by subjecting biomass to temperature levels between 200-300 °C in inert environment. During this process, hemicellulose is degraded and volatiles will be released from biomass. Decreasing of hemicellulose and volatiles will increase calorific value and maximize mass yield and energy yield from solid product [5-8]

Uemura et al. [5] torrefied empty fruit bunch at 225-275 °C and stated that temperature played a significant role toward solid product. Chin et al. [7] investigated OPT at 200-300 °C for 15-45 minutes. They found that time also affected torrefaction but was not as signifant as temperature. Lu et al. [6] torrefied oil palm fiber in nitrogen and air at 250-350 °C and found that torrefaction in air resulted in lower solid quality than in nitrogen. Chen et al. [8] investigated oil palm fiber at 250-350 °C and with varieties superficial velocities or carrier gas flow rate in 100-1000 ml/minutes. They stated that increasing carrier gas flow rate graphically did not give any effect for solid product.

All of the findings of the above-mentioned studies were reported relatively quick succession. However, the effect of interaction between variables has not been explored yet. If those variables like as temperature, time and carrier gas flow rate is investigated its effect to calorific value and proximate analysis, and processed statistically using Response Surface Methodology (RSM), then OPF utilization through torrefaction will be result the advantage of torrefaction as biomass pretreatment. RSM will give design and analysis of model so the affecting factor significantly will be known.

In this paper, torrefaction of OPF was carried out in a fixed bed horizontal reactor in inert condition. The effects of temperature, time and nitrogen flow rate on calorific value and proximate analysis of solid product were investigated using RSM.

2. Materials and methods

2.1 Material preparation

Oil palm frond was cut into blocks with the dimensions of 1x1 cm. Biomass samples were dried by sun for few hours and then also dried in an oven at 105 °C in order to remove residual water remaining in the biomass. The moisture content was measured for 24 hours.

2.2 Torrefaction process

The samples (50 g) was placed in a reactor and torrefied by heater under the conditions as follows: temperature 225, 250 and 275 °C for 15, 30 and 45 minutes. Torrefaction was performed in a fixed bed horizontal reactor of 6 cm diameter and 60 cm length, which is shown in Figure 1. They were torrified by flushing carrier gas with flow rate of 50-150 ml/minutes. The carrier gas was continuously blown into the reactor to keep the system in an inert environment and to remove volatiles produced in the reactor. After torrefaction, the heater was turned off and the reactor was left to cool down to the ambient temperature. Then the torrified sample was recovered, kept in desicator and weighed.

2.3 Result analysis

2.3.1 Calorific value (HHV)

The calorific value was determined according to ASTM standard D-5865-13 with a bomb calorimeter, in which 0.5 g of dried biomass was completely combusted under a presurized oxygen atmosphere.

2.3.2 Proximate

Moisture content was measured by drying in an oven at 105 °C for an hour according to ASTM standard D-3173-13. For the ash content, a prescribed amount (1 g) of the sample was weighed in a crucible and calcined in an electric furnace at 750 °C for 2 hours, according to ASTM standard D-3174-13. Volatile matter was also measured using a furnace which 1 g of sample was calcined at 950 °C for 7 minutes according to ASTM standard D-3175-13.

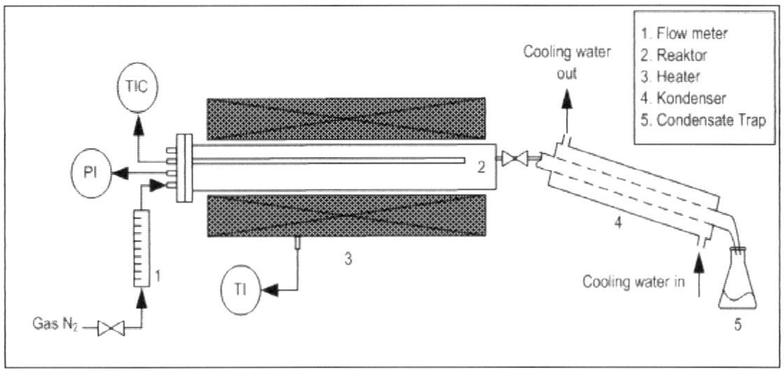

Figure 1. Fixed bed horizontal reactor.

3. Results and discussion

The results of calorific value (HHV) and proximate of OPF are listed in Table 1.

Table 1. Characteristic of torrefaction raw and product

Characteristic	Raw	Product
HHV (kJ/kg)	16,800	17,700-19,600
Moisture (%)	6-8	3-4
Ash (%)	0.95	1.5-4
Volatile (%)	62	45-55
Fixed Carbon (%)	16.9	37-46

3.1 Appearance of torrefied sample

In Figure 2, pictures of all the samples in every operation condition are exhibited. The color of OPF becomes darker from brown as the torrefaction temperature increases. The darker the biomass, the more weight loss in biomass [5,10]. When the time increased, the color also changes from dark to darker, not as significant as temperature affecting the samples. Meanwhile, sample does not show a significant change of color when nitrogen flow rate is increased.

3.2 The influence of time (a. 5; b.30; c.55 minutes) in operation condition 250°C of temperature and 100 ml/minute nitrogen flow rate

Calorific Value (HHV): The HHV ranged from 16,800-19,600 kJ/kg depending on torrefaction conditions. This implies that the HHV in the torrefied biomass increased by 5,4-16,5% when compared to the untreated material. The highest HHV obtained at temperature of 275 °C for 45 minutes reaction times and with 100 ml/minutes of nitrogen flow rate. The HHV significantly increased with torrefaction temperature than reaction time. This tendency is consistent with the previous work [7, 9], and is mainly due to the carbon content increase meanwhile the oxygen and hydrogen decrease. This decrease in hydrogen and oxygen is due to dehydration and de-carbondioxide from the biomass during torrefaction [10].

(a) (b) (c)

The influence of temperature (a. 208; b.250; c.292 °C) in operation condition of 30 minutes time and 100 ml/minute ntrogen flow rate.

(d) (e) (f)

The influence of time (a. 5; b.30; c.55 minutes) in operation condition 250 °C of temperature and 100 ml/minute nitrogen flow rate.

(g) (h) (i)

The influence of nitrogen flow rate time (a. 16 ml/minute; b.100 ml/minute; c.184 ml/minute) in operation condition 250 °C of temperature and 30 minutes of time.

Figure 2. Appearance of torrified sample.

3.3. Proximate

Moisture is one of important value in determination of solid fuel. High moisture will cause a lot of energy lost in combustion process. The higher the moisture, the more the calor needed to release water in biomass. Moisture decreased when temperature of torrefaction increased, this is due to dehydration reaction during torrefaction [11]. In contrast, ash content increased when temperature of torrefaction increased. Ash in inorganic component in biomass. The ash of biomass is not lost during torrefaction, though the overall mass of the biomass is reduced. The absolute amount of ash in the feed does not change while the overall mass reduces when temperature increased. That is, the ash percentage changes after torrefaction. Volatile matter is closely related with fixed carbon in biomass.

Devolatilization was affected by torrefaction temperature and torrefaction time. The result was volatile matter decreased when temperature and time increased. This due to torrefaction temperature which is also the range of hemicelluloce degradation temperature, so that it will be maximally degraded and produce volatiles that characterized by higher proportion of solid products [8]. This decrease causes an increasing for fixed carbon content in biomass due to ratio O/C and H/C decrease so C will increase toward total yield.

3.4. Respon surface analysis

HHV was fitted to the response surface model provided by the mathematic model in order to analyze the effect of torrefaction factor on HHV, that is shown in Equation 1 :

$$Y_1 = 18782.5 + 534.5X_1 + 378.2X_2 \qquad (1)$$

Y_1 is the predicted HHV (kJ/kg), where X_1 and X_2 denote temperature and time. To fit the response surface function and experimental data, regression analysis was performed and the model for HHV was evaluated by analysis of variance which is presented in Table 2. Before analyzing, confidence level (α) was determined which is 5%. This value represents the maximum probability of error that may be resulted during experiment. Data processing was started by using second order model to see the curvature.

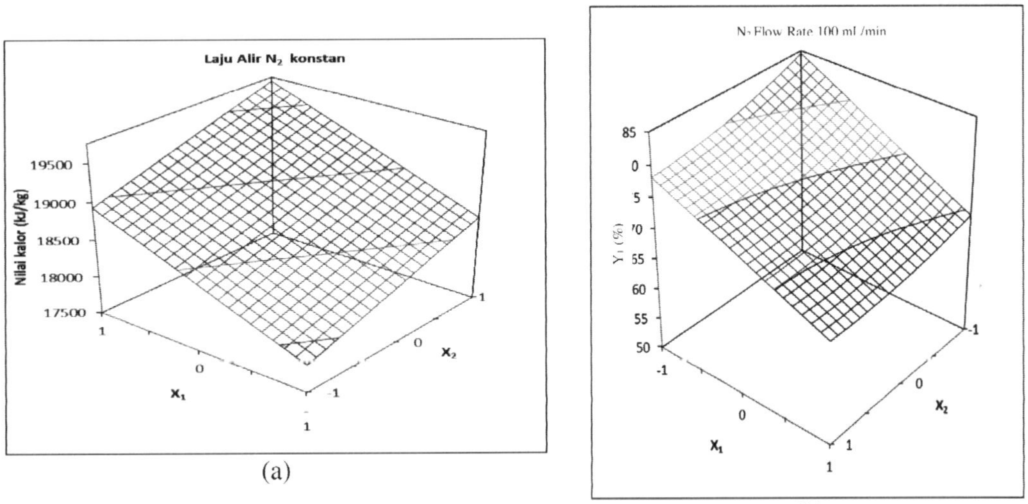

(a) (b)

Figure 3. Response surface graphic for (a).HHV; (b) mass yield and in nitrogen flow rate of 100 ml/minutes

Table 2. Analysis of variance (ANOVA) for HHV of biomass during torrefaction

Source	DF	Adj SS	Adj MS	F-Value	P-Value
Model	7	5487050	783864	7.31	0.000
Linear	3	220771	1740257	16.23	0.000
X_1	1	3368923	3368923	31.42	0.000
X_2	1	1527060	1527060	14.24	0.001
X_3	1	324789	324789	3.03	0.097
2-Way Interactions	3	258420	86140	0.80	0.507
X_1X_2	1	54666	54666	0.51	0.483
X_1X_3	1	195371	195371	1.82	0.192
X_2X_3	1	8384	8384	0.08	0.783
3-Way Interactions	1	7858	7858	0.07	0.789
$X_1X_2X_3$	1	7858	7858	0.07	0.789
Error	20	2144459	107223		
Lack-of-Fit	1	513051	513051	5.98	0.024
Pure Error	19	1631408	85864		
Total	27	7631508			
R^2 = 71.90%	R^2 (adj)= 62.06%		R^2 (pred) 33.02%		

The HHV and mass yield results obtained for all runs of the experimental design were represented by response surface in Figure 3. The responses for HHV from the torrefaction were depicted as three dimensional surface plots of three factors; temperature, reaction time and nitrogen flow rate with their corresponding contours plots. The strength of the effect of these three parameters on HHV is revealed more clearly by the surface plot. Temperature had a higher impact on the HHV of torrefied lignocellulosic biomass, while the effect of reaction time was considerable lower. Nitrogen flow rate insignificantly affected the HHV. Table 2 shows ANOVA of the linear model adjustment, where the total error was classified into lack of fit and pure error. The F-value estimated using the experimental data corresponded to the total residual and lack-of-fit values, respectively, and was lower than the tubular F value. This indicates that the model was significant in the region studied. The mathematic models (Equation 1) shows in good correlation with the actual data as justified by the relatively high R-squared value. For the model, the p-value was <0.0001, which shows that the models were strongly significant at the 99% confidence level.

4. Conclusion

Biomass can be upgraded and used as a solid fuel by torrefaction process. The torrefied biomass is more suitable than raw biomass in terms of HHV, physical and chemical properties. While the reaction temperature had a strong impact on the HHV yield of torrefied lignocellulosic biomass, the effect of reaction time and carrier gas flow rate were considerably lesser. As a whole, the torrefaction at 220 °C just degraded small amount of holocellulose and gave a mediocre effect on improving the energy properties. When biomass underwent the torrefaction at temperatures above 250 °C, large amount of hemicelluloses and cellulose were degraded which contributed to the increment of acid-insoluble material. The highest HHV obtained was 19,600 kJ/kg and it increased about 16.5% from the raw biomass.

IOP Conf. Series: Materials Science and Engineering **345** (2018) 012016 doi:10.1088/1757-899X/345/1/012016

References

[1] Kementrian Pertanian 2014 Pertumbuhan areal sawit meningkat *http://ditjenbun.pertanian.go.id/berita-362-pertumbuhan-areal-kelapa-sawit-meningkat-html.* Accessed 25 Maret 2015

[2] Warandi, L, Massijaya M Y, Machdie M F 2012 Pemanfaatan limbah batang kelapa sawit and plastik daur ulang sebagai bahan baku papan plastik komposit *Jurnal Ilmu and Teknologi Kayu Tropis* **10** (1) 52-59

[3] Simanihuruk K, Junjungan and Tarigan A 2007 Pemanfaatan Pelepah Kelapa Sawit sebagai Pakan Basal Kambing Kacang Fase Pertumbuhan *Prosiding Seminar Nasional Teknologi Peternakan and Veteriner 2007*

[4] Abduh S 2014 Peran masyarakat dalam pengelolaan energi nasional *Dewan Energi Nasional* Palu 17 November 2014

[5] Uemura Y, Omar W N, Tsutsui T and Yusup Z 2011 Torrefaction of oil palm wastes *Fuel* **90** 2585-91

[6] Lu K M, Lee W J, Chen W H, Liu S h and Lin T C 2012 Torrefaction and low temperature carbonization of oil palm fiber and eucalyptus in nitrogen and air atmospheres *Bioresour. Technol.* **123** 98-105

[7] Chin K L, H'ng P S, Go W Z, Wong W Z, Lim T W, Maminski M, Paridah M T and Luqman, A C 2013 Optimization of torrefaction condition for high energy density solid biofuel from oil palm biomass and fast growing species available in Malaysia *Ind. Crop Prod.* **49** 768-774

[8] Chen W, Lu K, Liu S, Tsai C, Lee W and Lin T 2013 Biomass torrefaction characteristics in inert and oxidative atmospheres at various superficial velocities *Bioresour. Technol.* **146** 152-160

[9] Poudel J, Ohm T I and Oh S C 2015 A study on torrefaction of food waste *Fuel* **140** 275-281

[10] Basu P 2013 *Biomass Gasification, Pyrolysis and Torrefaction*, 2nd Edition (New York: Elsevier Inc.)

[11] Prins J, Ptasinskri K and Janseen F 2006 *J. Anal. Appl. Pyrolysis* **77** 35-40

IOP Conf. Series: Materials Science and Engineering **345** (2018) 012017 doi:10.1088/1757-899X/345/1/012017

Fabricaton of brick without burning process

Zultiniar, A Fadli, S R Yenti, Drastinawati and M R A Thariq
Department of Chemical Engineering, Universitas Riau, Pekanbaru, 28293, Riau, Indonesia

E-mail: fadliunri@yahoo.com

Abstract. Brick is a material with many applications on building materials. The purpose of this research is to study the effect of fly ash addition and hardening time on mechanical properties of brick produced. Brick was formed by mixing cement, clay, water and fly ash with addition point was 0;0,3;0,6;0,9;1,2 kg weight. Brick was dried at room temperature for 7, 14 and 21 days. The result showed that compressive strength with addition 0,6 kg fly ash for 21 days was the the strongest where that's value was 9,375 N/mm^2. The highest porosity was get in on 21,34 % with 1,2 fly ash addition. The result of TCLP showed that for Pb, Cu, Cr content was still below the quality standards based on PP No.85 TH 1999 and still considered harmless. Leachate value for Pb of 3,10 mg/L, 1,213 mg/L for Cr and 4,374 mg/L for Cu.

1. Introduction

The using of bricks in the construction world as structural and non-structural elements cannot be replaced. The production of bricks was required a burning process. The useful of bricks in the community, not in line with environmental issues regarding air pollution and global warming due to increased production of carbon dioxide gas in conventional combustion bricks process. On the other hand, more industries are using coal that can contaminate the air. One of the waste generated from coal combustion is fly ash as 2-10%. Coal fly ash is generally disposed of in lagoon ash or stacked away in an industrial area. This fly ash buildup created problems for the environment, such as air pollution, waters and declining ecosystem quality [1]. The waste of fly ash can be utilized for mixed materials in the brick production process because it contains pozzolan chemical compounds such as alumina and silica, so it is suitable to be used as construction raw material [2].

Stabilization / solidification is one of the methods used in the handling of solid waste before disposal to a land. This method involves mixing waste with a binder to immobilize the heavy metals contained so that the hazardous and toxic waste (B3) can be converted and acceptable to the environment. Besides being disposed to landfills, the results of solidification can be utilized as construction materials [3]. The solidification study of alumina waste from petroleum processing is capable of providing compressive strength of 10.8 N / mm2, 12.8 N / mm^2 and 16.2 N / mm^2 [4]. While in addition of cement in brick making yield compressive strength equal to 1,312 N / mm^2, 2,124 N / mm^2, 2,905 N / mm^2, 4,475 N / mm^2 and 4,595 N / mm^2 [2]. The mechanical properties, in particular the compressive strength will increase significantly with the addition of coal waste, where as in the chemical properties, in particular leachet it is known that coal waste mixture can immobilize heavy metals well [5]

This paper reported about the effect of drying time on the manufacture of bricks made from fly ash without burning process. Brick making without combustion process is intended to reduce CO2 emissions (carbon dioxide) and the use of wood with clay for 5 minutes and water when stirring occurs. Then the mixture is inserted into the brick molding and pressed with a load of 25 kg. The result dried at atmospheric temperature for 7, 14 and 21 days. The resulting product was analyzed by compressive the

2. Material and method

The materials used in this study are coal fly ash, cement and clay. The process of making bricks using solidification techniques [4]. Fly Ash and cement was stirred perfectly. Then, the mixture was stirring the strength analysis, TCLP (Toxicity Caracteristictic Leached Procedure) and porosity test.

3. Result and discussion

3.1 Compressive strength analysis

The compressive strength is the ability of a brick product to accept a broad united press force, so the compressive strength identifies the quality of the brick product. The compressive strength of this study was tested on a variety of drying times of 7 days (A sample), 14 days (B sample) and 21 days (C sample). The results of the compressive strength test for 7-day drying time on each of the mixed compositions are shown in Table 1.

Table 1. Strong analysis of batubata press at 7 days drying time

Samples Code	*Fly Ash* (kg)	Max load (kN)	Compressive Strength (N/mm^2)	Compressive Strength Standart (N/mm^2)
A0	0	7	4,375	2,5
A1	0,3	8	5,0	2,5
A2	0,6	12	7,5	2,5
A3	0,9	5	3,125	2,5
A4	1,2	3	1,875	2,5

Base on Table 1, the addition of coal fly fly ash in the manufacture of bricks will initially increase the compressive strength of the brick products, but more fly ash will decrease the compressive strength of the bricks obtained. In this study the addition of 0.6 kg fly ash increased the compressive strength of the brick products, but the addition of 0.9 kg to 1.2 kg fly ash decreased the compressive strength of the brick products. The addition of fly ash to brick making can improve the mechanical characteristics of the resulting bricks [2]. The addition of fly ash above 50% in brick making can decrease the compressive strength of the resulting product [3]. Influence The comparable strength strength results were also obtained in samples with a drying time of 14 and 21 days as shown in the following table.

Table 2. Strong analysis results press bricks at drying time 14 days and 21 days

Samples Code	*Fly Ash* (kg)	Max load (kN)	Compressive Strength (N/mm^2)	Compressive Strength Standard (N/mm^2)
B0	0	9	5,625	2,5
B1	0,3	10	6,25	2,5
B2	0,6	14	8,75	2,5
B3	0,9	6	3,75	2,5
Samples Code	*Fly Ash* (kg)	Max load (kN)	Compressive Strength (N/mm^2)	Compressive Strength Standard (N/mm^2)
B4	1,2	4	2,5	2,5
C0	0	10	6,25	2,5
C1	0,3	11	6,875	2,5
C2	0,6	15	9,375	2,5
C3	0,9	7	4,375	2,5
C4	1,2	5	3,125	2,5

Table 2 shown that the highest compressive strength was obtained in the addition of 0.6 kg fly ash and decreased in addition to 1.2 kg. The effect of coal fly ash adds to the free lime-binding reaction produced in the cement hydration process by the silica contained in fly ash coal. In addition, the much smaller coal fly ash grains make the bricks more dense because the cavities between the aggregate granules are filled by coal fly ash so as to minimize the existing pores and utilize the existing pozzolan properties of fly ash to improve rock quality brick. Fly ash coal is an active ingredient when mixed with cement, and bricks made with fly ash coal mixture have higher compressive strength than normal bricks [1]. The same can be seen from the results of the compressive strength test in Table 2. From Table 2 showns that the addition of 0.3 kg fly ash increased the compressive strength from 5,625 N / mm^2 to 6.25 N / mm^2 or increased by 0.625 N / mm^2 for the drying time is seven days, while for the drying time of 14 days the addition of 0.3 kg fly ash increases the compressive strength from 6.25 N / mm^2 to 6.875 N / mm^2 or up 0.625 N / mm^2. The increasing continues until the addition of fly ash as much as 0.6 kg. While the addition above that number causes a decrease in compressive strength for various drying variations. The decrease in the strength of the bricks is due to the reduction in the value of the masonry of large brick-making materials due to the large amount of fly ash use of the brick material, resulting in a lack of binding strength between clay and fly ash. Between 10% -40% the use of fly ash can be utilized depending on the value of the keplastisan of the material used [4].

3.2 TCLP (Toxicity caracteristictic leached procedure) analysis

The aim of TCLP testing is to determine the level of protection of the nature of B3 in coal fly ash material after undergoing a solidification process with a brick product. The TCLP test was performed after 7 days, 14 days and 21 days old brick by Atomic Absorption Spectrofotomete (AAS) method. Further testing of the heavy metal content on bricks that have been tested the compressive strength. By doing this test we will know the ability of the bricks to immobilize the heavy metals contained in the sample. In the TCLP method the solidified material is crushed into grain particles which are further destructed until the process is obtained filtrate of filtrate which is then tested with AAS. The TCLP test on bricks was performed on samples with the highest compressive strength. TCLP test results of heavy metal content in the sample are as follows.

Table 3. Results of heavy metal content analysis on bricks

Component	TCLP Analysis of Brick (mg/L)			Standard of PP 85/1999 (mg/L)
	7 days	14 days	21 days	
Pb	5.03	4.93	3.10	5
Cu	6.34	5.551	4.374	10
Cr	2.864	1.600	1.213	5

From Table 3 it can be seen that brick as a solid waste asphalt produces maximum result in immobilizing heavy metals contained in the waste. From the picture above can also be seen that the longer the drying process will be more difficult heavy metal released to the environment due to the number of fewer in the sample of bricks. For 21-day drying Pb metals will decrease heavy metal content to 3.10 mg / L from 5.03 in seven-day drying. Likewise, in heavy metals Cu and Cr which decreased after concentration in solidified into bricks, this decrease caused by trapping heavy metals in the matrix of bricks making it difficult to decompose the environment. From the results of TCLP test above also shows that in some brick products produced are still below the quality standard of PP. 85 of 1999 so that its use is safe for the environment.

3.3 Porosity of brick analysis

Porosity is the ratio of pores or empty space in the bricks to the volume of bricks. There is the result of porosity analysis

Table 4. Porosity calculation result on brick

Samples	Fly ash	Drying Times	W1	W2	W3	Porosity
A0	0		87.46	97.26	30	17.06
A1	0.3		94.43	102.63	30	12.73
A2	0.6	7 days	98.02	104.72	30	9.85
A3	0.9		92.8	104.87	30	19.22
A4	1.2		88.98	103.32	30	24.31
B0	0		81.22	87.42	30	12.11
B1	0.3		87.87	93.57	30	9.85
B2	0.6	14 days	91.15	95.25	30	6.70
B3	0.9		85.95	94.35	30	15.01
B4	1.2		82.07	92.31	30	19.67
C0	0		75.2	80.5	30	11.73
C1	0.3		81.83	86.43	30	8.88
C2	0.6	21 days	85.03	86.53	30	2.75
C3	0.9		79.75	87.15	30	14.87
C4	1.2		75.77	84.67	30	19.45

The result data above will be compared with the compressive strength obtained. The compressive strength is inversely proportional to the porosity of a brick. The higher of the compressive strength of the porosity of the brick is getting smaller. This means that the highest compressive strength will result in low porosity. This is in accordance with the results of porosity test conducted, where from Figure 4. above can be seen that the highest porosity found in the addition of fly ash as much as 1.2 kg (compressive strength 1,875 N / mm^2) and drying time of sample for seven days ie 24.31%. The lowest porosity was obtained by adding fly ash 0,6 kg (9,375 N / mm^2) and 21 days drying time 2,73%.

4. Conclusion

Ash waste (fly ash) can be used as raw material for brick making without burning and can reduce pollution process. The best compressive strength bricks of 9,372 N / mm^2 were produced at the addition of 0.6 kg fly ash and drying time of 21 days. The resulting porosity is 1.875 N / mm^2. The TCLP test of the resulting bricks showed Pb metal content of 3.10 mg / L, Cu of 4.37 mg / L, Cr of 1.21 mg / L after the solidification process.

References

[1] Hughes E Randall 1996 *Bricks Manufacture with Fly Ash from Illionois Coals* (USA: ICCI Project)

[2] Garrabrants A C and Kosson D S 2005 *Leaching Processes and Evaluation Tests for Inorganic Constituent Release from Cement-Based Matrices, Stabilization and Solidification of Hazardous, Radioactive, and Mixed Wastes* (Florida – USA: CRC Press)

[3] Spence R and Shi C 2005 *Stabilization and Solidification of Hazardous, Radioactive, and Mixed Wastes* (Florida – USA: CRC Press)

[4] Hardjito D and Rangan B V 2005 *Development and Properties of Low Calcium Fly Ash Based Geopolymer Concrete* Faculty of Engineering Curtin University of Technology Perth - Australia.

[5] Stegemann J A 2005 *Interactions between Wastes and Binders Stabilization and Solidification of Hazardous, Radioactive, and Mixed Wastes* (Florida – USA: CRC Press)

Parametric study on the compressive strength geopolymer paving block

Aman[1], A Awaluddin[2], A Ahmad[1] and M Olivia[3]
[1]Department of Chemical Engineering, Universitas of Riau
[2]Department of Chemistry, University of Riau
[3]Department of Civil Engineering, University of Riau

E-mail: aman_syam@yahoo.co.id

Abstract. This paper reported about the investigated of sodium hidroxida concentration, effect of ratio liquid to solid (L/S), temperature and time on the compressive strength of geopolymer paving block using fly ash and fine aggregate as base material and combination of sodium hidroxida and sodium silicate as alkaline activator and the ratio of $Na_2SiO_3/NaOH$ was 2 and fly ash to aggregate of 1: 3. The experiments were conducted with variation of the sodium hidroxida concentration of (10-16 M) liquid to solid (L/S) 0.1- 0.7 ratio, curing temperature 30-100 °C and curing time (7-28 day). The main evaluation techniques in this experimental were Compressive strength, X-ray diffraction (XRD),and Scaning Electron Microscope (SEM). The result showed that the compressive strength of Geopolymer Paving block has increased with an increasing of concentration, liquid to solid ratio, curing temperature and curing time.

1. Introduction

Fly ash is composed a mixture of metal oxides and non-metal oxides. In addition, the fly ash also contain toxic substances such as chromium, arsenic, fluorine, boron and selenium [1]. Toxic substances are deposited in the soil (landfill) can be spread through the medium of water and air that can pollute the environment. One method for the management of fly ash waste is environmentally friendly with solidification way. In this research, the solidification method is the geopolymer process. Geopolymer produced by reacting solid aluminosilicate material contained in the coal fly ash with a strong base activator produce the binder/adhesive such as cement [2], binders have the same properties as

Factors that affected to the properties of the concrete/paving blocks geopolymer, the type and concentration of activators, activators modulus, temperature and duration of treatment, as well as the levels of alkaline solution in solids. The previous studies reported that each of these factors in the right amount into parameters to produce concrete/paving blocks geopolymer with a high compressive strength [4]. In this research the aims of this study is to check the parameters that influence the compressive strength of geopolymer paving blocks including the concentration difference, the ratio of liquid to solid, temperature and duration of treatment.

2. Materials and methods

In this study, fly ash is came from Paiton Probolinggo East Java as the source material geopolymer paving blocks. The chemical composition of fly ash are listed in Table 1.

Table 1. Chemical composition of fly ash

Oxida	(%) by mass
SiO2	53.616
Al_2O_3	28.47
Fe_2O_3	5.525
CaO	4.974
P_2O_5	0.569
SO_3	0.584
K_2O	0.857
TiO_2	0.724
LOI	4,681

Next, is by varying one variable to another variable fixed at the same time. There are 16 parameters shown in Table 2 below.

Table 2. Parameters used in this study

MIX DESIGN	[NaOH] M	Alkalin liquid to Fly Ash (L/S)	Curing Temperature(C)	Curing time (days)
M1	10	0,5	80	7
M2	12	0,5	80	7
M3	14	0,5	80	7
M4	16	0,5	80	7
M5	10	0,1	80	7
M6	10	0,3	80	7
M7	10	0,5	80	7
M8	10	0,7	80	7
M9	10	0,5	100	7
M10	10	0,5	80	7
M11	10	0,5	60	7
M12	10	0,5	T.room	7
M13	10	0,5	80	7
M14	10	0,5	80	14
M15	10	0,5	80	21
M16	10	0,5	80	28

The concentration of activator were used in this study with the range between 10-16 M, the ratio of liquid to solid(L/S) between 0.1 to 0.7, temperature between RT/Room temperature (30°C) – 100 °C, and length of treatment between 7-28 days, with Mix 1 as control and variable fixed modulus activator (Ms) = 2 and ratio of fly ash to aggregate (1: 3). Paving blocks geopolymer was started with mixing a solution of NaOH and Na_2SiO_3 activator, it was made for 24 hours before stirring. Aggregates (sand) saturated surface dry condition (SSD); aggregate and fly ash

is mixed for 3 minutes. The aggregate, fly ash, and activator were stirred until homogeneous approximately for 5 minutes. Then the mixture is poured into molds or molding machine paving blocks. Paving blocks was formed and then stored in room temperature for 3 days, then put in the oven for treatment with the variation of temperature , i.e. (30), 60, 80, and 100 ° C for 24 hours, after which the compressive strength test for the treatment time 7,14, 21 and 28 days. This was followed by XRD and SEM analysis.

3. Result and discussion

3.1. Variations of the activators concentration

Activator concentration is an important parameter to produce geopolymer material with a high compressive strength, in addition to the type of activator, the temperature and time of treatment [5]. Differences activator concentration will have effect to the difference in the mechanical properties of geopolymer produced particularly strong press.

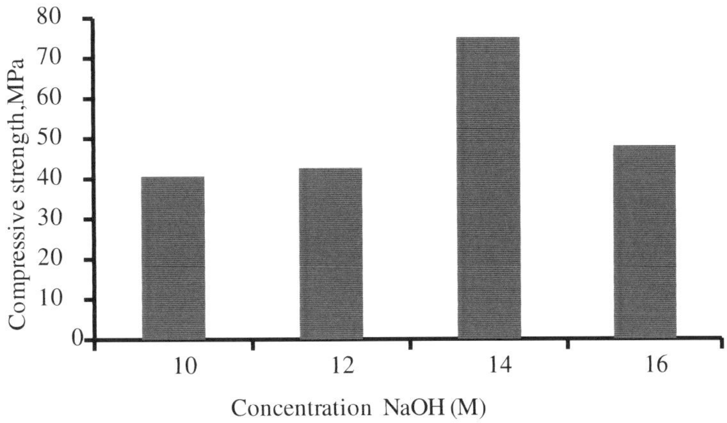

Figure 1. Effect of concentration activators on compressive strength.

The influence of concentration activator of compressive strength NaOH with a concentration of 10 to 16 M can be seen in Figure 1. It shown that there is an increase in the compressive strength of the concentration activator between 10 and 14 M, the higher is the concentration of NaOH. OH^- ion useful in controlling the leaching process Al^{3+} ions and Si^{4+} to accelerate the geopolimerisation reaction. In addition, it also have an effect on the solubility of silica-alumina. The high concentration cut off (breaks) of glassy material origin (fly ash) structure can helps atoms Si and Al become in the geopolimerisation reaction on the formation of the order geopolymer. It made the bond strength between the molecules become more reactive in order to generate monomers form, which affects the obtained compressive strength [6]. However, the compressive strength was decreased when the activator at 16 M due to the addition of alkaline activator concentration on the mixture which led to the increased coagulation of silica. It can reduce the mixture workability so that the mixture was hard to stirred and blend which caused the decreasing of the compressive strength of geopolymer [7].

3.2. Variations liquid to solid (L/S)

In Figure 2 below shows the effect of the ratio of liquid to solid (L/S) of compressive strength, where the compressive strength has increasing for the ratio is 0.1, 0.3 to 0.5. Because the mixture workability increased by the raised of liquid to solid ratio until ratio optimum at 0.5 (L/S). At 0.7 L/S ratio, compressive strength was decreased since the addition of the excess liquid activator which produced OH^- ions. It would leave the system thus weaken the geopolymer's structure. Besides that, the high liquid content caused the interaction among the fly ash particles become lessen and the distance between particles was getting wider [8].

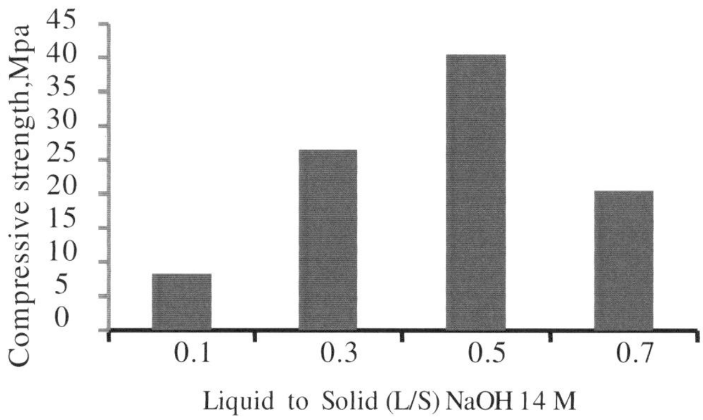

Figure 2. Effect of liquid to solid ratio (L/S) on compressive strength.

3.3. Variation of curing temperature
In figure 3 shown the effect of the temperature treatment of the compressive geopolymer strength, it seen that the paving block of room temperature to 80 °C compressive strength will be increasing with the increase of the temperature (> 80 °C).

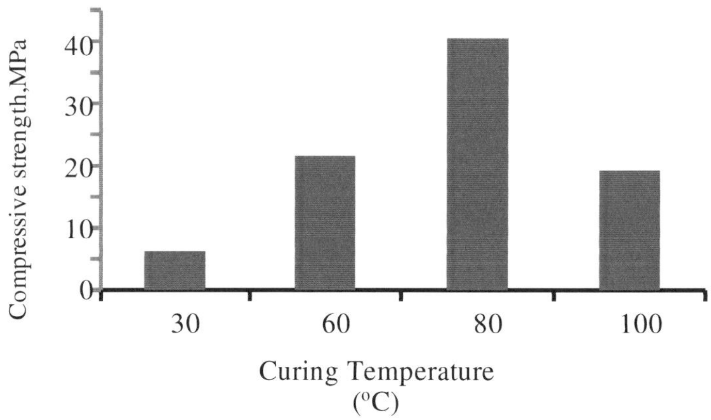

Figure 3. Effect of curing temperature (°C) on compressive strength.

A decrease in compressive strength, which is in line with previous researchers that the increase in temperature affects the mechanical strength of the concrete / paving blocks geopolymer but rise higher temperature and curing time longer can decrease the compressive strength due to lack of moisture in the structure caused easily occur cracking and also shrinkage of paving blocks / geopolymer concrete so that it can decrease the compressive strength [9, 10].

3.4. Varian duration of treatment
Figure 4 shows the compressive strength is influenced by the length of t h e treatment time at 7, 14, 21 and 28 day, it also will increased the compressive strength.

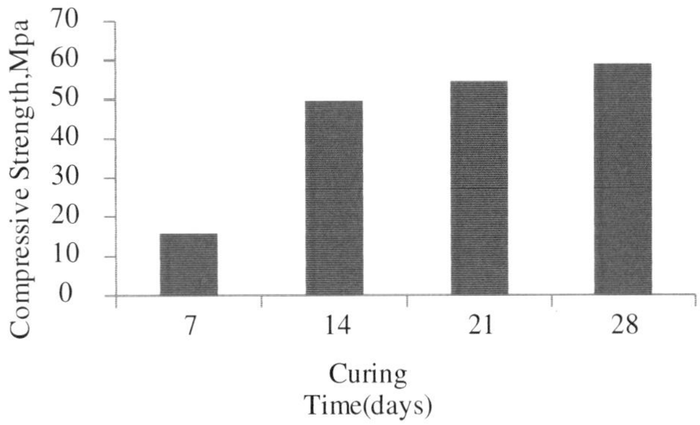

Figure 4. Effect of curing time (days) on compressive strength.

This is because the rate of chemical processes (solidification) takes on the mechanical strength of the paving blocks geopolymer, and according to the previous researchers also reported that the treatment the compressive strength increases [11-12].

3.5. X-Ray diffraction (XRD) analysis

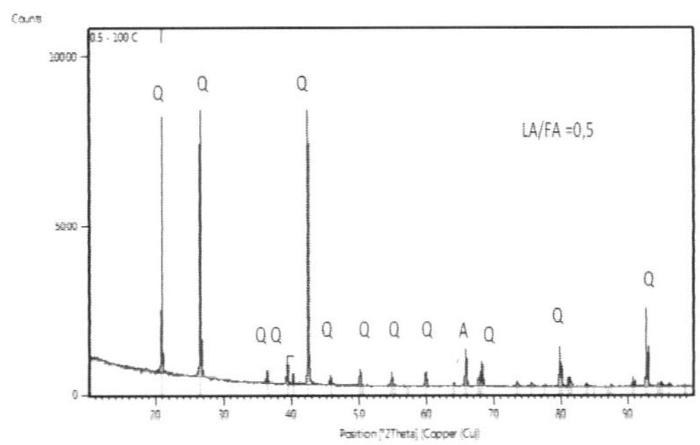

Figure 5. X-Ray Diffraction Analysis (XRD) on Paving Block geopolymer ratio L / S 0.5 NaOH 14 M (Q= quartz , A = Albite).

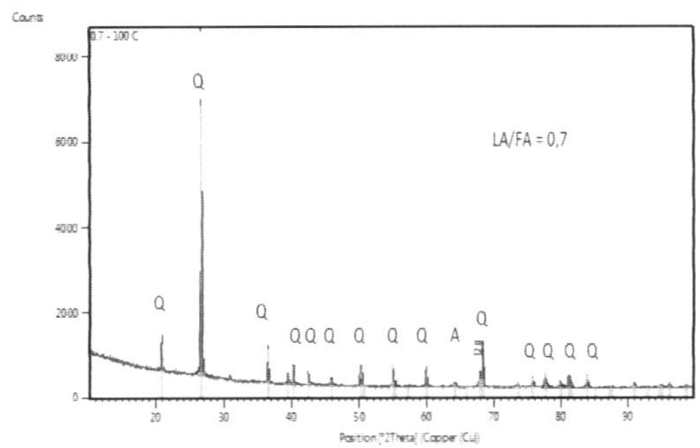

Figure 6. X-Ray Diffraction Analysis (XRD) on Paving Block geopolymer ratio L/S= 0.7 ,NaOH 14 M (Q=quartz,= Albite).

Figure 5 and 6 show the X ray diffraction (XRD) analysis for the L / S ratio of 0.5 and 0.7, respectively with the maximum compressive strength. the highest peak for all designs of the mixture is shown with quartz (Q) quartz. its shown that the level of violence. At the L / S ratio of 0.5, there are three highest Q peak (quartz), and there is a detected A (albite) peak indicating the strength of the geopolymer paste from the crystalline formation of the aluminasilicate gel [13]' while in the L / S ratio of 0,7 is not detected peak A (albite) and there is only one highest Q quartz peak so it can be concluded that L / S 0,5 has an optimum compressive strength because it has peak Q with highest intensity [14].

3.6. Scanning electron microscope (SEM) analysis

(a) L/S 0.5 (b) L/S 0.7

Figure 7. Geopolymer with difference ratio liquid to solid (L/S).

Figure 7a illustrates a geopolymer sample with ratio L/S 0.5 it shows complete geopolymerisation and fly ash reacts homogeneously with alkaline activator thus leading to maximum compressive strength. Geopolymer sample with ratio L/S 0.7 as in Figure 7b showed incomplete geopolymerisation the quantity of alkaline acktivator affects the saturation rate of the geopolymerasation process and the strength of geopolymer. In this sample, pores and microcrack was detected and probably due to the sample preparation for the SEM analysis.

4. Conclusion

Some of the parameter has been carried out for fly ash geopolymer paving block using various concentrasion activator, liquid to solid ratio, temperatur and time. It shown that the compressive strength of geopolymer paving block increased with an increasing of the concentration activator, liquid to solid ratio, curing temperatur and curing time. The presence of albite during XRD analysis was also detected in the geopolymer sample and the morphology of geopolymer samples will changes in the matrix with the increasing of the temperature.

References

[1] Ekaputri J, Maekawa K and Ishida 2011 The use of geopolymer process for boron fixation in fly ash Jinan China *Proc. 7th Intl. Symp. Cement and Concrete*

[2] Davidovits J 1994 Global warming impact on the cement and aggregates industries *World Res. Rev.* 263-278

[3] Olivia M and Damayanti P 2014 Parametric study on the compressive strength of palm oil fuel ash (POFA) geopolymer mortar *Intl. Conf. Environ. Friendly Civil Eng. Constr. Mater.* Manado Indonesia

[4] Hardjito D and Rangan 2004 Properties of geopolymer comcrete with fly ash as source material: Effect of mixture composition *Seventh CANMET/ACI Intl. Conf. Recent Adv. Concrete Technol.* Las vegas, USA

[5] Yao et al. 2009 Geopolymerization process of alkali-metakaolinite characterized by isothermal calorimetry *Thermochim. Acta* **493** (1) 49-54

[6] Ubolluk and Prinya 2009 The geopolymerization of alumina-silicate minerals Intl. J. Miner. Proc. **59** (3) 247-266

[7] Bergna H E and Roberts W O 2005 *Colloidal Silica: Fundamental and Applications* vol **131**, Taylor & Francis

[8] Sathonsaowaphak A et al. 2009 *Workability and strength of lignite bottom ash geopolymer mortar J. Hazard. Mater.* **168** (1) 44-50

[9] Muniz et al. 2011 *The effect of temperature on the geopolymerization process of a metakaolinbas Mater. Lett.* 65 (6) 995-998

[10] Wang and Cheng 2003 Production geopolymer materials by coal fly ash. *7th Intl. Symp. on East Asian Resources Recycling Technology*, Taiwan

[11] Mishra A and Choudhary 2008 Effect of concentration alkalyn liquid and curing time on strength and water absorption of geopolymer concrete *ARPN J. Eng. Appl. Sciences* **3** (1)

[12] Memon et al. 2011 Effect of curing condition on strength of fly ash based selfcompacting geopolymer concrete Intl. J. Civil Environ. Sci. Constr. Architect. Eng. **5** (8)

[13] Garcia-Lodeiro L 2010 Effect of Fresh C-S-H gels of the simultaneous addition of alkali and aluminium *Cement Concrete* **40** 27-32

[14] Alvarez-Ayuso E 2008 Environmental, physical and structural characterisation of geopolymer matrixes synthesized from coal (co-) combustion fly ashes *J. Hazard. Matter.* **154** 175-183

IOP Conf. Series: Materials Science and Engineering **345** (2018) 012019 doi:10.1088/1757-899X/345/1/012019

Synthesis of geopolymer from rice husk ash for biodiesel production of *Calophyllum inophyllum* seed oil

E Saputra[1,2], M W Nugraha[2], Z Helwani[2], M Olivia[3] and S Wang[1]
[1]Department of Chemical Engineering, Curtin University, GPO Box U1987, Perth, WA 6845, Australia
[2]Department of Chemical Engineering, Riau University, Pekanbaru 28293, Indonesia
[3]Department of Civil Engineering, Riau University, Pekanbaru 28293, Indonesia

E-mail: edysaputra@unri.ac.id

Abstract. In this work, geopolymer was prepared from rice husk ash (RHA) made into sodium silicate then synthesized by reacting metakaolin, NaOH, and water. The catalyst was characterized using Scanning Electron Microscopy (SEM), Energy-dispersive X-Ray analysis (EDX), Brunaeur Emmet Teller (BET), and basic strength. Then, the catalyst used for transesterification of *Calophyllum inophyllum* seed oil in order to produce biodiesel. The variation of process variables conducted to assess the effect on the yield of biodiesel. The highest yield obtained 87.68% biodiesel with alkyl ester content 99.29%, density 866 kg /m^3, viscosity 4.13 mm^2/s, the acid number of 0.42 mg-KOH/g biodiesel and the flash point 140 °C. Generally, variations of %w/w catalyst provides a dominant influence on the yield response of biodiesel. The physicochemical properties of the produced biodiesel comply with ASTM standard specifications.

1. Introduction

Currently, petroleum is still the main dependent energy source in the world. However, along with population growth, fuel consumption from petroleum continues to increase every year. In addition to the fact, petroleum-based fuels can pollute the environment (e.g. particulate matter, carbon monoxide and sulfur oxides), encouraging the development of environmentally friendly alternative energy i.e. biodiesel. Besides its renewability, biodiesel also has high degradability and produces far fewer emissions (e.g. nontoxic, lack of aromatic compounds and sulfur, low CO_x, SO_x, and particulate matter) thereby reducing environmental impact [1-3].

Biodiesel is produced from transesterification reactions of short-chain alcohols with triglycerides derived from vegetable oils or animal fats. The most commonly used catalysts for biodiesel are strong base homogeneous catalysts such as KOH and NaOH [4]. This high catalysts prices can cause biodiesel today to be more expensive than fuel derived from petroleum. It is necessary to develop an economical heterogeneous catalyst i.e. the geopolymer from rice husk ash. Large surface area and pore size, as well as strong base properties in geopolymer, made this material highly potential to serve as a catalyst for transesterification.

2. Experimental

2.1 Catalyst preparation

Rice husk was furnaced for 3 hours at 500 °C. Rice husk ash is filtered using 100 mesh strainer, then the silica from RHA was extracted which was carried out in the batch reactor. The extraction reacted RHA with 0.5 N NaOH (with SiO_2 / Na_2O mole ratio 1.9), at 100 °C for 60 min with 200 rpm stirring rate) [5]. After that, the mixture is filtered through the vacuum pump to separate the rice husk ash and filtrate as sodium silicate solution (Na_2SiO_3).

Metakaolin, sodium silicate, NaOH, and aqua DM were fed into a reactor with a weight ratio metakaolin: sodium silicate: NaOH: water = 10: 12.2: 1.1: 0.7. Polymerization was performed in a closed vessel at 60 °C for 48 h. After maturing of the geopolymer for one week, the geopolymeric monolith was ground [6]. Then, the catalyst was characterized by Scanning Electron Microscopy (SEM), Energy-dispersive X-Ray analysis (EDX), Brunaeur Emmet Teller (BET), and basic strength.

2.2 Transesterification process
First, pre-treatment of Calophyllum inophyllum oil using 0.3% phosphoric acid from the weight of the oil, heated to 60 °C with stirring. Then, the transesterification was carried out with a mol ratio of oil: methanol 9:1, the temperature of 60 °C, 2 h reaction time with the variation of process variables %w/w catalyst 1-2%, stirring rate 200-600 rpm.

3. Result and discussion
The basicity ($H\_$) of geopolymer catalyst was determined using Hammett indicators with the result of $H\_>9.3$, the catalyst is feasible to be used as a base catalyst in the transesterification reaction [7]. Then, for the surface area was determined by SAA (Surface Area Analyzer) instrument Brunauer Emmet Teller (BET) method with a result of 31.327 m^2/g. Thus, the larger surface area has more contact area with reactant molecule to produce biodiesel and larger catalytic activity.

The silica and aluminio content in geopolymer was determined using Energy Dispersive X-ray (EDX) spectrometry. The results are shown in Table 1. It can be seen that geopolymer has Si content of 14.17% and Al of 10.35%. Chemically, the geopolymer structure is similar to a low-silica zeolite with a Si/ Al ratio between 1-3 [8]. The ratio of Si/ Al in this research is 1.4. This indicated that the silica molecules were attached or incorporated into geopolymer structure during the impregnation process.

Table 1. Composition of geopolymer component elements

Element	Weight %	Atomic %
C	24.86	34.36
O	46.63	48.48
Na	2.82	2.04
Al	10.35	6.36
Si	14.17	8.37
K	1.17	0.50
Total	100.00	

Furthermore, the catalyst was analyzed using SEM morphology catalyst in Figure 1.

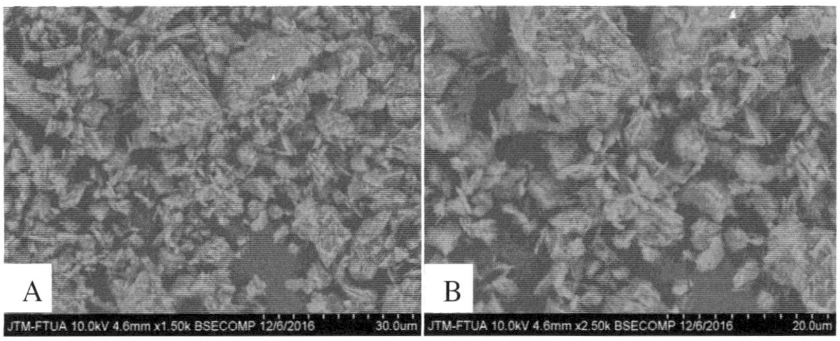

Figure 1. Surface morphology of geopolymer.

Based on the results of the research is seen from Figure 2. that the use of catalysts affects the yield of biodiesel produced in the transesterification process. The use of catalyst concentration of 1% (400 rpm stirring rate, 2 h time, mol ratio oil: methanol 9:1, temperature of 60 °C), yield is 83.6%, while at catalyst concentration 1.5% biodiesel yield increased 86.56%, then at catalyst concentration 2% yield of biodiesel produced decreased i.e. 84.45 %. This suggested that the increasing of the catalyst concentration may increase the yield of biodiesel due to the increasing area that accelerated the reaction of reactant conversion into the product until the optimum condition reaches with the highest yield obtained at 87.68% at 1.5% catalyst, 400 rpm stirring rate.

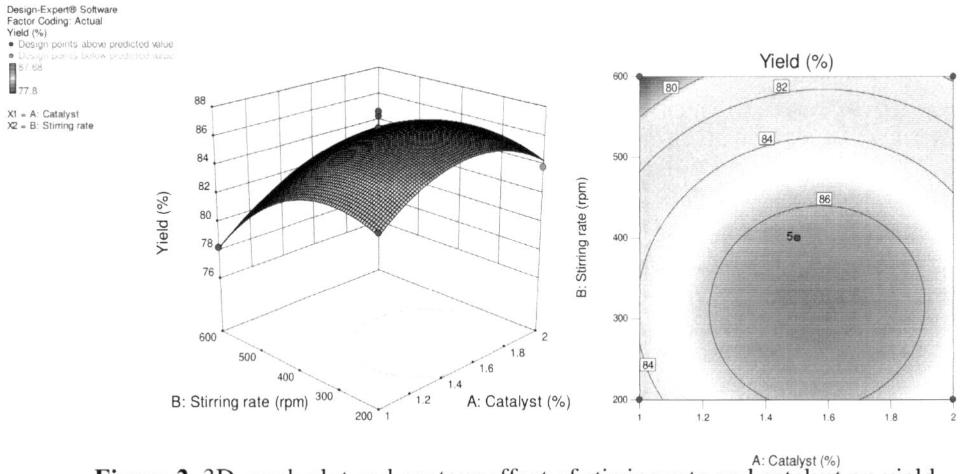

Figure 2. 3D graph plot and contour effect of stirring rate and catalyst on yield

The biodiesel obtained was determined its physicochemical properties. The content of the biodiesel product produced from this study was analyzed using GC-MS with 99.29% ester alkyl content presented in Figure 3. and Table 2.

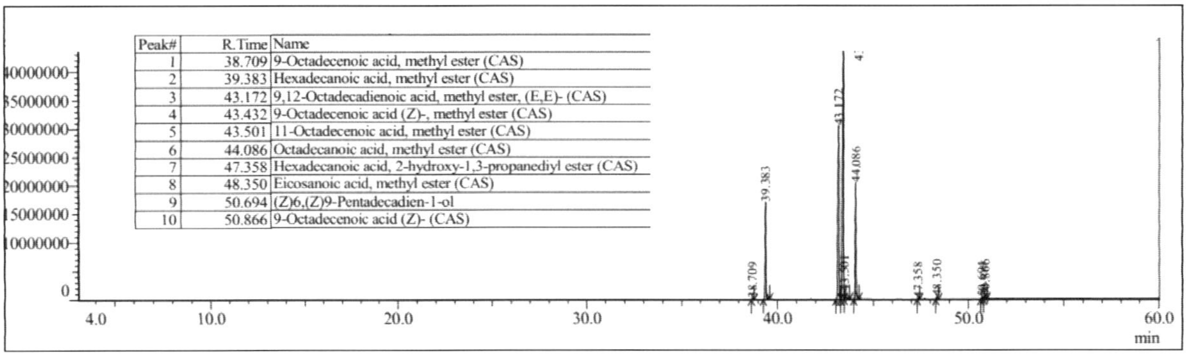

Figure 3. Calophyllum inophyllum biodiesel GC-MS analysis.

Table 2. Calophyllum inophyllum biodiesel composition content

R Time	Component	%-weight
39.383	9-Hexadecanoic acid, methyl ester	11.94
43.172	9-12-Octadecanoic acid, methyl ester	25.71
43.432	9-Octadecanoic acid, methyl ester	45.57
43.501	11-Octadecanoic acid, methyl ester	1.45
44.086	Octadecanoic acid, methyl ester	14.23
47.358	Hexadecanoic acid, 2-hydroxy-1,3-propanediyl ester	0.11
48.350	Eicasonoic acid, methyl ester	0.39
50.694	9-Pentadecadien-1-ol	0.23
50.866	9-Octadecanoic acid	0.37

The comparison of characteristic results of produced biodiesel in this study was compared with SNI 7182:2015 (Indonesia Standard) and ASTM D6751 presented in Table 3. The physicochemical properties of the produced biodiesel comply with SNI and ASTM standard specifications.

Table 3. Produced biodiesel comparison

Parameter, unit	ASTM D6751	SNI 7182:2015	Produced biodiesel
Ester alkyl content, %		Min 96.5	99.29
Density 40°C, kg/m^3		850-890	866
Kinematic Viscosity 40°C, mm^2/s	2,3-6,0	2,3-6,0	4,13
Acid Number, mg-KOH/g	Max. 0,5	Max. 0,5	0,42
Flash Point, °C	Min 100	Min 100	140

4. Conclusions

In this study, rice husk ash (RHA) geopolymer was prepared from metakaolin then by impregnation methods in sodium silicate solution from RHA extraction. The analyses of basicity, EDX, SEM, and BET showed the possibility of this economical catalyst being effective heterogeneous base catalysts for the manufacture of biodiesel. The RHA geopolymer catalyst provided the maximum biodiesel yield at 87.68% obtained in 2 h reaction time at 1.5% catalyst, 400 rpm stirring rate, a reaction temperature of 60 °C, methanol to oil molar ratio of 9:1.

References

[1] Ma F and Hanna M A 1999 *Biosource. Technol.* **70** 1-15
[2] Schuchardt, U, Sercheli R and Vargas R M *1998 J. Braz. Chem. Soc.* **9** 199-210
[3] Michael S G and McCormick R L 1998 *Prog. Energ. Combust. Sci.* **24** 125-164
[4] Meher L C, Sager D V and Naik S N 2006 *Renew. Sustain. Energ. Rev.* **10** 248-268
[5] Folleto E L, Gratieri E, Oliviera L H and J. Sergio 2006 *Mater. Res.* **9** 335-338
[6] Sazama, P, Bortnovsky O, Dedecek J, Tvaruzková Z and Sobalík Z *Cat. Today* **164** 92-99
[7] Ho W S, Hoonkiat N, Suyin G and SangHuey T 2014 *Energ. Conver. Man.* **88** 1167-1178
[8] Sharma S, Medpelli, D, Chen S and Seo D K 2015 *RSC Adv.* **5** 65454-65461

Effect of process variables on the calorific value and compressive strength of the briquettes made from high moisture *Empty Fruit Bunches* (EFB)

Z Helwani[1], W Fatra[2], L Arifin[1], M R Othman[3] and Syapsan[4]
[1]Departement of Chemical Engineering, Riau University, Pekanbaru, 28293, Indonesia
[2]Departement of Mechanical Engineering, Riau University, Pekanbaru, 28293, Indonesia
[3]School of Chemical Engineering, Universiti Sains Malaysia, 14300 Nibong Tebal, Pulau Pinang, Malaysia
[4]Economic Faculty, Riau University, Pekanbaru, 28293, Indonesia

E-mail: zuchra.helwani@lecturer.unri.ac.id

Abstract. In this study, the manual hydraulic press was designed to prepare the briquettes from selected biomass waste. Each biomass was sun-dried and milled into small particle sizes before mixing with crude glycerol that used as a biomass binder. The effects of applied pressure levels of 100, 110, 120 bars, the particle size of 60, 80 and 100 mesh and the binder composition on the density, compressive strength and calorific heating value of the prepared briquettes were investigated using response surface methodology (RSM). Results showed that the briquettes have an average inside diameter, average outside diameter, and height of 12, 38, and 25-30 mm, respectively. The density of the briquettes increased with increasing the applied pressure, was in the range of 623-923 kg/m^3. The densest briquettes were obtained at 80 mesh of particle size, 53:47 binder composition ratio and 110 bars of pressurizing. The heating value of the briquette reached up to 28.99 MJ/kg obtained on the particle size of 80 mesh, 53:47 binder composition, and 110 bars and the best compressive strength of 6.991 kg/cm^2 obtained at a particle size of 100 mesh, 60:40 binder composition, and 120 bars. Process conditions influence the calorific value significantly.

1. Introduction

In the last few years, energy has become a crucial issue in the world owing to economic development, which has been accompanied by population growth, the depletion of oil reserves, and the problem of emissions from these fuels. Consumption of fossil energy is constantly increasing, which has caused the fuel crisis. Limitations in the procurement of fuel oil and gas push to thedevelopment of a renewable fuel.

Indonesia is a rich country in agricultural products, such as biomass, that could beused as the sources of renewable energy. The amount of energy from biomass is about 50,000 MW, of which only 320 MW recently used, or approximately 0.64% of the amount of existing biomass [1]. The potential of biomass in Indonesia includes palm oil, rice mill waste, wood, bagasse, and other agricultural waste. The land area of oil palm plantations in Riau Province in 2015, there were approximately 2.39 million hectares with the production of approximately 7.44 million tons FFB [2]. On average, one ton of FFB produces 230-250 kg of empty fruit bunches of oil palm (EFB), 130-150 kg of fiber, 60-65 shells, and 55-60 kg of seed production of oil [3]. Of the wastegenerated, only EFB has not been utilized optimally. EFB by palm processing plants is still very limited. Most of the oil factories in Indonesia are still burning EFB in the incinerator, although this has been disallowed by the Indonesian government. An alternative treatment is to hoard (open dumping), which is then used as much in oil palm plantations or processed into compost [4].

EFB waste is processed into alternative fuel, maybe through the carbonization process, which is then followed by densification. Carbonization aims to minimize the elements forming the smoke, so the exhaust gas can be cleaner [5]. Generally, biomass has poor physical properties, such as the density is quite low, storage, and transport. Densification of biomass to be briquettes aims to increase the density and reduce handling issues, such as storage and transport. In general, biomass densification has several advantages. Among the other things, it can increase in density to be stored easily and transported and have the same size and quality [6].

The calorific value of the results of briquetting could be increased by adding a binder. One of the binders that can enhanced the calorific value of the briquette is crude glycerol. Crude glycerol is a byproduct of biodiesel results that can be useful, such as cosmetic products. However, it needs further purification by high costs [7]. The idea of combining the crude glycerol waste byproduct of biodiesel production with biomass is still relatively new. Crude glycerol usehas a calorific value of 25175.98 kJ/kg, as an enhancer combustion heat values) is one of the alternatives by using glycerol without purification [8]. By this way, reducing purification costs can also be done in an integrated manner by the manufacturer of small and medium-scale biodiesel [9].

EFB has a calorific value of 19,600 kJ / kg. This value is higher than the calorific value that is possessed by palm fronds and durian skin. Accordingly, this study will use raw materials of EFB as a source of biomass briquetting and crude glycerol by-product of biodiesel as binder. The research was conducted by varying the particle size, the composition of the binder, and the pressing pressure. Based on the research of Ilham et al., (2016) [10], these conditions have a significant effect on increasing the calorific value and the compressive strength of the product briquettes produced. Optimization of data processing will be done using the response surface method (RSM). Therefore, the manufacture of solid fuel from EFB and crude glycerol is a byproduct of biodiesel production that is expected to be an alternative solid fuel because of its rich source and because it is still not optimal in its use.

2. Materials and methods

2.1 Materials

Manufacturing of solid fuels in this study used the raw materials of palm empty fruit bunches (EFB) and the glycerol by-product of biodiesel as binder. The supply of EFB was obtained from PTPN V Sei Galuh while crude glycerol was obtained from biodiesel byproduct at PT. Wilmar Bioenergy Indonesia. The equipment used consisted of hydraulic press units, a universal testing machine, and a bomb calorimeter.

2.2 Preparation of empty fruit bunches

The raw material, such as EFB, cleaned, downsized, and then dried under the sun. Furthermore, EFB carbonized in the tube furnace at a temperature of 400 °C for two hours [11]. Next, the raw carbonized product was sieved using a size 60, 80, and 100 mesh. Then, the EFB charcoal powder was mixed with crude glycerol in the ratio of 80:20; 70:30; and 60:40 and then printed by using a hydraulic press at a pressure of 100, 110 and 120 bar for 10 seconds. Briquettes were already printed and dried under the sun to reduce their water content.

2.3 Samples testing
The calorific values of the briquettes products were tested the by using the bomb calorimeter and compressive strength using a universal testing machine.

3. Results and discussion

3.1 Characteristic of EFB, carbonized and briquettes
The results regarding the EFB characteristic, carbonized EFB and briquettes is shown in Table 1.

Table 1. Characteristics of EFB, EFB charcoal, and EFB briquettes

No.	Characteristics	Unit	EFB	Carbonized product	Briquettes
1	Calorific value	kJ/kg	18,200.9	21,591.359	22,991.4 – 28,999.4
2	Water content	% - b	10.36	4.6	5.04
3	Volatile mater content	% - b	87.08	24.16	21.48
4	Dust content	% - b	1.18	0.56	0.34
5	Carbon Content	% - b	1.38	70.68	73.14
6	Compressive Strength	kg/cm^2	-	-	1.851 – 9.5316
7	Density	gr/cm^3	-	-	0.623 – 0.932

3.2 Design and analysis of results by using response surface methodology

Table 2. Summary of test results of the curvature of order 1

Responses	*p-value Prob>F*	*p-value Lack of fit*	*p-value curvature*
Calorific value	< 0.0001	0.1916	0.5480
Compressive Strength	< 0.0001	0.1329	0.1631

Based on Table 2, it can be seen that the p-value of both responses has qualified the regression test p-value smaller then $\alpha = 0.05$ and the p-value for the lack of fit responsesgreater than $\alpha = 0.05$. The models are in accordance with the data. The p-value for the curvature of each response is greater than $\alpha = 0.05$, which means there is no curvature. From the analysis of the three tests above, it can be concluded that the first-order model was appropriate because the three parameters were met.

Table 3. Summary p-value for each response

Source of variants	*Calorific value*	*Compressive strength*
Model	< 0.0001	< 0.0001
A-Particle Size	0.6904*	0.1249*
B-Filler Composition	< 0.0001	<0.0001
C-Pressing Pressure	< 0.0001	0.0034
AB	0.6476*	0.0167
AC	0.0029	0.0076
BC	0.0027	0.0010
R^2	0.9394	0.9007

Table 3 shows the component model has an influence on the response variable (Y_i). The level of probability was used $\alpha = 0.05$, and then the variable that has a p-value smaller than $\alpha = 0.05$ are variables that have a significant impact on the model. While the p-value has greater than 0.1, it does not have a significant effect on the model. The statistical analysis of the results can be used to obtain the coefficient of determination (R^2). R^2 is the amount of variability in the data obtained or calculated based on the regression model. The R^2 value is in the interval from 0 to 1. The value of $R^2$is close to 1, which indicates a high degree of correlation between the results of observation of the resulting model [12]. After all statistical tests were performed on the models, the model was obtained to study the effect of operating conditions (X_i) of the response variable (Y_i).

Testing the hypothesis on a simultaneous test was done by calculating the F_0 (F_{test}) for each response variable. The F_0value can be determined by using the ratio of the mean square regression to the mean square error, if the ratio of F_{test} > F (α, DoF, DF2), H_0 is rejected.
The hypothesis applied for testing:
H_0: β1 = β2 =...= βk (there is no conformity to the model)
H_1: Not all βi there is conformance in the model.

Table 4. Summary F_0 value for each variable responses (Y_i)

Responses	Sources	SS	DoF	MS	F_0	F_{table}
Calorific Value	Regression	4.781E+007	7	6.830E+006	24.37	3.01
	Error	3.084E+006	11	2.803E+005		
	Total	5.100E+007	19			
Compressive Strength	Regression	52.57	7	7.51	14.25	
	Error	5.80	11	0.53		
	Total	59.55	19			

The F_{table} value is F (α, DoF, DF2) with a probability level of $\alpha = 0.05$ with the DoF being a degree of freedom. The F_{table} value for distribution of F (0.05,7,11) is 3.01. Therefore, if the result of F_0>F (α, DoF, DF2) and $H_0$is rejected, suggesting that the model is used significantly affectsthe responses. After all statistical tests are performed on the model, the model was obtained to study the effect of operating conditions (X_i) of the response variable (Y_i).

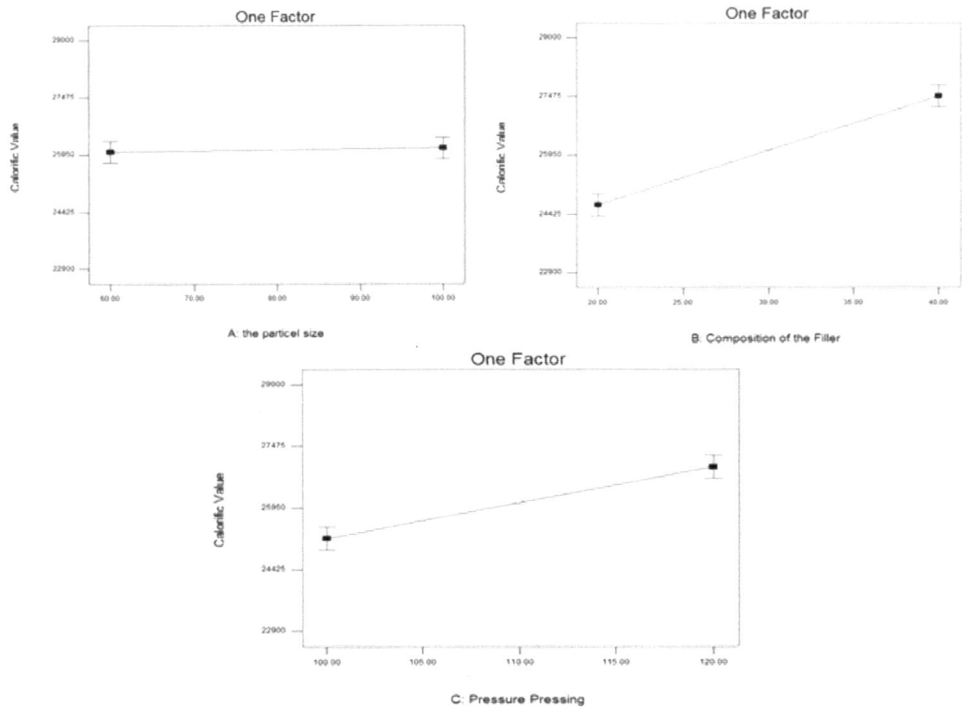

Figure 1. Effect of process conditions on the calorific value: a) The size of the particles; b) The composition of the filler; c) Pressure pressing

3.3 Analysis response calorific value

Figure 1 shows that the composition of the filler and the pressing pressure havea significant effect on the value of the heat produced because it can increase the calorific value of the briquettes. In this study, the raw material EFB has a calorific value of 18200.975 kJ / kg. The calorific value of briquettes produced by the densification process ranged from 22991.4 to 28999.4 kJ / kg. The addition of filler in the form of crude glycerol significantly increases the calorific value by 62.76% of the calorific value of the raw material. This is consistent with the results obtained by Asavatesanupap and Santikunaporn (2012) [7], which states that the addition of crude glycerol can improve the calorific value.

Stamping pressure also influences the calorific value of the briquettes. The greater the pressure, the density of the briquette presses becomes bigger so more charcoal gets in contact with the filler in the form of crude glycerol toincrease the calorific value [13]. In addition to the filler's composition, and the pressing pressure having a significant effect, the particle size does not affect the calorific value of the briquettes. This is evident from the p-value of the particle size of 0.6904. The p-value for the variable particle size exceeds the probability value is> $\alpha = 0.05$, so it can be concluded that the particle size does not provide a significant effect on the calorific value.

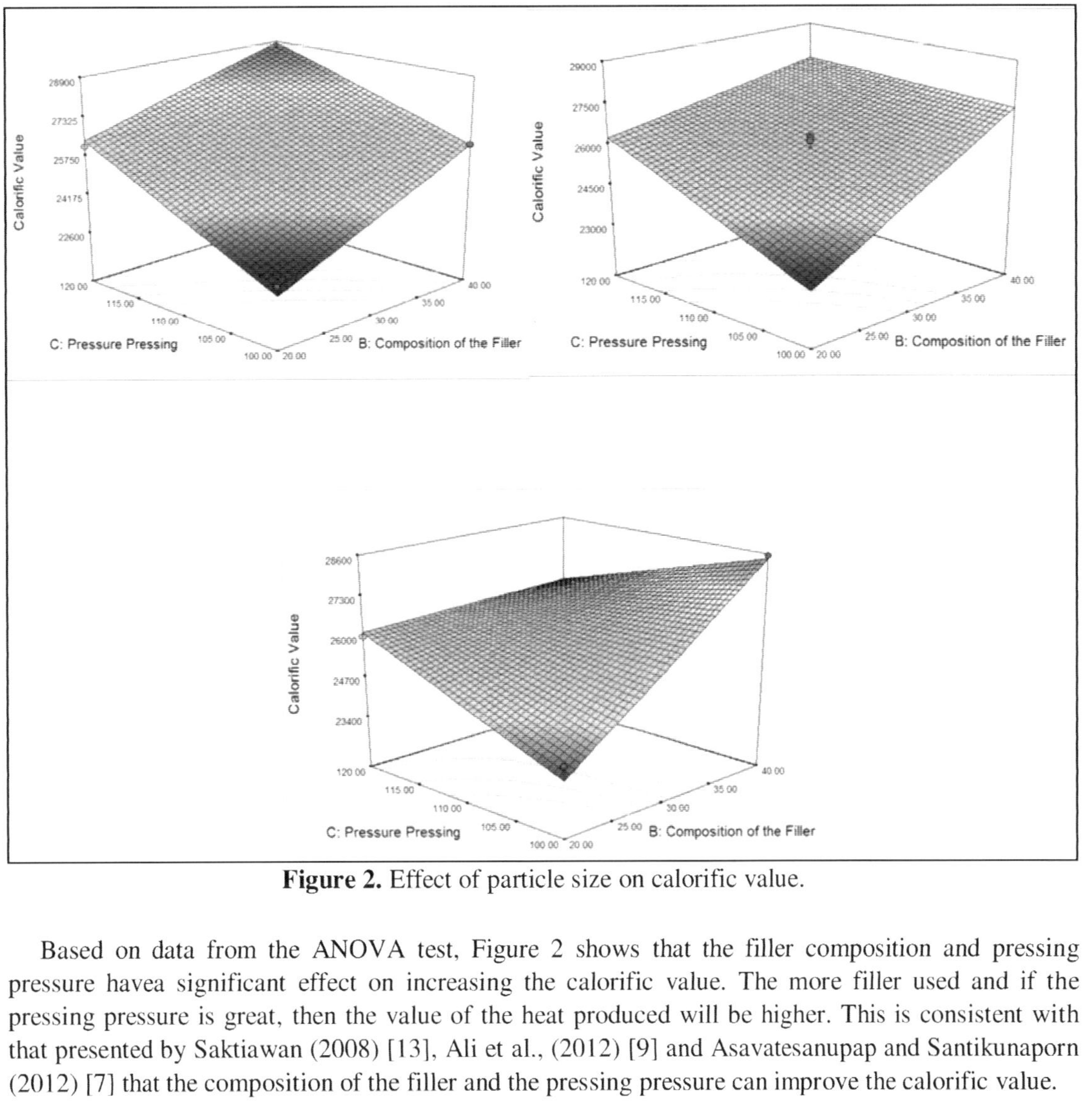

Figure 2. Effect of particle size on calorific value.

Based on data from the ANOVA test, Figure 2 shows that the filler composition and pressing pressure havea significant effect on increasing the calorific value. The more filler used and if the pressing pressure is great, then the value of the heat produced will be higher. This is consistent with that presented by Saktiawan (2008) [13], Ali et al., (2012) [9] and Asavatesanupap and Santikunaporn (2012) [7] that the composition of the filler and the pressing pressure can improve the calorific value.

3.4 Response analysis on compressive strength

Figure 3 indicates that the composition of the filler and the pressing pressure significantly influence compressive strength. From the p-value of the composition of the filler and the pressing pressure in Table 3.5, the more filler used, the compressive strength also increases. Likewise, if the pressing pressure is large, the pressure pressing densities and compressive strength will also be increased. The effects of particle size are not very significant, because of the p-value of the particle size in Table 3. The compressive strength of the briquettes produced ranges from 1.851 to 9.5316 kg/cm^2. The best compressive strength of 6.991 kg/cm^2 was obtained at a particle size of 100 mesh, 60:40 filler composition, and a pressing pressure of 120 bar.

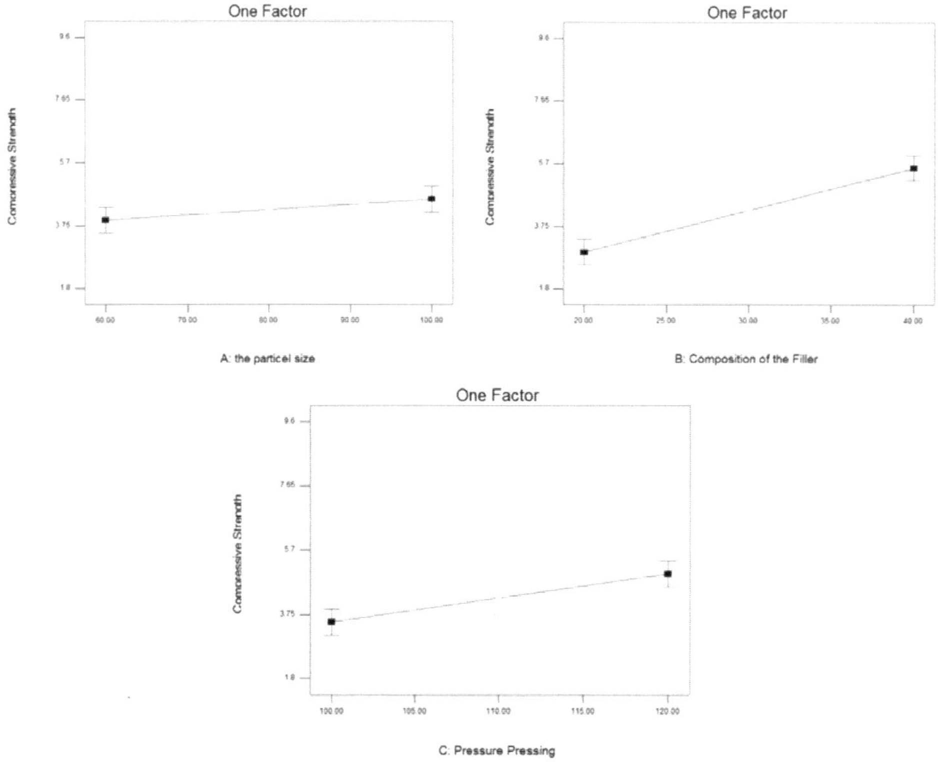

Figure 3. Effect of compressive strength of process conditions a) particle size b) the composition of the filler c) pressure pressing.

Figure 4 shows that the interaction between the filler composition and the pressing pressure has a significant influence on the compressive strength. The relationship between the composition of the filler and the pressing pressure is directly proportional to the compressive strength. The greater the pressure exerted when the densification process accompanied by the composition of the filler increasesthe compressive strength. The purpose of the pressing pressure is to producethe contact between the surfaces of the charcoal with crude glycerol as filler, so the briquettes are not easily broken [14].

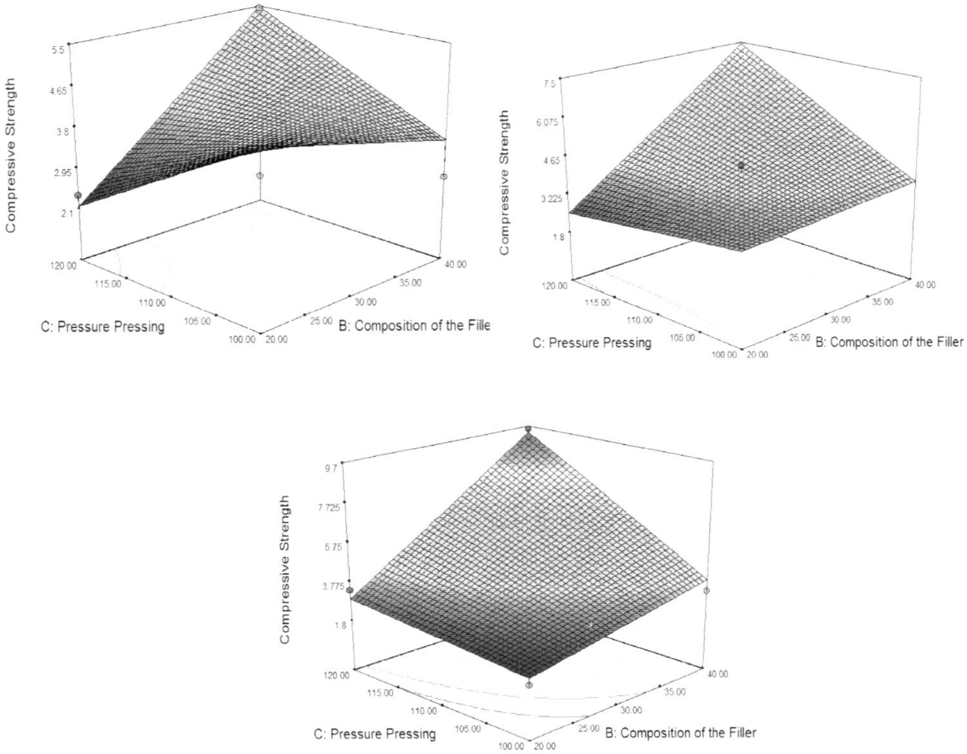

Figure 4. Effect of particle size on compressive strength.

3.5 Density of the briquette

The density of the briquettes produced in this study amounted to 0.623 to 0.932 g / cm^3. The density value greatly affects the quality of the briquettes. Therefore, if the density is large, the quality of the briquettes will be high. The density of the briquettes is affected by the composition of filler, particle size, and pressure pressing. The best density is 932 kg/cm^3obtained onaparticle size of 80 mesh, 53:47 filler composition, and pressure of 110 bar.

4. Conclusions

The composition of filler and pressure of pressing have a significant influence on the response of the calorific value and compressive strength. The best calorific value of 28999.4 kJ/kg was obtained on the particle size of 80 mesh, 53:47 filler composition, and the pressing pressure of 110 bar. The best compressive strength of 6.991 kg/$cm^2$was obtained at a particle size of 100 mesh, 60:40 filler composition, and a pressing pressure of 120 bar.

Acknowledgement

The support from the University of Riau under Hibah Guru Besar Riau University Grant is gratefully acknowledged.

References

[1] Nuriana W, Anisa N and Martana 2013 Synthesis preliminary studies Durian peel bio briquettes as an alternative fuels *Energ. Procedia* **47** 295-302

[2] Direktorat Jenderal Perkebunan RI. 2014. *Statistik Perkebunan Indonesia Komoditas Kelapa Sawit 2013-2015* Dirjen Perkebunan, Jakarta

[3] Nurdiawati A, Novianti S, Zaini I N, Nakhshinieva B, Sumida H, Takahashi F and Yoshikawa K 2015 Evaluation of hydrothermal treatment of empty fruit bunch for solid fuel and liquid organic fertilizer co-production *Energ. Procedia* **79** 226-232

[4] Nur M S 2014 *Karakteristik Kelapa Sawit Sebagai Bahan Baku Bioenergi* IFMN Kutai Timur

[5] Novianti S, Biddinika M K, Prawisudha P and Yoshikawa K 2013 Upgrading of palm oil empty fruit bunch employing hydrothermal treatment in lab-scale and pilot scale *Procedia Environ. Sci.* **20** 46-54

[6] Surono U B 2010 Peningkatan kualitas pembakaran biomassa limbah tongkol jagung sebagai bahan bakar alternatif dengan proses karbonisasi dan pembriketan *Jurnal Rekayasa Proses* **4** (1) 13-18

[7] Asavatesanupap C and Santikunaporn M 2012 A feasibility study on production of solid fuel from glycerol and agricultural wastes *Intl. Trans. J. Eng. Manage.* **1** (01) 43-51

[8] Umam M C 2007 *Optimasi Penambahan Limbah Gliserol Hasil Samping Transesterifikasi Minyak Jarak Pagar dan Perekat Tapioka pada Pembuatan Biomass Pellets Bungkil Jarak Pagar (Jatropha Curcas L.)* Thesis Institut Pertanian Bogor

[9] Ali A, Fortuna A D and Restuhadi F 2012 Kajian pemanfaatan biomassa limbah industri minyak picung (Pangium Edule Reinw) untuk biobriket sumber energi alternatif di desa Pulau Picung, Kampar *Agric. Sci. Technol. J.* **11** (1) 1-11

[10] Ilham M A, Helwani Z and Fatra W 2016 Proses densifikasi produk karbonisasi pelepah sawit menjadi briket menggunakan gliserol produk samping biodiesel sebagai filler. *Jurnal Online Mahasiswa Fakultas Teknik Universitas Riau* **3** (2) 1-4

[11] Yemita S, Helwani Z and Fatra W 2016 Karbonisasi pelepah sawit *Jurnal Online Mahasiswa Fakultas Teknik Universitas Riau* **3** (1) 1-6

[12] Montgomery C D 2013 *Design and Analysis of Experiments*, 8th ed. (New York: John Wiley & Sons, Inc.)

[13] Saktiawan. 2008 *Identifikasi Sifat Fisis dan Kimia Briket Arang Dari Sabut Kelapa* Thesis Institut Pertanian Bogor

[14] Thabout M, Pagketanang T, Panyachareon K, Mongkut P and Wongwicha P 2015 Effect of applied pressure and binder proportion on the fuel properties of holey bio-briquettes *Energ. Procedia* **79** 890-895

Evaluation of the effect of roasting process on the energy transition and the crystalline structures of *Arabica*, *Robusta*, and *Liberica* coffee from Jambi Indonesia

B M Perdana, R Manihuruk, R Ashyar, Heriyanti and Sutrisno
Department of Chemistry, Faculty of Science and Technology
University of Jambi, Indonesia
Kampus Pinang Masak, UNJA Mendalo Darat, Jambi, Indonesia

E-mail: herasutrisno@gmail.com

Abstract. The effect of the roasting process has been evaluated to determine of the energy transition and the crystalline structure of three types of coffee, Arabica, Robusta, and Liberica coffee both green and roasted coffee with the roasted temperature at 200°C and 230°C. The crystalline structure of the coffee was evaluated with X-ray powder diffraction (XRD). The result exposes that the three types of green coffee showed that an amorphous structure whereas the roasted coffee denotes a crystal structure of sucrose. The varied temperature in the roasting process leads to changes in the crystal structure shown by the peak shift of 2θ for all types of coffee. The added cations, such as Fe^{2+}, Ca^{2+}, and Mg^{2+} ions on Liberica coffee induced of changes in the crystal structures, which are assigned by the peak shift, that imply of metal ions of the sucrose complexes happened in the solution, except for the addition of Mg^{2+} ion.

1. Introduction

Roasting coffee is a complex process of heat transfer in closed conditions, where coffee beans will experience weight loss, volume increase and the decrease in density during roasting. It is well known that roasting process causes the chemical and physicochemical characteristics of green coffee beans. The coffee beans will experience changes such as colour, pH level, taste and aroma associated with the Millard reaction [1][2]. Thermal reaction in the roasted coffee beans is decarboxylation, dehydration, fractionation, isomerisation, polymerization, and sugar complexes [3][4]. Besides, the roasting process causes improve profile thermal and crystal behaviour of the coffee [5]. Although a few recent studies about thermal behaviour and identification of a chemical change of Arabica and Robusta coffee in the roasting process have established [6][7][8] but there is very limited information for Liberica coffee beans. Meanwhile, the extracts of C. liberica have higher antioxidant capacity compared to those of C. Arabica and C. Robusta [9].

The thermal and crystal behaviour of coffee can be evaluated by using thermal analytical techniques, among which differential scanning calorimetry (DSC) and by X-ray diffraction. Variation of temperature on the roasting process can be applied to study the phase transition through the change of structure of the coffee using XRD analysis. While the change of thermal properties will be monitor by DSC [1], [10]. The experiment of green coffee using DSC shows a high and fast decomposition after 200 °C until 289 °C [11]. The XRD spectra of spent coffee grounds (SCG) coffee silverskin (CS) show a similar trend, indicating the crystal regions of both coffee residues [12].

We focused on the comparative study of thermal behaviour and crystal structure formed during roasting of arabica, robusta, and liberica coffee. We have recently described the thermal behaviour and characterization of liberica coffee beans using DSC, spectroscopic methods and XRD. The XRD analysis shows that there is a formation of polysaccharide crystal, but the structure is not confirmed yet. There have limited investigations of crystal structure formed of roasted coffee, although the knowledge of its thermal changes and chemical characterizations of coffee beans may be of importance to compare the characteristic of a variated coffee bean.

2. Materials and methodology

All experiments were performed using a Fike 500i drum roaster, 2016 made in Indonesia and heating gas propane. For all trials 1 kg batches green beans were roasted. Green coffee beans of liberica, arabica, and robusta samples from Jambi Province, Indonesia with a moisture content of around 12% were roasted at temperature 200 °C and 230 °C, roasting time up 21 minutes were selected from trials using as criteria the visual colour (light, medium and dark) of the roasting process coffee beans. Green bean and roasted coffee beans than they were ground to be powder.

To study the transformations experimented by the coffee beans after roasting, three techniques were utilized to analyse powder samples of green and roasted material. Differential scanning calorimetry (DSC) was carried out with DSC Perkin Elmer DSC 4000. To study the transformations experimented by the coffee beans after roasting, two techniques were utilized to analyse powder samples of green and roasted material. To determine the crystal structure were used an X-ray diffractions (XRD) PAnalytical Xpert3 powder goniometer PW 3050/60 type of Bragg-Brentano. Data diffraction of XRD was collected using a target atom Cu (1.54 Å) at 40 kV and 30 mA at range 2θ 15o - 70o, step 0.040 and measurement time of 0.5 s per step. The calculation of the diffraction pattern using a database Crystallography Open Database (COD) on HighScorePlus (HSP) analysis software. Rietveld method was used for phase identification that analyses and separate peaks overlapping diffraction patterns and complex. The principle Rietveld analysis method is to match the peak profile calculation of the peak profile observations.

The Liquid Chromatography-Mass Spectrometer (LC-MS) analysis was using LC-MS Waters Quarto Micro Triple quadrupole with a retention time of 45 minutes and an analysis of a mass range of 500 to 1200 g/mol. The column temperature and sample were constant at 40 °C and 20 °C. The mobile phase used consisted of 90% water and 10% methanol with a flow rate of 0.2 ml/ min and a maximum pressure of 300 bar. The preparation of samples for LC-MS analysis were as follows: 500 grams of green coffee powder and roasted coffee were macerated with 150 ml of 96% ethanol solvent for 24 hours, and then it separated. The coffee bleach was then squeezed four times over four days. The obtained filtrate was evaporated using a rotary evaporator at a temperature of 70 °C until a coffee extract was obtained.

To study the effect of metal ions to coffee, three types of metal ions were added to roasted liberica coffee at 200 °C. The metal ions obtained from CaCl2, MgCl2 and FeSO4.7H2O (Merck). Each metal (0.8 grams) dissolved in 4 ml of distilled water and then mixed into a coffee of liberica (2 grams). Mixture is heated at 75 °C and oven until dry coffee powder at 100 °C. Samples were characterized using XRD.

3. Results and discussion

3.1 Roasting process

Roasting coffee is influenced temperature and time. The roasting profiles of arabica, robusta, and liberica coffee beans with the varied temperature 200 °C and 230 °C is presented in Figure 1. The roasting profiles of arabica, robusta, and liberica coffee beans show that there is no significant difference for all type of coffee during the roasting process. It means that the three types of coffees have same characteristic. In the beginning of the roasting process, the decrease in temperature indicates the occurrence of dehydration process in coffee. This dehydration process is the process of evaporation of water content in coffee. During the roasting process, the colour of the coffee will gradually change, from grey, greenish to brown, dark brown, and black. After the dehydration process, the roasting temperature rises steadily up to a temperature of 210 °C. The rise in temperature explains the occurrence of chemical and physical changes during roasting.

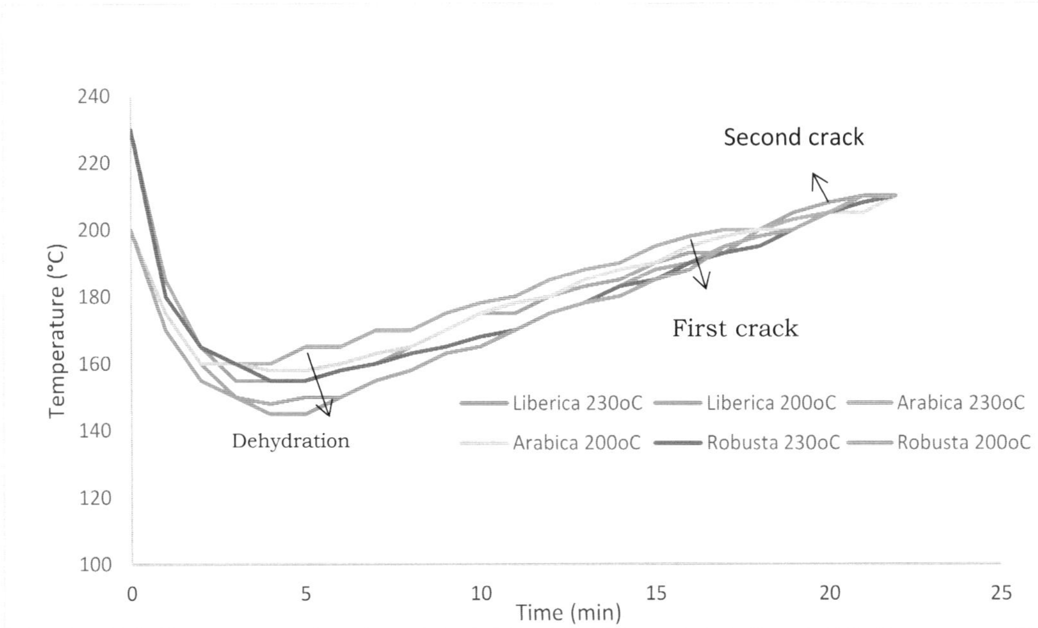

Figure 1. Roasting profile of Arabica, Robusta, and Liberica coffee.

The first stage was called the first crack that occurs in the 16th minute, temperatures around 190°C where the colour changes from yellow to brown, the beans sizes were doubled, and lose weight around 5%. The first cracking occurs at 6 minutes at 207 °C and the second cracking at 10 minutes with a temperature of 219 °C [8]. The first crack occurs due to the amount of gas that is formed that causes cracks in the cell wall of cellulose [13]. The occurrence of first crack signifies 75-80% of the total roasting time. The final stage in the roasting process was the second crack characterized by the colour change to dark brown and oily in coffee beans due to the breakdown of cellulose in the cell walls of coffee beans. The second crack occurs in the 21st minute, at temperatures around 200 °C. The roasting process causes the build-up of pressure inside the coffee bean due to the formation of the vapour and other gases, the second crack occurs when the pressure accumulation exceeds the strength of the cellulose wall [13]. The second crack occurs when the CO_2 pressure formed in the Stracke degradation exceeds the strength of the cellulose wall in the seed [2]. Loss of CO_2 indicates that excess sugar molecules begin to oxidize (caramelized reactions), producing non-cyclic aldehydes, ketones, and acids. As temperatures continue to rise, the organic molecules undergo thermal degradation that result in a chemical changes of coffee. The existence of the second crack was often used as a reference to the end of the roasting process in coffee beans.

3.2 DSC analysis

DSC was used to analyse thermal behaviour of green coffee beans and roasted changes under optimum and best auto hydrolysis condition. DSC can be used to determine the heat capacity and enthalpy of the materials. Figures 2 and Table 1 show a comparison of the DSC curve of three types of coffees. In general, all three types of coffee have similar DSC curves, except green liberica coffee. The green liberica coffee has two endothermic peaks, will be explained later. From the figure, the transition glass

(Tg) for all coffees did not have much difference, between -5.77 to -4.15 °C. In this condition, the amorphous portion of the crystalline polymer switches from the brittle/rigid to the soft/plastic state. This condition occurs due to the change of amorphous solid temperature in the form of heat capacity, but no phase change. The low value of Tg on the amorphous glucose due to the presence of fragmentation products with small molecular weight [14].

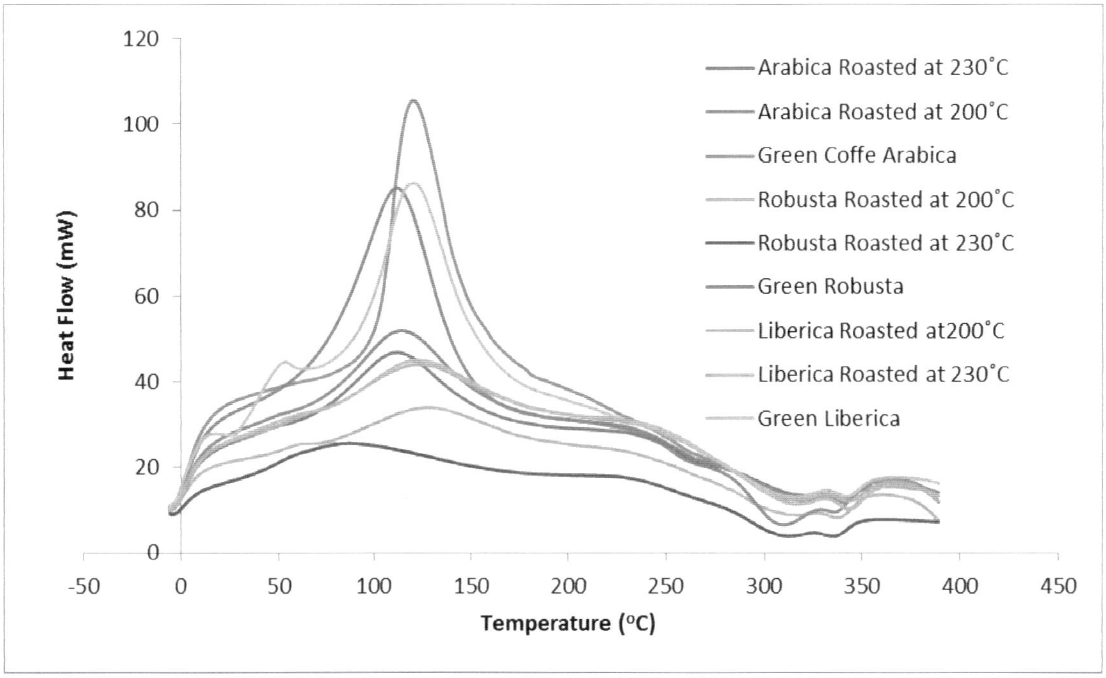

Figure 2. The DSC Curve of Arabica, Robusta, and Liberica Coffee.

For crystallization temperature (T_c), there was a considerable difference of green compared to roast coffee for all types of coffees. The green coffee has higher Tc compared to roasted coffee, which were 120.05 °C, 111.38 °C, and 120.25 °C for arabica, robusta, and liberica, respectively. The value of Tc explains about the transition process of the material from amorphous to crystal phase, known as an exothermic process. The high exothermic peak values possessed by green coffee compared due to lots of dirt and water (hydrophilic groups) contains in the coffee. The moisture content of arabica coffee was higher than robusta coffee, which is 69.7% and 60.0%, respectively [10]. The exothermic peak of roasted coffee have associated the loss of mass about 20% due to the depolymerisation reaction [4], breaking bonds, a release of some volatile compounds, and evaporation of water contains in the coffee [12].

Table 1. DSC analysis parameter of Arabica, Robusta, and Liberica coffee

Samples	T_g (°C)	T_c (°C)	T_m (°C)	ΔH_m (Jg^{-1})
Green Arabica	-4.15	120.05	388	1071.265
Roasted Arabica 200°C	-5.74	113.93	342	612.205
Roasted Arabica 230°C	-5.72	111.38	344	750.229
Green Robusta	-5.71	111.82	320	875.140
Roasted Robusta 200°C	-5.73	111.95	340	810.780
Roasted Robusta 230°C	-5.74	88.080	344	445.540
Green Liberica	-5.68	120.25	325	886.216
Roasted Liberica 200°C	-5.77	123.63	344	850.129
Roasted Liberica 230°C	-5.68	122.01	343	818.378

As for the value of T_m, the temperature of the melted crystals is called the endothermic transition. The Tm values of the three types of coffees have no much difference, except for green arabica and liberica. Green arabica coffee has the highest Tm value of 388 °C with the enthalpy of melting of 1071.265 Jg-1. The high Tm value means the high content of impurities and water in the arabica coffee. The lowest enthalpy of melting possessed by roasted robusta 230 °C is 445.540 Jg-1. The results obtained relate to the thermal decomposition of the sample due to the presence of impurities and water (hydrophilic groups) in the sample and the crystalline nature of the materials [15][9]. The endothermic transition shows the decomposition of compounds containing carbon, hydrogen and oxygen. The different phenomenon occurs on green liberica coffee, as two endothermic peaks appear. The first endothermic peak was at 63 °C, which explains that there was evaporation of water in green coffee [12]. The second endothermic peak was at 325 °C and an associated to melting process with an enthalpy change of 886.21 Jg-1. While both types of roasted coffee, the DSC curve was relatively similar. The emergence of endothermic peaks in coffee was associated with impurities and the evaporation of water in the sample (indicating the presence of hydrophilic groups) [12]. The melting temperature indicates the melting of some constituents of coffee such as amino acids, lipids and sugars such as sucrose, glucose, fructose, arabinose, galactose, maltose and polysaccharides [4].

3.3 XRD analysis
The results of x-ray spectra of green and roasted coffee for the three types of coffee (arabica, robusta, and liberica) can be seen in Figure 3. The XRD spectra for all types green coffees and roasted coffee have the same trend, it means arabica, robusta, and liberica coffee has the same characteristic. It appears that green coffee for all types of coffee does not have a peak (Figure 3.a) as in coffee that has been roasted (Figure 3.b). The green coffee diffraction patterns show an amorphous pattern. While coffee beans undergo thermal treatment (roasting), the XRD spectra of roasted coffee for the three types of coffees showed a peak. The coffee that exposed to thermal treatment has the change in the distribution of the crystalline phases [11]. The thermal treatment on the coffee beans causes the change at least a part of the crystallinity observed in SCG structure due to an elimination of some water molecules incorporated into the crystal fraction, transforming some α-polymorph to β-crystal phase structures [11], [12]. The amorphous phase changes to semi-crystal in XRD results in accordance with DSC where the DSC results show a large Tm point and energy change for the melting process (ΔH_m) in green coffee associated with the water elimination process (Table 1). The value of ΔH_m of green coffee for the Arabica, robusta, and liberica are higher than roasted coffee.

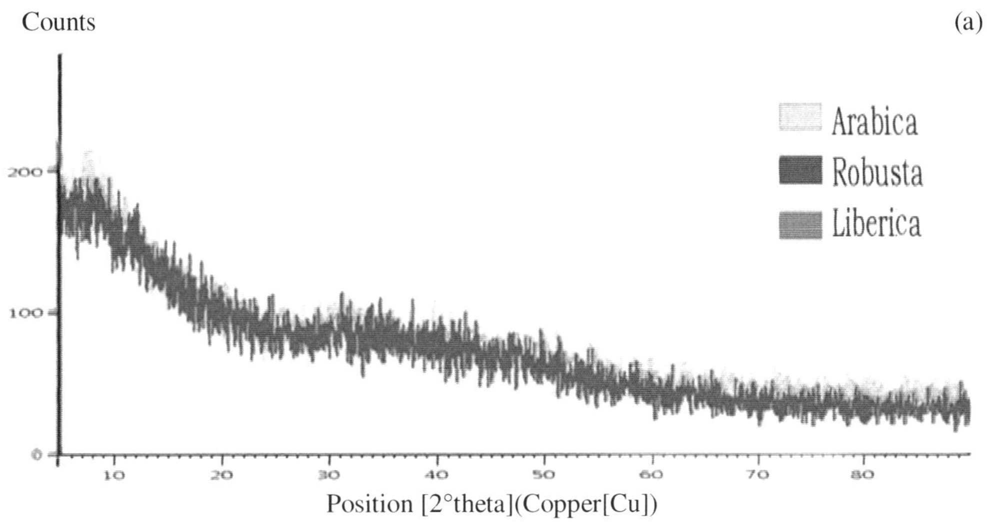

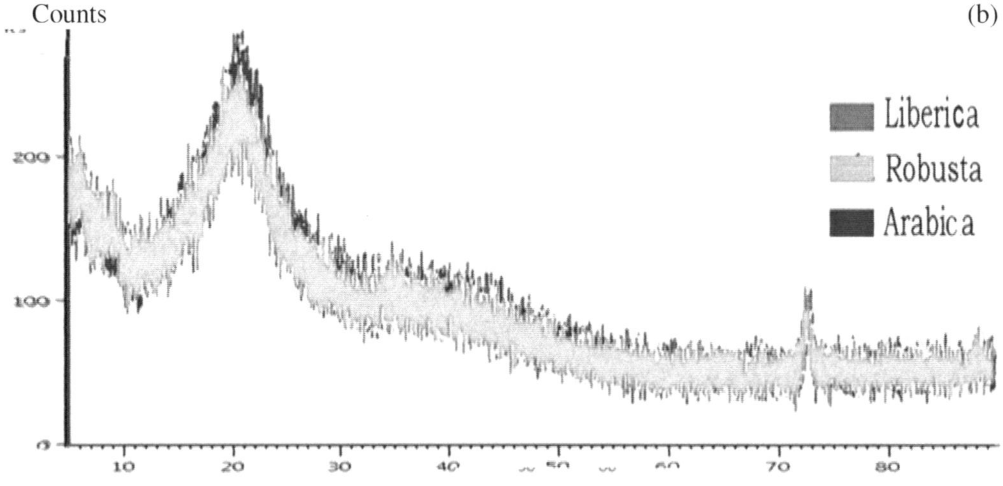

Figure 3. XRD Spectra of arabica, robusta, and liberica coffee (a) green (b) roasted at 200°C

In arabica coffee, the peak was identified at 2θ = 18.664° with d-spacing 4.7542 A and intensity of 100% and hkl = (111). In addition, there were two more peaks at 2θ = 11.5821° and 24.8565°. For robusta coffee, at 2θ = 24,0227° with d-spacing 4.1708 A and its intensity is 38.09% and hkl = (211). In the XRD results for liberica coffee, the peak appeared at 2θ = 18.737° with d-spacing 4.7319 A and the intensity of 90.9% and hkl = (111), and the other two peaks at 2θ = 11.6648° and 24.8401°. The peaks that appear for all roasting condition of the three types of coffees corresponds to the peaks possessed by the crystalline sucrose, which are 2θ = 11.4426°, 18.842°, and 24.773° (ICDD 00-024-1977). According to ICDD 00-024-1977, the crystal structure of sucrose is crystal sucrose with the monoclinic crystal system.

The comparison XRD spectra of green and roasted coffee with different roasting temperature, namely: 200 °C and 230 °C for arabica, robusta, and liberica coffee is shown in Figure 4 and the XRD data of 2θ and the intensity of the peak are presented in Table 2.

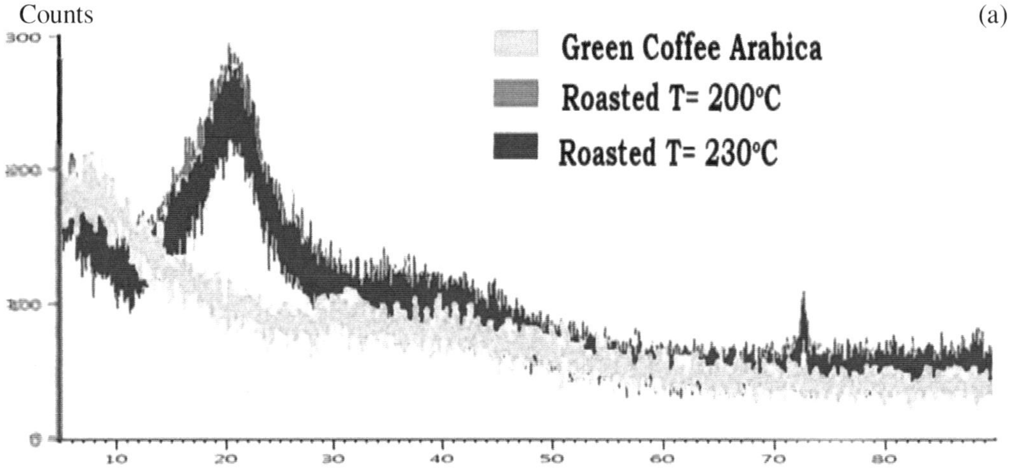

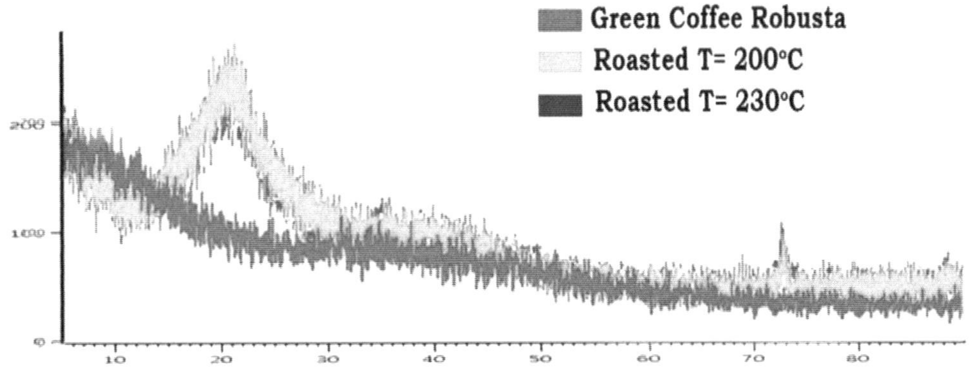

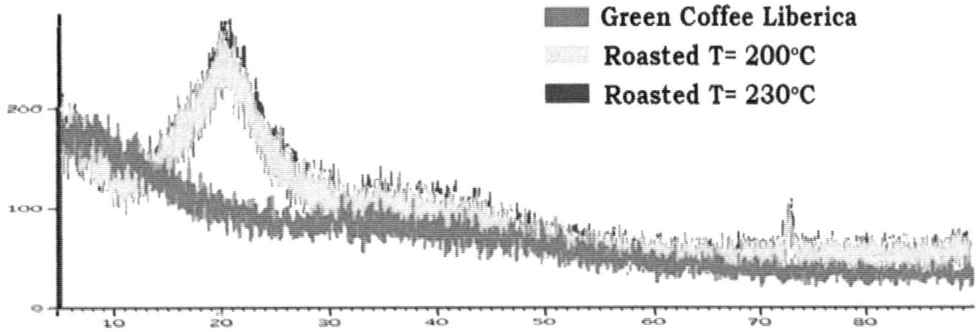

Figure 4. XRD spectra of green and roasted coffee (a) Arabica (b) Robusta (c) Liberica.

The three spectra show the trends are similar for all types of coffees. The green coffee, without thermal treatment, was in the amorphous phase, while the roasted coffees have amorphous and crystalline region. The significant difference of spectra is evidence of the change in the distribution of the amorphous to crystalline phases when the coffee is exposed to thermal treatment [11]. The thermal heating induces the formation of a crystal in the roasted coffee. It is known that the change in hydrated caffeine by the heating process is due to a water molecule incorporated into its crystal structure [11]. The thermal heating with temperature variations of 200 °C and 230 °C cause slightly change of 2θ (Table 2). Caffeine in the form of a-polymorph transforms into a b-crystal phase at 141°C and remains in this phase until melting process at 236 °C [16]. The crystal structure of sucrose hydrates is seen in Figure 5.

Table 2. The XRD data analysis of Arabica, Robusta, and Liberica roasted coffee at 200°C and 230°C

Type of coffee	200°c		230°C	
	2θ	Intensity (%)	2θ	Intensity (%)
Arabica	11.582	23.9	11.518	12.1
	18.664	100	18.265	64.9
	24.856	44.2	24.42	23.3
Robusta	11.442	25	11.805	26.5
	19.285	79.8	19.262	100
	24.022	38.09	24.001	32.1
Liberica	11.664	16.4	11.735	5.55
	18.737	90.99	18.149	72.3
	24.84	31.3	24.523	3.629

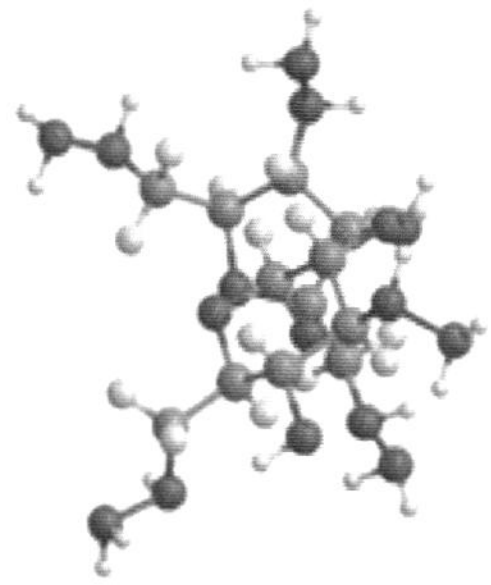

Figure 5. Sucrose hydrates crystal system.

3.4 LC-MS analysis

The LC-MS spectrum of liberica coffee is shown in Figure 6 and the data m/z is tabulated in Table 3. Caffeine compounds were identified in all coffee samples, either green or roasted green coffee. Caffeine compounds appear on m/z between 194-195 gmol^{-1}. Caffeine appears either on or after roasting green coffee because caffeine is stable to heat. Caffeine does not significantly change during roasting, but may slightly decrease due to sublimation. For the other chemical compound, there is chemical change due to the thermal treatment of the coffee.

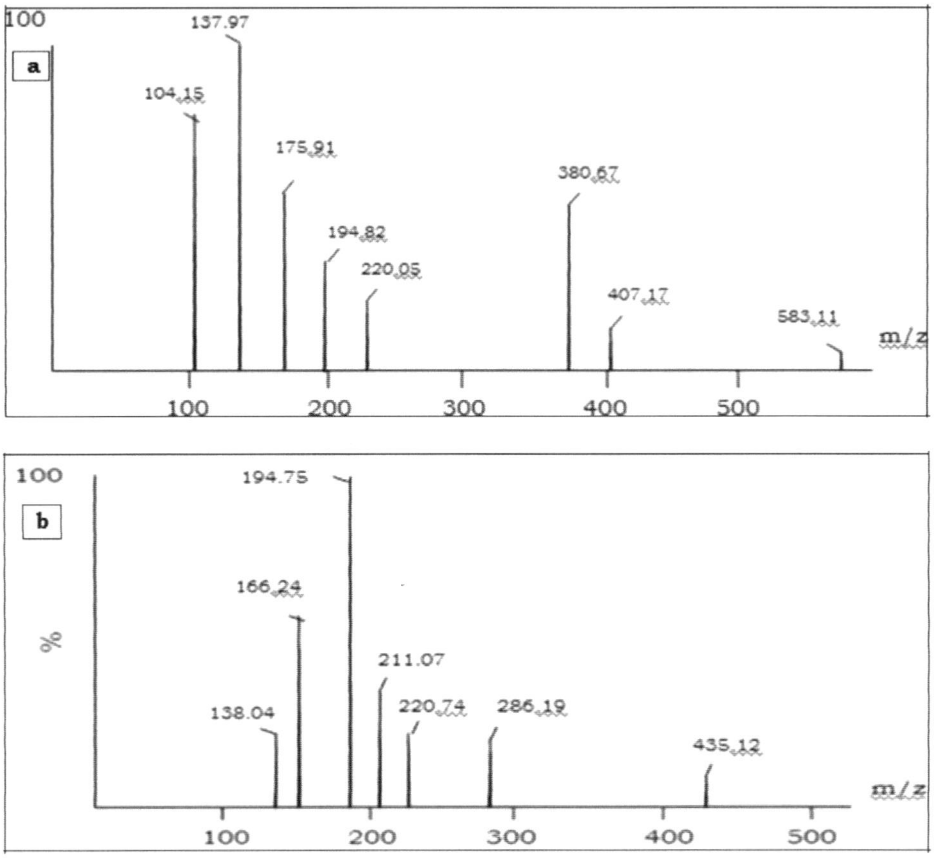

Figure 6. LC-MS spectra of liberica coffee (a) green (b) roasted at 200 °C.

Table 3. LC-MS data of Liberica Coffee

Compound	Formula	m/z	
		Green	Roasted 200 °C
Choline	$C_5H_{14}NO$	104.15	-
Trigonelline	$C_7H_7NO_2$	137.97	138.04
Caffeine	$C_8H_{10}N_4O_2$	194.82	194.75
Feruloylquinic acid	$C_{17}H_{20}O_9$	407.17	-
Sucrose hydrates	$C_{18}H_{20}O_{10}$	-	435.13
Diferuloylquinic acid	$C_{27}H_{28}O_{12}$	583.11	-

Choline, trigonelline, and caffeine are components of nitrogen compounds in addition to proteins and amino acids that are non-volatile compounds so they are not lost during roasting [2]. Choline ($C_5H_{14}NO$) was identified only in green liberica coffee at m/z 104.15. Trigonelline is a pyridine derivative that contributes to the aroma and bitter taste of coffee drinks. Trigonelin ($C_6H_5NO_2$) with a molecular weight of 123.111 gmol^{-1} was identified both in green and roasted liberica coffee at 137.97 and 138.04, respectively. Trigonelline will be identified on m/z 138 as $[M + H]^+$ [17]. Caffeine is one of the most important compounds in the coffee that determines the quality of coffee. In theory, caffeine has a molecular weight of 195.19 m/z. The results of this study indicate the presence of alleged caffeine with a near-theoretical molecular weight, where molecular weight in green liberica coffee and after roasting 200 ^{0}C is 194.82 m/z and 194.75 m/z. caffeine is detected at molecular weight of 195.08 m/z [18].

Chlorogenic acid ($C_{16}H_{18}NO_9$) is an ester group formed from a combination of quinic acid and some trans cinnamic acids, generally caffeic, p-coumaric and ferulic acid [19]. Chlorogenic acid can protect coffee plants from microorganisms, insects and UV radiation. While the benefits of chlorogenic acid for human health as an antioxidant, antiviral, hepatoprotective, and act as anti-spasmodic [5]. Chlorogenic acid in coffee can be found in the form of caffeoylquinic (CQA), pcoumaroylquinic (pCoQA), feruloylquinic acid (FQA), dicaffeoylquinic (diCQA) and caffeoylferuloylquinic acid (CFQA) [20]. In this study, chlorogenic acid is present in feruloylquinic form acid and diferuloylquinic acid in Robusta coffee extract [18], where the highest intensities are 406.73 m/z and 583.11 m/z respectively.

Sucrose hydrates identified only in roasted coffee with m/z 435.13. The fructose/glucose reacts with amino acids during the roasting process to form volatile compounds and melanoidin complexes [13].

3.5. Addition of metal

It well is known that coffee as natural antioxidant sources [11], [12]. As preliminary study pharmacological effects of coffee, three types of metals have been added to the liberica-roasted coffee, which is Fe^{2+}, Mg^{2+}, and Ca^{2+}. Iron sucrose complex (ISC) is effective in increasing haemoglobin, serum ferritin, and other haematological parameters on therapy for moderate anaemia in pregnancy [21]. Figure 7 and Table 4 shows the XRD spectra of addition metals on roasted liberica coffee. For the addition of three metals, the peak only appears in addition of Fe^{2+} with the activity of 2θ = 18,272°, d-spacing = 4.8553 A, and Intensity of 62%. The preliminary study can be predicted as an iron sucrose complex (ICD). While for Mg^{2+}, and Ca^{2+}, no sucrose peaks formed. It can be suggested for LCMS to make sure that ICD was observed.

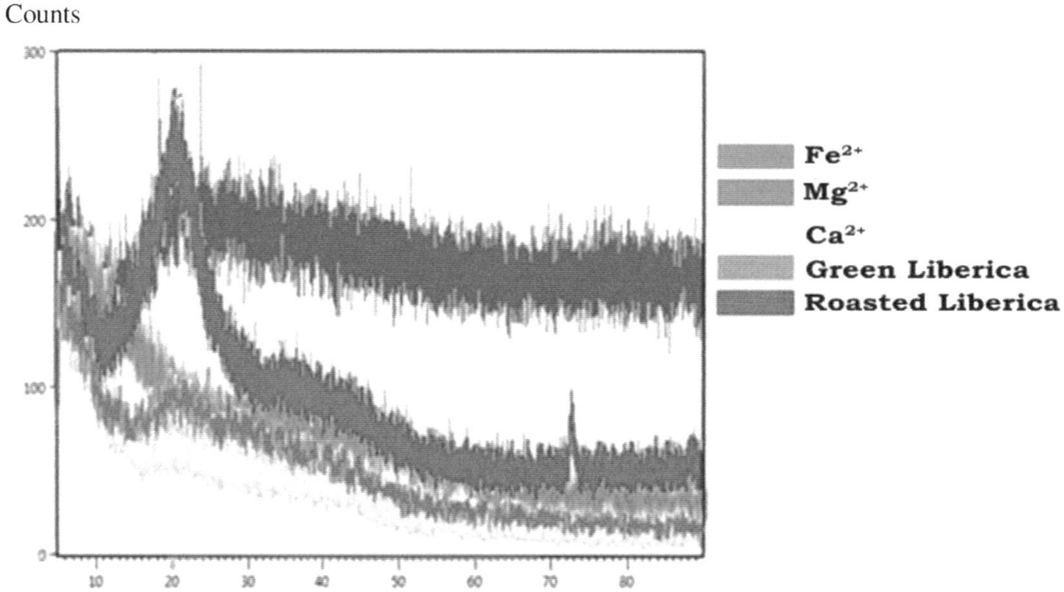

Figure 7. XRD Spectra of metal addition to roasted liberica coffee at 200°C (a) Ca^{2+} (b) Mg^{2+} (c) Fe^{2+}

Table 4. Comparison of XRD data of metal addition to roasted liberica coffee at 200 °C

	2θ(°)	
Without metal ion added	Fe^{2+} ion	Ca^{2+} ion
6.228	6.221	6.387
11.664	9.566	9.259
18.737	18.272	44.680
24.84	20.873	50.123
72.692	23.759	
88.316	27.551	
	41.912	

4. Conclusion

Thermal behaviour of arabica, robusta, and liberica green coffee associated with a glass transition around -5.77 to -4.15 °C for green and roasted coffee. The high exothermic peak of (Tc) appears for all type green coffee indicates that the high water and volatile compound content in the coffee for the occurrence of thermal decomposition processes in coffee. The green Arabica coffee has the highest exothermic peaks at 120.05 °C. The green liberica coffee has two endothermic peaks associated to the evaporation of water and the melting of some constituents of coffee such as amino acids, lipids and sugars such as sucrose, glucose, fructose, arabinose, galactose, maltose and polysaccharides. The green Arabica has the highest melting temperature at 388°C with energy change for melting process 1071.265 Jg-1. We have established experimentally by XRD that there is an increase of a crystalline phase of the sucrose with activity in 2θ = 18° with monoclinic crystal system for all types of roasted coffee. Addition of iron ion to the roasted liberica coffee forms iron sucrose complex that identified at 2θ = 18.272°. While the addition of Mg^{2+} and Ca^{2+} are not identified any peaks of sucrose.

References

[1] Calligaris S, Munari M, Arrighetti G and Barba L 2009 *J. Lipid Sci. Technol.* **111** 1270–1277
[2] Farah A 2012 2 Coffee Constituents *Emerg. Health Effects Dis. Prevent.* (United Kingdom:

Elsevier Inc.) p. 22-58.

[3] Franca A S, Mendonca J C and Oliveira S D 2005 *Swiss Soc. Food Sci. Technol.* **38** 709-715

[4] Brondi A M, Torres C, Garcia J S and Trevisan M G 2017 *J. Braz. Chem. Soc.* **28** 7 1308-1314

[5] Farah A and Donangelo C M 2006 *Braz. J. Plant Physiol* **18** 123-36

[6] Davis A, Chester M, Maurin O and Fay M 2007 *Am. J. Bot.* **94** 3 313-329

[7] Patui S, Clincon L, Peresson C, Zancani M, Conte L, Terra L D, Navarini L,Vianello A and Braidot E 2014 *Plant Sci.* **219-220** 19-25

[8] Edzuan A M F, Aliah A M N and Bong H L 2015 *Am. J. Chem.* **5** 56-60

[9] Saw A K C, Yam W, Wong K C and Lai C S 2015 *J. Essent. Oil Bear. Plants* **18** 64-73

[10] Keller G, Lavigne F, Forte L, Andrieux K, Dahim M, Lisel C, Ollivon M, Bourgaux C and Lesieur P 1998 *J. of Thermal Anal.* **51** 783-784

[11] Rivera W, Velasco X, Gálvez C, Rincón and Arango P 2011 *Ital. Oral Surg.* **1** 385-390

[12] Ballesteros L F, Silverskin C, Teixeira J A and Mussatto S I 2014 *Food Bioprocess Technol.* **7** 3493–3503

[13] Wang N and Lim L 2012 *J. Agric. Food Chem.* **60** 5446–5453

[14] Lee J W, Avenue S G, States U, Thomas L C and Schmidt S J *J. Agric. Food Chem.* **59** 3311-3319

[15] Elmer P 2014 *Guide to Selection of Differential Scanning Calorimetry (DSC) Sample Pans* (USA: PerkinElmer, Inc.)

[16] Emel'yanenko V N and Verevkin S P 2008 *J. Chem. Thermodyn.* **40** 12 1661-1665

[17] Stadler R H, Varga N, Hau J, Vera F A and Welti D H 2002 *J. Agric. Food Chem.* **50** 1192-1199

[18] Garrett R, Vaz B G, Hovell A M C, Eberlin M N and Rezende C M 2012 *J. Agric. Food Chem.*, **60** 4253-4258

[19] Monteiro M, Farah A, Perrone D, Trugo L C and Donangelo C 2007 *J. Nutr.* 2196-2201

[20] Clifford M, Johnston K, Knight S and Kuhnert N 2003 *J. Agric. Food Chem.* **51** 1 2900-2911

[21] Al Momen A K, Al Meshari A A, Al Nuaim L, Saddique A A, Abotalib Z, Khashogji T, Abbas M 1996 *Eur. J. Obstet. Gynecol. Reprod. Biol.* **69** 121-124

IOP Conf. Series: Materials Science and Engineering **345** (2018) 012022 doi:10.1088/1757-899X/345/1/012022

Extraction of citral oil from lemongrass (*Cymbopogon Citratus*) by steam-water distillation technique

P N Alam, H Husin, T M Asnawi and Adisalamun
Department of Chemical Engineering, Syiah Kuala University, Darussalam, 23111, Banda Aceh, Indonesia

E-mail: pocutnurulalam@unsyiah.ac.id

Abstract. In Indonesia, production of citral oil from lemon grass (*Cymbopogon Cytratus*) is done by a traditional technique whereby a low yield results. To improve the yield, an appropriate extraction technology is required. In this research, a steam-water distillation technique was applied to extract the essential oil from the lemongrass. The effects of sample particle size and bed volume on yield and quality of citral oil produced were investigated. The drying and refining time of 2 hours were used as fixed variables. This research results that minimum citral oil yield of 0.53% was obtained on sample particle size of 3 cm and bed volume of 80%, whereas the maximum yield of 1.95% on sample particle size of 15 cm and bed volume of 40%. The lowest specific gravity of 0.80 and the highest specific gravity of 0.905 were obtained on sample particle size of 8 cm with bed volume of 80% and particle size of 12 cm with bed volume of 70%, respectively. The lowest refractive index of 1.480 and the highest refractive index of 1.495 were obtained on sample particle size of 8 cm with bed volume of 70% and sample particle size of 15 cm with bed volume of 40%, respectively. The solubility of the produced citral oil in alcohol was 70% in ratio of 1:1, and the citral oil concentration obtained was around 79%.

1. Introduction

Indonesia is a country famous for its essential oil potential. Lemon grass (*Cymbopogon Cytratus*) is a plant that could be processed to produce essential oils. Oil from lemongrass is one of the most important and widely used essential oils to produce citral oil, which is one of major components in essential oil. Citral oil is widely used in cosmetic flavors and as a raw materials for the manufacture of ionons (a synthetic liquid aromatic compound). Besides, due to its strong lemon odor, lemon grass is often used as fragrant in soap, detergent and various technical products. As medicinal materials, essential oils used in industry and trade must meet the requirements as stated in official book [1].

Botanically, lemon grass falls in family of grasses (*Poaceae*), cultivated for its leaves as traditional herbal medicine, spices or refined to obtain its essential oil. Despite of aforementioned benefits, the utilization of lemon grass oil is relatively small compared to other essential oils derived from patchouli oil, ylang, nutmeg, cloves, sandalwood, eucalyptus, fragrant root, ginger and cubes that have been cultivated and used as a source of essential oil commercially. This fact causes the low production of Indonesian lemongrass oil [1-3].

Currently, the price of raw lemon grass in the market is around Rp 60.000/kg, while its extracted essential oil (*with a simple process*) reaches Rp 60.000/10 ml. This significant margin also occurs in other volatile production. The amount of selling value is certainly not very profitable if the post-harvest lemon grass directly sold to market as raw material. On the other hand, this condition will open up business opportunities in the industrial sector of processed products of lemon grass.

In general, the physical properties of chemistry of essential oils vary, so difficult to obtain a fixed quality. This is caused by many factors, including post-harvest handling of the plant, such as on the patchouli leaves are fermented before the essential oil is distilled. Another factor is caused by pretreatment of raw materials prior to distillation. Lemongrass leaves are usually distilled when they are fresh, but when large harvests of newly cut or trimmed clumps cannot be immediately distilled, so as a clump of crops spread over the field or left buried in a room that causes the clump to become semi-dried or wilted. This stockpiling is often done due to the lack of distillation equipment impediments or the absence of a distillation plant that can accommodate large quantities [1]. The aim of this research is to improve the rendement and quality of lemongrass oil by varying the effect of particle size and bed volume.

The objective of this research is to improve the distillation technique of lemongrass oil for enhancement of its yield and quality. It is expected that this research can help lemongrass farmers in Indonesia especially in Aceh to increase its selling price.

2. Materials and methods

2.1. Materials
In this work, leaves of the lemongrass, stem, ethanol 70% as solvent, $Na_2SO_4.H_2O$ and filter paper, erlenmeyer, beaker, measuring cup, burette, picnometer, water bath, spatula, raction tube, funnel, separator funnel, dropper pipette, volume pipette, scales and refractometer. In the steam distillation process the drying and distillation times are determined for 24 hours and 2 hours, respectively, for the bed volume and size varied as mentioned in the working procedure.

2.2. Methods
A quarter volume of the steam boiler filled with water and heated electrical steam. A 2: 1 proportions of mixed lemongrass leaf and stem put into the distillation kettle equipped with grid where the material was placed. Set the proportion with various density according to the varied. Flow the steam produced by direct contact with raw material to the shelter through the distillate pipe. Drain the oil steam mixture out of the distilled boiler to the condensate through the steam distillation pipe. Accommodated distillate produced on the column of essential oil storage. Repeat the above work by varying the volume of the bed (40%, 60%, 70%, 80%) and sample particle size (3 cm, 5 cm, 8 cm, 12 cm, 15 cm).

3. Results and discussion

3.1. Effect of packed bed volume and particle size on citral oil rendement
Distillation of lemongrass oil used bed volume with variation 40%, 60%, 70% and 80%. The effect of bed volume and sample size on citral oil yield as can be seen in Figure 1. It shows that with the increase of bed volume, the yield of lemon grass oil is less due to the larger volume of the lemongrass bed the harder the vapour produced from the distillation boiler to exstract its essential oil components. The vapors will be difficult to carry out the volatile oil components due to the density of the bed, and low vapour pressure generated from the distillation boiler. In Figure 1 it can be seen that the highest yield is on the sample particle size of 15 cm and the volume of bed 40% is 1.95%. In addition, there are several factors in determining the yield of lemon grass oil produced, such as the timber which has exceeded the time specified in this study it was 24 hours. As a consequence, many components of volatile oil that evaporated due to the timber. In addition, at the time of chopping will also affect the results, inappropriate cooking method also will damage the lemongrass so that the essential oil components contained will be more volatile before refining..

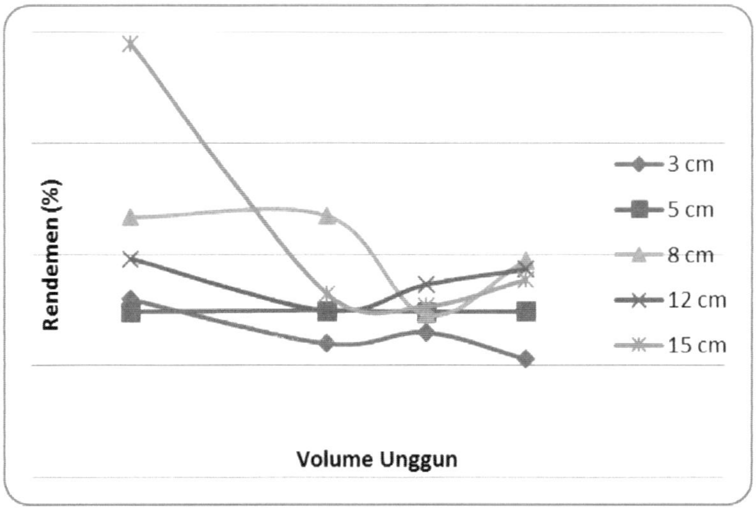

Figure 1. Influence of bed volume and sample size to yield.

3.2. Density

Weight values are shown in Figure 2 below for all bed volumes. It can be seen that at 70% bed volume with sample size of 3 cm, the specific gravity of 0.902 gram / mL, the sample size of 5 cm by 0.873 g / mL, by sample size 8 cm by weight of the type achieved 0.897 gram / mL, the sample size is 12 cm, the specific gravity is 0.905 gram / mL, and the sample size is 15 cm by 0.892 gram / mL. The specific gravity of kitchen sereh oil according to EOA standard ranges from 0.869 to 0.894. There are some samples of specific gravity that do not meet EOA standards, this is due to the presence of a mild fraction contained in lemon grass oil. It is also because at the time of distillation is not much fraction of coconut oil oil that distilled perfect. The weight fraction is referred to as sitral. As the sample size increases, the resulting density is lower. On the other hand, there are several factors that influence the low value of the weight of a component, such as the type and number of functional groups, the number and position of double bonds, the number of carbon atoms and others [4].

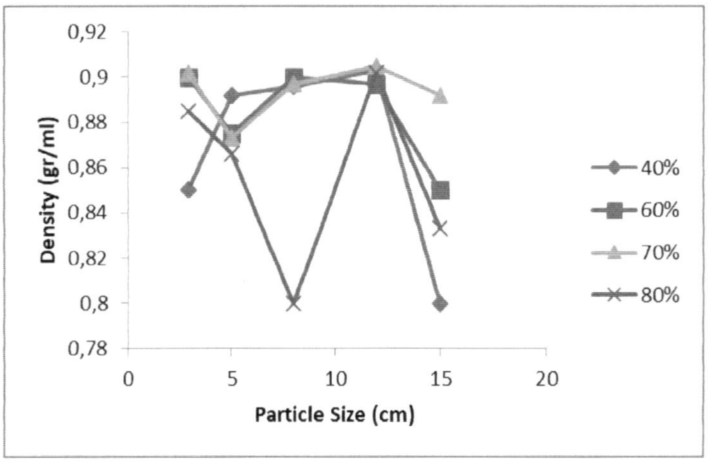

Figure 2. Effects of particle size and volume on density.

In Figure 2 it can also be seen that the lower volume of the essential oil bed of lemon grass is lower then the value of the specific gravity to be obtained on the lemon essential oil. This is due to the less volume of the lemon grass oil bed, the less weight or citral fraction can be distilled from the lemon grass oil. In Figure 2 it is found that the optimum density of the lemongrass oil is at 70% bed volume.

3.3. Refractive index

If a monochromatic light passes through a medium (A) to a more densely packed medium (B), then the change in velocity and refraction of the light approaches the normal line, or the angle of the incident ray (iA) greater than the angle of the bias ray (iB). The comparison of the incident angle of light with this refracted beam angle is called the refractive index. In this study to determine the refractive index of an oil sample used Abbe Refractometer. Figure 3 is an acquisition of refractive index for each lemon grass sample of at each bed volume.

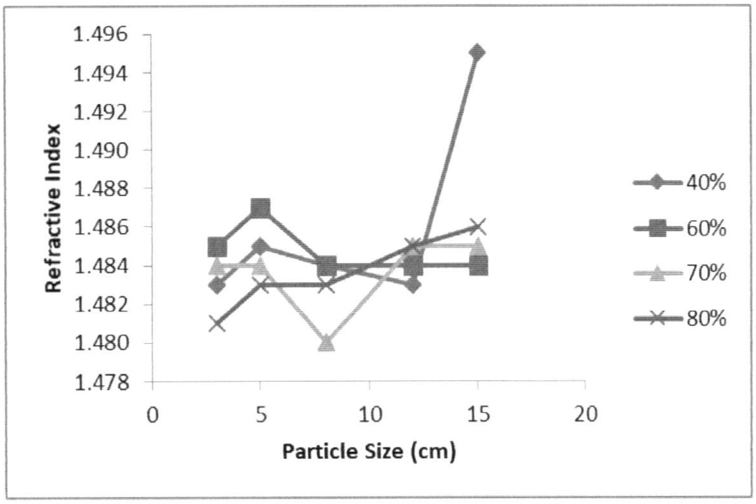

Figure 3. Effect of particle size and volume on refractive index.

Figure 3 shows that the refractive index obtained ranged from 1.480 to 1.495 for all samples. The refractive index obtained largely meets the established standards of EOA of 1.4830-1.4890. However, there are some that are still below and above the EOA provisions such as the lemon oil oil refractive index with sample size 15 cm and 40% bed volume is 1.495, sample size 8 cm and 70% bed volume is 1.480, sample size 3 cm and bed volume 80% is 1.481. This happens because there is still water contained in the lemon grass. The more water content, the refractive index value will exceed the predefined EOA standard. This is because the nature of the water is easy to refract the incoming light, but if the refractive index value is smaller below the EOA standard this means that the citric content contained in the lemon grass is little. From the results of the research obtained generally the refractive index obtained has met the EOA (Essential Oil of America) standard, only a few samples that do not meet the EOA standard due to the presence of water contained in the lemon grass. The high value of the refractive index of a material is one of the parameters to determine the purity of the material is no longer guaranteed, in other words the material has been mixed with other materials. In addition, chemical changes such as the occurrence of resinification due to polymerization, hydrogenation or isomerization of a substance can cause changes in the refractive index values [4].

3.4. Solubility of citronella oil kitchen in ethanol 70%

Since many essential oils dissolve in alcohol and are rarely water soluble, the solubility can be readily known by using alcohol at various concentrations. Determining the solubility of the oil depends also on the speed of solubility and the quality of the oil. Oxygenated oils are usually more soluble in alcohol than those rich in terpenes. In lower alcohol concentrations, the terpene fraction will separate as is the case with wax. Based on this, this method is widely used for the manufacture of free oil terpen and sesquiterpen, consentrate and extract. Oil solubility may also change due to the influence of oil age. Unfavorable storage conditions can speed up polymerization. Factors such as light, air and the presence of water usually cause unfavorable effects. The concentration of alcohol used to determine the solubility of essential oils is: 50% -60% -70% -80% -90%. In this study, alcohol (ethanol) used as a solvent in the analysis of lemon grass oil quality is 70% ethanol. The solubility of kitchen lemon oil obtained from the research in ethanol 70% still meets the EOA standard, that is, by an average ratio of 1: 1 for each sample of oil tested.

3.5. Gas chromatography

In this study used GC-MS in order to evaluate the components contained in the essential oil of lemon grass. From result of GC-MS analysis for lemongrass oil with particle size 3 cm and bed volume 40% hence obtained level of α-sitral equal to 28,18% and β-citral equal to 43,11%. And at the size of 3 cm and 60% volume of beds obtained α-sitral levels of 27.30% and β-cyral of 42.40%. As well as the size of 3 cm and the volume of bed 80% obtained α-sitral rate of 31.69% and β-cyral of 48.08%.

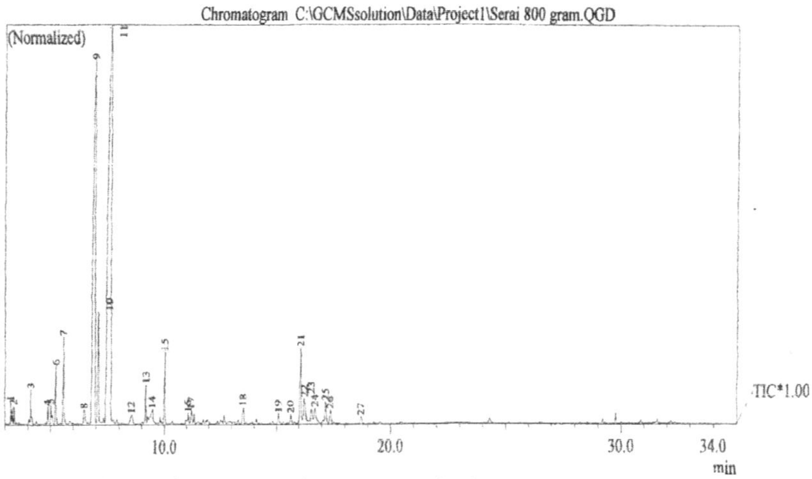

Figure 4. Gas data chromatography lemon grass 800 grams.

3.6. Color

Lemongrass oil obtained from this experiment is dark yellow color. This is in compliance with the EOA (Essential Oil of America) standard. The lemon oil oil color can be seen in Figure 5 below.

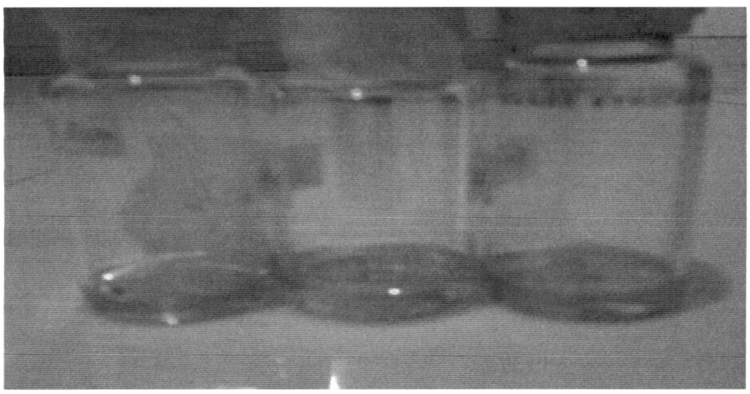

Figure 5. Citronella oil produced.

4. Conclusions

From the experimental study, the maximum yield obtained is at 40% bed volume and sample size 15 cm that is 1.95%. The quality of lemon grass oil obtained meet the EOA (Essential Oil Association) Standards, i.e the specific gravity ranged from 0.80 to 0.903, the refractive index is 1.484 and the solubility in alcohol 70% is average produced by 1:1. Citral oil content obtained on citronella essential oil based on GC-MS analysis of 79% on sample size 3 cm and bed volume of 80% and oil color is old yellow of already meets the EOA quality standards.

Acknowledgement

We gratefully acknowledge the financial support from the Syiah Kuala University Banda Aceh, Republic of Indonesia.

References

[1] Guenther E 1987 *The Essential Oils* vol. 1 (New York: Robert E. Krieger Publishing Company)
[2] Ketaren S 1985 *Pengantar Teknologi Minyak Atsiri* (Jakarta: Penerbit Balai Pustaka)
[3] Harris Ruslan 1987 *Tanaman Minyak Atsiri* (Jakarta: PT Penebar Swandaya)
[4] Fitriana and Djafar 2009 *Ekstraksi Minyak Atsiri dari Jahe dengan Metode Hidrodistilasi Sistem Kabohasi* Tesis Magister Teknik Kimia (Banda Aceh: Universitas Syiah Kuala)

Influence of tannin content in *Terminalia catappa* leaves extracts resulted from maceration extraction on decreasing corrosion rate for mild steel in 1M H_2SO_4

M Pramudita[1,2], Sukirno[1] and M Nasikin[1]
[1] Department of Chemical Engineering, Universitas Indonesia,Depok,16424,Indonesia
[2] Department of Chemical Engineering, University of Sultan Ageng Tirtayasa, Banten, 42124,Indonesia

E-mail: mnasikin@che.ui.ac.id

Abstract. The ability of natural compounds as corrosion inhibitors is necessary to obtain safe corrosion inhibitors for the environment. The tannin compounds derived from plant extract has the ability to decrease the corrosion rate. The purpose of this research is to find the ability of tannin compounds in *Terminalia catappa* leaves extracts to decrease corrosion rate on mild steel. *Terminalia catappa* leaves that have been mashed in ethanol solvent extraction using maceration with the variable time 2.4 and 6 days. Mild steel that has been on the sandpaper and cleaned then soak into the 1 M H2SO4. *Terminalia catappa* leaves extract concentration used is 0, 250, 500, 500, 750, 1000 ppm, the immersion time is 3.6 and 9 hours. Calculating of corrosion rate is used the weight loss method, the analysis of the tannin concentration using GC-MS. The results indicate that highest tannin content equal to 7.23% in 6 days maceration time. The result showed that the corrosion rate was reduced in the presence of tannin content in *Terminalia catappa* leaves extract.

1. Introduction

Mild steel is defined as alloys of iron (Fe) and other alloying elements, such as Cr, Ni, and C, etc. Mild steel has been widely used as the material for fluid transmission pipes, such as gas, water, and oil. However, corrosion has been identified as the main problem of mild steel application.

Corrosion is degradation or deterioration of metals caused by a reaction of materials (commonly metals) with the environment [1]. Corrosion cannot be prevented or stopped and caused damage, especially in the industrial sector. Nevertheless, the addition of corrosion inhibitor is known as a suitable method to control metal corrosion [2].

Corrosion inhibitor is defined as a substance with the ability to decrease corrosion rate when added to an environment in a small concentration [3]. Inhibitor works by precipitating on the surface of protected metals and forming a thin layer that will protect metal from corrosion.

Corrosion inhibitor generally comes from the organic compound and anorganic, containing groups with free electron pairs such as nitrite, chromate, phosphate, urea phenylalanine, imidazoline, and amines compounds. However, synthetic chemicals are toxic and not environmentally friendly [4]. Hence, further research is needed to explore eco-friendly corrosion inhibitor, for instance, natural products inhibitor.

Recently, research regarding natural products as corrosion inhibitor has received much attention. Natural products are not only eco-friendly and ecologically acceptable but also renewable and available abundantly [5]. Natural products notably come from plant extract. Plant extract is renewable sources of materials, readily available and inexpensive. In addition, plant extract is biodegradable and free of both heavy metals and toxic substances [6].

Phytochemicals (i.e.: alkaloids, tannins, glycosides, saponins, flavonoids, polyphenol, etc.) exist abundantly in plant extract. Phytochemicals consist of N, O, P, S and atoms with free electron pair, which can be promoted as ligands that will form complexes with metals [7]. Those are organic compounds and contain heteroatoms in their molecules [8].

Terminalia catappa is plant extract that can be used to produce corrosion inhibitor. It has been reported *Terminalia catappa* leaves embodies several compounds used to inhibit corrosion rate, and tannin, which presents in the high content, is one of them [9].

Tannin is the high potential organic compound to be utilized as corrosion inhibitor. Tannin exhibits ability to form complexes with metals and able to work under low pH environment. In the previous studies regarding mimosa tannin successfully inhibits corrosion rate of steel, mechanism of oxide films stable chelates formation with Fe (III) has been reported [10]. Tannin is polyphenol compounds and usually found in plants with a large structure. These types of plants form macromolecule and contain the considerable amount of hydroxyl (-OH) groups. It is inferred that these groups enable tannin to form complexes with metals [11].

The extraction process was conducted in order to determine tannin content in *Terminalia catappa* leaves. Maceration is employed as extraction method in the present study. Maceration extraction is extraction using a cold solvent (without heating process). An advantage of this method is to minimize the possibility of thermolabile compounds damage. Thus, many compounds can be extracted [12].

Simplisia materials were ground and mixed with extraction materials and kept not exposed to direct light. Maceration durations are varied from 4 – 10 days. It is theoretically impossible for an absolute extraction to occur in maceration process since solvent will reach its saturation point. The high ratio of simplisia to extraction solution, more results can be obtained [13].

The present study is required in order to investigate the influence of tannin in *Terminalia catappa* leaves extract on corrosion rate inhibition. This study is expected to provide important findings for the advancement of *Terminalia catappa* leaves extract as a bio corrosion inhibitor.

2. Materials and methods

2.1. Materials
The material used in this study is mild steel with composition (%wt) : 0.54Mn, 0.05Si, 0.01S,0.01P, 0.16C and the remainding Fe. Coupon (2.5 x 3 x 0.5 cm) were mechanically polished using grid 400# – 1200#, cleaned with acetone and dried. Afterwards, the coupon was weighed to determine initial weight [14].

2.2. Preparation of Terminalia catappa leaves extraction
Terminalia catappa leaves were obtained from University of Indonesia. Prior to maceration process, these leaves were dried, followed by size reduction using the blender. Maceration process was carried out with various duration (2, 4, and 6 days) and ethanol was used as a solvent.

2.3. Solutions
H_2SO_4 1M pa was simulated as corroding medium and corrosion inhibitor was applied in various concentration (0;250;500;750;1000 ppm). Immersion time was varied for 3, 5 and 9 hours [15].

2.4. Weight Loss Technique and analysis

Weight loss technique was applied in the present study. Coupon initial and final weights were accurately weighted before and after immersion in the corrosive medium. From the obtained data, weight loss was determined, and the corrosion rate was calculated as follows [16] :

$$C_{R\,(mmpy)} = \frac{87600 \times \Delta W}{A\rho t} \quad (1)$$

where C_R is corrosion rate; ΔW is weight loss (gram); A is coupon surface area (cm^2); ρ is the density of mild steel ($gram/cm^3$) and t is immersion time (hours).

3. Result and Discussion

3.1 Composition of tannin content in Terminalia catappa leaves in maceration extraction using ethanol as solvent

The influence of *Terminalia catappa* leaves extracts maceration extraction process on tannin content was investigated by various immersion time (2, 4 and 6 days). The GC-MS analysis was conducted in order to identify percentage tannin content from extraction maceration process. Figure 1 shows the relationship between tannin concentration and maceration time.

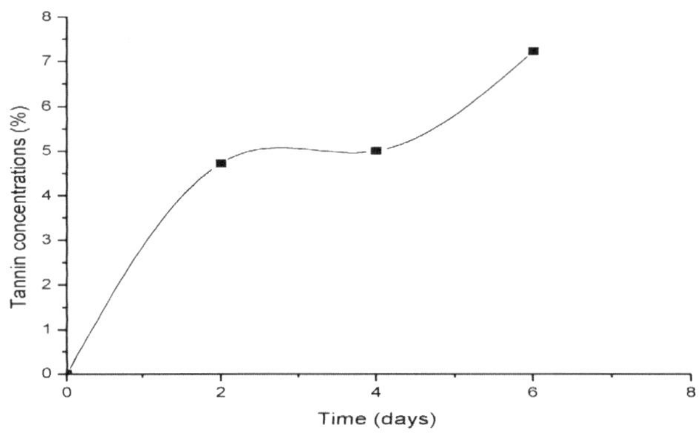

Figure 1. Graph plotting of tannin concentration in *Terminalia catappa* leaves extract vs maceration time.

From GC-MS analysis, tannin content percentages from immersion time of 2, 4 and 6 days in ethanol are 4.73%, 5.01%, and 7.23%, respectively. It is illustrated that the immersion time which obtained the highest tannin content (7.23%) was within 6 days. It is also inferred from Figure 1 that within 6 days, tannin concentration *Terminalia catappa* leaves extract was higher than 4 and 2 days. The ratio of *Terminalia catappa* leaves powder and ethanol were 1:4. Highest tannin content in maceration time of 6 days might be as a result of longer immersion time, which caused more tannin can be extracted from the process. However, it is theoretically impossible for an absolute extraction to occur in maceration process since solvent will reach its saturation point [15].

3.2 Effect of inhibitor concentration on corrosion rate

Corrosion rate was incited by addition of the inhibitor. Corrosion rates of mild steel decrease following the addition of inhibitor at a higher concentration, as depicted in Figure 2. Based on the amount of tannin in *Terminalia catappa* leaves extract, maceration 3 shows lower corrosion rate than maceration 1 and 2. Thus, tannin content in *Terminalia catappa* leaves extracts influenced corrosion rate. In addition, there are several compounds – as impurities - along with tannin in *Terminalia catappa* leaves extract act as corrosion rate inhibitor.

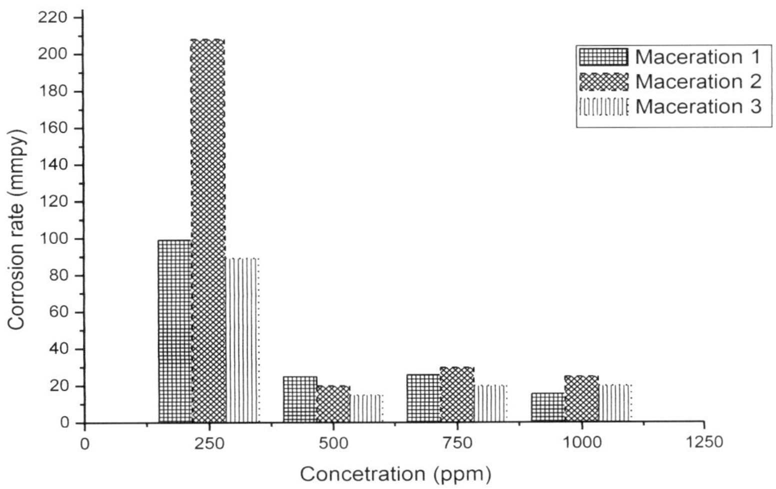

Figure 2. Graph plotting of corrosion rate vs concentration in maceration time 1, 2, and 3.

Addition of inhibitor at higher concentration corrosion rate might be originated from the presence of complex compounds (tannin compound) in inhibitor solution adsorbed on the surface of mild steel and form the protective layer. Hence, this layer plays the significant role in reducing corrosion attack within a short period of time. Corrosion rate reduced in the presence of tannin compounds is adsorbed on mild steel surface. Protection mechanism was carried out by reaction of Fe^{2+} and natural inhibitor results in the formation of complex compounds. Natural extract inhibitor which contains nitrogen act as an electron-pair donor to the surface of mild steel metal, whereas Fe^{2+} ion was diffused in the electrolyte solution, according to the following reaction scheme:

$Fe \rightarrow Fe^{2+} + 2e^-$ (electron donor) and
$Fe^{2+} + 2e^- \rightarrow Fe$ (electron acceptor).
Fe^{2+}+ tannin $\rightarrow$ [Fe(Tannin compound)]$^{2+}$(complex compounds)

Complex compounds formed in the reaction above have high stability, and consequently, mild steel samples with natural extract inhibitor will exhibit high corrosion resistance [16]. This phenomenon is in agreement with theory, which states that efficiency of inhibitor relies on the stability of chelate formation and inhibitor molecules should be occupied with the ability of its core to bond with the metal surface through electron transfer [17].

3.3 Effect of mild steel immersion time in 1 M H_2SO_4 on corrosion rate

The relationship between immersion time and corrosion rate was plotted into the graph to observe the influence of immersion time on corrosion rate.

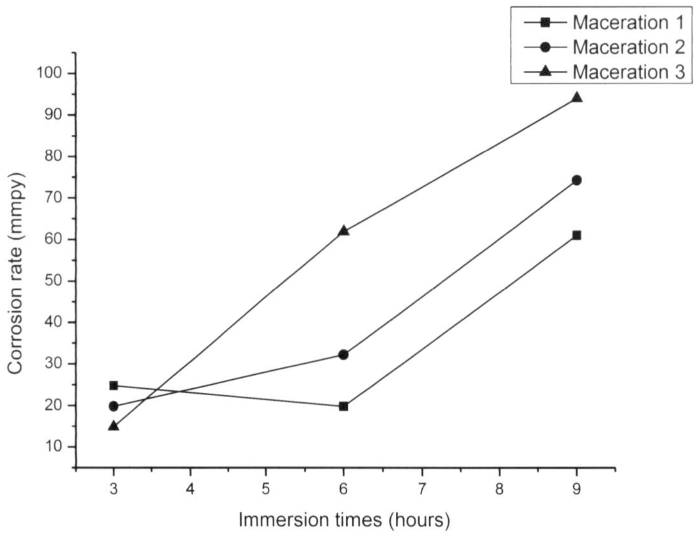

Figure 3. Graph plotting of corrosion rate as function of immersion time.

Figure 3 illustrates time maceration which gradually and cause to increase the rate of corrosion. This condition is a result of the longer interaction of mild steel surface with corroding medium H_2SO_4, which induced more corrosion occurs on the surface of mild steel due to H^+ ion attack. Decreasing corrosion rate within 6 hours inhibition period was originated from the natural extract inhibitor addition to corroding medium. Natural extract inhibitor generated passive layer on the surface of mild steel, and optimally resisted attack from corrosive ions. On the other hand, longer inhibition period (9 hours) shows increasing corrosion rate. This results may be caused by a large amount of H^+ ion was formed after interactions of the passive layer with corroding medium and accordingly, corrosion rate was increasing. It is evident with reduction of mild steel mass after inhibition process.
Furthermore, the passive layer of tannin-iron from *Terminalia catappa* leaves extract was unable to protect entire surface of steel, and consequently, unprotected parts can be ionized and experience corrosion. Corrosion rate was also influenced by accumulation of tannin impurities from *Terminalia catappa* leaves extract accumulated on the surface of the mild steel.

4. Conclusion

The process of maceration extraction affects the amount of tannin content in *Terminalia catappa* leaves extract. The highest content of tannin during the six-day maceration period is 7.23%. From the results of corrosion rate testing showed that the higher the concentration of the inhibitor, the corrosion rate decreases. Time maceration which gradually and cause to increase the rate of corrosion.

References

[1] Uhlig, Herbert H, Revie R and Winston 2008 *Corrosion and corrosion control* (New Jersey: John wiley & sons, inc., Publication)
[2] Kadhum A, Mohamad A, Hammed L, Al-Amiery A, San N and Musa A 2014 *Mater.* **7** (6) 4335-4348
[3] NACE Glossary of Corrosion Terms 1965 *Mat. Pro.* **4** (1) 79-80
[4] Umoren S A 2009 *Carbohydr. Polym.* **2** (1) 175-188
[5] Tan K W and Kassim M J 2011 *Corros. Sci.* **53** (2) 569-574
[6] Prabakaran M, Kim S H, Hemapriya V and Chung I M 2016 *J. Ind. Eng. Chem.* **37** 47-56
[7] Ramji K, Cairns D R and Rajeswari, S 2008 *Appl. Surf. Sci.* **254** (15) 4483-4493
[8] Zulfareen N, Kannan K, Venugopal T and Gnanavel S 2016 *Arab. J. Chem.* **9** (1) 121-135

[9] Eddy N O, Ekwumemgbo P A and Mamza P a P 2009 *Green. Chem. Lett. Rev.* **2** (4) 223-231
[10] Martinez S 2003 *Mater. Chem. Phys* **77** (1) 97-102
[11] Zelinka S L and Stone D S 2011 *Mater. Corros* **62** (8) 739-744
[12] Heinrich M B, Gibbons J, Williansom S 2004 *Fundamentals of Pharmacognosy and Phytotherapy* (Philadelphia: Elsevier)
[13] Voight R 1995 *Buku Pelajaran Teknologi Farmasi, diterjemahkan oleh Noerano Soendani* (Yogyakarta: Gajah Mada University Press) p. 56 159, 557
[14] ASTM G 01-03 *Standard Practice for Preparing, Cleaning, and Evaluating Corrosion Test Specimens.*
[15] ASTM G 31-72 *Standard Practice for Laboratory Immersion Corrosion Testing of Metals.*
[16] ASTM NACE/ASTMG31-12a, 2012, Standard Guide for Laboratory Immersion Corrosion Testing of Metals, ASTM International, West Conshohocken, PA
[17] Martell, M C 1953 Chemistry of Metal Chelate Compounds (New York: Prentice Hall) p. 28-34
[18] Slabbert N 1992 *Complexation of Condensed Tannins with Metal Ions*, ed R W Hemingway and P E Laks (New York: Plenum Press) p. 421-436

IOP Conf. Series: Materials Science and Engineering **345** (2018) 012024 doi:10.1088/1757-899X/345/1/012024

The application of vacuum redistillation of patchouli oil to improve patchouli alcohol compound

T M Asnawi, P N Alam, H Husin and M Zaki
Department of Chemical Engineering, Syiah Kuala University, Darussalam, 23111, Banda Aceh, Indonesia

E-mail: pocutnurulalam@unsyiah.ac.id

Abstract. Patchouli oil produced by traditional distillation of patchouli leaves and stems by farmers in Aceh still has low patchouli alcohol compound. In order to increase patchouli alcohol concentration, vacuum redistillation process using packed column was introduced. This research was conducted to fractionate terpene (*alpha-copinene*) from oxygenated hydrocarbon (patchouli alcohol) compound. The operation condition was conducted at two variables that was dependent variable and independent variable. The dependent variable was the 30 cm height distillation packed column, by using raschig ring with 8 mm x 8 mm dimension. And the independent variable was operating temperature 130 ^{0}C and 140 ^{0}C., vacuum pressure 143,61 mbar, 121,60 mbar and 88,59 mbar and operation time 2 hours, 3 hours and 5 hours. Total of treatments applied in this works were 3 x 3 x 3 or equal to 27 treatments. Patchouli oil used in this research was obtained from Desa Teladan-Lembah Seulawah, Aceh Province. The initial patchouli alcohol compound which analyzed with GC-MS contained 16,02% before treatment applied. After vacuum redistillation process treatment applied patchouli oil concentration increase up to 34,67%. Physico-chemical test of patchouli oil after vacuum redistillation is in accordance with SNI 06-23852006 standard.

1. Introduction

Patchouli plants (*Pogestemoncablin benth*) is one of the important essential oil-producing plants. Among 70 types of volatile oil which are traded in international market, about 9 to 12 essential oils are supplied from Indonesia. Therefore, Indonesia is categorized as an essential oil producing country and a top quality exporter of essential oil. Almost all patchouli plantations in Indonesia are planted by traditional farmers. As one of the largest supplier of patchouli oil in international market, Indonesia contributes 90% of market demand. Most of the patchouli oil products are being utilized in perfume, cosmetic, antiseptic, insecticide and currently aromatherapy industry (Anonimous, 2005). Patchouli oil is an essential oil which has a very important chemical compound namely fixating agent which is used in perfume-based industry, such as perfume, soap, deodorant, etc. This fixative property is caused by its main component of patchouli alcohol ($C_{15}H_{26}O$) belonging to oxygenated terpen. Patchouli oil extracted from patchouli plants (*Pogestemoncablin benth*) by distillation of patchouli leaf [1].

Basically the quality of patchouli oil is strongly influenced by patchouli alcohol content. In general, patchouli oil also contains oxygenated terpen hydrocarbon compounds and other compounds. For certain purposes, in the perfume industry, soaps, cosmetics, and medicines, patchouli oil should be free from terpen compounds due to its ability to the formation of resins in the presence of air. Separation of terpen from patchouli oil would increase its solubility in alcohol and inhibit oxidation process. The patchouli alcohol component has a relatively high boiling point of 150-160 °C (8mmHg), meanwhile, terpen and sesquiterpene components boiling point is ranging at 150-180 °C and 240-280 °C at atmospheric pressure, respectively [1-3].

Refer to this characteristics, it will be applied terpent separation for patchouli oil purification. The oxygenated compound has a better solubility in alcohol comparable to terpene component and sesquiterpen component. The reduction of terpen component and sesquiterpenes in volatile oil can be conducted by vacuum redistilation, column chromatography and polar solvent extraction. The main problems found in patchouli oil distillation from patchouli leaf which traditionally produced by farmers is still have low patchouli alcohol concentration (Dalimi, 1998). High levels of patchouli alcohol will result in low terpen content resulting in good quality patchouli oil. To produce patchouli oil with high concentration of patchouli alcohol, which is oxygenated terpen compound, it can be done patchouli alcohol component enrichment process with vacuum distillation method with packed column (packed column). In vacuum distillation process of patchouli oil, the fraction with low boiling point (terpen component) produced as a distillate product, whereas the fraction with high boiling point (oxygenated terpen) tends to remain inactive. It is expected that farmers can make changes to patchouli oil distillation equipment from used drums into non-oxidizable metal materials such as stainlesssteel.

The purpose of this research is to separate terpen component, to increase the patchouli alcohol level in patchouli oil by vacuum redistilation method using raschig ring type packed column and to evaluate the effect of temperature, pressure and distillation time to determining parameters quality of produced patchouli oil (refractive index, density, acid number, ester number, and patchouli alcohol level).

2. Materials and methods

2.1. Materials

In this work, patchouli oil obtained from traditional distillation unit in Desa Teladan, Aceh, Potassium hydroxide (KOH, 99,9%: Acros organics), hydrocloric Acid (HCl, 99,9%: Acros organics) , phenolthalein, ethanol (C_2H_5OH, 99,5% Merck), erlenmeyer flasks, beaker glass, measuring cup, burette, picnometer, water bath, spatula, reaction tube, funnel, separator funnel, dropper pipette, volume pipette, scales and refractometer (Reichart Leica). In the vacuum redistillation process the drying and distillation times are determined for 24 hours and 2 hours, respectively. Deionized water was used in whole experiments.

2.2. Methods

A quarter volume of the steam boiler filled with water and heated electrical steam. A 2: 1 proportions of mixed lemongrass leaf and stem put into the distillation kettle equipped with grid where the material was placed. Set the proportion with various density according to the varied. Flow the steam produced by direct contact with raw material to the shelter through the distillate pipe. Drain the oil steam mixture out of the distilled boiler to the condensate through the steam distillation pipe. Accommodated distillate produced on the column of essential oil storage. Repeat the above work by varying the volume of the bed (40%, 60%, 70%, 80%) and sample particle size (3 cm, 5 cm, 8 cm, 12 cm, 15 cm).

3. Results and discussion

3.1. Initial patchouli oli quality before redistilltion

The initial patchouli oil analysis results before redistilled can be seen in Table 1.

Table 1. Initial patchouli oil analysis results before redistilled

Parameters	Analysis results (Desa Teladan)
Colour	Brownish yellow
Density	0,9516 (mg/cm^2)
Ester number	17,8605
Acid number	6,0306
Refractive index	1,5074
Solubility in alcohol	1:5
Patchouli alcohol concentration (%)	16,0247

Initial analysis of patchouli oil can be seen that the patchouli alcohol content is very low that is 16.0247% while the standardized in SNI 06-2388-2006 (Table2.2) is at least 30%. The other parameters already similar to SNI 06-2388-2006 as the color produced brownish yellow. Patchouli oil color can be cleared back with the adsorption process. The higher ester numbers the better (max 20) while the smaller acid number the better (max. 8) quality of patchouli oil. The solubility in alcohol allowed is 1: 10, the more soluble in alcohol the better of its quality. Patchouli oil color after redistilation process changes from yellow brown to dark brown due to part of the light fraction has evaporated so that the concentration of heavy fraction increase. In addition, color changes may also occured due to a higher heating at low pressure so that the oil near by the kettle wall easily damaged. This also influenced by the burning of oil smells better known as distillate (Ketaren, 1985). If compared with SNI 06-2388-2006, the redistilled patchouli oil color become darker. In SNI 06-2388-2006, patchouli oil is standardized yellow to reddish brown while redistilated patchouli oil is in dark brown.

3.2. Density

The density for each treatment are shown in Figure 1 and Figure 2 for a 2 and 3 hours redistillation time, respectively.

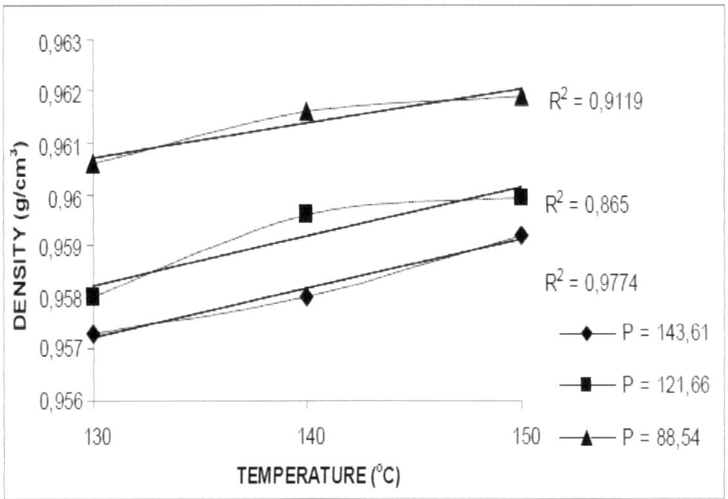

Figure 1. The relationship between temperature (°C) and patchouli oil density which was redistilled at various operating pressures for redistillation time 2 hrs.

Figure 1 shows that as the temperature increase the higher the density of redistilled pathchouli oil due to light fraction evaporated. Specific gravity is associated with the heavy fraction of the components contained therein. The larger the weight fraction contained in the oil, the greater the density. Often the heavier type of oxygenated terpen component is greater than that of oxygenated terpenes. The highest density obtained at 88.54 mbar and temperature 1500 ^{0}C is 0.9619 g/cm^3, and the lowest density is obtained at pressure 143.61 mbar and temperature 1300 ^{0}C is of 0.9573 g/cm^3. Fig.1 shown that the highest average density value obtained at a pressure of 143.61 mbar is 0.9774 g/cm^3.

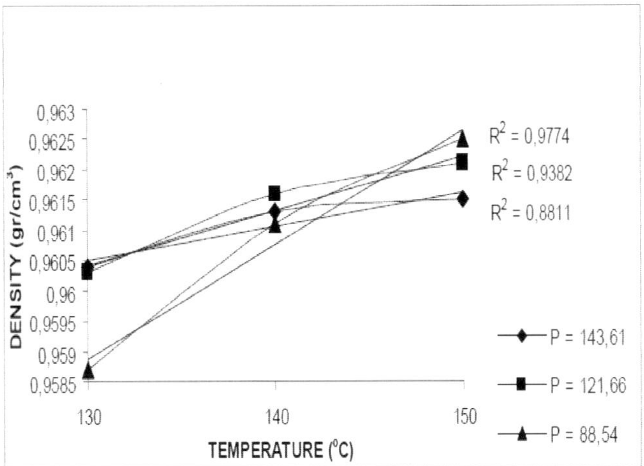

Figure 2. The relationship between temperature (°C) and patchouli oil density which was redistilled at various operating pressures for redistillation time 3 hrs.

Figure 2 shows that the highest density values are obtained under pressure 88,54 mbar and temperature 150 ^{0}C that is 0,9625g / cm^3, and the lowest value obtained at pressure 88,54 mbar and temperature 130 ^{0}C that is 0,9587 g/cm^3. The patchouli oil density increased from the initial state from 0.9516g / cm3 to 0.9573 0.9625g / cm^3. This increase is due to the process of concentrating heavy fractions on the residue affected by heating and decreasing pressure treatment.

The standardized density in SNI06-2388-2006 is 0.950 0.975 g / cm^3. The results showed that the density of patchouli oil from redistillation treatment was within the standard range. If compared between the two process (2 and 3 hours redistilation treatment) it can be seen that the lowest density is 0.9573g / cm3 for 2 hours while the highest density is 0.9625 g / cm^3 for 3 hours. On the process of redistilasi for 3 hours obtained the highest average value (mean value) at 88.54. This is due to the longer processing time then the light fraction that evaporates the greater of weight fraction increases and causes the density is also increasing.

The refractive index of essential oils is closely related to the components arranged in the volatile oil produced. Similarly, the specific gravity wherein the essential oil component may affect the refractive index. A long-chain components such as sesquiterpen or oxygen-compounds components involved in distillation process, the density of the essential oil will increase so that the incoming light will be harder to be biased. This causes the oil refractive index increase. The essential oil with a high refractive index is better than low one. The analysis results on refractive index are shown in Figure 3 for redistilation time 2 hours and Figure 4 for redistilation time 3 hours.

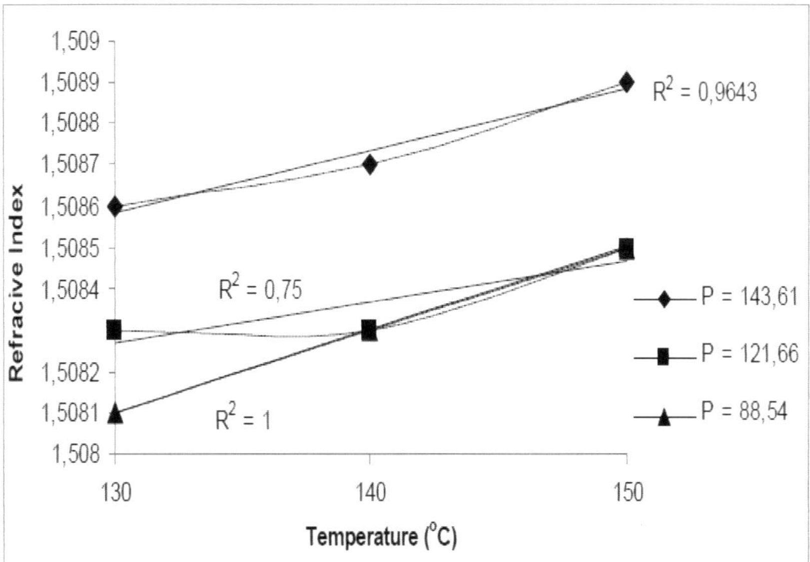

Figure 3. The relationship between temperature (°C) and the index of patchouli oil refraction resulted from redistillation at various operating pressures for a two-hour redistillation time.

If a monochromatic light passes through a medium (A) to a more densely packed medium (B), then the change in velocity and refraction of the light approaches the normal line, or the angle of the incident ray (iA) greater than the angle of the bias ray (iB). The comparison of the incident angle of light with this refracted beam angle is called the refractive index. In this study to determine the refractive index of an oil sample used Abbe Refractometer. Figure 3 shows that the highest refractive index values are obtained under pressure 88,54mbar and temperature 1500 °C that is equal to 1,5085 and the lowest refractive index value obtained at pressure 143,61 mbar that is equal to 1,5074.

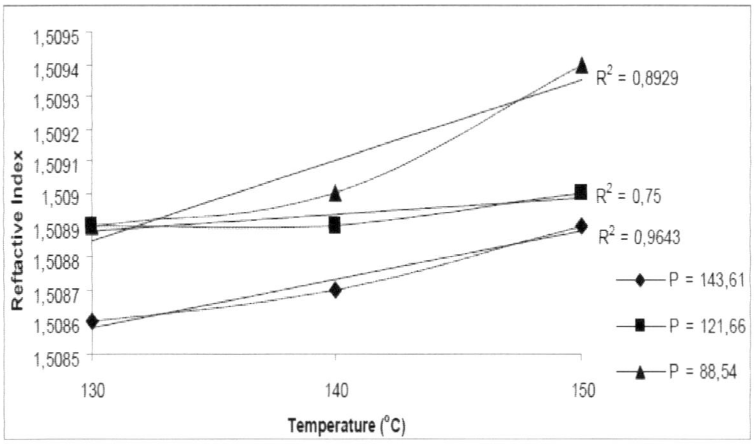

Figure 4. The relationship between temperature (°C) and the index of patchouli oil refraction resulted from redistillation at various operating pressures for a three-hour redistillation time.

Figure 4 shows the highest refractive index values are obtained under pressure 88,54 mbar and temperature 150 ^{0}C and the lowest refractive index obtained at pressure 143,61 mbar and temperature 130 ^{0}C that is equal to 1,5086, the initial refractive index of 1.5074 increased after the redistillation process. The refractive index required in SNI 06- 2388-2006 is 1,507 1,515, and the refractive index of patchouli oil of redistillation has been in that range. Comparing Figure 3 with Figure 4, it can be seen that the highest refractive index in the process lasted for 3 hours (Figure 4) and the lowest in the process that lasted for 2 hours (Figure 3). The refractive index on the redistillation process for 2 hours obtained the highest average value at 88.54 and the average value obtained was 1 and the highest average value on the redistillation process for 3 hours was obtained at 143.61 pressure. This is due to the increasing temperature and the lower pressure, the refractive index value will increase. This is because the longer the oil residue process becomes thicker due to the evaporation of mild fraction and water. Index value is also influenced one of them with the water in the oil content. The more the water content, the smaller the value of the bias index. This is because the nature of the water is easy to refract the incoming light.

3.3. Acid Number

Acid numbers show free acid levels in essential oils. The increasing number of acids may affect the quality of essential oils. Namely these acid compounds can change the peculiar smell of essential oils. This may be due to the duration of oil storage and the presence of contact between the volatile oil produced by the light and air around when it is in the oil sample bottle at the time of storage. Most essential oils contain small amounts of naturally occurring free organic acids or those produced from oxidation and hydrolysis ester processes (Ketaren, 1985).
Result of test of acid number for patchouli oil after redistilled for time 2 hours and 3 hours are shown in Figure 5 and Figure 6.

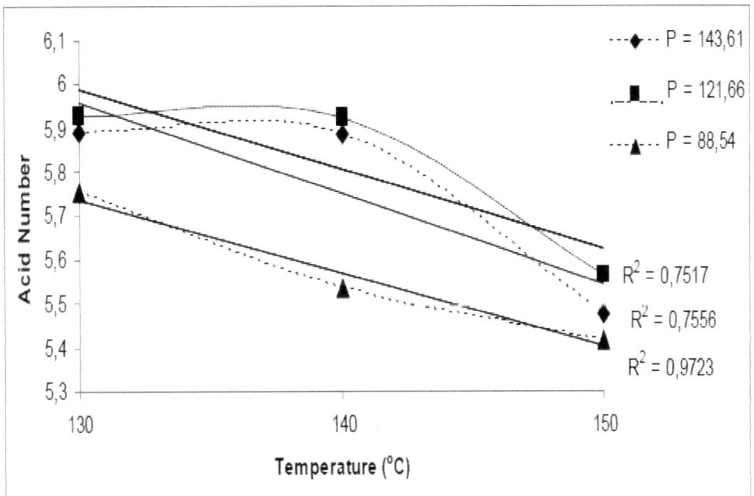

Figure 5. The relationship between temperature (°C) and acid number resulted from redistillation at various operating pressures for a two-hour redistillation time.

Figure 5 shows that the highest acid value is obtained at 121.66 mbar and 1300C at 5.9272 and the lowest acid value obtained at 88.54 mbar is 5,1476 Figure 6 The highest acid number is obtained at 143,61 mbar pressure and temperature 1300 °C that is equal to 5,7818 and the lowest acid number obtained at pressure 121,66 mbar and temperature 1500 °C that is equal to 5,4462. Overall, the patchouli oil acid number before it was redistilled by 6,0306 decreased after the redistillation process. This is due to the distilled patchouli oil that has dried up so that the organic acids vaporize, as suggested by (Guenther 1987) that the oil that has been dried and protected from air and light has a relatively smaller amount of free acid.

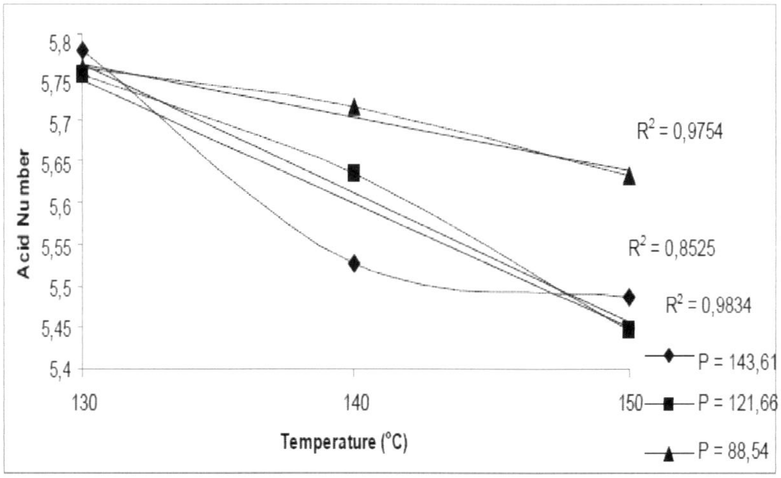

Figure 6. The relationship between temperature (°C) and acid number resulted from redistillation at various operating pressures for a three-hour redistillation time.

When compared between the 2-hour redistillation process (Figure 5) and a 3-hour process (Figure 6) it can be seen that the lowest acid number in the process is 2 hours. The average value obtained in 2-hour redistillation compared to 3 h of redistillation, it can be seen that the average value of 3-hour redistillation is better than the average redistillation value of 2 hours, because with the longer time the oil is likely to oxidize to increase the oil acid number. In SNI 06-2388-2006, it is required that the maximum acid number is 8, and the patchouli oil of the redistillation product is below the prescribed standard.

3.4. Esther number
Determination of the amount of esters is very important in determining the value of essential oils. The ester number for the redistilled patchouli oil is shown in Figure 7 for 2 hours refining time and Figure 8 for refining time 3 hours.

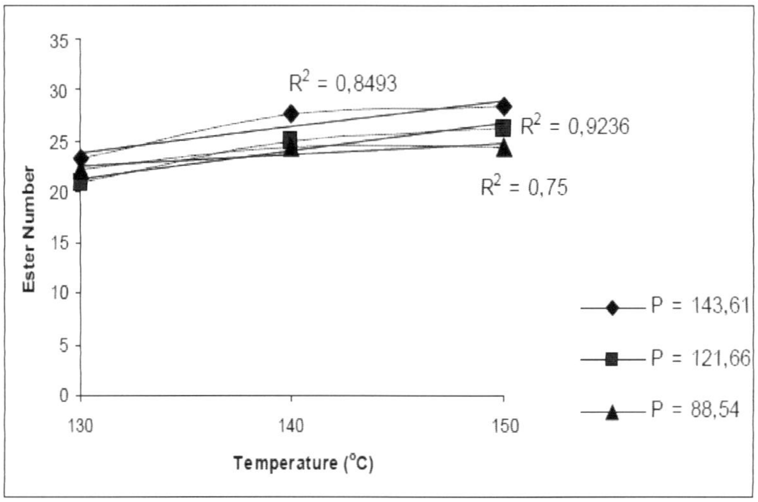

Figure 7. The relationship between temperature (°C) and esther number resulted from redistillation at various operating pressures for a two-hour redistillation time.

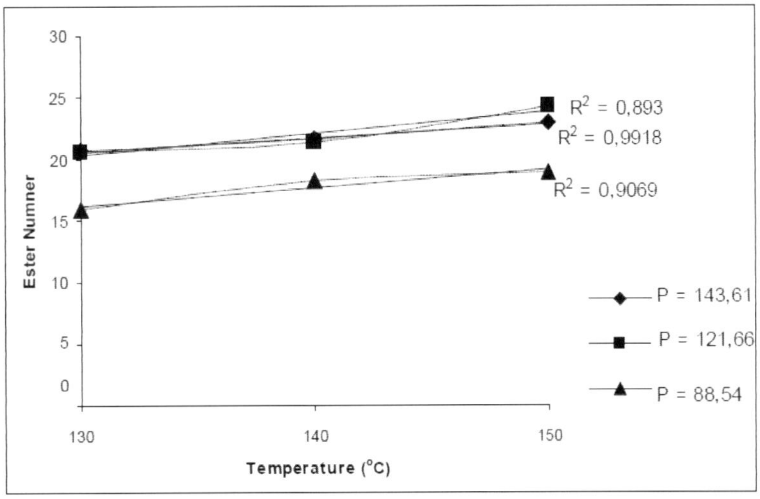

Figure 8. The relationship between temperature (°C) and esther number resulted from redistillation at various operating pressures for a three-hour redistillation time.

Figure 8 The relationship between temperature (°C) and esther number resulted from redistillation at various operating pressures for a three-hour redistillation time Figure 8 shows that at a pressure of 150 mbar, the ester number increases with increasing temperature. The highest este number is obtained under pressure 121,66 mbar and temperature 1500 °C that is equal to 24,2356 and lowest ester number obtained at pressure 364.44 mbar and temperature 1300 °C yaitu equal to 15,8776. The average value obtained on 2-hour redistillation compared to 3 h of redistillation, it can be seen that the average value of 3-hour redistillation is better than the average redistillation value of 2 hours, because with the longer time it is possible the oxidized oil is greater so it can increase the oil ester number .Overall, the value of patchouli oil ester before it was redistilled was 17.8605 lower than that of patchouli oil after it was redistilled. This is because the ester includes an oxygenated compound with a high boiling point of 2400 2800C at atmospheric pressure as revealed by Guenther (1987) that compounds, phenol, ether and some aromatic aldehyde compounds, sesquiterpen alcohols, esters and so on have a boiling point equal to the boiling point of sesquiterpenes, so that the mild fraction evaporated by the redistillation process increases the fraction of the ester in the residue. SNI 06-2388-2006 requires a maximum estimate of ester 20, but the patchouli oil of redistillation has an ester value higher than 20.

3.5. Solubility in Alcohol

Since many essential oils dissolve in alcohol and are rarely water soluble, the solubility can be readily known by using alcohol at various concentrations. Determining the solubility of the oil depends also on the speed of solubility and the quality of the oil. Oxygenated oils are usually more soluble in alcohol than those rich in terpenes. In lower alcohol concentrations, the terpene fraction will separate as is the case with wax. Based on this, this method is widely used for the manufacture of free oil terpen and sesquiterpen, consentrate and extract. Oil solubility may also change due to the influence of oil age. Unfavorable storage conditions can speed up polymerization. Factors such as light, air and the presence of water usually cause unfavorable effects. The solubility value in patchouli oil alcohol after the redistillation process has increased from 1: 5 to 1: 4. Increased solubility in alcohol occurs because in the process of redistillation, the conventional compound has largely evaporated. P This is because the terpene fraction is somewhat difficult to dissolve in dilute alcohol, while paraffin and sesquiterpen are insoluble at all. Oxygenation compounds generally have better solubility, for example alcohols, aldehydes, ketones and phenols, whereas the ester and phenol ester compounds have smaller solubility values (Guenther, 1987). In SNI 06-2388-2006 it is required that patchouli oil should have a solubility in alcohol of a maximum of 1: 10 in a clear solution, and patchouli oil of redistilated soluble in alcohol in a ratio of 1: 4.

3.6. Patchouli alcohol content

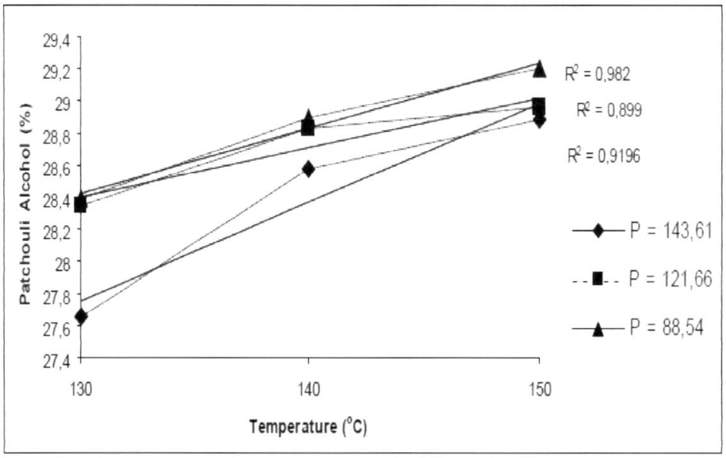

Figure 9. The relationship between temperature (°C) and patchouli alcohol content resulted from redistillation at various operating pressures for a two-hour redistillation time.

Patchouli alcohol is a group of oxygenation compounds is an important component that determines the quality of patchouli oil. The greater the level of patchouli alcohol, the better the quality of patchouli oil produced. Increased levels of patchouli alcohol can be done by the process of redistilasi on the vacuum to vaporize terpen compound. Test results on patchouli alcohol content for patchouli oil resulted from redistillation are shown in Figure 9 for a 2-hour redistillation time and Figure 10 for a 3-hour redistillation time. Figure 9 shows that the level of patchouli alcohol increases with increasing temperature and lower pressure. The highest patchouli alcohol level was obtained at 143,61mbar pressure and temperature 1500C that was equal to 29,12% and the lowest patchouli alcohol level was obtained at 88,54 mbar and temperature 1300C ie by 27.64%.

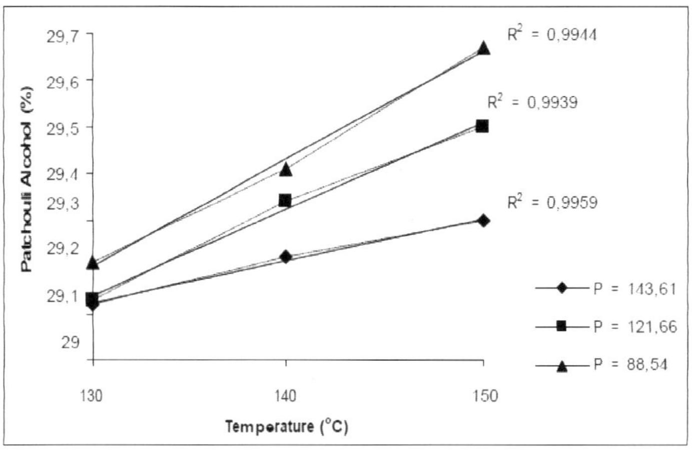

Figure 10. The relationship between temperature (°C) and patchouli alcohol content resulted from redistillation at various operating pressures for a three-hour redistillation time.

Figure 10 shows that patchouli alcohol levels are increasing with increasing temperature and lower pressure. The highest patchouli alcohol content was obtained at 143,61 mbar and temperature 1500 °C that was equal to 29,67% and the lowest patchouli alcohol concentration was obtained at 88,54 mbar and 1300C at 29,13%. Increased levels of patchouli alcohol is quite significant when compared with patchouli alcohol patchouli oil levels before redistilasi. The increase of patchouli alcohol level is above 11,6%, that is from 16,02% to above 27,63%. Comparison of the average value of patchouli alcohol level during redistilation 2 and 3 hours, the average value of redistilasi for 3 hours has a higher value than the average redistilasi 2 hours. This is because the patchouli alcohol level is very influential on pressure, temperature and time. The lower the pressure, the high temperature and the length of time will increase the level of patchouli alcohol in patchouli oil. SNI 06-2388-2006 requires that patchouli oil patchouli oil content is at least 30%, but patchouli oil from redistillation still has not reached the concentration. Patchouli oil from the distillation of the people has a low rpatchouli alcohol so that the process of redistilasi in a vacuum to increase the level of patchouli alcohol. In this research also do the process of redistilation without using the column as the comparison only. The result of patchouli oil redistilation without column Indicates that its physico-chemical properties do not differ greatly from the result of patchouli oil redistilation using the column.

4. Conclusions

From the experimental study the increase concentration of patchouli alcohol in patchouli oil can be conducted by re-distillation process under vacuum conditions. The optimum condition of the re-distillation process is obtained at a pressure of 88.54 mbar and temperature 1500 ^{0}C. Patchouli oil result for optimum condition is dark brown, density 0,9611g / cm^3, refractive index1,5074, alcohol solubility 1: 4, acid number 5,7129, ester number 18,9073 and patchouli alcohol level is 29,67%. These parameters of optimum conditions is in accordance with SNI 06- 2388-2006 standard.

Acknowledgment

We gratefully acknowledge the financial support from the Syiah Kuala University Banda Aceh, Republic of Indonesia.

References

[1] Guenther E 1987 *The Essential Oils* vol. **1**, ed E Robert (New York: Krieger Publishing Company)
[2] Ketaren S 1985 *Pengantar Teknologi Minyak Atsiri* (Jakarta: Penerbit Balai Pustaka)
[3] Ruslan H 1987 *Tanaman Minyak Atsiri* (Jakarta: PT Penebar Swandaya)
[4] Djafar F 2009 Ekstraksi minyak atsiri dari jahe dengan metode hidrodistilasi sistem kabohasi *Tesis Magister Teknik Kimia*, Universitas Syiah Kuala Banda Aceh
[5] Aprilina P and Silviana 2006 *Prosiding Konferensi Nasional Minyak Atsiri 2006*, Solo 18-20 September 2006
[6] Brown G G 1978 *Unit Operation* 14th ed. (Japan: Modern Asia Edition)
[7] Dalimi A 1998 *Monograf Nilam* No. **5** Badan Penelitian dan Pengembangan Pertanian Balai Penelitian Tanaman Rempah dan Obat, Bogor Indonesia
[9] Hermani and Marwati T 2006 *Prosiding Konfrensi Nasional Minyak Atsiri 2006*, Solo 18-20 September 2006.
[10] Ketaren S 1985 *Pengantar Teknologi Minyak Atsiri* (Jakarta: PN Balai Pustaka)
[11] Rusli M S and Ketaren S *Standar Peralatan Untuk Penyulingan Minyak Atsiri*
[12] Pocut N A 2006 *Pelatihan Analisis dan Pengolahan Minyak Atsiri* (Banda Aceh: Universitas Syiah Kuala)
[13] Panjaitan L 1993 Kajian tahanan gesekan tumpukan nilam terhadap aliran udara serta profil suhu tumpukan pada penyulingan dengan metode air dan uap *Skripsi* Fateta-IPB Bogor
[14] Rusli S 2001 *Pengaruh Padatan dan Penyulingan terhadap Rendemen dan Mutu Minyak Nilam* (Bogor: Lembaga Penelitian Tanaman Industri) p 17-18

IOP Conf. Series: Materials Science and Engineering **345** (2018) 012025 doi:10.1088/1757-899X/345/1/012025

Properties enhancement of cassava starch based bioplastics with addition of graphene oxide

A Amri, L Ekawati, S Herman, S R Yenti, Zultiniar, Y Aziz, S P Utami and Bahruddin
Department of Chemical Engineering, Universitas Riau, Pekanbaru, Indonesia

E-mail: amun.amri@eng.unri.ac.id

Abstrak. The properties of cassava starch based bioplastic have been successfully enhanced by additioning of graphene oxide (GO) filler. The composite was synthesized via starch intercalation method using glycerol plasticizer with variation of 5 – 15 % v/v GO filler and mixing time of 30 and 60 minutes. The effects of GO content and the mixing time to the mechanical, water uptake and biodegradation were studied. The synthesis of GO and its integration in the bioplastic composite were also elucidated. The increasing of the GO content and mixing time improved the mechanical properties of composite mainly due to of good homogeneity among the constituents in the composite as indicated by scanning electron microscopy (SEM) and Fourier Transfom Infrared (FTIR) spectroscopy. The bioplastic produced using 15% of GO and 60 minutes mixing time had the highest mechanical properties with tensile strenght of 3,92 Mpa, elongation of 13,22% and modulus young of 29,66 MPa. The water uptake and biodegradation increased as the increase of GO content and decreased as the increase of the mixing time. Graphene oxide is the promissing filler for further development of cassava starch based bioplastics.

1. Introduction

The industrial development which is accompanied by the increase in the number of inhabitants, will increase the consumption of natural resources and energy. It causes the creation of several problems including the buildup of waste in large numbers. One of the industrial products has penetrated all the aspects of life is plastic. Plastic is more acceptable, besides due to of its price aspect, plastic is also considered superior to other media such as metal or glass due to of its nature (lighter and not easy to break). Each passing year, the production of the plastic world reached 100 million tons per year [1]. In Indonesia the plastic consumption growth reached 6-7% per year [2, 3]. Significant improvement against the production and consumption causes buildup of plastic wastes.

Conventional plastics is made from petroleum which is limited natural resources (unrenewable) [4]. In addition, the petroleum plastic is difficult to be decomposed by the microorganisms leading to the environmental pollution [5]. It gives an overview of the potential development of biodegradable plastic as substitute of conventional plastic [1]. Biodegradable plastic or often called with bioplastic, it is plastic that can be used like as conventional plastic and has an ability easily to be degraded by microorganisms such as fungi and bacteria [6]. Raw materials used in making of bioplastic filmed are renewable and biodegradable materials such as starch, vegetable oils, and microbiota [7]. The potential of starch derived from cassava or better known as the cassava (*Manihot Esculenta Crantz*) as an environmentally friendly plastic materials are very promising. It contains around 90% of the cassava starch with other small content such as lipids, proteins and fibres [8]. Generally, the bioplastic from starch cassava has the nature of transparency and a good tensile strength [9].

Even though it has many advantages, the bioplastics still have drawbacks compared to the conventional plastics, namely in term of mechanical properties which is easily to torn and not as strong as conventional plastic. One of the strategies way for improving the mechanical properties of bioplastic is the filler addition in bioplastic matrix. One of the filler which can potentially be added into bioplastic is graphene oxide (GO). GO is one layer of oxidized graphite obtained via chemical synthesis or through others graphite oxidation methods. There are many advantages of the use of GO as filler including it is easily to dissolve in matrix of bioplastic, it has high surface area and biocompatibility, and it still inherit the graphene characters, i.e. strong and flexible [10]. Based on the nature of the cassava starch and the graphene oxide, it is predicted the combination of these materials can produce bioplastic with better properties. To the best our knowledge, no research like this has ever been made.

This work tries to understand the influence of graphene oxide addition to the properties of cassava starch based bioplastic. The main result showed that the addition of graphene oxide and the mixing time improved the mechanical properties of produced bioplastic. The water uptake and biodegradation increased as the increasing of GO content and decreased as the increasing of the mixing time. Graphene oxide is a promising filler for further development of cassava starch based bioplastics.

2. Experimental

2.1 Sintesis of graphene oxide.

Graphene oxide was synthesized using Hummer's method, it has been reported elsewhere [11]. The graphite powder (1 g) was stired with 23 ml of H_2SO_4, 0.5 grams of $NaNO_3$ and 3 grams of $KMnO_4$ in a beaker glass in an ice bath container with the temperature was kept below 20^0C until homogen. The solution was then stirred at a temperature of 40^0C in a heat bath. After 14 hours of stirring, the solution was moved to the ice bath and as many as 50 ml of H_2O were added gradually into the aqueous solution with the temperature maintained under 90^0C. The solution was then heated at a temperature of 50 ^{o}C and stirred for 1 hour. Afterthat 150 ml of H_2O were added into solution, and as much as 5 ml of H_2O_2 was added into the solution and stirred for 30 minutes [11].

The phase separation of graphene oxide solution used filter paper and was washed with H_2O repeatedly up to neutral pH was approached. The deposit of graphene oxide was heated at a temperature 80^0C using oven for 4 hours, so it gained a solid graphite oxide. Then graphite oxide mixed with H_2O using ultrasonic dispersion to peel layers to form the graphene oxide dispersed in solution [11, 12].

2.2 Sintesis of bioplastic

The manufacturing of bioplastic was carried out using solution intercalation method. Tapioca flour (10 g) were dissolved into the 100 ml aquadest and heated at 60^0C while stirring for a while. The glycerol (as plasticizer) and graphene oxide corresponded to variable (0%, 5%, 10%, 15%) were then mixed with the tapioca solution and stirred with a stirring time variation (30 minutes and 60 minutes). A homogeneous solution was then poured into the glass mold and dried during 3 days in room temperature [13]. The produced bioplastic was separated from the mold and the characterizations were performed.

2.3 Chracterizations

The identification of the produced graphene oxide used via X-ray diffractometer (XRD) and scanning electron microscopy (SEM) evaluated using JSM-6390A SEM. Furthermore, the mechanical properties (tensile strength, elongation, and elasticity modulus) and surface morphology of bioplastic samples, biodegradation test and the resilience of water test (swelling test) were also investigated. The cristalline phase identification of the produced graphene oxide was carried out using Bruker - D822 X-

Ray Diffraction with Cu Kα radiation. The mechanical properties were performed via ASTM D822-02 using RG 3010 tensile test equipment with specimen size of 100 mm x 15 mm. The rate of biodegradation was carried out by burying the samples in the soil and measured their each mass in a week. The swelling test was carried out by cutting the samples in size of 2cm x 2 cm then was soaked in water with controlling of weight.

3. Result and discussion

3.1. Characterizations of graphene oxide (GO)

The XRD patterns of the produced graphene oxide obtained via Hummers method were shown in Figure 1.(a,b). Figure 1.(a) shows the XRD pattern of the raw graphite before the process, while Fig. 3.1.(b) shows the produced graphene oxide synthesized via Hummer method. From Fig. (a), it can be seen that the graphite has a diffraction peak at 26.60^o and it is in accordance with which has been reported by many researchers [14, 15, 17]. After the oxidation, this peak disappears and new peak appears at 2Θ of arround 10° (Fig. 3.1.b). It has been reported by many researchers that the diffraction peak at GO lies at 9.10 [18], 10.3 0 [19] and 10.5 0 [20].

Figure 2 shown SEM micrograph image of raw graphite used, while Figure 3 shown the SEM image of the produced graphene oxide after the synthesis process. The morphology of raw graphite consists of very dense stacks of graphene, while after being oxidized and passing through the peeling stages the piles appear more loosely as seen in Figure 3. This occurs because the SO_4^- ions penetrate on the interlayer of graphite and break the Van der Walls bonds [21]. The sonification may cause the oxidized graphite to exfoliate further. The addition of water and oxygen molecules in between the layers causes the spacing between d-spacing layers of graphite to become large (Shen et al, 2010). This is supported by XRD data analysis from the raw graphite showing d-spacing of 0,35 nm while the produced graphene oxide having d-spacing of 0,71 nm. Similar case was also reported by Spyrou and rudolf (2012) with d-spacing of 0.35 nm and 0.68 nm for their graphene and graphene oxide,

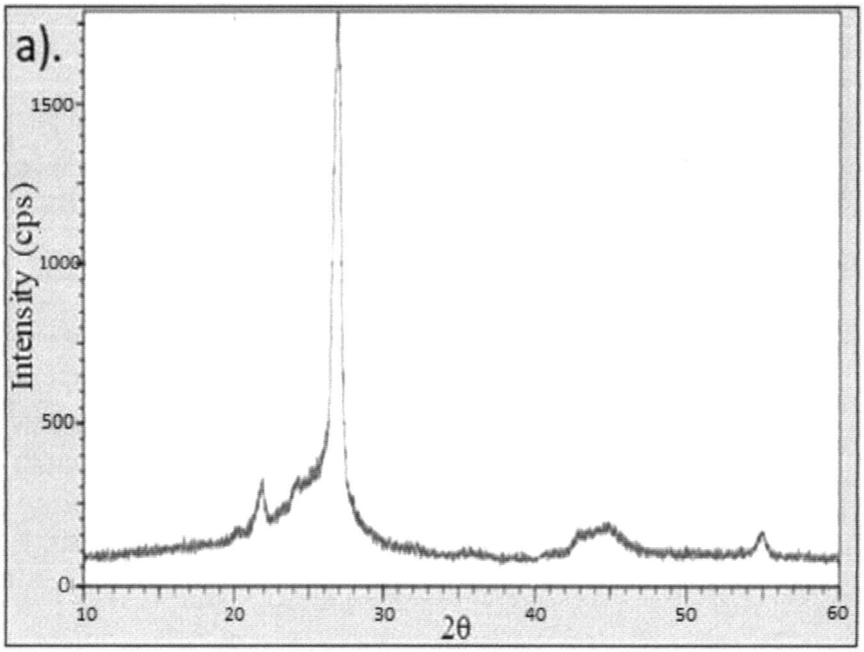

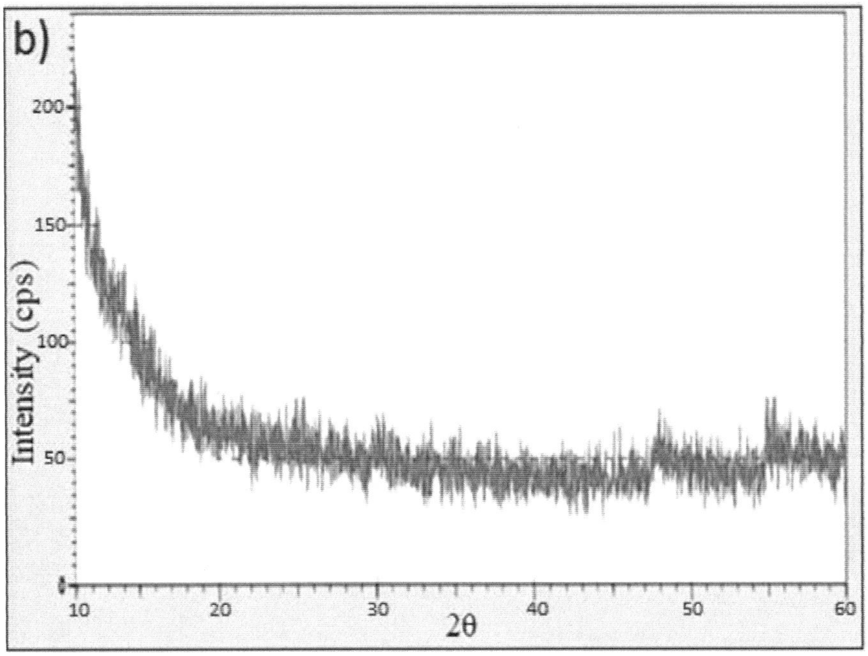

Figure 1. XRD pattern of (a). Graphite before process and (b) produced graphene oxide

respectively [22]. These XRD and SEM analyses confirm the formation of graphene oxide performed in the synthesis process [23].

Figure 2. SEM image of the raw graphite used (2000x)

Figure 3. SEM image of produced graphene oxide (2000x)

3.2. Caracterizations of bioplastic – graphene oxide composites

3.2.1. Mechanical properties of composites

The tensile tests applied to the samples based on the standard of ASTM D882-92 using the Universal Testing Machine (UTM) were presented in Table 1 and Figure 4 - 6. From Table 1 and Figure 4, it can be seen that the higher the graphene oxide concentration, the higher tensile strength and Young modulus will be, while the elongations decrease. The similar trend is also showed by the increase of the stirring time. This phenomenon indicated that there is a strong influence of graphene oxide to the composite matrix. The highest mechanical properties is exhibited by sample synthesized by using graphene oxide as much as 15% with 60 minutes stirring time, namely tensile strength of 3.92 Mpa elongation value of 13.22 Mpa and young modulus value of 29.66 Mpa.

Table 1. The mechanical properties of bioplastic – graphene oxide composites

Concentration of GO	Time (minutes)	Sample	*Tensile Strenght* (Mpa)	Elongation (%)	Young modulus (Mpa)
0%	30	1	1,38	28,09	4,88
	60	2	2,25	25,03	9,00
5%	30	3	2,55	20,62	12,36
	60	4	2,65	18,90	13,99
10%	30	5	2,74	15,35	17,88
	60	6	2,84	14,96	19,00
15%	30	7	3,63	13,45	26,95
	60	8	3,92	13,22	29,66

The increase in tensile strength due to the addition of graphene oxide is caused by the strong interaction between the bioplastic compiler molecules with graphene oxide layers. In more detail, it is associated with the interaction between the hydroxyl (OH) group, either from the bioplastic matrix or from the graphene oxide to form oxygen bridge [24]. The stirring time affects the process of mixing between the composite matrix and the filler graphene oxide. This stirring time correspond to the filler particles distribution in the matrix cavity. The longer the stirring time (60 min.), the more GO particles will spread evenly within the matrix of bioplastic. As the result, the better mechanical properties of bioplastic will be gained.

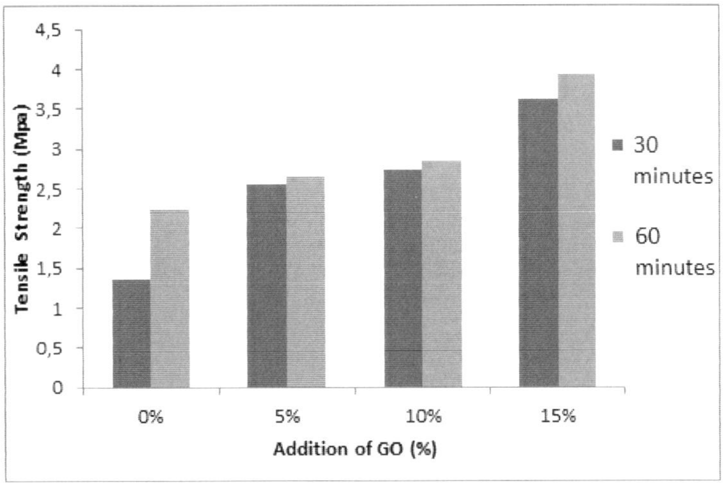

Figure 4. The effects of GO addition and stirring time to the tensile strength

Figure 4 shown the influence of graphene oxide addition and mixing time against the elongation. The percentages of elongation decrease with the increasing of the GO percentages and the increase of mixing time. Elongation is the maximum change of bioplastic length before breaking up when it gets a tensile load. The higher the elongation value, the lower the quality of bioplastic will be. The decrease of elongation is because the filler will reduce hydrogen bonds internally so that there is weakening of the polymer chain before breaking up [25].

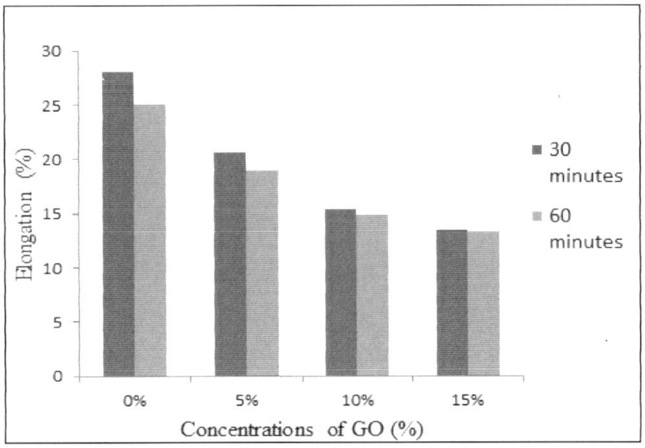

Figure 5. The effects of GO addition and stirring time to the Elongation

The influence of GO addition and mixing time of composites against the Young's modulus can be seen in Figure 6 Young's modulus can be said to be a measure of the stiffness of a material. It can be seen the Young's modulus increase significantly as the increase of GO concentration and mixing time. This confirms good integration of GO into the bioplastic matrix which enlarges the flexibility level of bioplastic composites. The bioplastic synthesized using GO addition of 15% and mixing time of 60 minutes shows the highest Young's modulus of 29.66 Mpa.

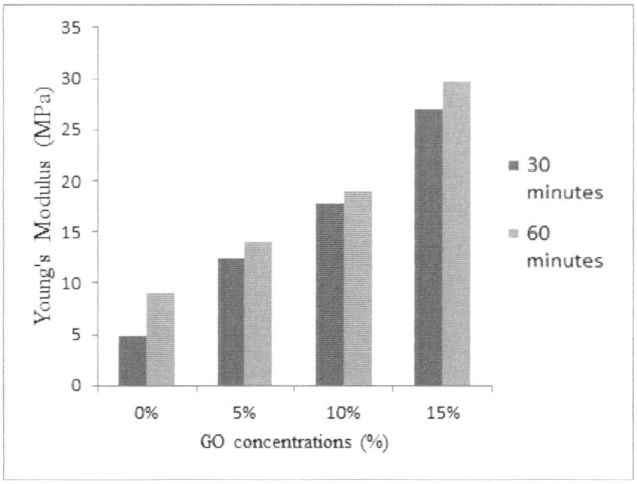

Figure 6. The influence of GO addition and mixing time to the Young's Modulus

3.2.2. Morphology of composites

Figure 7 shows the morphology surface of bioplastic synthesized by various of GO addition (%) and mixing time (t) investigated by scanning electron microscopy (SEM). It can be seen that the surface of composites is relatively smooth with some blemishes when they were stirred for 60 minutes, either with or without GO addition (Figure 7.a-b). It indicated that the GO filler is capable to have good interaction in the composites, and the bioplastic itself (without GO) also contributes to the formation of blemishes. However, when the composites was mixed at a lower strring time (30 minutes), the surface is relatively inhomogen (Figure 7.d) compared to a longer stirring time. It can be concluded that the stirring time greatly affects the homogeneity of the sample.

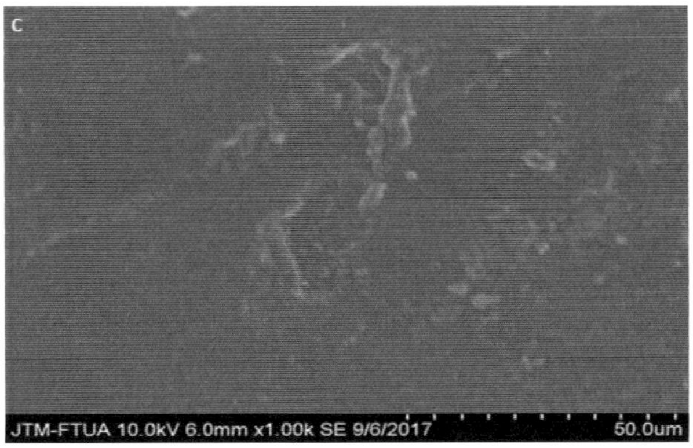

Figure 7. SEM micrograph of bioplastic composites (a) 0% GO, t = 60 min, (b) 15% GO, t = 60 min, (c) 15% GO, t = 30 min.

3.2.3 FTIR Spectra of bioplastic-graphene oxide composites

FTIR analysis is useful to look at the presence of functional groups in bioplastics, graphene oxide and composites as well as their behavior when the varibles are changed. The FTIR spectra of bioplastics, graphene oxide and their composites are presented in Fig. 3.8. It can be seen that the FTIR spectrum of graphene oxide consists of some functional groups formed during the synthesis process. The oxygen functional group peak appears strongly at wavenumber around of 3200-3300 cm^{-1} indicating that the presence of an O-H bond from water molecules and hydroxyl functional groups in the graphene oxide solution. There is also a vibration at the wavenumber of 2850 cm-1 corresponding to the presence of the C-H bond. The presence of vibrations around the wavenumber 1540-1640 cm^{-1} corresponds to the C=O bonds associated with the presence of carboxyl groups at the edges of the graphene oxide layer. In addition, there is a vibration around the wavenumber around of 1200 cm^{-1} corresponding to the presence of the C-O bond. The presence of a C-O group signifies the success of the graphite oxidation process and it also indicates that graphene oxide has a hydrophilic property [14].

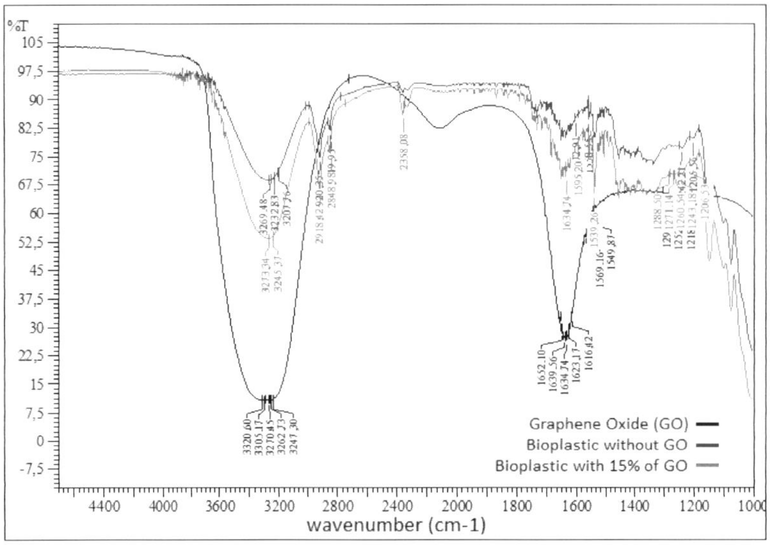

Figure 8. FTIR spectra of graphene oxide, Bioplastic and Bioplastic – GO composites

Figure 8 also describe the FTIR spectrum for bioplastic without graphene oxide addition. It can be seen that generally the bioplastic FTIR spectrum pattern has lower intensity of absorption especially for O-H, C-H and C=O bonds compared with the FTIR spectrum of GO. The oxygen bond is probably derived from the bonds between the starch, glycerol and water. When the bioplastic is added with the graphene oxide, the O-H bond is strengthened due to of hydroxy functional groups from GO increasing the hydrophilic properties of bioplastics. Furthermore, there is a change in the bioplastic structure in the wavelength range 2853-2962 cm^{-1} where there is an increase in the C-H grade in bioplastic by the addition of GO. The similar case with the vibrations in the range around of 1600 cm^{-1}. The vibrations at around 1200 cm^{-1} in the FTIR spectrum of composite indicate an interaction between the bioplastic matrix and GO that creates both intermolecular and intramolecular hydrogen bonds.

3.2.4 Interaction mechanism in composite

The bioplastic composite synthesized consisted of the starch as the matrix and the constituent fillers. The starch is composed of amylose and amylopectin. Basically the heated starch granules will break when it reaches the gelatination temperature (75°C). The heat energy will cause the hydrogen bond to break and the water molecules come into the starch granules. The process of entering the water into the starch will cause the granules to expand and eventually break resulting the cavities in the starch structure and cause the amylose and amylopectin to diffuse out of the granules [5,13].

The composite process begins by mixing the cracked starch granules with glycerol plasticizer and graphene oxide filler. At first, the glycerol molecule will enlarge the formed cavity so that the matrix becomes more elastic which affects the elongation value. Furthermore, the graphene oxide will fill the cavities to form a more dense and strengthening the mechanical properties of bioplastics. It is estimated that the hydroxyl groups from the starch, glycerol and graphene oxide interact each other forming the oxygen bridge and water byproducts. The interaction mechanism in the formation of bioplastic - graphene oxide composites is outlined in Figure 9.

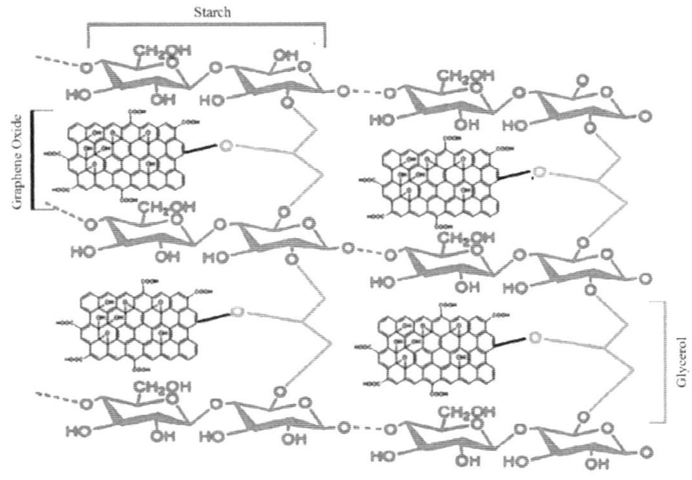

Figure 9. Interaction mechanism in bioplastic-GO composites

3.2.5. Water uptake

The water uptake is the ability of bioplastic absorb water. Bioplastic has good water resistance if it has small percentage of water uptake. Figure 10 shows the water uptake of bioplastic synthesized using various GO addition and mixing time. It can be seen that the water uptake increase monotonically as the increase of GO addition. Darni and Herti [26], stated that the water resistance properties of a molecule depended on the nature of the molecule. In this case, the ingredients of starch and glycerol used are hydrophilic, while the GO is also hydrophilic. Therefore, the addition of GO in bioplastic certainly enhances the ability of bioplastic to absorb the water due to the nature of hydrophilic GO [22]. However, it is different with the mixing time where a longer mixing time (60 minutes) has lower water uptake compared to the mixing time of 30 minutes. It could be attributed to the completeness of reaction between the starch, glycerol and GO resulting H_2O as elucidated in previous section. The longer mixing time then more complete the reaction and more water will be resulted. Eventually, this will disembogue to the water saturated condition which decrease the ability of composites to receive water from external.

The highest water uptake is showed by the sample synthesized using GO addition of 15% and the stirring time of 30 minutes, with the percentage of absorbed water of 30.77%.

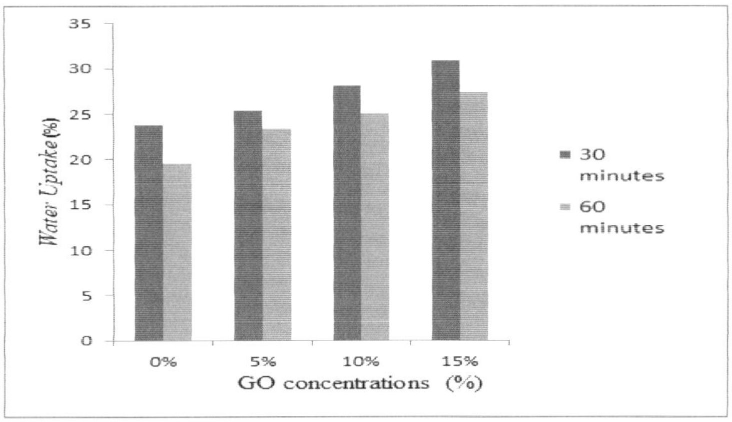

Figure 10. The effect of GO addition and mixing time toward the water uptake

3.2.6. The degradation test of bioplastic – GO composites

Biodegradation test aims to find out how long the bioplastic to be broken down in the environment. Figure 11 shows the results of the composites biodegradation test. It can be seen that more the GO added, more the rate of mass will be reducing. This is related to the hydrophilic properties of GO that improves the hydrophilic properties of composite making the bioplastic more easily degraded [12, 26]. According to Wypich [27], water can break down the structure of the material and help the activation of biology (microbial) during the biodegradation process.

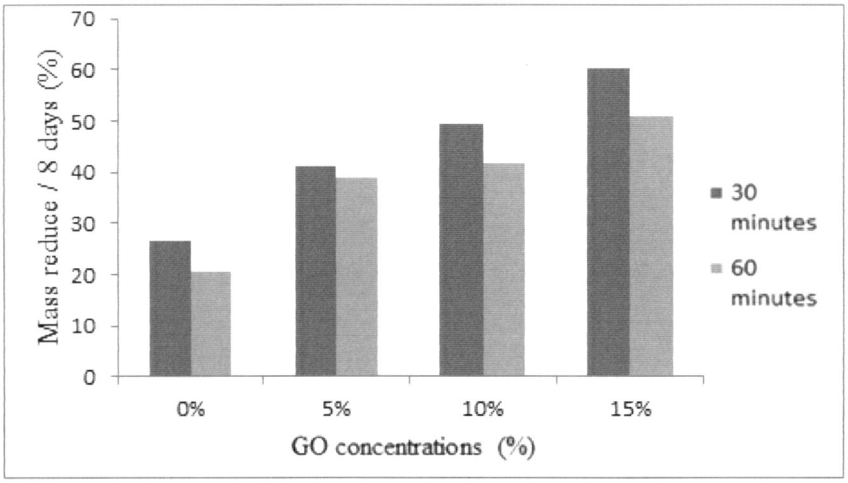

Figure 11. The biodegradation test of bioplastic – GO composites.

Base on Fig. 3.11, it can be seen that the mixing time also influence the rate of biodegradation. The shorter mixing time (30 minutes) has the higher of biodegradation rate. This may be due in the shorter mixing, the interactions among the constituent has not occurred completely yet and there is an empty space between matrix and GO filler forming the intramolekul hydrogen bonds with the water.

4. Conclusions

The graphene oxide filler has been successfully synthesized and integrated in the cassava starch based bioplastics forming the bioplastic-graphene oxide composites. The addition of graphene oxide with various concentrations improved the mechanical properties of bioplastic composites. The more graphene oxide and mixing time the more tensile strength and Young's modulus and the less elongation would be. This was be attributed to the homogeinity of components in the composite as indicated by SEM micrograph images and the FTIR spectra. The estimation of interactions mechanism in the formation of bioplastic - graphene oxide composites was proposed. The water uptake increased as the increase of GO addition and decreased with the longer mixing time attributed to the completeness of interactions among the components. The more GO added, more the biodegradation rate will be. This was atributted to the hydrophilic properties of GO. However, the shorter mixing time, the higher of biodegradation rate due to of empty space in composites as conducive site for biodegradation activities.

Acknowledgment

We gratefully acknowledges to Kemenristek-dikti *via* PUPT research grant 2017.

References

[1] Setiawan H, Lutfi M and Masuroh 2014 Optimasi plastik biodegradable berbahan jelarut (*Marantha Arundinaceae L*) dengan variasi LLDPE untuk meningkatkan karakteristik mekanik *Jurnal Keteknikan Pertanian Tropis Dan Biosistem* Vol **2** 124-130

[2] Supriadi A 2016 Inaplas soroti manajemen sampah dan pengelolaan dana plastik. CNN Indonesia. http://www.cnnindonesia.com/ekonomi / 2016022 2123332-92-*112525/inaplas-sorot - manajemen-sampah-dan- pengelolaan- dana-plastik* / Accessed10 April 2017

[3] LIPI 2016 Indonesia penghasil sampah plastik kedua terbesar di dunia. *http://lipi.go.id/lipimedia/konsumsi-plastik-indonesia-tertinggi-kedua-di dunia/ 15182* Accessed 10 April 2017

[4] Atifah N 2010 *Pemanfaatan hidrolisat pati sagu sebagai sumber karbon pada produksi bioplastik polihidroksialkanoat secara fed-batch oleh Ralstonia Eutropha.* Tesis Institut Pertanian Bogor.

[5] Marbun S 2012 Sintesis bioplastik dari pati ubi jalar menggunakan penguat logam Zno dan penguat alami selulosa. *Skripsi Fakultas Teknik.* Universitas Indonesia, Jakarta.

[6] Avella M 2009 Eco-challenges of bio-based polymer composites *Mater.* **2** 911-925.

[7] Stevens E S 2002 *Green Plastic: An Introduction to the New Science of Biodegradable Plastic* (New Jersey: University Press)

[8] Liu Q 2005 *Food Carbohydrates.* 978-0-203-48528-6.

[9] Arrieta A A, Piedad F G, Samith E M and Robin Z 2011 Electrically conductive bioplastics from cassava starch *J. Brazilian Chem. Soc.* **22** (6) 1170–1176

[10] Rahman M M, Jiang Z T, Yin C Y, Amri A, Chuah L S, Goh B M, Wood B J, Mondinos N, Altarawneh M and Dlugogorski B Z 2016 Structural thermal stability of graphene oxide-doped copper-cobalt oxide coatings as a solar selective surface *J. Mater. Sci. Technol.* **32** (11) 1179-1191

[11] Rafitasari Y, Suhendar H, Imano N, Luciana F, Radean H and Santoso I Sintesis graphene oxide dan reduced graphene oxide *Prosiding Seminar Fisika* FMIPA. Universitas Gadjah Mada, Yogyakarta.

[12] Widjaja H, Jiang Z T, Altarawneh M, Yin C Y, Goh B M, Mondinos N, Amri A and Dlugogorski B Z 2016 Double-sided F and Cl adsorptions on graphene at various atomic ratios: Geometric, orientation and electronic structure aspects *Appl. Surf. Sci.* **373** 65-72

[13] Putri S L 2016. *Pembuatan bioplastik dari pati singkong (Manihot Escullenta Crantz) dengan menggunakan gliserol dan polivinil alkohol (PVA).* Skripsi Universitas Andalas.

[14] Syakir 2015 Kajian pembuatan oksida grafit untuk produksi oksida grafena dalam jumlah besar *Jurnal Fisika Indonesia* 55 Vol XIX, Edisi November 2015

[15] Shao et al. 2012 Graphene oxide: The mechanisms of oxidation and exfoliation *J. Mater. Sci.* **47** (10) 4400-4409

[16] Huang N et al. 2011 Simple room-temperature preparation of high-yield large-area graphene oxide *Intl. J. Nanomed.* **6** 3443-3448

[18] Bayhaki M 2015 Kristalinitas graphene oxide (GO) dari karbon aktif menggunakan metode difraktometer sinar-X UNESA *J. Chem.* **4** 3

[20] Naebe M et al. 2014 Mechanical property and structure of covalent functionalised Graphene/Epoxy nanocomposites *Scientific Reports* **4** (3) 4375

[21] Chen et al. 2013 An improved hummers method for eco-friendly synthesis of graphene oxide *journal homepage www.elsevier.com/locate/carbon*

[23] Tiannan et al. 2011 A simple and efficient methode to prepare graphene by reduced of graphite oxide withh sodium hydrosulfite *Nanotechno.* **22** 045704

[24] Indrayati L, Indrarti and Rahmi E 2006 Pengaruh carboxymethyil cellulose (CMC) dan gliserol terhadap sifat mekanik lapisan tipis komposit bakterial selulosa *Journal Sains Material Indonesia* **8** (1) 40-44

[25] Amni C et al. 2015 Pembutan bioplastik dari pati ubi kayu berpenguat nano serat jerami dan ZNO *Journal Universitas Syiah Kuala*

[26] Darni Y and Utami H 2010 Studi pembuatan dan karakteristik sifat mekanik dan hidrofobisitas bioplastik dari pati sorgum *Jurnal Rekayasa Kimia Dan Lingkungan* **7** (4) 88-93

[27] Wypich G 2003 *Plasticizer Use and Selection for Specific Polymers* (Toronto: ChemTec Laboratories)

IOP Conf. Series: Materials Science and Engineering **345** (2018) 012026 doi:10.1088/1757-899X/345/1/012026

Conductive composites of tapioca based bioplastic and electrochemical-mechanical liquid exfoliation (emle) graphene

A Amri[1,2], H Rahmana[1], S P Utami[1], R S Iriyanti[1], Z T Jiang[2] and M M Rahman[2,3]
[1]Department of Chemical Engineering, Universitas Riau, Pekanbaru, Indonesia
[2]School of Engineering and Information Technology, Murdoch University, Australia
[3]Department of Physics, Jahangirnagar University, Bangladesh.

E-mail: amun.amri@eng.unri.ac.id

Abstract. The conductive composites of tapioca based bioplastic and the electrochemical-mechanical liquid exfoliation (EMLE) graphene have been successfully synthesized via the solution intercalation method for conductive bioplastic applications. The synthesized EMLE graphene quality, the mechanical properties, the functional group interactions and the conductivity of bioplastic composites, respectively, were analyzed using Raman spectroscopy, Universal Testing Machine (UTM) via ASTM D882-92, Fourier Transform Infrared (FTIR) spectroscopy, Multitester via Four Probe Method. Raman spectroscopy analyses revealed that the graphene used is multi layer graphene (~ 3-10 layer) with deffects and minor impurity of graphene oxide (EMLE graphene). The tensile strength and the Young's modulus increased with the increasing of the EMLE graphene content in the composites, while the elongation decreased. The bioplastic synthesized using the 9% EMLE graphene content and the mixing time of 50 minutes exhibited the best mechanical properties with the tensile strength of 4.116 Mpa, the Young's modulus of 75.476 Mpa, and the elongation of 5.453%. The FTIR spectra indicated that there was a good interactions of EMLE graphene in the bioplastic matrix due to the hydrophylic properties and the secondary bonds between the EMLE graphene and the starch and glycerol plasticizer. The higher amount of graphene added, the higher conductivity of bioplastic would be, and vice versa for the resistivity. The best electrical properties of 1.57 $x10^{-1}$ /ohm.cm (conductivity) and 6.34 ohm.cm (resistivity) was reached by the bioplastic synthesized with addition of 9% EMLE graphene and 50 minutes stirring time. EMLE Graphene is the promissing filler for further development of Tapioca based conductive bioplastics.

1. Introduction

Plastics play an important role in human life, ranging from simple needs to complex and sophisticated needs. The need for plastic from time to time increases continously [1]. Conventional plastics that are still used today are derived from synthetic polymeric materials made from petroleum or natural gas whose feedstocks are increasingly limited and non-renewable as well as difficult to recycle and non-degradable. This can lead to the environmental pollution [2]. Different from the conventional plastic, the bioplastics is a biodegradable plastic, which is in turn, it will break down by microorganisms enzymatically in soil and produce water and carbon dioxide gas. Bioplastic is a solution to reduce the environmentally problem due to of the non-degradable plastic wastes.

Bioplastics can be synthesized from renewable natural materials such as starch [3,4], protein [5], and bacteria [6]. Generally, among the polysacharides and biopolymers, starch is considered the most potential and promising natural ingredient for use in biodegradable plastics [7]. Because it is easy to obtain and the amount is abundant as well as it is easy processing so it is very promising to be developed. One source of starch is tapioca flour from cassava plants which has the highest starch content reaching 90% [8]. Generally, the bioplastics from cassava starch have good tensile strength and transparency properties [3].

The lack of bioplastic applications in the electronics sector is due to the isolator properties of bioplastic. Whereas the biodegradable properties of bioplastics are very useful in electronic devices. The strategy of converting the non-conductive to conductive properties of bioplastic is the brilliant thought. One of the most suitable alternatives and up to date is by the integration graphene into the bioplastic matrix. Graphene is the thinnest, strongest and most powerful material in the world today formed from the arrangement of two-dimensional carbon monolayer atoms that make up the hexagonal crystal structure resembling a honeycomb [9]. Graphene has electronic properties with high electrical conductivity and good flexibility. Graphene has unique and superior properties compared to other materials such as high electron mobility reaching 200,000 cm^2/Vs, high electrical conductivity ($0.96 \times 10^6\ \Omega^{-1}\ cm^{-1}$), high thermal conductivity (5000 W/mK), good optical transparency (97.7%), and has a tensile strength of 1 TPa or 200 times harder than steel and 20 times harder than diamond and 130 Gpa stiffness [10]. The electronic structures of graphene is also affected by the relative orientation of the adsorbed sites (zigzag or armchair) and possibly the relative size / mass of the adatoms and carbon [11,12].

This work investigates the properties of tapioca based bioplastic with the addition of graphene and study their interactions. The discussed variables are the effects of graphene content and mixing time in the synthesis process of composite against the mechanical and electrical properties of bioplastic composites. In this work, Graphene is obtained from combination of electrochemical and mechanical liquid exfoliation (EMLE) method. The identity of graphene used (EMLE graphene) and its interactions in the composite are also discussed. The tensile strength, Young's modulus and the conductivity of biocomposite increase around 2.8 times, 10,5 times and 14.4 times, respectively, after the addtion of 9% v/v multi layers (~3-10 layers) graphene. The remarkable improvement of mechanical and electrical properties accompanied by the low cost and environmentally friendly in the synthesis process, make this bioplastic-graphene conductive composite is very promising to be explored further for industrial scale.

2. Experimental

2.1 Synthesis of electrochemical-mechanical liquid exfoliation (EMLE) graphene

The electrochemical-mechanical liquid exfoliated (EMLE) Graphene made via electrochemical process of graphite rod of 2B pencil continued by the mechanical liquid turbulence shear exfoliation. In the electrochemical process, the graphite was applied as working electrode, while the steinless steel was as counter electrode. Both electrodes were then dipped in Na_2SO_4 solution. Electrochemical peeling was started by applying a positive voltage of 10 V in the system. When the exfoliation process was complete, the produced graphite flakes were filtered and washed using aquadest repeatedly until netral pH was reached [14]. Then, it was dried forming the graphite flakes powder. Mechanical liquid exfoliation process was applied to the 10 of grams graphite flakes powder in rotor blade household blender with addition of 5 ml FL (fairy liquid) surfactant and addition of aquadest until total volume of 500 ml. Blender was operated at maximum speed for 60 minutes and keep the process in the blender cold [12]. The product of this electrochemical and mechanical exfoliation process was called as EMLE graphene.

2.2 Synthesis of conductive bioplastics composite

Preparation of bioplastic was carried out using solution intercalation method. 10 grams of tapioca flour were dissolved into 100 ml aquadest and heated over with water bath with temperature of 60^0C while stirring. Thereafter, glycerol plasticizer and EMLE graphene were added according to the variables (0%, 3%, 6%, 9% v/v) of EMLE graphene, then stirred until homogen with stirring time variation (30 minutes and 50 minutes). The homogeneous solution was then poured into the flexi glass mold and dried for 1 day at room temperature. The bioplastic was separated from the mold and ready for characterizations.

2.3 Characterization

The produced graphene was analyzed using Raman spectroscopy indetified using excitation energy at wavelength 532 nm. The mechanical properties (tensile strength, elongation and Young's modulus) were tested based on the ASTM D882-92 standard using Universal Testing Machine (UTM). The qualitative interactions of graphene with the organic/functional groups in bioplastics were investigated using FTIR spectrophotometer. The electrical conductivity of bioplastic was probed based on the four probe method by using multitester.

3. Results and discussion

3.1 Raman spectroscopy analysis

Raman spectroscopy is used to study the quality and the number of graphene layers produced. Fig. 1 shown the Raman spectrum of sample generated from the process of electrochemical and mechanical liquid exfoliation graphite. From the spectrum looks there are three major peaks, namely peak of G band at the wavenumber of 1583.64 cm^{-1}, peak of 2D band at 2690.88 cm^{-1} and peak of D band at 1353.46 cm^{-1}. Normally, for a single-layer graphene, the G band appears at wavenumber around of 1580 cm^{-1} [15]. This vibrational involves sp^2 of carbonized carbon hybridization consisting of graphene sheets. While the 2D and D bands appear at around wavenumber of 2680 cm^{-1} and 1350 cm^{-1}, respectively [15].

The size and shape of G and 2D band peaks can be analyzed for identification purposes. The single-layer Graphene has higher 2D band spectrum with a symmetrical and sharp compared to the G band [17], and it is vice versa exhibited in Fig. 1. Another analysis, from Fig. 1, the peaks intensity ratio of 2D and G bands (I_{2D}/I_G) is 0.48 which is far from the pristine (without deffects) and single layer graphane where the I_{2D}/I_G value of 2 [15,16, 17]. The wider and the higher peak of the 2D band indicate the higher multi layer of graphene and vice versa. The thicker the layer, the 2D band form is more asymmetric [15,18]. These elucidations indicate that the produced graphene in this work is not a single layer.

Quantitatively, the number of layers in graphene can be counted via the analyses of G and 2D bands, while the presence of D band indicates the existence of deffects in the graphene [15]. The position of G band will shift to a lower wavenumber if the number of graphene layer increase. It can be predicted through the following equation [15, 16]:

$$\omega_G = 1581{,}6 + 11/(1+ n^{1,6}) \quad (1)$$

where the ω_G is the G band wavenumber detected on the sample. From this equation, it is found the number of layers in the EMLE graphene is ~ 3 layer. The 2D band can also be used to determine the graphene layer qualitatively. However, the position and shape of the 2D bands can differ significantly depend on the excitation energy applied. Therefore, it is important to use the same excitation laser frequency when the comparisons are carried out (532 nm or 633 nm) [15]. [19] have The identification of graphene layers based on the G and 2D band shapes has been elucidated [19]. Based on this view, the G and 2D bands in this work have a high similarity with the G and 2D bands of ~5-10 layers of graphene [19].

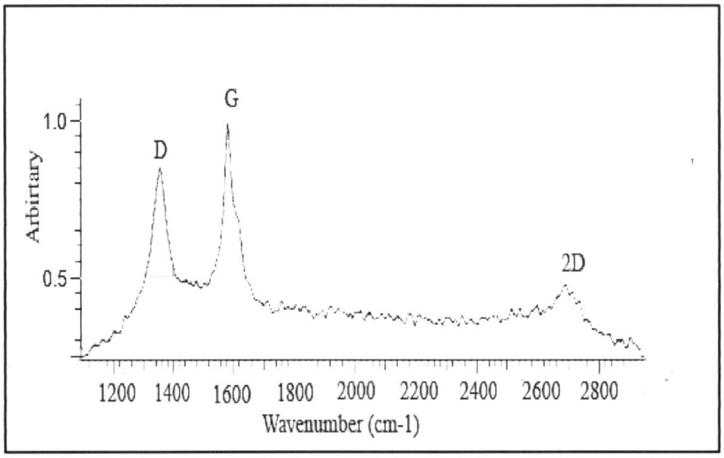

Figure 1. Raman spectrum of sample obtained from EMLE graphite.

The appearance of band D in this study indicates the presence of defects in graphene produced and the addition of sp 3 hybridization by covalent bonds. This peak is very weak in graphite or in high quality graphene. If the value of band D is high or significant, it can be interpreted that a lot of damages or impurities exist in the sample [15,16]. If the deffect level in graphene is very high, indicated by the very high D band peak, then there will be the D' band at around 1620 cm^{-1} and the D+G band at around 2940 cm^{-1} as reported in [16]. However there is no these extra bands in the graphene produced in this work.

The graphene oxide could be exist in the sample even though the synthesis process did not involve the oxidation reactions directly. It is possibly due to the C^+ from the graphene sheet edge (when deffects occured) captures the electron from hydroxyl (OH^-) forming the graphene oxide. The shoulder at the higher wavenumber of G band peak as seen in Fig. 1, it indicated that the presence of graphene oxide (C-OH) bonds in the EMLE graphene produced [20]. However the C-OH bonds are minority in the sample as indicated by the lower intensity of D band peaks than G band peak and the absence of D' and D+G extra bands [20]. From the analyses either from band position, band shapes, comparisons or from the calculations performed, it can be concluded that the graphene produced in this research has defects with the number of layers are ~ 3-10 layer (multi-layer) and with minor impurity of graphene oxide. We call this product as electrochemical and mechanical liquid exfoliation (EMLE) graphene or EMLE graphene.

3.2 Mechanical properties

The mechanical properties of bioplastic with several of EMLE graphene content and mixing time are presented in Table 1 and Fig. 2-4. From Figure 2, it can be seen that the tensile strength increase with the increasing of the EMLE graphene content, and the increasing of the mixing time. This phenomenon may be elucidated as follow: the graphene will fill the cavities formed when the galatinization of starch (in the synthesis process) occured. This composite forms a solid and compact structure due to the interaction of graphene with the matrix physically or chemically via the secondary bonding especially the adjacent oxygen and hydrogen atoms between the graphene oxide and the matrix as elucidated in FTIR analysis section. Indriyanti et al. [21] reported that the higher filler content in bioplastics will increase the stiffness and tensile strength values of bioplastics. The increase in tensile strength is due to the tensile forces interactions between thin-film molecules. The stirring time will affect the distribution of graphene filler within the matrix cavity, the longer the agitation (50 minutes) the graphene particles will spread evenly within the bioplastic matrix, and the mechanical properties of bioplastic tensile strength will increase.

Table 1. Testing results of bioplastic mechanical properties

Sample (v/v)	*Tensile Strenght* (Mpa)		Elongation (%)		*Young Modulus* (Mpa)	
	30 min	50 min	30 min	50 min	30 min	50 min
0%	1.274	1.47	27.58	20.617	4.619	7.129
3%	2.254	2.548	18.684	18.613	12.063	13.689
6%	2.848	3.038	16.12	11.782	17.63	25.783
9%	3.92	4.116	8.516	5.453	46.031	75.476

The addition of larger quantity of graphene in this work (9%) results the larger tensile strength values compared with no graphene addition (0%) and fewer graphene additions (3%). A longer mixing time (50 minutes) has the higher tensile strength than a shorter mixing time (30 minutes). The largest value of tensile strength is 4.116 MPa showed by sample synthesized using graphene addition of 9% and stirring time of 50 minutes.

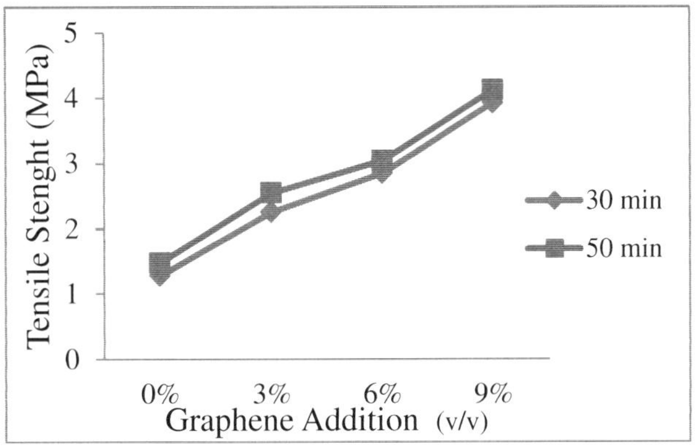

Figure 2. Effects of graphene addition and mixing time to the bioplastic tensile strength.

The effect of graphene filler and mixing time to the elongation properties of bioplastic produced can be seen in Figure 3. It can be seen that the elongation value greatly decreases as the increase of graphene content and the length of stirring time. Elongation is a maximum change in length of the bioplastic before it is broken. The higher the elongation value the less of quality of the plastic [22]. In this work, the lowest elongation value showed by sample at 9% graphene content and 50 minutes agitation time namely 5.453%. The addition of graphene filler will decreases the bonding distance between molecules and reduces the internal hydrogen bonds to weaken the tensile force on the polymer chain before breaking up [23]. The same thing was also reported by Sriprachuabwong et al., [24] which obtained good flexibility results along with the increase in the amount of graphene in their polymer [24].

The Young's modulus is obtained from the comparison between the tensile strength to percent of elongation (elongation at break). Young modulus can also be regarded as a measure of the strength of a material [25]. Fig. 4 shows the effect of graphene filler and stirring time to the Young's modulus. It can be seen that the Young's modulus increases with the increase of graphene content and stirring time. The greatest Young's modulus (75.476 Mpa) is exhibited by sample synthesized at addition of 9% graphene and 50 minutes of stirring time. The similar trend was also reported by Susanti et al., [26] who conducted research to synthesize tapioca starch bioplastic with bamboo fiber reinforcement. This trend is due to the value of the Young's modulus is directly proportional to the value of tensile strength and inversely proportional to the value of elongation. Thus, the addition of graphene and the duration of mixing time will affect the value of tensile strength and Young's modulus.

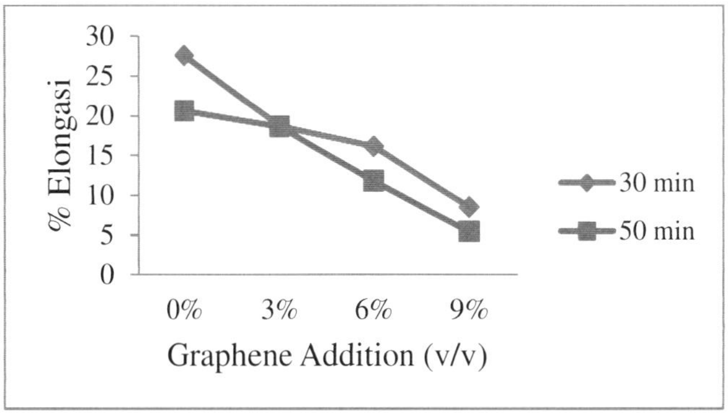

Figure 3. Effect of graphene addition and stirring time against the bioplastic elongation.

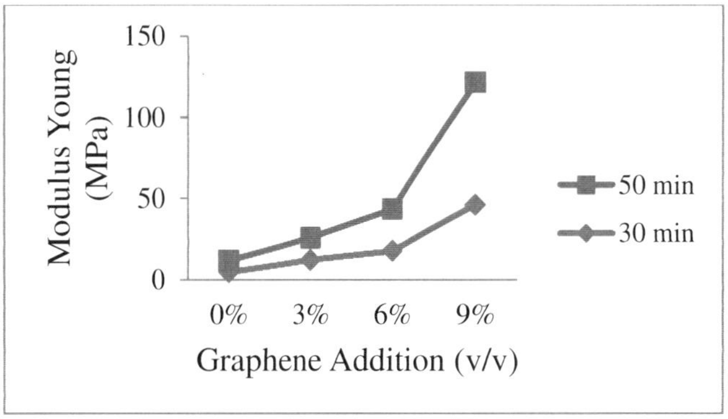

Figure 4. Effect of graphene addition and mixing time to young's modulus of bioplastic.

3.3. FTIR analyses and interactions mechanism

Figure 5 is the FTIR spectra of bioplastic without and with graphene content, respectively. It can be seen that the FTIR spectrum of bioplastic without the addition of graphene shown the presence of O-H bonds at 3321.56 cm^{-1}, C-H bonds at 2917.46 $cm^{-1}$and C-O bonds at 1057.75 cm^{-1} (Fig. 5.a). They are from the starch (glucose) and glycerol plasticizer. The bonds from O-H group indicated the hydrophilic property of bioplastic. Plasticizers in bioplastic make hydrogen bridges with starch molecules that weaken the strength of both the intra and intermolecular bonds of hydrogen, thereby reducing the interaction between amylase and amylopectin and making the film flexible. Fig. 5b shows the wavenumber shifting of bioplastics after the addition of graphene. It can be seen the significant change ocures at spectra around 1057 cm^{-1} (C-O bonds) and the bonds at around 2400 cm^{-1} which could be associated with the C=C bonds in graphene. These indicate a good interaction between the bioplastic matrix and graphene filler.

The interactions in the biocomposites are elucidated as the following. Basically the heated starch granules will be break when the gelatination temperature (60°C) is reached and water come into the starch granules [27]. This will cause the granules expand and eventually break. The cracked starch granules will open the cavities in the starch crystalline structure and also cause the amylose and amylopectin as the component of the starch diffuse out of the granules [27,28]. The amylose and the

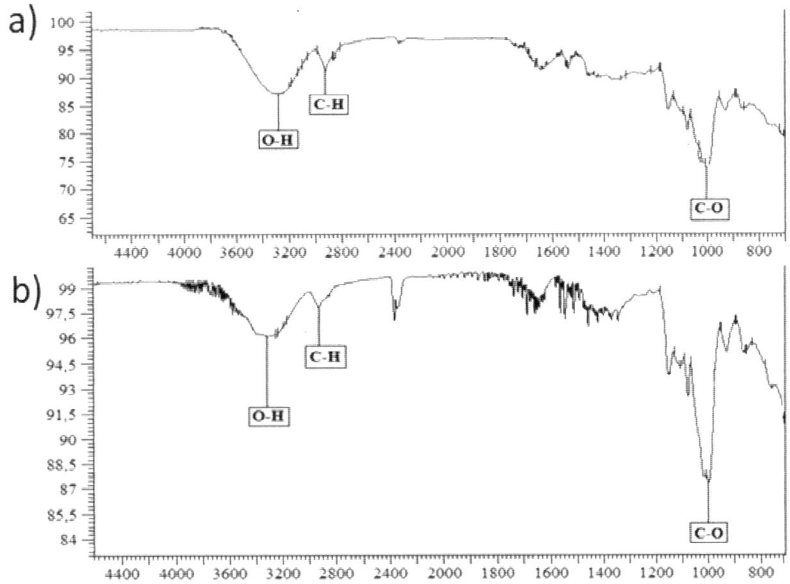

Figure 5. FTIR spectra for bioplastic: a). without graphene and, b) with 9% of graphene.

amylopectin then have a good interaction with the glycerol structure and the graphene filler due the hydrophylic properties among them and due to the secondary bonds (hydrogen bonds, O--H) between the oxygen from the graphene and the hydrogen from the matrix, and vice versa. The interactions mechanism among the starch, glycerol and the graphene in the biocomposite can be estimated as seen in Figure 6.

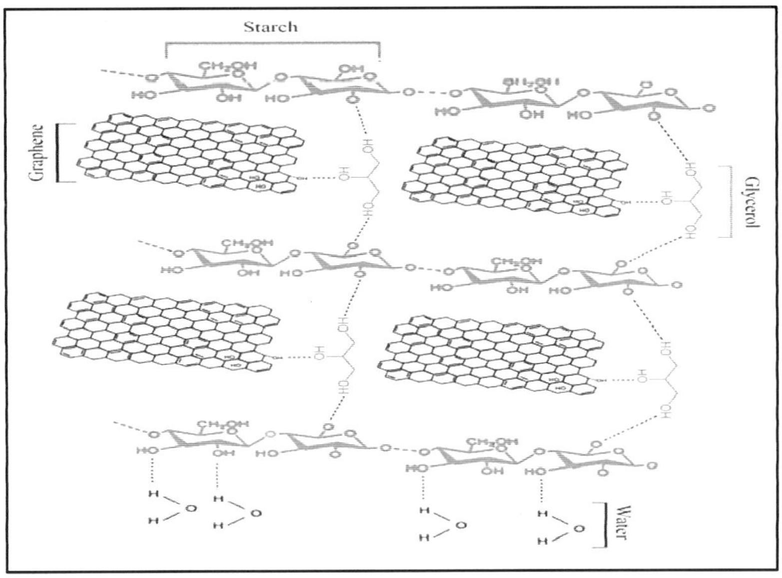

Figure 6. Estimation of starch-glycerol-graphene interactions mechanism.

3.4 Resistivity and conductivity of bioplastic composite

Bioplastic electrical properties were measured using the Four Probe Method. Electrical resistance (resistivity, Rs) and conductivity (σ) are calculated by the following equation:

$$Rs = \frac{\pi t}{ln2} x \text{ R} \quad (2)$$

$$\sigma = \frac{1}{Rs} \quad (3)$$

The results of the experiment are listed in Table 2 and Fig. 7. Generally, it can be seen that when the

Table 2. Resistivity and conductivity of bioplastic composite

Sample (v/v)	0%		3%		6%		9%	
	30 min	50 min	30 min	50 min	30 min	50 min	30 min	50 min
R (ohm)	200	202	38	34	30	28	19	14
t (cm)	0.1	0.1	0.1	0.1	0.1	0.1	0.1	0.1
Resistivity/Rs (ohm.cm)	90.60	91.51	17.21	15.40	13.59	12.68	8.61	6.34
Conductivity (/ohm.cm)	0.011	0.011	0.058	0.065	0.074	0.079	0.111	0.158

EMLE graphene content increase, then the resistivity decrease or conductivity increase. Likewise the mixing time, where the longer mixing time, the resistivity decrease or the conductivity increase.

The best conductivity value was found in sample with graphene content of 9% v/v and mixing time of 50 minutes namely 1.57 x10^{-1} /Ωcm or resistivity of 6.34 Ωcm. As a comparison, Arrieta et al., [3] reported the best conductivity value for their bioplastic-lithium perchlorate system is 8.1 x 10^{-3} /Ωcm. For their bioplastics, the similar trend was also found where the conductivity increased when the lithium perchlorate filler was increased [3]. Even though the filler added was quite high, but the conductivity was still very low [3]. The type of fillers will affect the properties of the resulting bioplastic composites including the electrical property. Therefore, the structure and functional properties of the composite can be prepared in accordance with the desire by selecting the filler materials [2]. In this case, graphene has unique and superior properties compared to other materials, i.e. high electrical conductivity of 0.96 × 10^6/Ωcm [10].

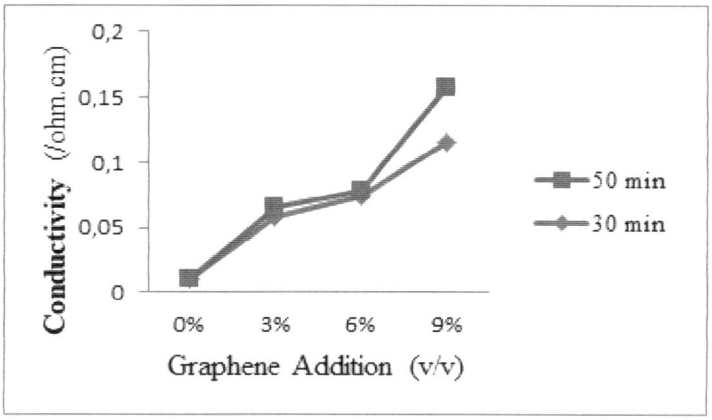

Figure 7. Effect of graphene content and mixing time on electrical conductivity of bioplastic composite.

4. Conclusions

The conductive composites of tapioca based bioplastic and the electrochemical-mechanical liquid exfoliation (EMLE) graphene have been successfully synthesized via the solution intercalation method. Raman spectroscopy analyses showed that the graphene used is multi layer graphene (~ 3-10 layer) with deffects and minor impurity of graphene oxide (EMLE graphene). The tensile strength and the Young's modulus increased with the increase of the EMLE graphene content in the composites, while the elongation decreased. The bioplastic synthesized using the 9% EMLE graphene content and the mixing time of 50 minutes exhibited the best mechanical properties with the tensile strength of 4.116 Mpa, the Young's modulus of 75.476 Mpa, and the elongation of 5.453%. The FTIR spectra indicated that there was a good interactions of EMLE graphene in the bioplastic matrix due to the hydrophylic properties and the secondary bonds between the EMLE graphene and the starch and plasticizer. The higher amount of graphene added, the higher conductivity of bioplastic would be, and vice versa for the resistivity. The best electrical properties of 1.57 $x10^{-1}$ /ohm.cm (conductivity) and 6.34 ohm.cm (resistivity) was reached by the bioplastic synthesized with addition of 9% EMLE graphene and 50 minutes stirring time.

Acknowledgments

We gratefully acknowledges to DIKTI *via* PUPT research grant 2017. Thanks are also for Fathiyah Zulfahni, Rifati Hanifa and Lestari Ekawati for their help and valuable discussions.

References

[1] Gosta D R 2016. Inaplas: Konsumsi plastik tumbuh 5,2%. http://industri.bisnis.com /read/20161227/257/615083/inaplas-konsumsi-plastik-tumbuh-52. Accessed 12 April 2017
[2] Avella M, Aleksandra B, Maria E E, Gennaro G and Anita G 2009 Eco-challenges of bio-based polymer composites *Mater.* **2** (3) 911-925
[3] Arrieta A A, Piedad F G, Samith E M and Robin Z 2011 Electrically conductive bioplastics from cassava starch *J. Brazilian Chem. Soc.* **22** (6) 1170-1176
[4] Sagnelli D, Kim H H, Eric L, Agnes R S, Sophie G, Jacob J K K, Kell M, Denis L, Andreas B 2016 Plant crafted starches for bioplastics production *Carbohydrate Polym.* **152** 398-408
[5] Perez V, Manuel F, Alberto R and Antonio G 2016 Characterization of pea protein-based bioplastics processed by injection moulding *Food and Bioprod. Proc.* **97** 100–108
[6] Singh S and Mohanty A K 2007 Wood fiber reinforced bacterial bioplastic composites: Fabrication and performance evaluation *Comp. Sci. Technol.* **67** (9) 1753-1763
[7] Vieira M G A, Mariana A D S, Lucielen O D S and Marisa M B 2011 Natural-based plasticizers and biopolymer films: A review *Eur. Polymer J.* **47** (3) 254-263
[8] Cui S W 2005 *Food Carbonhydrate: Chemistry Physycal Properties and Applications* (CRC Press) p. 1-8
[9] Fikri A A and Dwandaru W S B 2016 Pengaruh variasi konsentrasi surfaktan dan waktu ultrasonikasi terhadap sintesis material graphene dengan metode liquid sonification exfoliation menggunakan tweeter ultrasonication graphite oxide generator *Jurnal Fisika* **5** 3.
[10] Casero E, Parra-Alfambra A M, Petit-Dominguez M D, Pariente F, Lorenzo E and Alonso C 2012 Differentiation between graphene oxide and reduced graphene by electrochemical impedance spectroscopy (EIS) *Electrochem. Comm.* **20** (1) 63-66
[11] Hartanto W, Zhong T J, Mohammednoor A, Chun Y Y, Amri A, Bee M G, Nicholas M, Bogdan Z D 2016 Double-sided F and Cl adsorptions on graphene at various atomic ratios: Geometric, orientation and electronic structure aspects *Appl. Surf. Sci.* 373 65-72
[12] Varrla E, Keith R P, Claudia B, Andrew H, Ronan J S, Joe M C and Jonathan N C 2014 Turbulence-assisted shear exfoliation of graphene using household detergent and a kitchen blender *Nanoscale* **6** (20) 11810-11819

[13] Rahman M M, Jiang Z T, Yin C Y, Amri A, Chuah L S, Goh B M, Wood B J, Mondinos N, Altarawneh M, Dlugogorski B Z 2016 Structural thermal stability of graphene oxide-doped copper-cobalt oxide coatings as a solar selective surface *J. Mater. Sci. Technol.* **32** (11) 1179-1191
[14] Parvez K, Wu Z, Li R and Graf R 2014 Expoliation of graphite into graphene in aqueous solutions of inorganic salt *J. American Chem. Soc.* 136 6083-6091
[15] Wall M 2011 The raman spectroscopy of gaphene and the determination of layer thickness *Thermo Scientific* 1-5
[16] Childres I, Luis A J, Helin C, Wonjun P and Yong P C 2013 Raman spectroscopy of graphene and related materials *Birck Nano Tecnology Center* **19** 4-5
[17] Yan Z and Andrew R B 2010 Characterization of graphene by raman spectroscopy raman spectrum of graphene the G - band *OpenStax CNX* 6-9
[18] Papageorgiou D G, Kinloch I A and Young R J 2015 Graphene/elastomer nanocomposites *Carbon* **95** 460-484
[19] Ferrari A C, Meyer J C, Scardaci V, Casiraghi C, Lazzeri M, Mauri F, Piscanec S, Jiang D, Novoselov K S, Roth S and Geim A K 2008 Raman spectrum of graphene and graphene layers *ResearchGate* 2-3
[20] Yazici E, Yanik S and Yilmaz M B 2016 Graphene oxide nano-domain formation via wet chemical oxidation of graphene *Carbon* Elsevier Ltd.
[21] Indriyati L, Indrarti and Rahimi E 2006 Pengaruh carboxymethyil cellulose (CMC) dan gliserol terhadap sifat mekanik lapisan tipis komposit bakteri selulosa *Jurnal Sains Materi Indonesia* **8** (1) 4044
[22] Amni C, Marwan and Mariana 2015 Pembuatan bioplastik dari pati ubi kayu berpenguat nano serat jerami dan ZnO *Jurnal Litbang Industri* **5** (2) 91-99
[23] Zuraida A, Yusliza Y, Anuar H and muhaimin M K 2012 The effect of water and citric acid on sago starch bio-plastics *Intl. Food Res. J.* **19** (2) 715-719
[24] Sriprachuabwong C, Sorawit D, Komkrit S, Anurar W and Adisorn T 2015 Electrolytically exfoliated graphene–polylactide-based bioplastic with high elastic performance *J. Appl. Polymer Sci.* **132** (6) 2-9
[25] Rifaldi A 2017 *Pembuatan bioplastik berbasis pati sagu dengan filler clay dan gliserol* Skripsi Fakultas Teknik Universitas Riau
[26] Susanti S, Jasruddin D M and Subaer J 2015 Sintesis komposit bioplastik berbahan dasar tepung tapioka dengan penguat serat bambu *Jurnal Sains dan Pendidikan Fisika* **11** (2) 173-178
[27] Jacobs H and Delcour J A 1998 Hydrothermal modifications of granular starch with retention of the granular structure: Review *J. Agric. Food Chem.* **46** (8) 2895-2905
[28] Yuniarti L I, Hutomo G S and Rahim A 2014 Synthesis and characterization of bioplastic based on sago starch (Metroxylon Sp) *e-J Agrotekbis* **2** (1) 38-46

Lauryl Amine as heavy metal collector of boiler ash from pulp and paper mill waste

M P Sembiring[1,2], J Kaban[1], N Bangun[1] and E Saputra[2]
[1]Chemistry Department, Facultyof Mathematic and Natural Science, University of Sumatera Utara
[2]Chemical Engeenering Department, Faculty of Engeenering, University of Riau

E-mail: m_peratenta@yahoo.com

Abstract. Theincreasing of demand of pulp and paper products, will following with the growing the pulp and paper industryand generate significant mill waste. The total waste reached 1/3 of the amount raw materials used and ash boiler is the waste with the largest percentage of 52%. For that it takes effort to manage the existing waste. The boiler ash contained the chemical elements, it can be utilized such as fertilizer, because it also contains transition metals in form of heavy metal such as Cadmium (Cd), Cobalt (Co), Chrome (Cr), Cupprum (Cu), Ferrum (Fe), Nickel (Ni), and Zinc (Zn), the use of boiler ash must follow the threshold specified by the Government. Several studies have been undertaken to reduce and extract heavy metals from ash and sand of the boiler by using carbon dioxide as its ligand. Eelectrochemical method was used to remove and recovery of heavy metals from the incenerator. This study focused on removal of heavy metals using Lauryl Amine as collector and three solvents namely Dichloromethane, Ethanol and n-Hexane. The treatmentswas able to extract the heavy metal and generally reduce the heavy metal content of ash boiler pulp and paper mill waste. The combination treatment used toreduce the heavy metal content of 5 gram Lauryl Amine collector in Dichloromethane solvent for 4 hours process time.

1. Introduction

The pulp and paper industry generate significant mill waste such as boiler ash and lime by-products (lime mud, grit, and dregs) daily. As example, a pulp and paper company namely PT. Riau Andalan Pulp and Paper (RAPP) in 2012 produce 225,410 ton/adt of boler ash, 83,380 ton/adtof dregs and grits, 66,138 ton/adt of lime mud, 24,529 ton/adt of screen reject[1]. Most of these products are placed in landfills. RAPP is examining the feasibilty of these solid waste in road making, brick production and explore their use to be including in the construction and concrete manufacture. Land application of these solid waste especially boiler ash, for its potential as soil liming materials may also be conducted further as long as it meet Indonesia's regulatory requirement.The research on the applications of waste is necessary.

Several studies shown that the nutrient content of these boiler and lime ash is generally low, especially for N, P and K, they have high pH and the ability to increase soil alkalinity [2]. Application of boilers and other wood ash to soil can improve the growth or yields[3-5]. Evaluation on boiler ash as soil amelioration on pulp wood plantation *Acacia crassicarpa* grown in peat areas have been also conducted and in addition itincreasesthe plant growth and yield, boiler ash application at 10 kg/tree do not change the peat properties[6]. However, considering its heavy metal content such as Cd, Co, Cr, Cu, Ni, Fe, and Zn that may potentially toxic to the plant when its concentration exceed the threshold that plant can tolerate and protecting environment by preventing soil contaminaton, the effort to explore the possibilityto reduce or removeheavy metal content of boiler ash to meet the Indonesia's regulatory requirement is important [7-9].

Some studies have been undertaken to reduce and extract heavy metals from ash and sand of the boiler using carbon dioxide as its ligand[10], absorption of Cs^{+}, Co^{2+}, Eu^{3+} on zirconium silikat [11], extraction [12,13], biosorption [14],pythoremediation[15]and separationkromatography sodium, cobalt and europium [16]. This study focused on removal of heavy metals that can be treated by converting the heavy metal into complex compounds with organonitrogen or amine compounds. In this studyLauryl Amine isused as a collector to reduce the heavy metals .

The removal of heavy metals found in boiler ash can be done by converting the metal transition into complex compounds with organonitrogen or amine compounds [17].The formation of complex compounds with heavy metals by two phases (solid-solution) is determined by various factors such as collector type, collector concentration, particle size, stirring rate, reaction temperature, and contact time of the ligand solution with waste.Amines have distinctive properties that allow them to be absorbed in solids and form cationic films. In this research, the Lauryl Amine was used as collector with three different solvents including Dichlorometane, Ethanol and n-Hexane.

2. Material and method

This research was conducted in Organic Chemical Laboratory, Faculty of Matematic and Natural Science, University of Sumatera Utara,Medan and in Soil Laboratory of PT. Riau Andalan Pulp And Paper Pangkalan (RAPP) Kerinci, Pelalawan, Riau. The samples are obtained from the combustion boiler ash pulp and paper mill of PT. RAPP.

The boiler ash sample was smoothed to 400 mesh sizes, and then the sample was placed in an oven with 100°C for 30 minutes. A 5 g ash is prepared and separated then used as a control.5 g ash was added to 250 ml Erlenmeyer which is filled with 5 g of lauryl amine collector, 30 ml of Dichloromethane solvent. This mixture wass stirred for 4 hours, thenit is left until the upper and lower layers are formed, the two layers were separated using a separation funnel. The bottom layer as slurry was dried using oven. Once the dried sample is obtained, the sample thenanalyzed using Inductively Couple Plasma (ICP) to determine the content of heavy metal. The same procedure was applied to LaurylAmine with extraction time 8 hours. The same procedure was applied for Lauryl Amine 15 gram in Dichloromethane solvent with extraction time of 4 hours and 8 hours.

For ethanol, the same procedures also conducted for 30 ml ethanol with the following procedure. The 5 g ash was added to 250 ml Erlenmeyer which was filled with 5 g of Lauryl Amine collector, 30 ml ethanol. This mixture was stirred for 4 hours, then it is left until the upper and lower layers are formed, the two layers were separated using a separation funnel. The bottom layer as slurry was dried using oven. Once the dried sample is obtained, then the sample was analyzed using Inductively Couple Plasma (ICP) to determine the heavy metal content. The same procedure was applied to Lauryl Amine with extraction time 8 hours and also for Lauryl Amine 15 gram in Ethanol solvent with extraction time of 4 hours and 8 hours.For n-hexane the same procedure was also applied.

3. Results and discussion

3.1. Cadmium

All Lauryl Amine collector treatments combination with three solvents able to reduce Cadmium content compared than control (Figure 1 and Table 1). In term of dosage, 15 gram treatment combination is slightly better than 5 gram, the length of process time 4 hours is better than 8 hours except for n-Hexane with 5 gr Lauryl Amine. The highest Cadmium reduction at 42.5% shown by treatments with Dichloromethane both for 4 hours and 8 hours. The 5 gram Lauryl Amine in

Dichloromethane solvent for 4 hours process time is the best treatment that effecively and efficiently reduce Cadmium content from boiler ash.

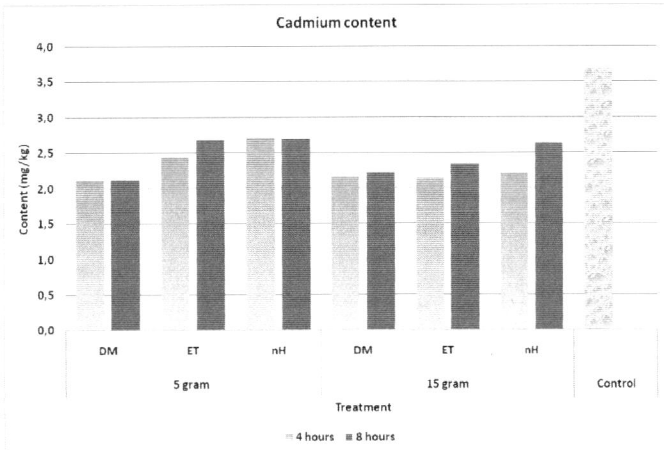

Figure 1. Cadmium content on Lauryl Amine collector and threedifferent solvents at 4 and 8 hours process length.

Table 1. Cadmium content on Lauryl Aminecollector and three different solvents at 4 and 8 hours process length; and its reduction to control

Lauryl Amine Dose (gr)	Solvent	Concentration (mg/kg)		Reduction to control (%)	
		4 hours	8 hours	4 hours	8 hours
5	Dichloromethane	2,1	2,1	42,5	42,5
	Ethanol	2,5	2,7	33,6	27,4
	n-Hexane	2,7	2,7	26,8	27,1
15	Dichloromethane	2,2	2,2	41,5	39,8
	Ethanol	2,1	2,3	42,0	36,6
	n-Hexane	2,2	2,6	40,1	28,7
Control		3,69			

3.2. Cobalt

Similar to Cadmium, Lauryl Amine collector and all three solvents treatments shown the reduction on Cobalt content compare than Control (Figure 2 and Table 2). Dosage of 15 gram show higher reduction compare than 5 gram, except in Dichloromethanesolvent.Lauryl Amine collector at 15 gram with Dichloromethane solvent in 8 hours process time show the highest Cobalt reduction at 41% followed by ethanol solvent in 4 hours process time at 35.7%.

3.3. Chromium

Lauryl Amine collector and all three solvents treatments show reduction on Chromium content compare than Control (Figure 3), with range of 19.5–34.3% and 13.3 – 34.9% for length of process time 4 hours and 8 hours respectively (Table 3). Lauryl Amine collector at 15 gram dosage is better than 5 gram, while 4 hours process time better than 8 hours, except for Dichloromethane solvent.The highest Chromium reduction is shown by 15 gram Lauryl Amine in Dichloromethane solvent for 8 hours process time at 34.9% it is not so much different with 5 gram for 4 hours process time at 32.2%. Considering less Lauryl Amine and time used, treatment with 5 gram Lauryl Amine collector in Dichloromethane solvent for 4 hours process time is the best option.

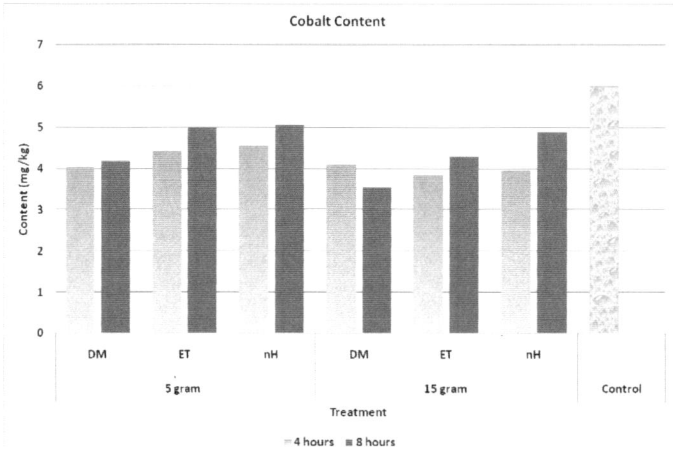

Figure 2. Cobalt content on Lauryl Amine collector and three different solvent at 4 and 8 hours process length

Table 2. Cobalt content on Lauryl Amine collector and three different solvent at 4 and 8 hours process length; and it reduction to control.

Lauryl Amine Dose (gr)	Solvent	Concentration (mg/kg)		Reduction to control (%)	
		4 hours	8 hours	4 hours	8 hours
5	Dichloromethane	4,03	4,18	32,7	30,2
	Ethanol	4,44	4,99	25,9	16,7
	n-Hexane	4,55	5,05	24,0	15,7
15	Dichloromethane	4,11	3,54	31,4	40,9
	Ethanol	3,85	4,29	35,7	28,4
	n-Hexane	3,96	4,88	33,9	18,5
Control		5,99			

3.4. Cuprum (copper)

Reduction of Cuprum content occursin all treatments combination Lauryl Amine collector and all three solvents (Figure 4 and Table 4). The highest Cuprum reduction is on Lauryl Amine collector at 15 gram with Dichloromethane solvent in 8 hours process time at 39.2%. The 15 gram dosage produce higher reduction in all solvent combination compare than 5 gram dosage. 4 hours processing time produce higher reduction compare to 8 hours, except for Dichloromethane solvent. It is more efficient to use 5 gram Lauryl Amine in Dichloromethane solvent for 4 hours process time.

3.5. Ferrum

Ferrum content of boiler ash is reduce after treated with Lauryl Amine collector with all three solvents (Figure 5). Overall the Lauryl Amine at 15 gram dosage produce higher reduction compared than 5 gram in both 4 hours and 8 hours process time, except in Dichloromethane solvent (Table 3). Process time of 4 hours will be reducing the heavy metal higher than 8 hours, except in Dichloromethane solvent. The highest Ferrum reduction at 41.5% it shown that 15 gram Lauryl Amine in Dichloromethane solvent with 8 hours process time.

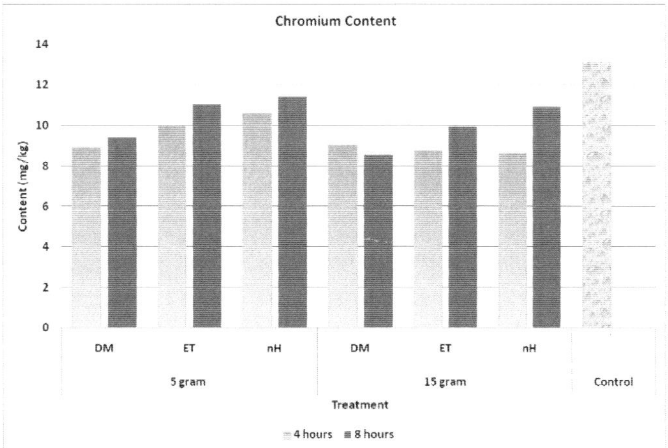

Figure 3. Chromium content on Lauryl Amine collector and three different solvent at 4 and 8 hours process length.

Table 3. Chromium content on Lauryl Amine collector and three different solvent at 4 and 8 hours process length; and it reduction to control

Lauryl Amine Dose (gr)	Solvent	Concentration (mg/kg)		Reduction to control (%)	
		4 hours	8 hours	4 hours	8 hours
5	Dichloromethane	8,9	9,37	32,3	28,7
	Ethanol	10	10,99	23,9	16,4
	n-Hexane	10,58	11,39	19,5	13,3
15	Dichloromethane	9,03	8,56	31,3	34,9
	Ethanol	8,77	9,93	33,3	24,4
	n-Hexane	8,63	10,91	34,3	17,0
Control		13,14			

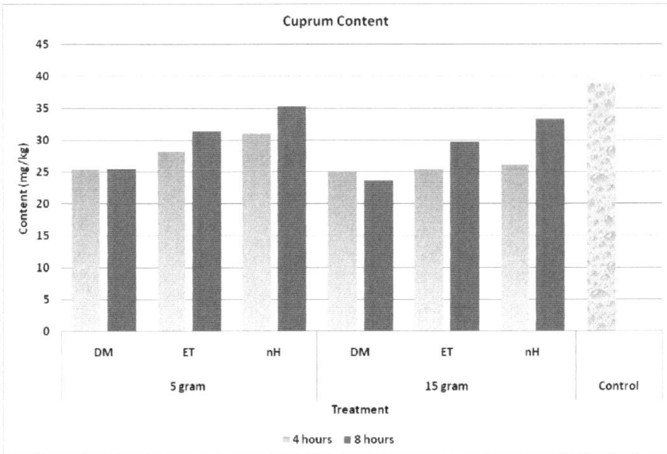

Figure 4. Cuprum content on Lauryl Amine collector and three different solvent at 4 and 8 hours process length.

Table 4. Cuprum content on Lauryl Amine collector and three different solvent at 4 and 8 hours process length; and its reduction to control.

Lauryl Amine Dose (gr)	Solvent	Concentration (mg/kg)		Reduction to control (%)	
		4 hours	8 hours	4 hours	8 hours
5	Dichloromethane	25,35	25,44	34,8	34,6
	Ethanol	28,19	31,39	27,5	19,3
	n-Hexane	31,07	35,22	20,1	9,5
15	Dichloromethane	24,98	23,65	35,8	39,2
	Ethanol	25,41	29,72	34,7	23,6
	n-Hexane	26,18	33,28	32,7	14,4
Control		38,9			

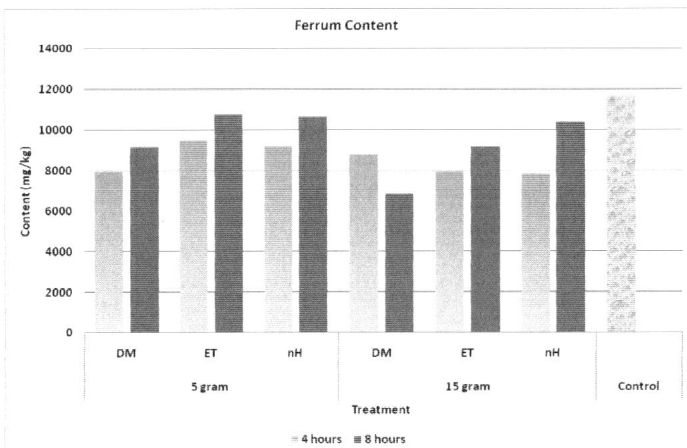

Figure 5. Ferrum content on Lauryl Amine collector and three different solvent at 4 and 8 hours process length.

Table 5. Ferrum content on Lauryl Amine collector and three different solvent at 4 and 8 hours process length; and it reduction to control.

Lauryl Amine Dose (gr)	Solvent	Concentration (mg/kg)		Reduction to control (%)	
		4 hours	8 hours	4 hours	8 hours
5	Dichloromethane	7939	9131	31,8	21,6
	Ethanol	9475	10725	18,6	7,9
	n-Hexane	9166	10615	21,3	8,8
15	Dichloromethane	8777	6814	24,6	41,5
	Ethanol	7952	9169	31,7	21,2
	n-Hexane	7841	10363	32,6	11,0
Control		11642			

3.6. Nickel

Nickel reduction occur in all treatments and the pattern of Nickel reduction figureis quite similar with Ferrum reduction figure, where except in Dichloromethane solvent, the Lauryl Amine at 15 gram dosage produce higher Nickel reduction compared than 5 gram in both 4 hours and 8 hours process time (Figure 6). The 4 hours process time reduce higher Nickelcontent than 8 hours, except in Dichloromethane solvent (Table 6). The highest Nickel reduction at 41.7% shown by 15 gram Lauryl Amine in Dichloromethane solvent with 8 hours process time.

3.7. Zinc

Both 5 gram and 15 gram Lauryl Amine collector in Dichloromethane, Ethanol and n-Hexane solvents will be reducing Zinc content after the treatment (Figure 7 and Table 7). The 15 gram Lauryl Amine at 4 hours process time give the highest reduction of Zinc content at 48.4%, 46.3%, 37.9% for n-Hexane, Ethanol and Dichloromethane solvents respectively.

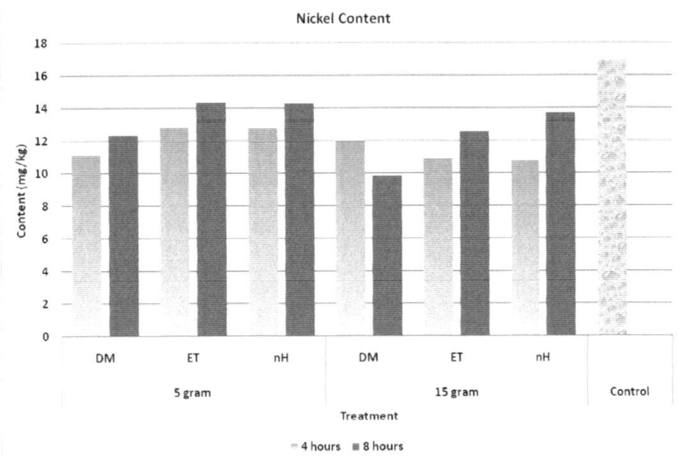

Figure 6. Nickel content on Lauryl Amine collector and three different solvent at 4 and 8 hours process length.

Table 6. Nickel content on Lauryl Amine collector and three different solvent at 4 and 8 hours process length; and it reduction to control.

Lauryl Amine Dose (gr)	Solvent	Concentration (mg/kg)		Reduction to control (%)	
		4 hours	8 hours	4 hours	8 hours
5	Dichloromethane	11,09	12,31	34,3	27,1
	Ethanol	12,82	14,3	24,1	15,3
	n-Hexane	12,76	14,27	24,5	15,5
15	Dichloromethane	12,04	9,84	28,7	41,7
	Ethanol	10,86	12,55	35,7	25,7
	n-Hexane	10,74	13,7	36,4	18,9
Control		16,89			

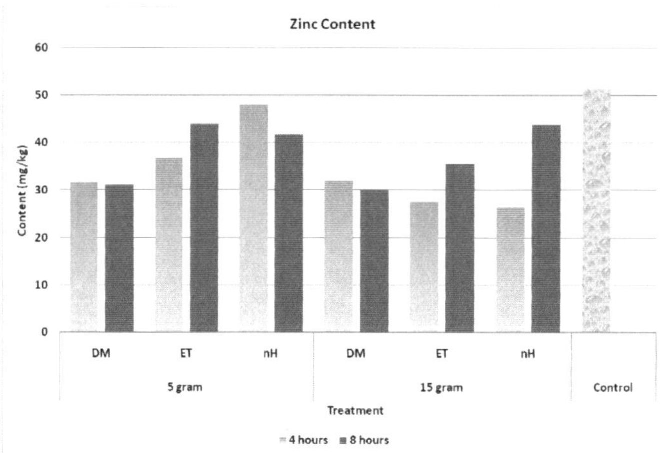

Figure 7. Zinc content on Lauryl Amine collector and three different solvent at 4 and 8 hours process length.

Table 7. Zinc content on Lauryl Amine collector and three different solvent at 4 and 8 hours process length; and it reduction to control.

Lauryl Amine Dose (gr)	Solvent	Concentration (mg/kg)		Reduction to control (%)	
		4 hours	8 hours	4 hours	8 hours
5	Dichloromethane	31,68	31,11	38,4	39,5
	Ethanol	36,86	43,99	28,3	14,4
	n-Hexane	47,965	41,73	6,7	18,8
15	Dichloromethane	31,89	30	37,9	41,6
	Ethanol	27,59	35,6	46,3	30,7
	n-Hexane	26,52	43,76	48,4	14,8
Control		51,39			

Within the 5 gram dosage of Lauryl Amine collector, the Dichloromethane solvent consistently showed higher heavy metal reduction in all elements including Cd, Co, Cr, Cu, Fe, Ni and Zn both in 4 hours and 8 hours process time (Table 8). Within the 15 gram dosage, only the Dichloromethane solvent at 8 hours process time consistently as the highest treatment toreduce heavy metal, 4 hours process time, it shown that not consistently as the highest solvent in reducing heavy metal, the percentage reduction of Dichloromethane solvent is not so different with other solvents. In term of process time, it is obvious that 4 hours reduce more heavy metal than 8 hours. Based on these results, the most effective and efficient treatment combination is 5 gram Lauryl Amine collector in Dichloromethane solvent for 4 hours process time.

Table 8. Summary of heavy metal reduction.

Lauryl Amine Dose (gr)	Solvent	Cd		Co		Cr		Cu		Fe		Ni		Zn	
		4h	8h	4h	8h	4h	8h	4h	8h	4h	8h	4h	8h	4h	8h
5	Dichloromethane	42,5	42,5	32,7	30,2	32,3	28,7	34,8	34,6	31,8	21,6	34,3	27,1	38,4	39,5
	Ethanol	33,6	27,4	25,9	16,7	23,9	16,4	27,5	19,3	18,6	7,9	24,1	15,3	28,3	14,4
	n-Hexane	26,8	27,1	24,0	15,7	19,5	13,3	20,1	9,5	21,3	8,8	24,5	15,5	6,7	18,8
15	Dichloromethane	41,5	39,8	31,4	40,9	31,3	34,9	35,8	39,2	24,6	41,5	28,7	41,7	37,9	41,6
	Ethanol	42,0	36,6	35,7	28,4	33,3	24,4	34,7	23,6	31,7	21,2	35,7	25,7	46,3	30,7
	n-Hexane	40,1	28,7	33,9	18,5	34,3	17,0	32,7	14,4	32,6	11,0	36,4	18,9	48,4	14,8

4. Conclusion

Lauryl amine collector in Dichloromethane, Ethanol and n-Hexane solvent may reduce heavy metal such as Cd (26,8%-42,5%), Co (15,7%-40,9), Cr (13,3%-34,9%), Cu (9,5%-39,2%), Fe (7,9%-41,5%), Ni (15,3% 41,7%), and Zn (6,7% 48,4%) from boiler ash. the most effective and economically efficient treatment used to reduce the heavy metal content of boiler ash from pulp and paper mill waste is 5 gram Lauryl Amine collector in Dichloromethane solvent for 4 hours process time.

References

[1] Anonim 2012 *Sustainability Report PT. Riau Andalan Pulp Paper Waste Management* p. 52-53
[2] Demeyer A, Nkana and Verloo 2001 Characteristics of wood ash and influence on soil *Bioresour. Technol.* **77** 287-295
[3] Cynthia J H and Raymond G P 1985 Recycling mill waste for agricultural use *Tappi J.* **68** (10) 70-73
[4] Makela M, Watkins G, Dahl O, Nurmesniemi H and Poykio R 2010 Integration of solid residue from the stell and pulp and paper industries for forest soil amendment *J. Residuals Sci. Technol.* **7** (4) 191-198
[5] Muse J K and Mitchell C C 1995 Paper mill boiler ash and lime by-products as soil liming materials *J. Agron.* **87** 432-438
[6] Purwati S, Soeetopo R and Setiawan Y 2007 Feasibility of utilization of boiler ash from pulp and paper industry as peat soil conditioner in industrial plantation forest *J. Selulosa* **42** (1) 8-17
[7] Kep 04/bapedal/09/1995 about *Hazardous Waste*
[8] Government Regulation **101** 2014 about *Hazardous Waste*

[9] Government Regulation **82** 2001 on *Waste Management*

[10] Kersch C M, Roosmalen J E Woerlee G F and Witkamp G J 2000 Extraction of heavy metals from fly Ash and sand with ligands and supercritical carbon dioxide *American Chem. Soc.* 4670-4872

[11] Naggar I M, Mowafy E A and Wahed M G A 2007 Sorption mechanism for Cs^+, Co^{2+}, and Eu^{3+} on amorphous zirconium silicate as cation exchanger *J. Solid State Ionic* 741-747

[12] Poykio R, Nurmesniemi H, Peramaki P, Kuokkanen T and Valimaki I 2005 Leachability of metal in fly ash from a pulp and paper mill complex and environmental risk characterization for eco-efficient utilization of the fly ash as fertilizer *Chem. Speciation Bioavail.* **17** (1) 1-9

[13] Nurmesniemi H, Poykio R, Keiski R L and Matsuto T 2008 Total and extractable heavy metal and sulfur concentrations in fly ash from a large-sized (90MW) fluidized bed boiler at pulp mill incinerating biofuels *J. Residuals Sci. Technol.* **5** 171-179

[14] Verma V K, Gupta R K and Rai J P N 2005 Biosorption of Pb and Za from pulp and paper industry effluent by water hyacinth (Eichhorniacrasipes) *J. Sci. Ind. Res.* **64** 778-781

[15] Kumar B A, Jothiramalingam S, Thiyagarajan S K, Hidhayathullakhan T and Nalini R 2014 Phytoremediation of heavy metals from mill effluent soil using Croton spariflorus *Intl. Lett. Chem. Phys. Astron.* **36** 1-9

[16] Gamma El B and Shady S A 2006 Chromatofigureic separation of sodium, cobalt and europium on the particles of zirconium molybdate and zirconium silicate ion exchangers *J Colloid Surf. Physicochem. Eng.* 132-138

[17] Stoica L, Jitaru I., Razvan. A. Micu O and Plapcianu C 2002 The use of recipitate-flotation as new route for mixed oxides precursors obtaining *J. Environ. Protect. Ecol.* **3** 191-195

IOP Conf. Series: Materials Science and Engineering **345** (2018) 012028 doi:10.1088/1757-899X/345/1/012028

Biopolymers to improve physical properties and leaching characteristics of mortar and concrete: A review

M Olivia, H Jingga, N Toni and G Wibisono
Department of Civil Engineering, Faculty of Engineering, Universitas Riau, Pekanbaru 28293 Indonesia

E-mail: monita.olivia@lecturer.unri.ac.id

Abstract. The invention of environmentally friendly, high performance, and green material such as biopolymers marked an emerging trend for sustainable construction over the past decades. Biopolymer comprises of natural monomers and synthesized by plants or other organisms. The sustainable, biodegradable, and renewable biopolymers were used in concrete mixes to improve their physical and mechanical properties and durability. The aim of this paper is to provide a brief an overview of the impact of biopolymer addition into concrete and mortar mixes. Many studies on the influence of biopolymer on the properties of concrete and mortar by adding biopolymers at a certain proportion (usually less than one wt.%) to the concrete or mortar mixes, and the heavy metal leaching, rheological, and mechanical properties of the mixes were conducted. Biopolymers included in this review are chitosan (CH), xanthan gum (XG), guar gum (GG), lignosulphonate (LS), and cellulose ethers (CE). Data from previous studies showed that the addition of certain types of biopolymer into concrete and mortar mixes improve workability, water retention, and compressive strength by up to 30 percent. Chitosan strengthens heavy metal encapsulation in the mortar and neutralizes the negative impact of heavy metal on the mortar properties and environment. To sum up, the use of biopolymers improve physical properties and leaching characteristics of mortar and concrete.

1. Introduction

The construction industry is one of the most significant pillars that support the development of every country. According to Plank [1], the world spent approximately 3 trillion USD to construct home buildings, industrial buildings, and other infrastructures such as bridges, highroads, and railroads in 2000. Some countries with a massive population such as India has invested 1 trillion dollars in infrastructures between 2012-2017, and in the USA, rehabilitation of public infrastructures such as bridges will cost 1.6 trillion dollars in the next 5 years [2, 3]. Concrete is an essential infrastructure material. The primary binder of concrete, cement, is the second most used material with production reach 6000 million tons annually [4]. On the other hand, the processes involved in the cement production have a detrimental impact on the environment. The production of cement worldwide contributes to approximately 5% of global CO_2 emission [5]. CO_2 is one of the greenhouse gasses that accelerates global warming. Due to these facts, researchers continuously study and create inventions in the construction field. New construction methods and materials are intensively investigated to open the window for better sustainability and efficiency with less detrimental impact on the environment.

Bio-based materials such as vegetable fat, animal oil, egg white, fish oil, olive oil, linseed oil, and blood have been used as historical construction materials for many centuries [6]. It was used mostly as an admixture of mortar and plaster to improve viscosity or retarder agent for gypsum. Builders in Brazil used whale oil for concrete workability and durability (waterproofing). The Romans used blood as an air-entraining agent and the Portuguese added vegetable oil in the mortar [7, 8]. In modern building construction, approximately 500 different bio-based admixture products such as lignosulphonate to modify concrete fresh and hardened state have been utilized [7].

The bio-based admixtures can be classified as biopolymers since it is consisted (partly) of bio-based raw materials and biodegradable. The biopolymers are mostly produced by a sustainable process using renewable agricultural and biotechnological biomass residues. It has non-toxic, easy extraction, biocompatibility, and biodegradability characteristics. Biodegradability is an important characteristic that allows decomposition of polymeric chains into smaller molecules [9]. These characteristics indicate a tight relationship between biopolymer and sustainable construction concepts. Since it comes from raw materials such as plants and animals, the material has less cost in producing it, and very popular due to demand for sustainable materials in the construction industry.

The usage of more environmentally friendly material such as biopolymers in concrete and mortar mixtures has been studied intensively. Biopolymers such as polysaccharides, cellulose, starch, chitin, chitosan, alginates from nature are entirely sustainable and biodegradable. In this review paper, the influences of the biopolymer to the rheological, mechanical, and leaching properties of concrete and mortar are discussed. This paper also presents the impact of the biopolymer in mortars which encapsulate heavy metal. The biopolymers in this review paper are limited to chitosan, xanthan gum, guar gum, lignosulphonate, and cellulose ethers.

2. Biopolymers as additive to concrete and mortars

Biopolymers are polymers were synthesized by plant or organism with more complex chemical structure than human-made polymer [10]. Recently the definition of biopolymer has been expanded to any polymer synthesized using natural monomers such as polylactic acid which comprises of lactic acids.

Biopolymers are used in cement mortar and concrete as admixtures to improve properties, as matrices in biocomposites with high mechanical strength, and insulator or high-temperature retardant properties [11, 12]. Two sources of biopolymers for mortar and concrete are from agro-resources extraction and biotechnology through microorganism fermentation and conventional synthesis. Extraction of natural materials or agro-resources materials produce biopolymers such as polysaccharides, cellulose, starch, chitin, chitosan, and alginates. Alginates are extraction polysaccharide from brown algae and act as a self-healing agent for crack repair in concrete without reducing the strength [13]. Biotechnology through microorganism fermentation produces polyhydroxyalkanoates. Biotechnology using conventional synthesis could produce polylactides, and biopolyethylene (PE). Both sources give different dosage levels, product quality, price and environmental safety. Microbial biopolymers such as welan gum are more expensive hence just could be added at 0.002-0.1 wt% from the total formulation of the building material [7]. Lignosulphonate from lignin could be included more in the mixture at 0.1-0.4 wt% since lignin is available abundantly and cost of processing is relatively cheaper than biopolymer from microbial fermentation. Raw materials, process, and equipment cost of biopolymers determine the price and significant contribution of biopolymers to the material properties.

Three basic forms of biopolymers can be included in concrete, i.e., powder, liquid, and fiber [6]. Powder biopolymer, such as chitin, chitosan, and starch is added to cement and water in concrete manufacturing preparation. Chitosan increased viscosity, reduced porosity and improved the final strength of cement pastes, however when added to the concrete mixture it gave adverse impact to the properties [14]. Chitosan is more beneficial to absorb heavy metals although it was included in the mix in small amount (0.4wt.%) [15]. Rubber, xanthan, rubber, duar, gelan, and gutta-percha are the liquid biopolymer. While natural fiber biopolymer without any biopolymerization process and added to the cement mix such as coconut, sisal, jute, hemp, kenaf, and pineapple was used to increase stiffness and tensile strength of concrete. Natural fiber must have several criteria such as high thermal stability, excellent adhesion of fibers and matrix, good long time behavior, high elongation at failure, dynamic behavior and low price and processing cost. Fibres as a part of biopolymer composite matrix are necessarily required to have good fiber-matrix adhesion to allow optimum impregnation in the system [16].

3. Chitosan, Xanthan Gum, and Guar Gum as admixtures to heavy metal encapsulating mortar mixes

Biopolymers such as chitosan, xanthan gum, and guar gum are good viscosizers. Chitosan is derived from chitin by a process known as deacetylation [17]. Chitin is from crustacean shells (crabs and lobsters). Figure 1 shows chemical structure of chitosan and chitin. Xanthan gum and guar gum are usually used to thicken paint materials, soups, and drilling muds. Xanthan gum is produced by the bacterial: Xanthomonas campestris, whereas guar gum is acquired from legume seed: Cyamopsis tetragonolobus which is abundant in India and Pakistan.

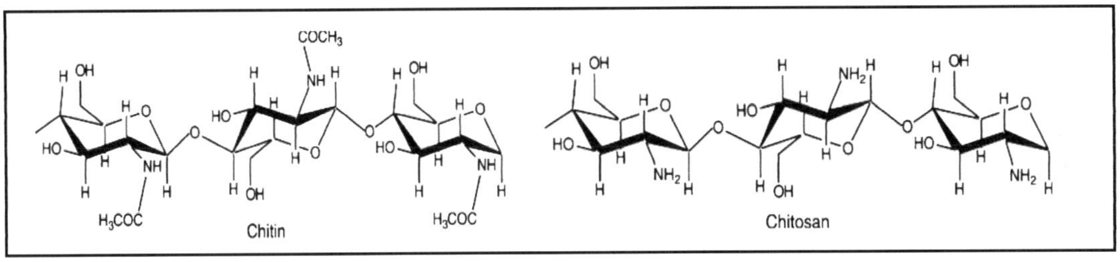

Figure 1. Chemical structure of chitin (left) dan chitosan (right) [18].

Heavy metal is high-density metal which is toxic to the environment. Heavy metal includes lead (Pb), chromium (Cr), zinc (Zn), and mercury (Hg). Heavy metal is not degradable and might contaminate human body at high concentration. Heavy metal enters the human body through food, water, and air. The use of lead-based pipe in clean water distribution pipeline might dissolve lead substance into the water by the process of corrosion [19]. Industrial activity produces heavy metal that adversely impacts nearby air quality. Despite these facts, heavy metal such as zinc is still required in a specific concentration to support metabolism in human body.

Heavy metal is hazardous due to a process called bioaccumulation. Bioaccumulation is defined as the steadily increasing concentration of the chemical substance in an organism over a long time [20]. If the accumulation speed of particular chemical substance is faster than the ability of the organism to degrade or dispose of it, then bioaccumulation occurs. Related to lead exposure, people that work in demolition, welding, or lead-based paint material combustion are at risk of lead poisoning. In 1989, eight workers suffered from lead poisoning while working on nuclear reactor's protective structure in Monroe, Louisiana. Another issue related to lead contamination is the accumulation of cathode ray tube (CRT) wastes which contained a high concentration of lead.

A study conducted by Lasheras-Zubiate et al. [15] showed that chitosan added to mortar contains heavy metals Pb, Cr, and Zn to reduce their detrimental impact on the environment. This process is known as heavy metal encapsulation or immobilization in a mortar. The experiment results showed that the addition of chitosan improves the rheological, mechanical, and leaching characteristics of mortar. Slump test results indicate that chitosan reduces the deleterious impact of heavy metal encapsulation (Pb and Cr) to the workability characteristic of fresh-state mortar. The addition of chitosan neutralizes the influence of heavy metal encapsulation to the setting time of mortar but does not significantly affect water retention characteristic of mortar. The XRD analysis results also showed that the heavy metals were not merely encapsulated inside the mortar matrix system; instead, they formed new silicate and aluminosilicate structures which are not quickly released from the mortars.

Heavy metals encapsulated in mortar might be released to specific liquid medium. This process is called leaching. The leaching test results showed that mortar system efficiently encapsulates heavy metals. The cumulative leaching of Pb and Zn at 64th day are 58 mg/m^2 and 15 mg/m^2, which are less than the maximum allowable value of 400 mg/m^2 and 800 mg/m^2, respectively for Pb and Zn. For Cr, the retention value is lower at 70-75% due to the formation of weaker chromatic compounds compared to the silicate compounds formed by Pb and Zn. On the other hand, the addition of other types of chitosan produces a lower retention value compared to chitosan-free mortar samples.

Kim et al. [21] also performed a study related to heavy metal encapsulation in a mortar by biopolymer. The heavy metal Pb was acquired from cathode ray tube (CRT) wastes. CRT is the component of television which is considered toxic due to its high lead content. The lead content in CRT is 0-3 wt.% for the panel, 22-25 wt.% for the funnel, 30-40 wt.% for the neck and 60-85 wt.% for the solder. In the study, CRT was mixed with mortar that contains biopolymer, which is further called CRT-Biopolymer-Concrete (CBC) composite. The CRT was shattered into pieces of glassy aggregates and used as a replacement for silicate sand. The CRT aggregates were then submerged in biopolymer solution. The biopolymer solution was made by mixing xanthan gum and guar gum into the pure water at 0.1% concentration. For specific variations, 0.1% boric acid was also added as a crosslinking agent. The CBC sample casting process was conducted by mixing Paste I and Paste II. Paste I is the mixture of water, cement, and sand, whereas Paste II is the mixture of water, cement, sand, CRT aggregates, and biopolymer solution. The resulting CBC samples were 10.16 x 20.32 cm^2 cylinders which were cured for 7 days. The CBC samples were then put into compressive strength tests and toxicity characteristic leaching procedure (TCLP) tests. The mortar sample variations and laboratory test results are tabulated in Table 1.

Table 1. Mortar sample variation dan laboratory test results [21]

Variation	Ordinary Mortar (Control)		Mortar + CRT		Mortar + XG 0.1%		CBC 1 GG + BA 0.1% (70 gram)		CBC 2 GG + BA 0.1% (417 gram)		CBC 3 XG + GG 0.1% (70 gram)		CBC 4 XG + GG 0.1% (417 gram)	
Composition	Weight		Weight		Weight		Weight		Weight		Weight		Weight	
	gram	%	gram	%	gram	%	gram	%	gram	%	gram	%	gram	%
Water	417	11.2	417	11.2	347	9.34	347	9.34	-[a]		347	9.34	-[a]	
Cement	1050	28.3	1050	28.3	1050	28.3	1050	28.3	1050	28.3	1050	28.3	1050	28.3
Sand	2250	60.5	1900	51.1	2250	60.5	1900	51.1	1900	51.1	1900	51.1	1900	51.1
CRT Aggregate	N/A		350	9.42	N/A		350	9.42	350	9.42	350	9.42	350	9.42
Boric Acid Solution	N/A		N/A		70	1.88	70	1.88	417	11.2	70	1.88	417	11.2
Biopolymer	N/A		N/A		0.07	$1.93x10^{-5}$	0.07	$1.93x10^{-5}$	4.17	$1.12x10^{-4}$	0.07	$1.93x10^{-5}$	4.17	$1.12x10^{-4}$
Laboratory Test Results														
TCLP (mg/L)	N/A		4.6		N/A		N/D		0.148		N/D		0.2529	
Compressive Strength (MPa)	28.5		28.3		35.4		34.8		34.6		37.6		21.2	

N/A: Not available **XG: Xanthan gum** **GG: Guar gum** **BA: Boric Acid**

N/D: Not detected (< 0,001 mg/L) **-[a]: no extra water for cement hydration other than that from biopolymer solution**

The laboratory test results is shown in Table 1 indicate that the addition of xanthan gum and guar gum improves the compressive strength of CBC samples by approximately 30%. For the case of CBC 4, compressive strength reduction of 26% relative to control samples was noticed. This is due to the large water usage by xanthan gum and guar gum in the biopolymer solution; thus the remaining water was less for cement hydration (lower w/c). On the other hand for CBC 3, water was added after the biopolymer was mixed with cement and aggregate mixture. Therefore, more water was available for cement hydration. Judging from CBC 1 and CBC 2 test results, it is clear that the addition of boric acid as the crosslinking agent was more successfully produced specimens with high compressive strength. The crosslinking between guar gum and boric acid is shown in Figure 2.

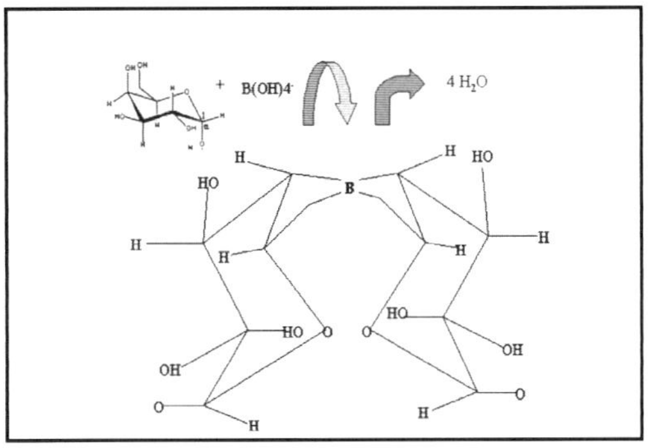

Figure 2. Crosslinking between guar gum and boric acid [21].

TCLP test results show that mortars are good Pb encapsulating agents. For mortar + CRT without biopolymer samples, the leachable lead measured was only 4.6 mg/L, which is far less than the leachable lead of CRT aggregates that reach hundreds of mg/L. This value is also less than the allowable leachable lead of 5 mg/L [22]. For CBC samples, the results were much more remarkable with the leachable lead of 0.2529 mg/L. Specifically for CBC 1 and CBC 3, the leachable lead was not detected (< 0.001mg/L). According to Olmo et al. [22], cement matrix alone is not capable of retaining Pb permanently due to its porous microstructure. However as stated earlier, the crosslinking of biopolymer molecules: xanthan gum and guar gum in the mortar matrix system creates nano-sized barriers which retain Pb particles effectively and provide long-term Pb encapsulation. Thus, it can be concluded that the addition of chitosan, xanthan gum, and guar gum is capable of modifying and improving the rheological, mechanical, and leaching characteristics of mortars which encapsulate heavy metals.

4. Lignosulphonate to increase compressive strength

Lignosulphonate is one of macromolecule organic and polyelectrolyte [23]. Lignosulphonate produced from lignin chemical process, that is a natural biopolymer from wood and grass. Wood cell wall layers are a complex mixture of polymers which are 70-80% holocellulose (contained in cellulose and hemicellulose), and the rest is lignin [1]. Lignin and hemicellulose are then dissolved in a hot acid liquid (usually calcium bisulfite $Ca(HSO_3)_2$) to produce sulfite liquor that containing lignosulphonate [24]. Sulfite liquor composition is shown in Table 2.

Table 2. Sulfite liquor composition obtained from soft wood [1]

Component	[% (wt/wt)]
Lignosulphonate	55
Sugar Hexose	14
Sugar Pentose	8
Hemicellulose, sugar acids	12
Acetic-, formicacid	4
Resin, tall oil	2
Ashes	5

One of the most popular characteristics of lignosulphonate is its ability as dispersant or plasticizer. Lignosulphonate does not contain toxic at a severe level and not causing any health problem to human, animals, and plants. In concrete, lignosulphonate is used as plasticizer mixture to separate cement particle and another fine particle, thus increasing the workability of fresh concrete. Results from Petersen and Gundersen [23] research about the effect of lignosulphonate on mortar mixture workability indicated that fresh mortar workability increases exponentially with the number of lignosulphonate added.

Plank [1] researched lignosulphonate effect to slump, water reduction effect, and concrete compressive strength. Table 3 and 4 shows the experimental results. In general, concrete with low wcf has higher strength due to smaller porosity but reducing concrete workability. With the lignosulphonate addition, can be applied low wcf value in the mixture (water reduction effect), with higher compressive strength while keeping maintaining fresh concrete workability (plasticizer effect) at a certain rate. From Table 4, can be seen that water mixture reduction by 7% can be reached by adding 0.14% lignosulphonate, meanwhile adding 0.28%, lignosulphonate can reduce water usage by 10%.

Table 3. Sodium lignosulphonate effect to concrete slump [1]

Characteristics	Without additive material	0,1% Sodium Lignosulphonate*
Water cement factor (wcf)	0,68	0,68
Slump (mm)	95	135
Slump increase (mm)	-	40

* Percentage of cement weight (= 300 kg/m^3)

Tabel 4. Lignosulphonate water reduction effect in concrete mixture [1]

Lignosulphonate dose (% weight)	wcf	Air content (% volume)	Compressive strength (% from concrete compressive strength control without additive material)	
			3 days	28 days
-	0,58	1,3	100	100
0,14	0,54	2,4	125	112
0,28	0,52	2,9	130	115

Compared to lignosulphonate, several types of plasticizer synthetics such as β-naphthalene sulfonate (BNS), melamine formaldehyde sulfonate (MFS), polycarboxylate, and amphoteric polycarboxylate have better water reduction effect. For example, amphoteric polycarboxylate has water reduction effect up to 60% (maximum). The plasticizer such as polycarboxylate selected by the construction industry to produce high-quality concrete, although it has a higher price than lignosulphonate. The usage of lignosulphonate reduces due to lower performance compared to new type plasticizer, in addition to the price that relatively increases due to wood reduction for lignosulphonate production process related to environmental issues [24]. However, lignosulphonate have a higher sustainability because it is made from the wood material that can sufficiently renewable, compared to polycarboxylate that based on crude oil (resources that are not renewable).

5. Cellulose ethers to increase mortar water retention

Cellulose ethers (CE) is a biopolymer of cellulose, the most abundant biopolymer in nature. For more than 60 years, CE plays an essential role in several products from construction products, ceramics, paints, to food, cosmetics, and pharmacy [25, 26]. CE that commonly used in construction can be seen in Table 5.

Usually, CE product applied to building product to increase water retention and viscosity. Water retention is fresh mortar mixture ability to maintain its water content when in contact with a substrate (rocks that can absorb water). For example, an installation of a red brick wall, the water content in mortar can be absorbed by red brick (substrate) thus affecting mortar workability and mechanical properties. This condition can be prevented by improving mortar water retention capability.

Table 5. Cellulose ethers that used in construction [1]

Ether	Abbreviation
Methyl cellulose	MC
Methyl hydroxyethyl cellulose	MHEC
Methyl hydroxypropyl cellulose	MHPC
Hydroxyethyl cellulose	HEC
Hydroxypropyl cellulose	HPC
Na-carboxymethyl cellulose	CMC, PAC
Na-carboxymethyl hydroxyethyl cellulose	CMHEC
Ethyl cellulose	EC
Ethyl hydroxyethyl cellulose	EHEC

Patural et al. [27] researched CE effect to mortar water retention and workability. In that research, three types of cellulose ether were examined, i.e., MHEC, MHPC, and HEC (Table 5). Mortar water retention level was investigated using the standard procedure from ASTM C1506-09 and Standard DIN 18555-7 which mimics a process of fresh mortar water absorption by the substrate. Results showed that mortar with CE (0.27% weight) have water retention rate up to 98.8%, that means water lost from that test only reach 1.2%. Besides, the most important fact is the greater molecular mass of CE is added, the higher consistency and mortar water retention level.

6. Conclusions

Rapid development in the construction industry has recently signaled engineers for the needs of sustainability and the use environmentally friendly materials. The renewable, sustainable, and biodegradable biopolymers have great potential to be used as admixtures in the mortar and concrete mixtures. Previous studies showed that the use of chitosan, xanthan gum, and guar gum as admixture improves the properties of heavy-metal-encapsulating mortars. The addition of chitosan neutralizes the adverse impact of heavy metal encapsulation to the mortar's properties, such as workability, setting time, and compressive strength. The crosslinking features between xanthan gum and guar gum (+ boric acid) in mortar matrix system escalate the compressive strength of mortar by up to 30% and reduce heavy metal cumulative leaching characteristic to less than 0.001 mg/L. Lignosulphonate as plasticizer improves the workability of fresh-state mortar. The addition of cellulose ethers in mortar mixture improves water retention capability of fresh-state mortar by up to 98.8%. Based on the facts stated above, it could be summarized that the use of biopolymers certainly improve physical and leaching characteristic of mortar and concrete.

References

[1] Plank J 2005 *Biopolymers Online* 10
[2] Chakraborty S, Iyer N, Khrisna P and Thakkar S 2011 *Assessment of Civil Engineering Inputs for Infrastructure Development* (New Delhi: Indian National Academy of Engineering)
[3] Davalos J F 2012 *Advanced Materials for Civil Infrastructure Rehabilitation and Protection Seminar* at the City College of New York (New York)
[4] Flatt R, Roussel R and Cheeseman C R 2012 *J. Euro. Ceram. Soc.* **32** 2787
[5] Hosseini M M, Shao Y and Whalen J K 2011 *Biosys. Eng.* **110** 351
[6] Bezerra U T 2016 Biopolymers with superplasticizer properties of concrete *Biopolymers and Biotech Admixtures for Eco-Efficient Construction Materials* ed Pacheco-Torgal F, Ivanov V, Karak N, and Jonkers H (Amsterdam: Woodhead Publishing Series in Civil and Structural Engineering)

[7] Plank J 2004 *Appl. Micr. Biotech.* **66** 1

[8] Pacheco-Torgal F and Jalali S 2011 *Eco-efficient Construction and Building Materials* (London: Springer Verlag)

[9] Nakajima-Kambe T, Shigeno-Akutsu Y, Nomura N and Nakahara T 1999 *Appl. Micr. and Biotech.* **51** 134

[10] Ebnesajjad S 2013 *Handbook of Biopolymers and Biodegradable Plastics* (Oxford: William Andrew)

[11] Kim J-H, Shim B, Kim H, Lee Y-J, Min S-K, Jang D, Abas B and Kim J 2015 *Int. J. Precis. Eng. Manufac. Green Tech.* **2** 197

[12] Nguyen T, Feng J, Ng S, Wong J, Tan V and Duong H 2014 *Coll. Surf.* A 445 128

[13] Mignon A, Snoeck D, D'Halluin K, Balcaen L, Vanhaecke F, Dubruel P, van Vlierberghe S and De Belie N 2016 *Con. Build. Mat.* **110** 169

[14] Bezerra U T, Ferreira R M, and Castro-Gomes J P 2011 *Key. Eng. Mat.* **466** 37

[15] Lasheras-Zubiate M, Navarro-Blasco I, Fernandez J M and Alvarez J I 2012 *J. Haz. Mat.* **233-234** 7

[16] Riedel U and Nickel J 1999 *Die Angewandte Makromolekulare Chemie* **272** 34

[17] Kim D 2004 *Study of Biopolymer-Modified Concrete System: Geopolymerization of Biopolymers* (Los Angeles: University of Southern California)

[18] Mohanasrinivasan V, Mishra M, Paliwal J S, Singh S Kr, Selvarajan E, Suganthi E and Devi C S 2013. *3 Biotech* **4** 167

[19] Environmental Protection Agency (EPA) United States. 2014. Basic Information about Lead in Drinking Water. http://water.epa.gov/drink/contaminants/basic information/lead.cfm Accessed 2 August 2014

[20] Dimitrakakis E, and Gidarakos E 2014 *Closed-loop CRT Glass Recycling Alternatives* (Crete: University of Crete)

[21] Kim D, Quinlan M and Yen T F 2009 *Waste. Man.* **29** 321

[22] Olmo I F, Chacon E and Irabien A 2003 *J. Env. Eng.* **129**

[23] Petersen B G, and Gundersen N L 2004 *Ann. Trans. Nordic Rheology. Soc.* **12** 39

[24] Adams J W 1988 *Environmental Effect of Applying Lignosulfonate to Roads* (Chicago: Daishowa Chemicals)

[25] Gargulak J D and Lebo S E 2000 Commercial use of lignin-based materials *Lignin: Historical, Biological and Materials Perspectives* ed Glasser W G, Northey R A and Schultz T P (Washington DC: American Chemical Society)

[26] Cellulose Ethers Technical Overview and Product Guide. 2012. http://www.dow constructionchemicals.com/na/en/pdfs /832-00226.pdf. Accessed 5 August 2014

[27] Patural L, Marchal P, Govin A, Grosseau P, Ruot B and Deves O 2011 *Cem. Con. Res.* **41** 46

IOP Conf. Series: Materials Science and Engineering **345** (2018) 012029 doi:10.1088/1757-899X/345/1/012029

Computational fluid dynamic on the temperature simulation of air preheat effect combustion in propane turbulent flame

Elwina[1], Yunardi[2] and Yazid Bindar[3]
[1]Chemical Engineering Dept., State Polytechnics of Lhokseumawe, Lhokseumawe, Indonesia
[2]Chemical Engineering Dept., Syiah Kuala University, Banda Aceh, Indonesia

[3]Energy and of Processing System Chemical Engineering Dept., Faculty of Industrial Technology, Bandung Institute of Technology, Bandung, Indonesia

E-mail: elwina@pnl.ac.id, yyunardi@yahoo.com; y.bindar@yahoo.com

Abstract. this paper presents results obtained from the application of a computational fluid dynamics (CFD) code Fluent 6.3 to modelling of temperature in propane flames with and without air preheat. The study focuses to investigate the effect of air preheat temperature on the temperature of the flame. A standard k-ε model and Eddy Dissipation model are utilized to represent the flow field and combustion of the flame being investigated, respectively. The results of calculations are compared with experimental data of propane flame taken from literature. The results of the study show that a combination of the standard k-ε turbulence model and eddy dissipation model is capable of producing reasonable predictions of temperature, particularly in axial profile of all three flames. Both experimental works and numerical simulation showed that increasing the temperature of the combustion air significantly increases the flame temperature.

1. Introduction

Combustion modelling has played an important role not only in the design, but also in the operation of a turbulent combustion system. In order to accurately predict the temperature in such system, an integrated model that at least represents turbulence and combustion phenomena, as well as radiation heat losses is required. The ability to apply such an integrated model, of demonstrated accuracy, aiming at reaching maximum combustion efficiency and minimizing heat losses, emission of pollutants in relation to safety and environmental considerations, would represent a major step-forward in our ability to design and manage combustion processes. However, it is important to note that the modelling could not stand alone, since modelling requires verification. Therefore, experimental works are required the modelling not only for the purpose validation but also for providing better understanding on the phenomena being studied.

Numerous experimental works have been performed in an attempt to study the effects of air preheat on the combustion process. Gupta [1] conducted an experimental study on the structure of turbulent propane – air diffusion flames with preheated air and found out that the emission NOx was observed to be less with high preheat temperature and low oxygen concentration in the oxidizer. Sadakata [2] also confirmed that by increasing the temperature of combustion air up to 573 K, the emissions of thermal and fuel NO were not significantly increased. Although previous works mentioned that air preheating did not significantly affect the NO concentrations in a flame, recent experimental works emphasize different findings. Lim [3] investigating the effects of air preheat on the structure and emission of counter-flow methane –air diffusion flame have shown that the peak concentrations of CO and H_2 increased by 20 per cent, and that of NO increased by 70 per cent, as the air temperature was increased from 300 to 560 K. In another study, Fuse [4] also observed that the increase in air temperature increased the NO formation in diffusion flames. However, they found out that NO formation can be decreased, even with high air temperature, if concentration of oxidizer is significantly reduced. With respect to the temperature of the flame, all above studies confirmed that the flame temperature increases due to the use of air preheat. Since air preheat could be produced from waste heat sources, the use of air preheat will not only improve the combustion efficiency but also could reduce the heat loss in an industry. Therefore, it is important to investigate if the modelling could also produce similar increase of flame temperature when the combustion is supplied with air preheat.

A number of combustion models, ranging from simple to complex approaches, have been proposed in the literature, and research to develop new models and improve the existing ones is on-going. Those approaches that have been proposed for modelling combustion in turbulent flames have their own advantages and disadvantages. Although numerous models are available, modelling of non-premixed turbulent combustion using the eddy dissipation model (EDM) still receives great attention due to its simplicity and low computer resource usage. Therefore, for predictions of temperature in the propane flames being studied, this model was employed for the sake of keeping the computational cost at an acceptable level without losing an acceptable degree of accuracy.

In this paper, the results of the application of a computational fluid dynamics (CFD) code Fluent 6.3 to modelling of air preheated propane non-premixed flames are presented. The objective of the present work is the assessment of the EDM combustion model, available directly in the code to represent the temperature profile in the turbulent propane flames, with and without air preheat. Turbulence of the flame is calculated using a standard k-ε model. Validation is achieved by comparing the computed temperature predictions against available experimental data of propane flames with and without preheated air [5,6].

2. Methodology

2.1 Target flame

Three turbulent non-premixed propane flames with and without air preheats considered in the present study have been experimentally reported by Nishida and Mukohara [5], and Jurng et al [6], respectively. The important characteristics of each of these flames are presented in TABLE I. Further details with regards to the flame geometry, methods of data collection and processing can be found in the relevant references.

Table 1. Operating conditions for air-propane and preheated air propane flames

Air exit temperature	298 K	323 K	773 K
Absolute pressure/atm	1	1	1
Fuel exit velocity/ms^{-1}	16	30.0	30.0
Reynolds number	11000	13000	13000
Fuel exit temperature/K	298	298	298
Nozzle diameter/mm	3.0	2.0	2.0
Co-flow air velocity/ms^{-}	0	0.4	0.96

2.2 Turbulent flow field calculation

Since the flow of the flames being studied is turbulent in nature, the flow is characterized by fluctuating velocity fields. These fluctuations affect other transported quantities within the flame such as momentum, energy, and species concentration, and cause the transported quantities to fluctuate as well. Since these fluctuations can be of small scale and high frequency, they are too computationally expensive to simulate directly in practical engineering calculations. Instead, the instantaneous (exact) governing equations can be time-averaged, ensemble-averaged, or otherwise manipulated to remove the small scales, resulting in a modified set of equations that are computationally less expensive to solve. However, the modified equations contain additional unknown variables, and turbulence models are needed to determine these variables in terms of known quantities. For the purpose of propane flames calculation, a standard *k-ε* turbulence model was used to represent the flow field within the flame. As the *k-ε* turbulence model is categorized into two-equation models, it requires the solution of two transported equations for *k* and *ε* [7], in conjunction with the mass and momentum conservation equations.

Calculation of turbulent kinetic energy, *k* and turbulent dissipation rate, ε is required during the combustion calculation under EDM approach. The turbulent kinetic energy and turbulent dissipation rate can be calculated using Equation 1 and 2, respectively.

$$\frac{\partial}{\partial t}(\rho k)+\frac{\partial}{\partial x}(\rho k u_i)=\frac{\partial}{\partial x_j}\left[\left(\mu+\frac{\mu_t}{\sigma_k}\right)\frac{\partial k}{\partial x_j}\right]+G_k+G_b-\rho\varepsilon-Y_M \quad (1)$$

$$\frac{\partial}{\partial t}(\rho\varepsilon)+\frac{\partial}{\partial xi}(\rho\varepsilon u_i)=\frac{\partial}{\partial x_j}\left[\left(\mu+\frac{\mu_t}{\sigma_\varepsilon}\right)\frac{\partial\varepsilon}{\partial x_j}\right]+ C_{1\varepsilon}\frac{\varepsilon}{k}(G_k C_{3\varepsilon} G_b)-C_{2\varepsilon}\rho\frac{\varepsilon^2}{k} \quad (2)$$

In these equations, G_k is turbulent kinetic energy in average velocity gradient, G_b is kinetic energy related in driving force. Y_M is contribution in compressible turbulent dilatation contribution to overall dissipation rate. $C_{\varepsilon 1}$ and $C_{\varepsilon 2}$ is constant, σ_k and σ_ε is Prandtl number turbulent for *k-ε*.

Standard turbulence modeling constants in the standard *k-ε* modelling appropriate to asymmetric flows were employed, although to ensure the accurate prediction of the spreading rate of the jets, an adjustment was made to the value of $C_{\varepsilon 2}$ from 1.92 to 1.86 to affect an increase in the dissipation rate of turbulence kinetic energy

2.3 Combustion calculation

The Eddy Dissipation Model (EDM), a combustion model originally developed by Magnussen and Hjertager [8], is based on the assumption that the chemical reactions occur in regions where the dissipation of turbulence energy takes place. The region of the flow where this dissipation occurs is in fine structures or small turbulence scales, which are treated as perfectly stirred reactors. To find the amount of fine structures in the flow, Magnussen [9] uses an energy cascade model which accounts for finite Reynolds number effects and the influence of molecular dissipation processes on the reaction rates of combustion. Combustion is treated as a single-step reaction. In later developments, EDM models have been developed to account for finite-rate chemistry [9]. The main advantage of the EDM approach clearly lies in its simplicity and low computational cost due to the use of a conserved scalar concept, e.g. mixture fraction, which allows decoupling of the chemistry terms and the flow field.

Although Hjertager et al. [10] reported that the standard EDM model is not suitable for liquid-phase reactions; it has found extensive application in the simulation of turbulent reacting flows. Magnussen [9] applied detailed chemistry to investigate the performance of the EDM for the prediction of soot formation and destruction in an acetylene diffusion flame. Gran and Magnussen [11] used the model to simulate a non-premixed bluff body stabilized flame, and good agreement with experimental data was reported, and its performance was comparable to that of a transported PDF model. Chakraborty et al. [12] also used the model to simulate H2/air combustion in a scramjet combustor. Here, the EDM model with detailed chemistry and finite rate single-step chemistry was found to capture well all essential features of the reaction rate profile distribution, with similar order of magnitude peak values. In Giacomazzi et al. [13], the EDM model is used to describe single-step combustion chemistry close to the dissipation scales in the context of LES and RANS predictions of a bluff-body C3H8/air premixed flame.

The eddy dissipation model (EDM) proposed by Magnussen and Hjertager [8] was developed for the prediction of gaseous combustion reactions in turbulent flows. The combustion reaction is treated as a single-step irreversible reaction with finite rate for fuel F, oxidant O and product P, as illustrated in Equation 3.

$$1\text{ kg F} + \text{s kg O} \rightarrow (1 + \text{s})\text{ kg P} \tag{3}$$

The EDM assumes reaction to be much faster than mixing, and therefore the rate of combustion is entirely determined by the turbulent mixing rate ε/k and by the limiting ingredient needed for reaction, either the fuel or oxidizer or product. The model requires the solution of two variables, the mass fraction of species and the mixture fraction, where the source term in mass fraction of species equation is solved using Equation 4.

$$\overline{w_\alpha} = A\bar{\rho}\frac{\varepsilon}{k}\, min\left[\bar{Y}_F, \frac{\bar{Y}_O}{s}, B\,\frac{\bar{Y}_P}{(1+s)}\right] \tag{4}$$

Where k and ε are the turbulence kinetic energy and its dissipation rate obtained from the turbulence flow field calculation, and s is the stoichiometric coefficient of the reaction given as Equation 3. The constants A and B are often set to 4.0 and 0.5, respectively, for gaseous combustion reactions.

2.4 Numerical computation

With respect to numerical simulation, the flame geometry was created using GAMBIT software. The calculation of flow and mixing fields was achieved by equations under the standard k-ε turbulence modeling with the use of the commercial CFD code FLUENT ver. 6.3. The combustion calculation which solved the energy and species concentration was performed by utilizing EDM that is embedded in the FLUENT code. Computations were performed using a laptop computer of 2.00 GB RAM having a 2.13 GHz of speed. The computation time for each flame took around 4 to 6 hours.

3. Result and discussion

Figure 1 demonstrated axial and radial predictions of axial temperature for the propane flame without air preheating compared to experimental data [6]. The solid line represents the simulations resulting from the use of k-ε standard turbulence model and EDM model. The centreline axial temperature predictions in the flame display qualitatively good agreement with the experimental data, particularly further downstream of the flame. As a consequence, the predicted temperature in radial direction in the area closer to the nozzle significantly deviated from the experimental data. However, the radial temperature predictions further downstream are in line with measurements. This could be due to adjustment of spreading rate through modifying the constant $C_{\varepsilon 2}$ of the standard k-ε turbulence model.

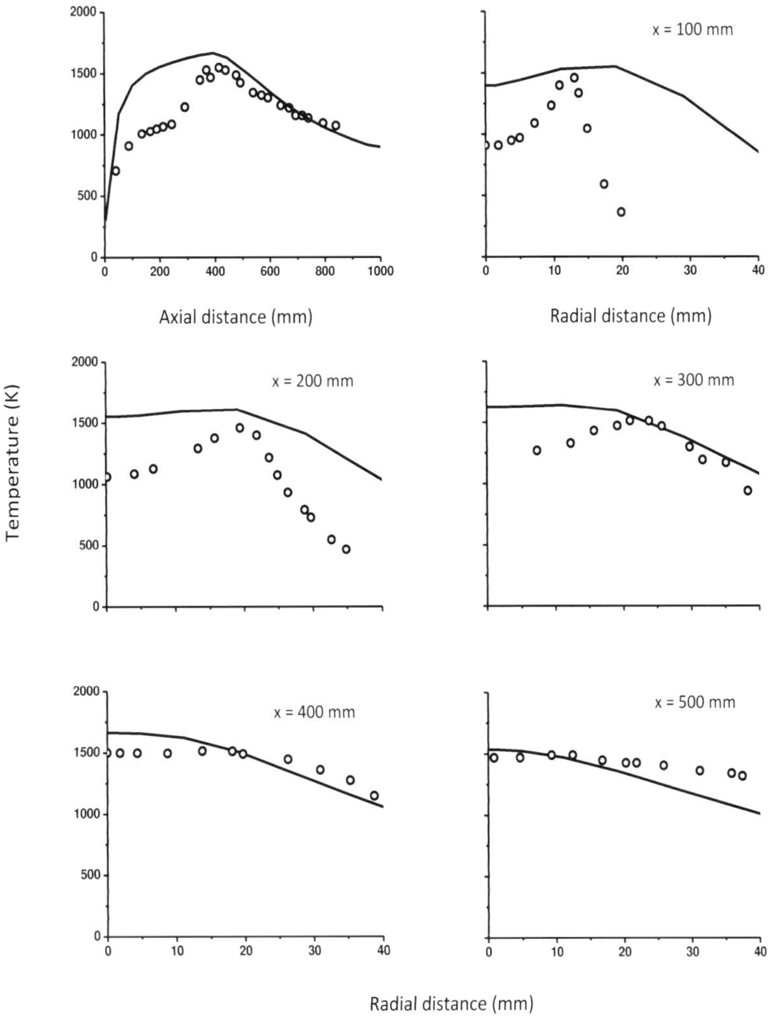

Figure 1. Axial and radial profiles of temperature for propane flame, Tair = 298 K (symbol – measured, solid line – predicted with EDM model)

Temperature predictions calculated using Eddy Dissipation Model showed that the centreline temperature predictions in the propane flame without preheating display qualitatively good results in comparison to the experimental data. The peak temperature in axial profile slightly over predicted. The evolution of the computed axial temperature is in line with the measurements along the core of the flame. Overall, the position and value of peak temperature is qualitatively and qualitatively well captured by the EDM model. This agreement is achievable not only due to the conformity of flow field predictions with experimental data but also due to the correct selection radiation model to represent the combustion.

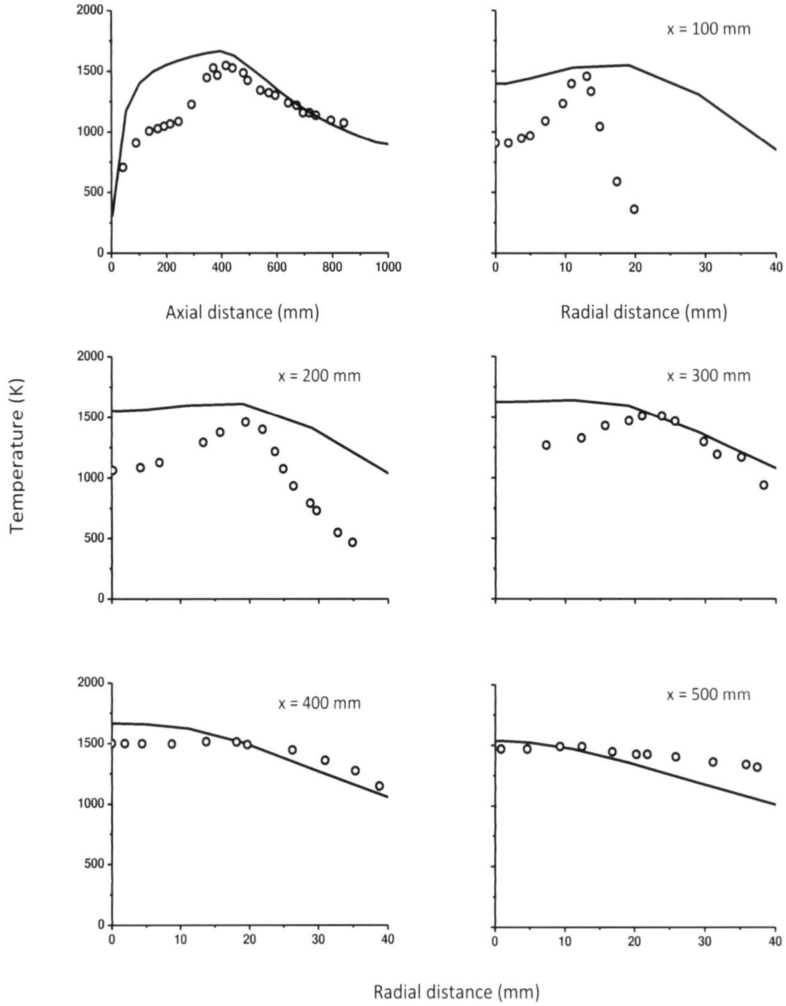

Figure 2. Axial and radial profiles of temperature for propane flame, Tair = 323 K (symbol – measured, solid line – predicted with EDM model)

Figure 2 presents graphs of temperature predictions compared with experimental data for flames with air preheating of 323 K. The calculated axial temperature profile predictions for the propane flame with air preheating of 323 K are in good agreement with the experimental data, indicating that the mixing field the lower air preheated propane flame is well represented by the application of the k–ε turbulence closure, as in the case of propane flame without preheating. In connection with temperature predictions, the results of upstream mean temperature predictions of the air preheated flames are not as satisfactory compared with experiment data, although the temperatures in the mid-flame region and its peak are well represented. Radial temperature profiles along with measurements for the lower air preheated propane flame in the position x = 500 mm are in good agreement with the experiments. However, closer to the nozzle, radial predictions of the temperature are not in agreement with the data.

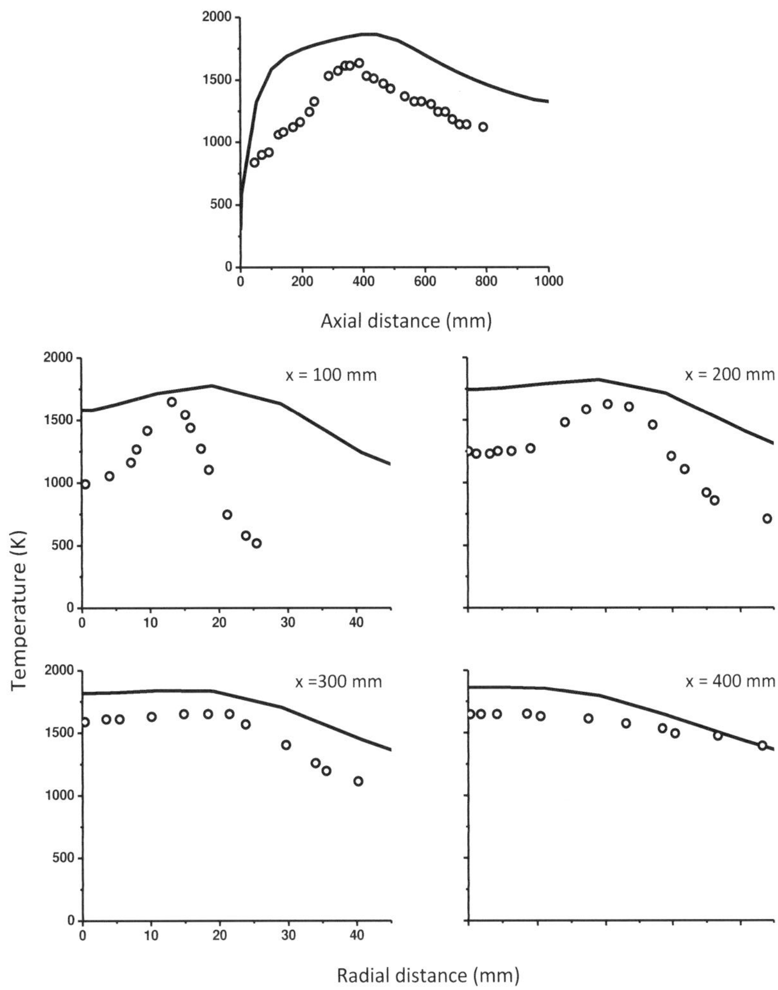

Figure 3. Axial and radial profiles of temperature for propane flame, Tair = 332 K (symbol – measured, solid line – predicted with EDM model)

Figure 3 represented predicted centerline and radial temperatures of the propane-preheat flame with air preheating of 772 K. Solid line and symbol represent temperature predictions and measurement, respectively. It is clearly seen that air preheating leads to an increase of centerline flame temperature, particularly up to the mid-flame zone. The temperature of air preheat flames increases sharply in core of the flame, leading to a shift of peak temperature and the predictions result is over prediction in the peak temperature. Further downstream, the radial profile of temperature predictions deviated significantly from measured data.

At this stage it is useful to compare the peak temperature in the three flames being studied. In the axial profile, it is clearly seen that the peak temperatures are as follows: 1862 K in the flame of 773 K air preheat, 1667 K in the flame of 323 K air preheat and 1652 K in the flame with no air preheat. As the temperature of the air preheats increases, the peak temperature also increases. As a consequence, the highest peak temperature among the three flames was observed in the flame of 773 K air preheat, reaching its peak at 450 mm and the lowest peak temperature among the three flames was in the flame with no air preheat having its peak at 400 mm.

It is seen that in all three flames, over prediction in temperature exists in the zone closer to the nozzle or in the rich fuel area, x = 100 mm. However, in the fuel lean area x = 400, 500 and 600 mm the result of prediction is getting better, reaching in closer agreement in measured data. On the contrary, Yunardi et al [14] could produce good result in rich fuel area but not in the lean fuel area when they used Conditional Moment Closure approach to calculate profile temperature coupled with a kinetic reaction scheme. However, it should be noted that Yunardi et al [15] demonstrated excellent temperature predictions in methane flame with the use of EDM model. It seems that the EDM model assuming single reaction step to represent the chemistry of combustion is suitable for simpler fuel like methane. With a more complex fuel, like propane, a single step reaction assumption is no longer applicable.

4. Conclusions

From the results and discussions, the following conclusions are drawn:

1. Increasing the temperature of the combustion air significantly increases the flame temperature. As a consequence, the flame of 773 K air preheat has the highest peak flame temperature of 1862 K and the flame of 298 K air preheat has the lowest peak flame temperature.
2. Although the position and the value of peak temperature in each flame is qualitatively predicted well by the EDM combustion model couple with *k-ε* turbulence model, the radial temperature predictions are not satisfactorily represented which much is caused by inability of the model to include finite rate chemistry to represent the combustion.

Acknowledgment

The first author would like to express her gratitude to the State Polytechnic of Lhokseumawe - Indonesia for the financial support in the forms of travel grant to enable the presentation and publication of this research.

References

[1] Gupta A K, Bolz S and Hasegawa T 1999 Effect of air preheat temperature and oxygen concentration on flame structure and emission *J. Energ. Resour. Technol.* **121** 209-216

[2] Sadakata M, Fujioka F and Kunii D 1981 Eighteen symposium (international) on combustion *Proc. Combust. Inst.* vol. **18** 65-72

[3] Lim J, Gore J and Viskanta R 2000 A study of the effect of air preheat on the structure of methane/air counterflow diffussion flame *Combust. Flame 121* 262-274

[4] Fuse R, Kobayashi H, Ju Y, Maruta K and Niioka T 2002 NOx emission from high-temperature air/methane counterflow diffusion flame *Intl. J. Thermal Sci.* **41** 693-698

[5] Nishida O and Mukohara S 1982 Characteristic of soot formation and decomposition in turbulent diffusion flames *Combust. Flame* **47** 49-61

[6] Jurng J, Shin H D and Chun S L 1988 Prediction of structure with soot concentration of propane-air turbulent-diffusion flame *J. Inst. Energ.* **61** 201-208

[7] Jones W P and Launder B E 1972 The prediction of laminarisation with a two-equation turbulence model *Intl. J. Heat Mass Trans.* **15** 301-314

[8] Magnussen B F and Hjertager B H 1977 On mathematical modeling of turbulent combustion with special emphasis on soot formation and combustion *Proc. Combust. Inst.* **16** 719-729

[9] Hjertager L K, Hjertager B H and Solberg T 2002 CFD modelling of fast chemical reactions in turbulent liquid flows *Comput. Chem. Eng.* **26** 507-515

[10] Gran I R and Magnussen B F 1996 A numerical study of a bluff-body stabilized flame *Combust. Sci. Technol.* **119** 191-217

[11] Chakraborty D, Paul P J and Mukunda H S 2000 Evaluation of combustion models for high speed H2/air confined mixing layer using DNS data *Combust. Flame* **121** 195-209

[12] Giacomazzi E, Battaglia V and Bruno C The coupling of turbulence and chemistry in a premixed bluff-body flame as studied by LES *Combust. Flame* **138** 320-335

[13] Yunardi, Darmadi, Hisbullah and Michael F 2011 Conditional moment closure approach to modelling sooting turbulent non-premixed propane flames with air preheating *J. Thermal Sci.* **20** 548-555

[14] Yunardi, Elwina, Sylvia N, Wusnah and Bindar Y 2012 A comparative performance study of soot formation models in methane elevated pressure non-premixed flames *Appl. Mech. Mater.* **110-116** 18-22

IOP Conf. Series: Materials Science and Engineering **345** (2018) 012030 doi:10.1088/1757-899X/345/1/012030

Corrosion control of carbon steel using inhibitor of banana peel extract in acid diluted solutions

Komalasari, S P Utami, M I Fermi, Y Aziz and R S Irianti
Department of Chemical Engineering, Universitas Riau
Jl. HR Subrantas KM 12,5 Sp Panam Pekanbaru, Riau, Indonesia

E-mail: komalasari@lecturer.unri.ac.id

Abstract. Issues of corrosion happened in pipes, it was used as fluid transportation in the chemical industry. Corrosion cannot be preventing, however it could be controlled or blocked. Inhibitor addition is one of the method to control the corrosion inside the pipe. Corrosion inhibitors consisted of inorganic and organic compound inhibitors. Organic inhibitor is composed from synthetic and natural material. This study focused to evaluate the inhibition's efficiency from banana peel to carbon steel in different concentration of inhibitor and immersing time in acid solution variation. The research employed inhibitor concentration of 0 gram/liter, 2 gram/liter, 4 gram/liter and 6 gram/liter, immersed time of carbon steel for 2, 4, 6, 8 and 10 hours. It was immersed in chloride acid solution of 0.5 M and 1.5 M. Carbon Steel AISI 4041 was used as specimen steel. Results were analyzed using corrosion rate evaluation for each specimens and inhibitor efficiencies determination. It was found that the specimen without inhibitor yielded fast corrosion rate in long immersing time and high concentration of HCl. However, the specimens with inhibitor gave lowest corrosion rate which was 78.59% for 6 gram/litre and 10 hours in 0.5 M HCl.

1. Introduction

Metal is a fundamental raw material which supported in industry development thus it is very important role in human civilization. One of the uses of metal, as industry material is pipe production as fluid transportation tools. Generally, pipe channels are passed by various fluids types which caused the corrosion formed on the pipe wall.

Common metals which used for piping system are carbon steel because it considered more effective and economical. However, carbon steel has low resistance to corrosion. Metal utilization with better a corrosion resistance are a good option to choose. However, this method is rarely become the main choice for industries, since the increasing the durability of material means accompanied by increased cost. Therefore, protection methods is needed in order to decrease corrosion rate in carbon steel [1]. Method was used to protect internal carbon steel pipe are by adding corrosion inhibitor. Corrosion inhibitor is a chemical compound which added into the environmental, it able to decrease corrosion attack rate in metal. Inhibitors are divided into two types, organic and inorganic. Organic inhibitor is divided into synthetics and natural resources organic which also known as green inhibitor [2].

The aims of this research is to discover the inhibition effieciency in different concentration of banana peel's extract, determine the maximum banana peel's extract concentration as corrosion inhibitor of carbon steel in varied acid environment and immersing times and also to compare this study with the previos research regarding the banana's peel extract as corrosion inhibitor.

2. Literature review

Corrosion is degradation of metal due to redoks reaction between a metal to different substrance in their environment then later produce undesirable compound [3]. Corrosion is also known as metal depletion rate or coronation. Corrosion is defined as attack that damages the metal since metal reacted with environment in chemically. Corrosion is a nature condition where every metal has its corrotion rate. Metal corrosion in eletctrolyte environment has impact from electrochemical process. Based on the electrochemical process, the necessity of the corrosion consisted of four main component [4], as listed below:

a. Cathode
Generally, cathode does not undergo corrosion but it would be happen in certain conditions. Reactions in the cathode are reduction and it is depend to pH solution. As we can see the reaction below:

$$pH < 7: \quad H^{+} + e^{-} \rightarrow H\ (\text{atom}) \quad (1)$$
$$2H \rightarrow H_2\ (\text{gas}) \quad (2)$$
$$pH \geq 7: \quad 2H_2O + O_2 + 4e^{-} \rightarrow 4O \quad (3)$$

b. Anode
Commonly, anode have experience the corrosion by releasing electrons from its neutral metal atom in order to form the related ions. These ions still alive in the solution or reacted to form undissolved corrosion. Reaction for anode is listed as below.

$$M \rightarrow M^{Z+} + ze^{-} \quad (4)$$

c. Anode and cathode should be electrically connected
Between anode and cathode should be have the electrical connection so the current inside the corrosion cell able to flow. Physycal contact was unnecessary needed if anode and cathode has the same metal.

d. Electrolyte
Electrolyte is a solution which has conductive properties. Electrolyte categorized to acid solution, base and saline solution. Electrolyte solution has important role in metal corrosion since this solution made electricity current between anode and cathode.

Corrosion process happen if half cell reaction which released electron (oxidation reaction in anodic side) and half cell reaction which received electron (reduction reaction in cathodic side). Both reaction will continue until reached the equilibrium dynamic where the released electrons amount equal to receive electrons. There are mechanism reaction of corrosion in iron (Fe) as shown [5]:

$$Fe_{(s)} + H_2O\ (l) + \tfrac{1}{2}\, O_{2\,(g)} \rightarrow Fe(OH)_{2\,(s)} \quad (5)$$

Ferro Hydroxide [$Fe(OH)_2$] is temporary result which oxidized naturally by water and air into Ferri Hydroxide [$Fe(OH)_3$] then the next reaction become:

$$4\, Fe(OH)_{2\,(s)} + O_{2\,(g)} + 2H_2O_{\,(l)} \rightarrow 4Fe(OH)_{3\,(s)} \quad (6)$$

Ferri Hydroxide formed changed into Fe_2O_3 which has brownish red called stain. Stain reaction can be shown :

$$2\, Fe(OH)_3 \rightarrow Fe_2O_3 + 3H_2O \quad (7)$$

2.1 Corrosion rate

Corrosion rate has defined as amount of released metal for each time in certain surface. Generally, corrosion rate had noted in mils per year (mpy). Corrosion rate analysis has been held by using weight loss method as been seen [6]

$$CR = \frac{W \times 24 \times 365 \times 1000}{A \times T \times \rho} \quad (8)$$

Equation 8 has elaborate in CR as corrosion rate (mm/y), W is mass of steel loss (gram) where $W = W_o - W_f$, W_o is initial weight (gram), W_f is final weight (gram), ρ adalah steel density (gram/cm^3), A is surface area of steel (mm^2), and T is immersing time (days).

2.1.1 Affected factors of corrosion rate
In metal, formation of corrosion has been influenced by several factors, They are: temperature, fluid flow, pH, and concentrations of corrosive material. They are describes:

a. Temperature
Temperature increase caused the accelerated the corrosion reaction rate. Since more elevated temperature made kinetic of the reacted more increased so exceeds the activated energy values. Therefore, it made the corrosion reaction rate faster and vice versa. However for open system, the ambient with above 80 °C, the corrosion rate has been decreased since the oxygen would released while the closed system the corrosion rate will continue to increase due to oxygen dissolved content [7].

b. Fluid flow
Corrosion rate tend to increase if the fluid flow also getting raise. This is caused by contact between substance and metal get heighten so the metal ions will be loose which made metal experiencing fragility, its called corrosion [8]
c. pH
pH netral is 7, pH < 7 has properties like acidic and corrosive, pH > 7 with base and corrosive properties. However, for iron, low corrosion rate lies between 7 to 13 and it is increased below 7 and above 13 [4].
d. Concentration of corrosive material
Concentration of material corrosive related with pH or acidic and base of the solution. The acidic solution is highly corrosive to the metal where the metal in the acid solution medium will be more rapidly corroded because it is an anode reaction. Whereas, an alkaline solution can cause corrosion in the cathode reaction, since the cathode reaction always co-exists with the anode reaction and the reduction reaction always occurs simultaneously with the oxidation reaction known as the redox reaction.

2.1.2 Carbon steel
Generally, carbon steel is used for piping system because it is more effective and economic. Carbon steel was formed from main element of iron (Fe), second element was carbon compound which influenced to its properties while another element has influenced according to its percentages. Carbon steel contained three types, they are [9].
a. Low carbon steel
Low carbon steel has carbon content less than 0.3 and its micro structure devided into ferrit and perlite. Enhancement method used cold working method. This carbon has mild, firm and tough. Besides that, low carbon steel has good machinability and weldibility. It is commonly utilized because it is cheaper but it easy to corroded.

b. Medium carbon steel
Medium carbon steel has carbon content in range 0.3 – 0.6%. This steel can be enhanced its mechanic through heat of austenitizing, quenching, and tempering. It is widely used in tempering condition and has stronger properties than low carbon steels.

c. High carbon steel
High carbon steel has carbon content between 0.6-1.4%. This is the hardest, strongest, most brittle steel among the others and napless resistance. It is widely utilized in tempering condition so the

micro structure become martensite, and its mechanic properties is enhanced via heat of austenitizing, quenching, and tempering.

2.2 Corrosion control method
Metal protection technologies has been widely known to offer a better solution to solve corrosion problem. The utilized cost and method applications can be adjusted with condition and needs. There are several methods i.e. coating, cathodic protection, sacrificial anode method and inhibitor corrosion.

2.3 Types of corrosion inhibitor
Corrosion inhibitor was divided into inorganic and organic inhibitor. As describes as follows :

a. Inorganic inhibitor
Inorganic inhibitor has active group, they are negative anion which useful to decrease corrosion rate to metal [10]. Generally, inorganic inhibitor are Sodium Nitrite, Chromat, Phosphat, and Zinc. The weakness of Sodium Nitrite is it is needed high concentration (300-500 mg/l) so it is not economical. Others, Chromat and Zinc are toxic and phosphate is considered as pollution compound since it improve phosphorous content in water [11]

b. Organic inhibitor
Besides inorganic inhibitor, organic inhibitor has been developed. It is classified into synthethic organics and organic from nature's extraction [2]. Compounds used as organic inhibitors are compounds with consisted of nitrogen, sulfur or oxygen atoms having a pair of free electrons. According to Stupnisek-Lisac et al, [12], the more effective inhibitor for metal corrosion is organic compounds. This is caused the organic compound has free pair of electrons in their carbon chain or in the aromatic chain systems which able to bind to metal positive charge so the adsorption between metal surface and inhibitor happened. This adsorption formed protector layer in the metal surface due to physysorption or to generate chelate barrier that does not dissolve as a result of chemisorption process. It could inhibite metal from direct contact with corrosive media [13].

Organic inhibitor of natural material come from natural matter extraction which contained nitrogen, oxygen, phosphor and atomics with free electron pairs. Elements which has free electron pairs has function as ligan to generate complex compound with metal and will cover its metal from corrosion attacks. This is called as "green inhibitor" since it is benign to environment, inexpensive, easy to get and applicable to environment. Todays, a lot of researchers has been developed the green corrosion inhibitors [14].
Several plant has been investigated its ability as organic inhibitor corrosion such us mangoness peel, tea leaves, and acacia barks. Researches about mangoness peel extract as inhibitor for 200 ppm in H_2SO_4 0.02 M for 24 h generated inhibition efficiency around 48.79% [15]. Tea leaves *Camellia sinensis* produced 86.3% efficiency in HCl 3% medium for 4 days [16]. Accacia's bark generated 36.1% in 80 ppm to feed water boiler in 65°C[2]. Eddy and Ebenso [17] investigated banana's peel (*Musa Sapientum*) in H_2SO_4 2.5 M as corrosion inhibitor. Their research had obtained efficiency around 71.05% in 0.5 g/l in 303 ^{o}K for immersing temperature.

2.4 Tannin compound
Another natural were used as organic inhibitor corrosion is tannin. It is a polyphenol compound, and it is also able to form complex coumpund and indissolved to metal ions. Complex compound generated between tannin and metal ions will cover the metal from direct contact of corrosive fluid so the corrosion rate of metal has been decreased [2]. Figure 2.1 illustrates the tannin core molecule structure.

.

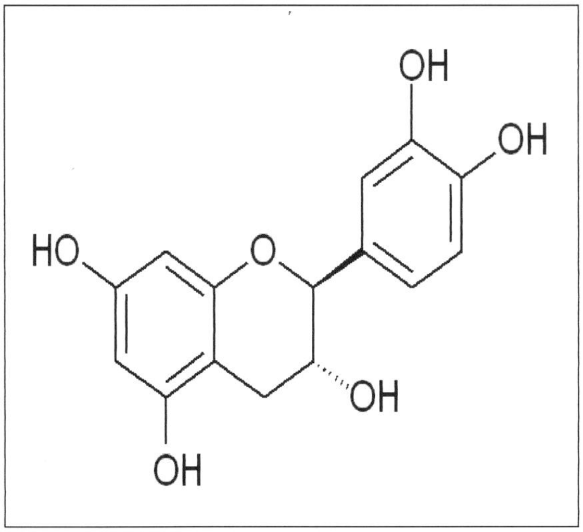

Figure 1. Sturcture of tannin molecule.

Tannin is a massive phenol compound with large molecule weight. Tannin is macromolecule compound with polar polyphenol group thus tannin extraction product is able to conduct by using polar solvent [18]. Tannin is found in almost every green flora with different content and qualitative [19]. Tannin can inhibit the corrosion since it generated a mixture complex compound of iron and tannin. This compound cover the surface of metal and useful to detain corrosion. Iron is metal transition, its properties in transition side is to tend produce complex ions or compounds. Ions from iron has empty orbital which able to receive electrin couple from tannin into d^2sp^3.

a. Qualitative analysis of tannin

Tannin qualitative analysis is an analysis to identify the presence of tannin plant extracts. The method is $FeCl_3$ of 1%. The presence of tannin is detected with the change color into greenish black if it is reacted with $FeCl_3$ since tannin in the extrac engender complex compound with ion Fe^{3+} [20].

b. Analysis quantitavie of tannin

Qualitative analysis is conducted to analyze the percentage of tannin in the extract's product. It is determined by using *Folin Ciocalteau* method. This method was using reagen *Folin Ciocalteau* since phenolic compound able to react with folin to form blue coloreed solution which measured the absorbance using spectrofotometre. Phenolic compound only reacted with *Folin Ciocalteau* in base ambient so the proton dissosiasion in phenolic compound into phenolic ion.. Therefore, it employed Na_2CO_3 15% for base condition in solution [21].

Next, the solution is measured by using UV-Vis Spectrofotometre. The absorb spectrum was shown in the outside electron energy change from a molecule which caused the emerging of bond or not chemical bond. Generally, energy change happen caused by the carbon double bond or nitrogen pairs with oxygen. Tannin compound contain conjungated aromatic system, therefore it depicted the strong absorption bands in ultraviolet and visible range [20].

3. Methods

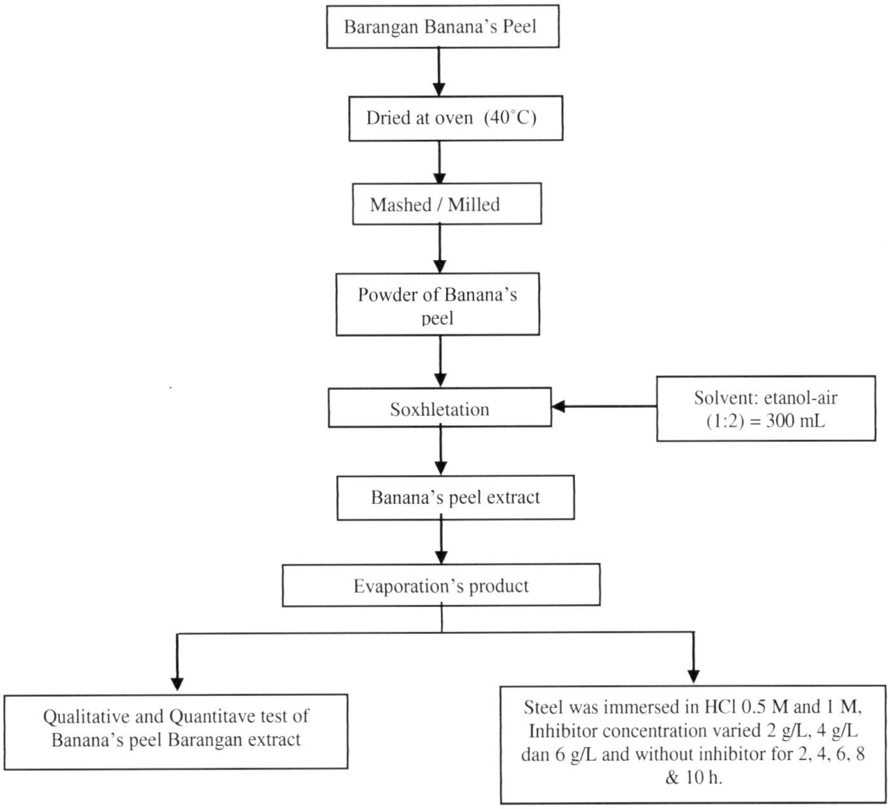

Figure 2. Diagram of banana peel extraction and corrosion rate analysis.

4. Results and discussion

Crude tannin extraction from banana's peel was held by soxhletaion method. It was evaporated to eliminate the solvent then condense the extract's product. In soxhletation process, ethanol: water ratio was 1: 2 for ± 8 h was conducted. Tannin is a polar compound, therefore etanol-water as polar compound is able to extract it. As in solvent side, ethanol-water mixture was used to obtain the amounts of extracted material [23]. The products extraction waere evaporated to omit the remaining solvent and to concentrate extract's product. Qualitative analysys or phitochemical test was held to assest the presence of tannin in extracts banana's peels. Phitochemical test was conducted by adding $FeCl_3$ reagen into guava leaves extract which showed the greenish or blue ink color change.

Phythochemical test has been done using $FeCl_3$, it is to determine either sample contained phenol group or not. After its change then the sample might contain phenol compound and it would might be proven the presence of tannin. This color changed occurred since the reaction between tannin and $FeCl_3$ produced the complex compound which made the mixture became greenish black color. Quantitave analysys of crude tannin was determined by using phenolic analysis method (Folin-CioCalteu) since tannin is one of phenolic compound which contained in plants. This method using regression linear equation as y = 0,005x + 0,021 from the calibration curve, y is absorbance value from Atomic Absorption Spectroscopy (AAS) analysis and x is crude tannin concentration in banana's peel (ppm). In soxhletation process, it was found that the absorbance was 0.1061 by twenty times dilution process then it was estimated that the crude tannin concentration in banana's peel sample was 340.4 ppm.

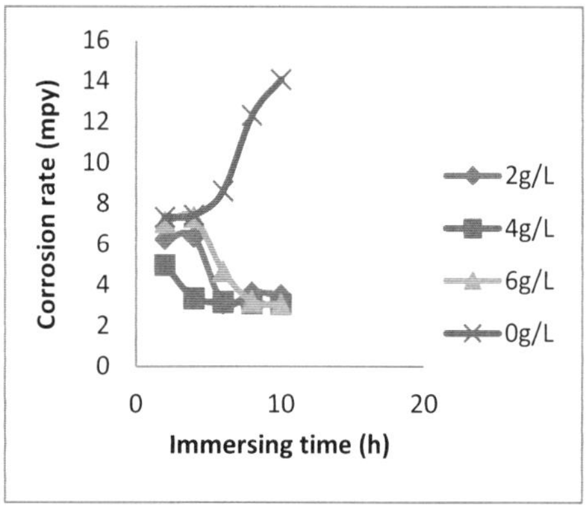

Figure 3. Immersing time vs corrosion rate in HCl 0.5 M.

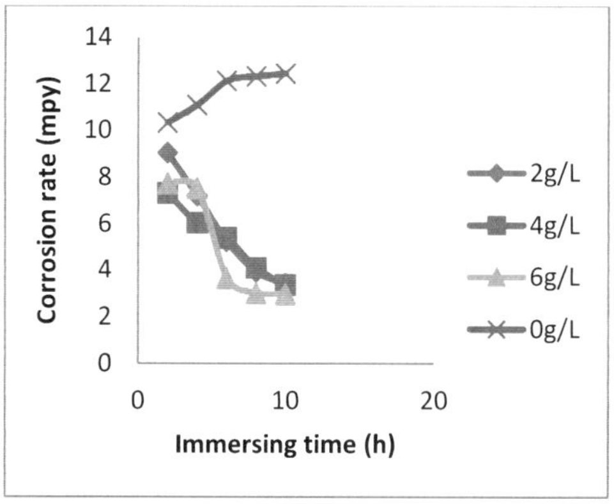

Figure 4. Immersing time vs corrosion rate in HCl 1 M.

Base on Figure 3, it shown that the lower corrosion rates were obtained by adding the inhibitor, it was contrary to the sample without inhibitor addition. It was occured because the presence of tannin in the banana's peel extract. Tannin is a complex poliphenol compound, where tannin molecule adsorpsed in the metal surface and formed a thin layer. This layer is unable to see by directed eyes, but it can detained the corrosion process in metal. Figure 4 depicted the difference of corrosion rate inside the HCl 0.5 M solution by adding and without the inhibitor. The lowest corrosion rate was reached at amount 3.0118 mpy in 6 g/l inhibitor concentration for 10 h while the highest corrosion rate reached out in 7.3143 mpy for 4 h in 6 g/l inhibitor concentration. It was proved that tannin has potential to decrease the corrosion rate thus it able to to use as inhibitor for shelving the corrosion rate.

Immersing time did influenced the corrosion rate. Figure 3 and 4 shown that the longer immersing time made the corrosion rate became more decelerate. It showed that more consuming contact time between tannin molecules with metal surface which formed thin layer protection then the corrosion rate was decreased and corrosion product $Fe(OH)_3$ able to cover the metal surface to develop pasive layer in cathode side then it affect reduction reaction in cathode. If reaction in cathode was inhibited, steel oxidation in anode would be inhibited.

In Figure 4 there is the difference of corrosion rate steel in HCl 1 M by adding and without inhibitor. The lowest corrosion rate was 2.9688 mpy in 6 g/l inhibitor concentration for 10 h while the highest corrosion rate was achieved 7.5294 mpy for 4 h and 6 g/l concentration inhibitor.

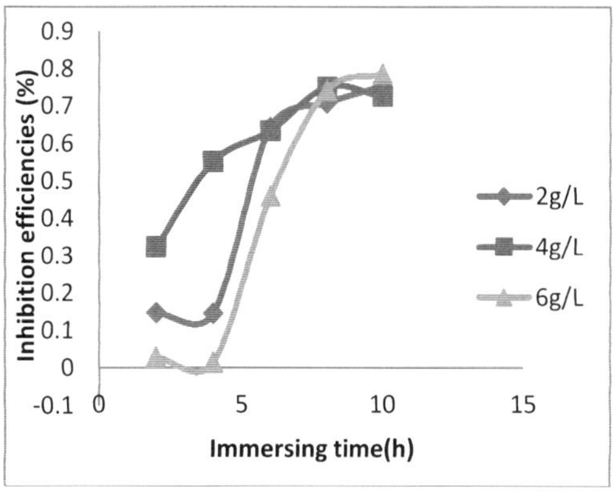

Figure 5. Immersing time vs inhibitor efficiencies in HCl 0.5 M.

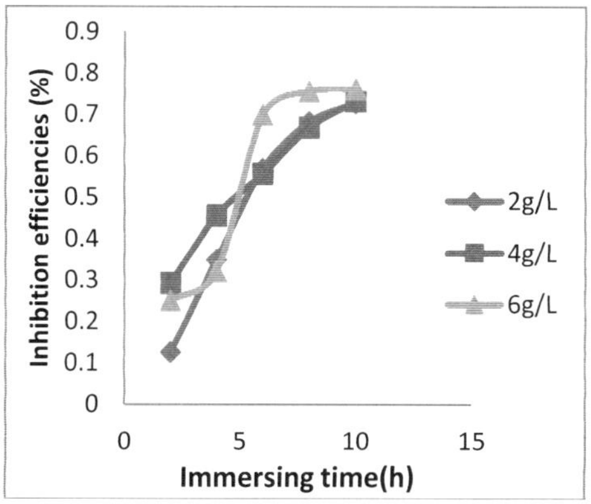

Figure 6. Immersing time vs inhibitor efficiencies in HCl 1 M.

Figure 5 shown the highest efficiency of inhibition 78.593% for 10 h in 6 g/l inhibition addition. It was proven in the theory, the inhibition concentration depended on immersing time. When the contact time take more prolonged between metal and corosive media made larger inhibition efficiencies, and also the oppposite ways. Figure 6 shown the highest efficiency inhibition at 76.125% in 6 g/l inhibitor addition for 10 h.
This results showed that the presence of tannin in the banana's peel has not fully formed complete complex compound then it was uncovered the entire surface of steel metal.

5. Conclusion

Banana's peel extract was proven as corrosion inhibitor because of the presence of tannin in the banana's peel. Lowest corrosion rate was obtained at 3.0118 mpy in 0.5 M HCl and 2.9688 mpy in 1 M HCl for 10 h and 6 g/l inhibitor concentration. Immersing time influenced the corrosion rate. The lower corrosion rate occured for longest immersing time. Highest inhibition obtained 78.593% and 76.125% for 10 h and 6 g/l inhibitor concentration in 0.5 M and 1 M HCl, respectively.

References

[1] Goffar A 2011 *Lembaran Publikasi Lemigas* **1** (45) 79-90
[2] Gusti D R 2013 *Prosiding Semirata FMIPA* Universitas Lampung
[3] Parker M E and Peattie E G 1988 *Pipe Line Corrosion and Cathodic Protection : A Practical Manual for Corrosion Engineers, Technicians and Field Personal*, Third Edition (Houston : Gulf Professional Publishing)
[4] Sidiq M F 2013 *Jurnal Foundry* **1** (3)25-30
[5] Haryono G 2010 *Prosiding Seminar Nasional Teknik Kimia "Kejuangan"* ISSN 1693-4393 p D09-1 – D09-6
[6] Baboian R 2005 *Corrosion Tests and Standars : Application anda Interpretation* Edisi ke 2 (ASTM International West Conshohocken).
[7] Fogler 1992 *Elements of Chemical Reaction Engginering* Edisi ke 2 (Prentice-Hall International Inc).
[8] Kirk and Othmer 1965 *Encyclopedia of Chemical Technology* Vol. **6** (New York: John Willey and Sons Inc) p 320
[9] Amanto H and Daryanto 2006 *Ilmu Bahan* (Jakarta : Bumi Aksara)
[10] Wiston R 2000 *Uhlig's Corrosion Handbook* Edisi ke 2 (New York : John Willey & Sons Inc) p 1091
[11] Marcus P and Mansfield F B 2006 *Analytical Methods In Corrosion Science and Engineering* (Boca Raton : CRC press)
[12] Stupnisek-Lisac E, Gazioda A and Madzarac M 2002 *Electrochim. Acta* **26** (47) 4189-4194
[13] Zhang D, Gao L and Zhou G 2004 *Appl. Surf. Sci.* **225** 287
[14] Raja P B and Sethuraman M G 2008 *Mater. Lett.* **62** 113-116
[15] Asdim 2007 *Jurnal Gradien* **2** (3) 273-276
[16] Sari D M, Handani S and Yetri Y 2013 *Jurnal Fisika Unand* **3** (2) 204-211
[17] Eddy N O and Ebenso E E 2008 *African J. Pure Appl. Chem.* **2** (6) 046-054
[18] Rosyda I K and Ersam T 2010 *Prosiding Kimia FMIPA* SK-01 Institut Teknologi Sepuluh November Surabaya.
[19] Rahim A A and Kassim J 2008 *Recent Patents on Materials Science* **3** (1) 223-231
[20] Harborne J B 1987 *Metode Fitokimia: Penuntun Cara Modern Menganalisis Tumbuhan* Edisi ke 2 (Bandung : Institut Teknologi Bandung)
[21] Ryanata E 2014 *Jurnal Ilmuah Mahasiswa Universitas Surabaya* **1** (4) 1-16
[22] Pambayun R, Gardjito M, Sudarmaji S, and Rahayu K 2007 *Majalah Farmasi Indonesia* **18** (3) 141-146

Geloina coaxans shell as calcium source on synthesis hydroxyapatite

P H Yanti and A Kamiah
Chemistry Department, Faculty of Mathematics and Natural Sciences
Universitas Riau Kampus Bina Widya Jl HR soebrantas KM 12, 5 Panam Pekanbaru Indonesia 28293

E-mail: pepihelzayanti@lecturer.unri.ac.id

Abstract. *Geloina coaxans* shell (GCS) is one of mullusc shell mainly composed by calcium carbonate. In this work, calcium carbonate has been converted to calcium oxide by calcination at 1000°C for 12 hours. The calcined of *geloina coaxans* shell were treated with HNO_3 to produce $Ca(NO_3)_2$ as calcium source on synthesis hydroxyapatite. Orthophosphoric acid (H_3PO_4) was used as phosphate donor. Reaction of Ca/P has been done by precipitation method at molar ratio of precursors of 1.67 and pH adjusted at 10 using NH_4OH. The XRD result revealed that hydoxyapatite can be prepared at 3 M of HNO_3 and stirring time for 240 minutes. Specific band of hydroxyapatite such as PO_4 and OH observed using FTIR instrument. Analysis of crystal size using Schererr equation proved nanosize of powder hydroxyapatite can be produced.

1. Introduction

Hydroxyapatite is one of calcium phosphate salt which have many application in several fields. Similarity of hydroxyapatite structure with inorganic component of teeth and bone will be the reasons to produce hydroxyapatite. Properties of hydroxyapatite or $[Ca_{10}(PO_4)_6(OH)_2]$ such as biocompatible, bioactive, osteoconductive and non toxic caused many applications in biomedics or ortopedics. Hydroxyapatite also can be used as catalyst [1,2] and adsorbent for removing the trace metal.

Hydroxyapatite have been produced by several sources of calcium carbonate. Usability of waste to obtain hydroxyapatite is become a way to reduce the enviromental problem. Waste kina (*Evechinus chloroticus*) shell converted to hydroxyapatite using pyrolisis method and produced hydroxyapatite with porous structure [3]. Hydroxyapatite also have been produced using *Mussel* shell with microwave method [4]. *Macoma balthica* was also a potensial material to prepare nano hydroxyapatite [5].

The objective of this work is to produce hydroxyapatite using waste of *Geloina coaxans* shell. *Geloina coaxans* shell (GCS) is one of mollusc shell with high calcium carbonte. It is potential to convert the hydroxyapatite material. *Geloina coaxans* shell as precursor is advantages due to low cost material. *Geloina coaxans* shell and H_3PO_4 are used as starting materials on synthesis hydroxyapatite.HNO_3 is used to produce $Ca(NO_3)_2$. Variation of HNO_3 and stirring time will be done to learn the crystalinity of hydroxyapatite sample. Synthesis is processed by precipitation method with ratio precursor 1.67 and pH is 10. Precipitation is one of simple and rapid method to produce hydroxyapatite. The obtained Powder is then characterized using XRD (X-ray Diffraction) technique to analyze mineral phase of hydroxyapatite synthesized and FTIR (Fourier Tranform Infra Red) instrument was then used to observe functional group of sample hydroxyapatite.

2. Experimental section

Geloina coaxans shell (GCS) were collected and washed to remove the impurities. The GCS were dried in sunlight and continued using oven for 2 hours at 100 °C. Sampel GCS was calcined at 1000 °C for 12 hours. The powder obtained was CaO. The calcined GCS was treated using HNO_3 to produce Ca(NO3)2 in various concentration of HNO_3. H_3PO_4 solution was added drop wise to $Ca(NO_3)_2$ solution. Stirring rate was adjusted at 300 rpm and pH was 10 using NH4OH. The obtained suspension was kept for 12 hours at room temperature. next step is the obtained suspension was filtered and dried for 2 hours at 105 °C. The Preparation of the powder was calcined for 1 hour at 900 °C and characterized using XRD and FTIR. Variation of HNO_3 and stirring time were chosen as reaction parameter of synthesis.

3. Results and discussion

3.1 Concentration of HNO[3]

Synthesis hydroxyapatite has been done using waste of Geloina coaxans shell as calcium source and H_3PO_4 as phosphate precursor. The reaction adjusted using rat io of precursor 1.67 at pH was 10 by precipitation method. Several reaction parameters such as concentration of HNO_3 and stirring time selected to learn crystalinity of hydroxyapatite synthesized. The XRD and FTIR analysis were done to examine the crystalline structure and functional group of hydroxyapatite synthesized. The result of various concentrations of HNO_3 is shown in Figure 1.

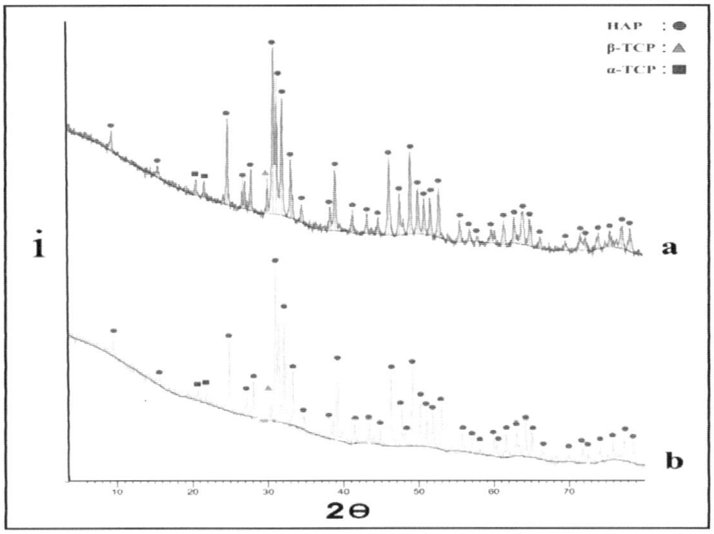

Figure 1. XRD pattern of hydroxyapatite in various concentration of HNO_3 (a) 3 M ; (b) 5 M

Based on Figure 1 Both of sample hydroxyapatite shown crystalline phase of hydroxyapatite. The specific peak of hydroxyapatite observed in both of produced hydroxyapatite.The highest intensity of XRD pattern appeared at 2θ =31.7°. Hydroxyapatite peaks also can be seen at 2θ =32.8°; 33,9°. These results are similar compared than JCPDS file of hydroxyapatite No 09-0432. Secondary phase of calcium phospate also observed at 2θ =30.5°; 2θ =21.67 ° and 2θ =22.75° with minor intensity (α-β TCP). Hydroxyapatite synthesized using 3 M of HNO_3 have higher intensity and proved better crystalinity of hydroxyapatite compared than 5M of HNO_3. The XRD patterns of sample correspond to crystal orientation plane of hyfroxyapatite [211]. Percentage of hyroxyapatite obtained was 71,39% and 65,06% for 3 and 5 M of HNO_3 respectively. The result of hydroxyapatite pattern, noted it shown that the concentration of HNO_3 influence crystalinity obtained, although both of sample prepared had similar diffraction patterns. Previous research, pure hydroxyapatite also produced using 3 M of $Ca(NO_3)_2$ and calcined at 900 °C specific peak appeared at 2θ= 31,8° by sol-gel method. Reaction of synthesis devided in several steps

$$CaCO_3 \longrightarrow CaO + CO_2 \quad (1)$$
$$CaO + 2HNO_3 \longrightarrow Ca(NO_3)_2 + H_2O \quad (2)$$
$$10Ca(NO_3)_2 + 6H_3PO_4 + 20NH_4OH \longrightarrow Ca_{10}(PO_4)_6(OH)_2 + 20NH_4NO_3 + 18H_2O \quad (3)$$

Debby Schereer equation also be used to estimate crystalite size of sample hydroxyapatite. Schererr equation can be seen below

$$L = \frac{0.89\,\lambda}{\beta\cos\theta} \quad (4)$$

L= crystal size; λ= Cu K α (0,154060); β=FWHM

The crystal size of hydroxyapatite at stirring time for 60 minutes and 3 M of HNO_3 was 37.48 nm. Hyroxyapatite *nano-wire and nano-rod* also produced with crystal size was 12 - 20 nm[4]. Analysis FTIR was also done to observe functional group of hydroxyapatite synthesized. FTIR spectra of hydroxyapatite at stirring time for 60 minutes and 3 M of HNO_3 is shown in Figure 2.

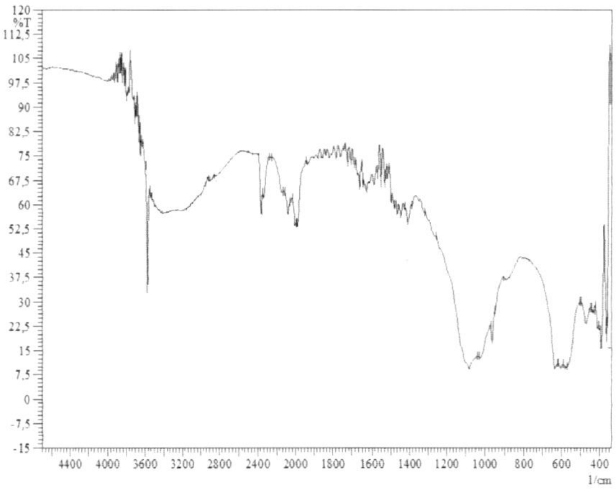

Figure 2. FTIR spectra of hydroxyapatite at stirring time for 60 minutes and 3M of HNO_3

Table 1. FTIR spectra of hydroxyapatite synthesized at stirring time for 60 minutes and 3M of HNO_3

Functional groups	Wavenumber (cm^{-1}) Results	Chen et al (2015)	Sasikumar et al (2006)
v O-H	3571	3567	3571
v PO_4	1025	1037	1045
v PO_4	568	567	571
v C-O	1410	1462	1550

The FTIR spectra of sample hyroxyapatite shown characteristic vibration of phoshate in two vibrational bands at 1025 cm^{-1} (v_3) and 568 cm^{-1} (v_4). Hydroxyl group (OH) of hydroxyapatite also assigned at 3571 cm^{-1} that revealed formation of hydroxyapatite phase. From the Previous research, it also observed that OH at 3567 and 3571 cm^{-1} respectively [6,7]. The band appeared at 1410 cm^{-1} corresponded to functional of CO_3^{2}

3.2 Stirring time

The XRD pattern of hydroxyapatite on variation of stirring time for 60; (b)120; (c)240; and (d) 360 minutes, it was depicted in Figure 3.

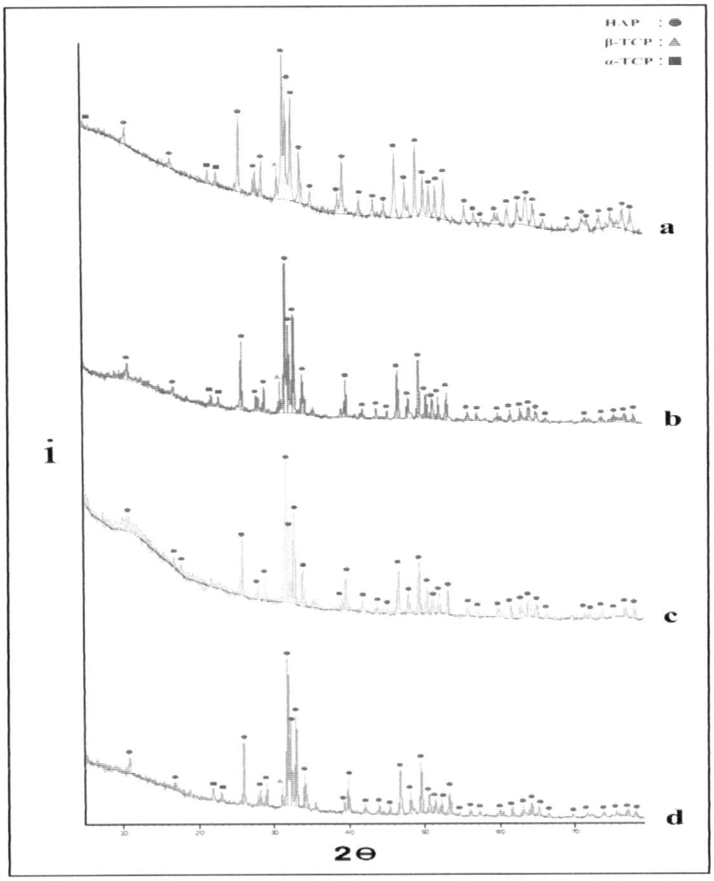

Figure. 3 XRD pattern of hydroxyapatite in various stirring time (a) 60; (b)120; (c)240 and (d) 360 minutes

The stirring time influenced the crystalinity and purity of hydroxyapatite synthesized. The characteristic peak of powder hydroxyapatite appeared at $2\theta = 31.7^{o}$, additional peak of hydroxyapatite also observed at $2\theta = 32{,}9\,^{o}$. These are suitable with JCPDS of hydroxyapatite file No 09-0432. The secondary phase of calcium phosphate salt also observed at $2\theta = 21{,}7\,^{o}$ with minor intensity, the mineral was TCP-tricalcium phosphate. The XRD pattern shown, the highest intensity of hydroxyapatite peak have been obtained at stirring time for 240 minutes, the percentage of hydroxyapatite prepared was 96 %. Analysis crystal size using Schereer equation, the crystal size of hydroyapatite was 53.24 nm. Synthesis hydroxyapatite using Ca(OH) and H_3PO_4 as precursors also fabricated hydroxyapatite nanopowder with crystalsize was 26 nm. The percentage of hydroxyapatite obtained was 86% at $2\theta = 31{,}7\,^{o}$ for stirring time 240 minutes[8], this result indicated that type of calcium as precursor also influence hydroxyapatite. The powder of the obtained hydroxyapatite was also analyzed using FTIR analysis. Specific band of hydroxyaopatite can be seen on Figure 4 and Table 2.

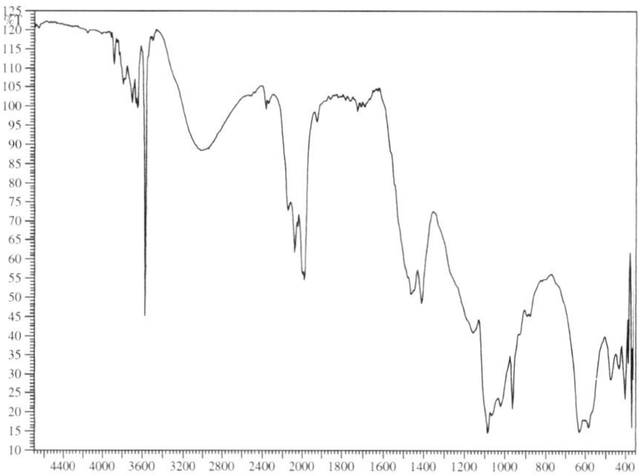

Figure 4. FTIR spectra of hydoxyapatite at stirring time for 240 minutes and 3M of HNO_3

Table 2. FTIR spectra of hydroxyapatite obtained at stirring time 240 minutes and 3M of HNO_3

Functional group	Wavenumber (cm^{-1})		
	Results	Wu et al (2013)	Shavandi et al (2014)
v O-H	3571	3580	3570
v PO_4	1035	1031	1100
v PO_4	569	554	565
v CO_3	1465	-	1478

Vibrasi of phosphate or PO_4^{3-} (v_4) appeared at 569; 613 cm^{-1} while PO_4^{3-} (v_3) observed at 1024 and 10535 cm^{-1} respectively. This group also observed at 1037 ; 960 dan 567 cm^{-1} using *abalone shell* as precursor [5]and at 603 dan 565 cm^{-1} [6]. OH *streching* also assigned at 3571 cm^{-1} . Usability of microwave method to *Mussel* shell as calcium source also obtained OH *streching* at 3570 cm^{-1} [6], 3580 cm^{-1}. [9]

4. Conclusion

The hydroxyapatite can be synthesized by precipitation method using Geloina coaxans shell as calcium source. Usability Geloina coaxans shell as precursor is advantages due to low cost material. The XRD result indicated that pure phase and high crystallinity of hydroxyapatite powder can be obtained. The result of FTIR also proved that specific band of hydroxyapatite appeared in characteristic band of hydroxyapatite.

References

[1] Boukha Z, Jonatan J and Guttirez O 2006 *J. Mater.* **190** 125-136
[2] Zahouily M, Abrouki Y, Bahlaouan B, Rayadh A and Sebti S 2003 *Catal. Commun.* **3** 521-4
[3] Shavandi A,Wilton V and Bekhit A 2016 *J. Taiwan Inst. Chem. Eng.* 1-7
[4] Shavandi A, Bekhit A, Ali Z and Sun Z 2014 *J. Mater. Chem. Phys.* 1-10
[5] Felsen T J 2015 *J. Adv. Powder Technol.* **300** 300-310
[6] Chen J, Wen Z, Zhong S, Wang Z, Wu J and Zhang Q 2015 *Mater. Design* **87** 445-449
[7] Sasikumar S and Vijayaraghavan 2006 *Trends Biomater.* **19** 70-73
[8] Yanti P H and Maylinda 2015 *Prosiding of International Conference on Oleo and Petrochemical Engineering* 77-82
[9] Wu S C, Tsou H, Hsu H and Liau S 2013 *J. Ceram. Intl.* 8183-8188

IOP Conf. Series: Materials Science and Engineering **345** (2018) 012032 doi:10.1088/1757-899X/345/1/012032

The effect of woven roving fiberglass total layers on resin infusion time in vacuum infusion

A H Saputra and R H Ibrahim
Department of Chemical Engineering, Faculty of Engineering, Universitas Indonesia Depok 16424

E-mail: sasep@che.ui.ac.id

Abstract. Composite material consists of reinforcement materials and resin as a matrix. Vacuum infusion isone of composite material manufacturing process. This process is to minimize the air cavity on composite material. The composite material will have good mechanical properties. There is a problem in vacuum infusion related to resin gelling time that must be considered. In this study, the area as well as the reinforcement layers are variated. Unsaturated polyester was used as resin and woven roving fiberglass was used as reinforcement. This study was obtained that resin infusion time data for woven roving, 15x20 cm of size, in two until six layers are 55 seconds to 78 seconds; whereas, the infusion times for 15x25 cm of size,in two until six layers are 119 seconds to 235 seconds; whereas the infusion time for 15x35 cm of size, in two until six layers are 181 seconds to 303 seconds. By data processing, the maximum fiber area that resin still can flow, for 6 layers, is 0,4391 m^2 (or 15 cm x 2.92m). Maximum fiber total layers for the specimen with 15x20cm^2, 15x25cm^2 and 15x35 cm^2 of areaare 147, 145 and 125 layers respectively.

1. Introduction

Composite materials are combination of two or more different components on macro scale, with these components having interfaces between them [1]. The material properties of the composite material will differ in the shape and chemical composition of the original substance [2]. There are several methods for manufacturing of compisite material, one of them is vacuum infusion. Vacuum infusion process is a process that use closed mold. This process is done by coating the reinforcement in the mold. Then the resin is fed into the mold. The resin infusion process is carried out using a vacuum pump, to allow the resin to be evenly distributed and to prevent the excess resin in the reinforcement. The vacuum infusion process can minimize the air cavities formation, so that the resulting composite material will have little free space and eventually be able to produce composite material products with good mechanical properties [3]. The reinforcement type manifested in this research is fiberglass with woven roving (WR), which is the most widely used fiberglass type in industrial composite materials. Resin that used in this research is unsaturated polyester, which is wide used resin in the manufacture industry.

However, there is a problem that often occur in vacuum infusion process namely a white spot. White spot is an area where the fiberglass was not filled by resin. White spot occurs because during the infusion, the resin begins to hardened. If the resin undergoes gelation when the infusion persists, some fiberglass will not be wetted by the resin. White spots should be avoided because it can reduce the mechanical strength. The fiberglass total layer must be considered as it may affect the resin infusion time. Therefore, to prevent white spots, this study is to obtain the effect of woven roving fiberglass total layers on resin infusion time, so that the resin can be flown among fiberglass.

2. Materials and methods

There are four stages that performed in this study. The first stage is to prepared all of the equipment and material that will used in this study, to make a vacuum infusion system. This research is focused on polymer matrix composite (PMC). PMC is one type of composite material in which composed of short fibers or continuous fibers combined into one by an organic polymer matrix [5]. The resin used in this study is Unsaturated Polyester Resin (UPR). UPR is an unsaturated condensation product or anhydrous acids and diols with or without diacids [6]. Fiber glass that used in this study is Woven Roving (WR) fiber glass.

After the tools and material are prepared, the vacuum infusion equipment system rearrangement can be done. All of the equipment and material must be rearranged because the vacuum infusion system is each separate device that must be combined to form a unified equipment system. The arrangement stage is done as well as to test the vacuum infusion equipment system in order to avoid air leakage. Vacuum infusion equipment schematic and arrangement can be seen in Figure 1.

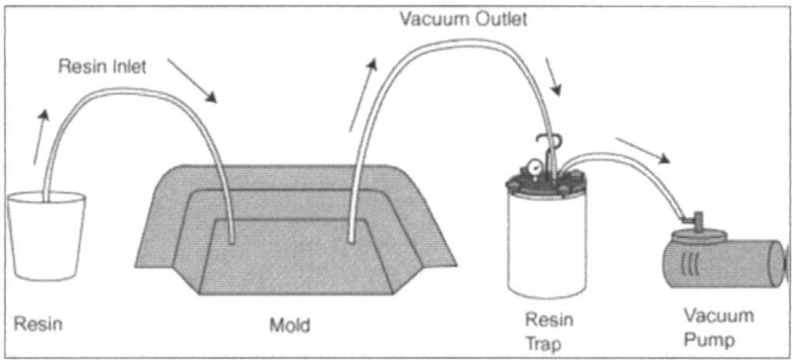

Figure 1. The vacuum infusion equipment arrangement

Second stage is, if the equipment system successfully arranged and can run without leakage, resin infusion process to fiberglass and research variations in the fiberglass total layers to the resin infusion time can be done immediately, and we get resin infusion time data in every total layers from 2 until 6 in three different fiberglass layer area. Finally, the data processing in order to get fiberglass dimension length and fiberglass total layer maximum.

3. Result and discussion

The results obtained the resin infusion time data on each woven roving fiberglass total layers from 2 to 6 layers, in the fiberglass area 15x20 cm, 15x25 cm, and 15x35 cm. The resin infusion data were 15 data from all composite material that successfully made. Test results can be seen in Table 1.

Table 1. Resin infusion time data in the fiber total layers' variation.

Total Layers	Woven Roving Fiberglass Area		
	15x20 cm	15x25 cm	15x35 cm
	Resin Infusion Time (s)		
2	55	69	97
3	59	75	104
4	66	80	110
5	71	87	118
6	78	92	123

Table 1 shows that in fiberglass total layer addition or the fibers thickness, resin infusion time becomes even greater. Research data is appropriate to Darcy law. In Darcy law, if the fiberglass layers L is enlarged then the resin flow rate (Q) will decrease. Small resin flow rate (Q) indicates it takes a long time to wet the entire fiberglass made. The dimension of the resin flow rate (Q) is m^3/s. Because of this, the resin infusion time is directly proportional to the fiberglass total layers. From Table 1 data, it is possible to determine the fiberglass maximum dimensional length that can be flown by the resin at each layer amount. The calculations performed by extrapolating the line equations from the graph between the fiberglass area to the resin infusion time. The graphs between the fiberglass area to the resin infusion time on two to six total layers can be seen in Figures 2.

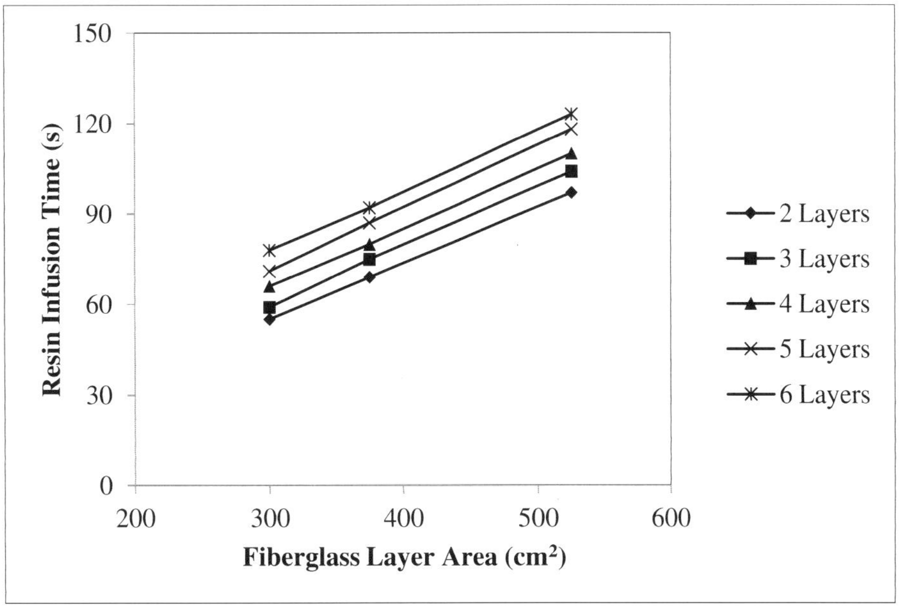

Figure 2. Relation of resin infusion time to fiber area at 2 to 6 layer.

Using linear regression, we obtain the line equation in the graphs in Figure 2, from 2 until 6 layers consecutively $y = 0.1867x - 1$, $y = 0.199x - 0.2857$, $y = 0.1982x + 6.8571$, $y = 0.2025x + 8.5714$ and $y = 0,201x + 17.286$. The x-axis shows the fiberglass area (cm^2), and the y-axis represents the resin infusion time (s). Line equation of each graph is then extrapolated by substituting the y value as the resin infusion time for 15 minutes or 900 seconds, obtained from UPR resin geltime for 20 minutes minus 5 minutes' preparation. From these extrapolations will be obtained the area dimensional maximum woven roving fiberglass that can be made on each layer. The woven roving fiberglass maximum length dimension can be calculated by dividing the maximum area by the fiberglass dimensional width. The fiberglass dimensional width in this case is the resin inlet and outlet in the fiberglass side. Since the fiberglass dimensional width that used into the composite material in this study is only 15 cm, the maximum area obtained is assumed to be applicable only to 15 cm. In the maximum fiberglass dimension length calculation, the maximum fiberglass area obtained is divided by the by 15 cm. Fromthe divided result obtained maximum woven roving fiberglass dimensional length which can be flown by resin in each layer. The maximum fiberglass dimension length can be seen in Table 2.

Table 2. Maximum fiberglass dimensional length (m)

Fiberglass Total Layer (Layer)	Fiberglass Maximum Dimension Length (m)
2	3,21
3	3,01
4	3,00
5	2,93
6	2,92

From Table 2 there is a woven roving fiberglass dimensional length which can be flown by resin in each layer from 2 to 6 layers, on 15 cm dimensional width. If the fiberglass dimensional length used does not pass through the maximum length data, the resin can still flow because it has not thickened or undergone gelation, in other words the resin has not passed its geltime. In contrast, the resin cannot flow if the fiberglass dimensional length used is greater than the data in Table 2. Because the resin has entered geltime. From the calculation it can be said that in order to avoid the gel-time resin problem in vacuum infusion which gives white spot on composite material, the fiberglass dimensional length made into composite material at 15 cm fiberglass width in every layer number from 2 to 6 layers should not pass the fiberglass dimensional length calculation data in Table 2.

Next, The Table 1 data is processed into graphs to see the relationship between the fiber total layer to the resin infusion time in a more easily visible form. The graphs are made up in three graphs for the three fiberglass areas made, which are 15x20 cm, 15x25 cm, and 15x35 cm. Below is a graph of testing the relationship between the woven roving fiber total layers against the resin infusion time on the fiberglass area of 15x20 cm, can be seen in Figure 3.

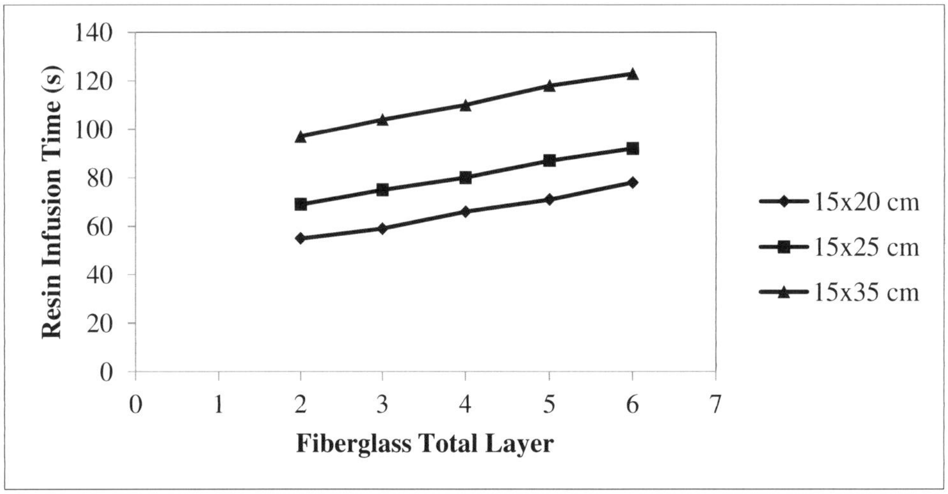

Figure 3. The relation of resin infusion time to the fiberglass total layers.

Using linear regression, we obtain the line equation in the graphs in Figure 3, from 15x20 cm until 15x35 cm consecutively $y = 5.8x + 42.6$, $y = 5.8x + 57.4$, and $y = 6.6x + 84$. In Figure 8 to 10, the x-axis shows the fiberglass total layers, and the y-axis shows the resin infusion time. Using linear regression, we will find the graph equation. The line equation is extrapolated by substituting the y value as the resin infusion time for 15 minutes or 900 seconds. From the y value of 900 seconds substitution will be obtained the maximum woven roving fiberglass total layer that can be made using the vacuum infusion equipment system in this study. From substitution result value y = 900 seconds then got result shown in Table 3.

Table 3. The fiber total layer maximum in every fiber area that used

Fiberglass Area (cm x cm)	Total Layer Maximum (Layer)
15x20	147
15x25	145
15x35	123

From Table 3 we obtained the woven roving fiberglass total layers that can be flown by resin in each fiberglass area 15x20 cm, 15x25 cm, and 15x35 cm. In each data the fiberglass maximum total layers for each fiberglass area, the resin can still flow because it has not thickened or undergone gelation, in other words the resin has not passed its gel time. Conversely, the resin cannot flow if the fiberglass layer is used more than the data in Table 3. because the resin has entered gel time. From the calculation it can be said that in order to avoid the gel time resin problem in vacuum infusion that cause white spot on the composite material, the woven roving fiberglass total layers made into composite material on 15x20 cm, 15x25 cm, and 15x35 cm fiberglass should not pass the calculation data in Table 3.

4. Conclusion

Based on test results, the longer resin infusion time, woven fiberglass roving total layer is increase. In other words, the relationship between time infusion resin with a number of woven roving fiberglass layers is directly proportional. The maximum woven roving area and length dimensional at 2 to 6 fiberglass total layers in 15 cm width dimensional is obtained as follows: 2 layers, Area: 0.4825 m^2, Length: 3.21 m. 3 layers, Area: 0.4524 m^2, Length: 3.01 m. 4 layers, Area: 0.4506 m^2, Length: 3.00 m, 5 layers, Area: 0.4402 m^2, Length: 2.93 m. 6 layers, Area: 0.4391 m^2, Length: 2.92 m. Then the maximum woven roving fiberglass total layers to be resin in each fiberglass area is as follows, fiberglass area of 15x20 cm is 147 layers, fiberglass area of 15x25 cm is 145 layers, fiberglass area of 15x35 cm is 123 layers.

Acknowledgement

Great gratitude to PT Solchem for providing materials needed in this research and PITTA Research Fund of Universitas Indonesia for the research financial support.

References

[1] Akovali G 2001 *Handbook of Composite Fabrication* (United Kingdom: RAPRA Technology ltd.)
[2] Smith W F 1996 *Principles of Materials Science and Engineering*, 2nd ed. (Singapore: McGraw-Hill Company)
[3] Poorzeinolabedin M, Parnas L and Dashatan S H 2014 *Mater. Des.* 64450-455
[4] Saputra A H and Setyarso G 2017 *Mater. Sci. Eng.* C1621
[5] Callister W D 2007 *Material Science and Engineering* (New York: John Willey & Sons Inc.)
[6] Holloway L 1994 *Handbook of Polymer Composites for Engineers* (Cambridge: Woodhead Publishing)

IOP Conf. Series: Materials Science and Engineering **345** (2018) 012033 doi:10.1088/1757-899X/345/1/012033

Characterization nano crystalline cellulose from sugarcane baggase for reinforcement in polymer composites: Effect of formic acid concentrations

N A S Aprilia, S Mulyati, P N Alam, Karmila and A C Ambarita
Chemical Engineering Departement, Engineering Faculty, University of Syiah Kuala, Banda Aceh, Indonesia

E-mail: sriaprilia@unsyiah.ac.id

Abstract. Nanocellulose from sugarcane bagasse for reinforcement in polymer composites has isolated from formic acid (FA) with different concentration. This research was conducted with three level concentration of FA ei. 15, 30 and 50%. The nanocellulose were successfully prepared with variations of total yields of 66.66, 67.33 and 69.33% respectively with increase of FA concentrations at 6 hours of hidrolysis time. The obtained nanocellulose were characterized by fourier transform infrared (FT-IR) spectroscopy confirmed the introduction of carboxyl goups on the surface of cellulose. The X-ray diffraction (XRD) spectra proved the existence of cellulose, with a highly crystalline of 62.466, 71.033, and 76.296% with increase of FA concentrations. The size of crystallinity of nanocellulose were decreased with increased of FA concentration. The result investigated that size of crystallinity of nano cellulose reduced from 4.37, 4.15 and 3.94 nm.

1. Introduction

The use of agriculture solid waste for reinforcement in polymer composites is one of reduce environment issue. In addition, the use of agricultural waste for various purposes also to increase the added value for the waste itself. The increasing interest in environment-friendly materials has motivated industrial in development and use of biomass for various reinforcement in polymer composites.

Sugarcane bagasse is one of biomass agriculture waste that potential for reinforcement in polymer matrixes because has cellulose content. Many researcher has developed nanocellulose from many bioresources such as kenaf [1], oil palm fronds [2] oil palm tree [3], bagasse pulp [4], dhaincha, corn stalks, rice straw and wheat straw [5], sugarcane bagasse [6-8], Gigantochloa *scortechinii* [9]. These nanocellulose developed in various method to find out nanocellulose in form of both nano fibril and nano crystalline.

Cellulose found in the main constituent of plant cell wall, is the mayor constituent and is a homopolysaccharide composed of β-1-4 glucopyranose units [10-11]. Many study have extracted nanocellulose from the plant cell wall throught chemical and mechanical treatment. Acid hydrolytic processes are also used to degrade amorphous cellulose in forming cellulose nanofibers. In common processes, cellulose crystallites from cellulosic materials were prepared using acid hydrolysis. During isolation process, the dimention of cellulose is mainly dependent on its origin as well as the preparation methods and condition implemented [12]. The acceptance of acid inside part of the amorphous is easier than in crystalline part due to the more compact strcture of crystals. Acid hydrolysis allows to break and removal of disodered and part of amorphous in cellulosic fibers, releasing single and well defined crystals in the form of nano crystaline cellulose [7]. Therefore, the extensive properties of nano crystalline cellulose from different bioresources such as mechanical properties, low density, biodegradability are very importent for the development of bionanocomposites.

In this study, to extract nano crystalline cellulose from sugarcane bagasse by using formic acid with various concentrations has been investigated. The developed of nano crystalline cellulose by acid hydrplysis having high crystallinity index, supposed to have good surface morphology for polymer reinforcement. The obtained nano crytalline cellulose were characterized using FT-IR and XRD analysis.

2. Experimental

Nano crystalline cellulose extracted from sugarcane baggase that were colected from seller sugarcane juice of Banda Aceh City, Aceh Province, Indonesia. After drying in sunlight, it was cut in size ± 2 cm, and that the sample wash with ditilated water for 3 times. Bagasse was dried in oven at 60 °C for over night and stored at room temperature.

2.1. Extraction of bleached pulp sugarcane bagasse

Approximately 10 grams of dried sugarcane bagasse was hydrolyzed using 300 ml NaOH 1 M for an hour and 30 minutes at temperature 80 °C. This process is to eliminate lignin content (delignification process). Then the sample was washed by distilled water in a centrifuge for 3 times. After that, added 150 ml of NaOCl 5% for 20 minutes at temperature of 80 °C. Sample washed by distilled water in a centrifuge for several times. The pulp was dried in an oven at 70 °C for 3 hours. This process was prepared base on previously describe methods [6].

2.2. Preparation of nano crystalline cellulose

Isolated nano crystalline cellulose from sugarcane bagasse used FA with ratio between bleaching pulp and FA 1:30. The three variations of FA concentrations were used 15, 30 and 50%. The hydrolysis process was quenched by sttirer at 400 rpm at temperature 90 °C. A colloidal suspension which produced was centrifuged at 4500 rpm for 30 minutes. After that, nano cryslline cellulose was washed with distilled water in the centrifuge as many times. The nano crystalline cellulose was dried in an oven at 70 °C for 3 hours. The percentage of yield was calculated according to Eq. 1 [3].

$$yield(\%) = \frac{M_1 x M_3}{M_2 x M_o} x100 \qquad (1)$$

Where M_o is the initial mass of sample, M_1 is the weight of nano crystalline cellulose finally obtained. M_2 is the mass of the suspension sample used to get the dry powder. M_3 is the mass suspension in the final preparation.

2.3. Characterization

2.3.1 Fourier transform infrared spectroscopy (FT-IR) analysis

The FT-IR spectra was used to identify the functional group of nano crystalline cellulose. The analyzed were recorded on Fourier Transform Infrared Spectrometer Shimadzu Prestige 6400. The amount of samples was blended with KBr powder and then pressed into thin pellets. The sample was measured in the wavelength range from 4000 cm^{-1} to 400 cm^{-1}.

2.3.2 X-ray diffraction (XRD) analysis
To detemine crystallinity index and crystal size of nano crystalline cellulose were used X-Ray diffraction analyses. XRD analyses were on an X-Ray diffractometer with measurement condition are 40 kV and current 30 mA. The scattering angle range (2θ) was from 10° to 80° with the rate 10°/min. The crystallinity index (CrI) was calculated using the Eq. 2 and crystal size was calculated using Scherrer Eq. 3 [11].

$$C_I(\%) = \left(1 - \frac{I_{am}}{I_{200}}\right) x100 \tag{2}$$

$$D = \frac{0.9\lambda}{\beta_{hkl}\cos\theta_{hkl}} \tag{3}$$

Crystallinity index (CrI) was calculated from the maximum peak intensity at lattice diffraction (200) and I_{am} is the intensity minimum between the 200 and 110 peaks. While for calculating crystal size, K is the Scherrer constant (K = 0.9), λ is the wavelength of X-ray radiation (λ = 1,5406 Å) and β1 / 2 is the full width at half-maximum of the Reflection, and θ is the Bragg angle.

3. Result and discussion

3.1. Yield of production nano crystalline cellulose
Production of nano crystalline cellulose with acid hydrolysis has investigated with three concentrations of FA and constant of time, acid ratio with fiber and temperature. The obtained yield of nano crystalline cellulose from sugarcane bagasse as concentration function show in Table 1. The basis of the cellulose of sample at 3 grams. The result show that the higher of FA concentration, the greater of the percent yield due to the less dissolved area of amorphous cellulose.

3.2 Functional group nano crystalline cellulose
The FT-IR spectra subjected with various concentrations of FA and cellulose functional group show in Figure 1. The functional group of cellulose could be determine from the wave length at 3247 cm^{-1} (C-H), 2895 cm^{-1} (C-H), 1647 cm^{-1}(N-H), 1369 cm^{-1} (C-H), 1315 cm^{-1}(C-O), 1155 cm^{-1}(C-H), 1055 cm^{-1}, dan 898 cm^{-1} (C-H), at these wavelengths are typical of cellulose groups.

From Fig. 1 it is clear that the nanoselulose spectrum is similar to the cellulose spectrum, the difference being in the transmittance intensity, meaning that for all three FA concentrations it does not affect the functional group, the similarity of the sample indicated the similar in chemical composition. The deferent sample of cellulose with nano crystalline cellulose, the wavelength appears new peaks at 1714 cm-1 (C ≡ O stretching) and 2360 cm-1 (H-C = O: C-H stretch). From FT-IR spectra are not different nano crystalline cellulose and cellulose. It indicates that there are not new bonds formed during hydrolysis process.

Nuruddin [5] has employed that removed the amophous cellulose on the surface by the acid hydrolysis, therefore, more COH, C-O-C and C-C bonds were exposed, resulting in increased stretching absorbency. Sensitive to crystal structure of the cellulosic part are the wavelength 850-1500 cm^{-1} (Kumar et al, 2014). Spectral bands at 1420-1430 cm^{-1} and 893-897 cm^{-1} are very important to elucidate to the crystal structure of cellulosic material and its (1420/893 cm^{-1}) spectral ratio and (1375/2900 cm^{-1}) spectral ratio show index of crystallinity or lateral order index (LOI) and total crystallinity index (TCI), respectively.

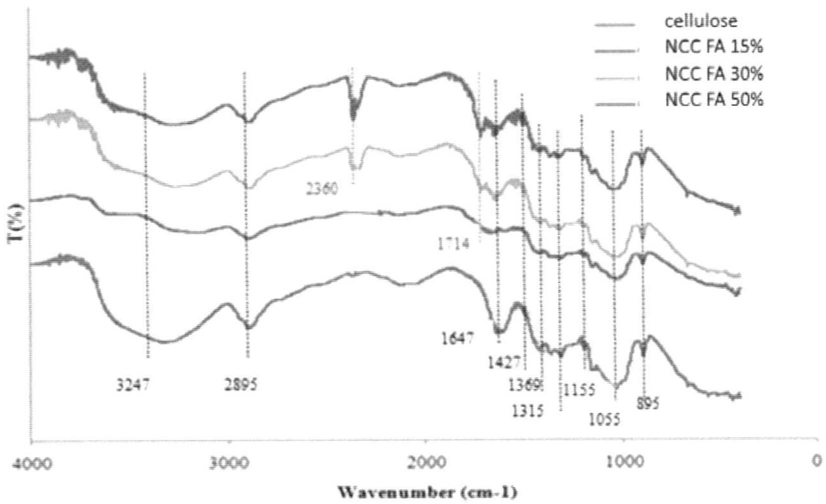

Figure 1. FT-IR spectra of cellulose and nano crystalline cellulose various FA concentrations.

3.3 Structure and crystallinity of nano crystalline cellulose

The structure and crystallinity index of isolated cellulose and nano crystalline cellulose were studied using XRD. The XRD patterns of cellulose and nano crystalline cellulose effect of FA concentrations are shown in Figure 2. The crystallinity percentage and crystal size, as calculated from Eq. 1 and Eq. 2, respectively, are summarized in Table 1. In four curves, the peak are observed at $2\theta = 15.25^{o}$, 20.29^{o} and 22.39^{o}.

Table 1. Yield, crystallinity and size of crystalline for sugarcane bagasse of nano crystalline cellulose

Concetration of FA (%)	Yield (%)	Crystallinity (%)	Size of crystalline (nm)
Cellulose	-	44.030	4.55
FA-15%	66.66	62.644	4.37
FA-30%	67.33	71.033	4.15
FA-50%	69.33	76.296	3.94

From Table 1 the crystallinity index increase with increasing concentration of FA. It has been reported that the cellulose is partly crystalline and and partly amorphous in molecule structure, meaning that the cellulose chains contains the crystalline regions and amorphous regions [2]. The intensity of the peaks was higher for nano crystalline cellulose, showing that nano crystalline cellulose are more crystalline. On the other hand, there is a decrease of diffraction intensities in the amorphous region [5]. The higher crystallinity is associated with the higher tensile strength of nano crystalline cellulose [3, 13]. When the amorphous regions that hold the crystalline parts dissolve during acid hydrolysis, single crystals are released and the crystallininity increase [1].

The size of crystalline of nano crystalline cellulose decrease with increasing of acid concentrations. From Table 1 show that size crystalline of cellulose higher than nano crystalline cellulose. The loose structure of celluloses favored alteration in the crystal size and that the degradation of smaller crystals as well as the growth of defective crystal was resposible for increasing the crystallite size [1].

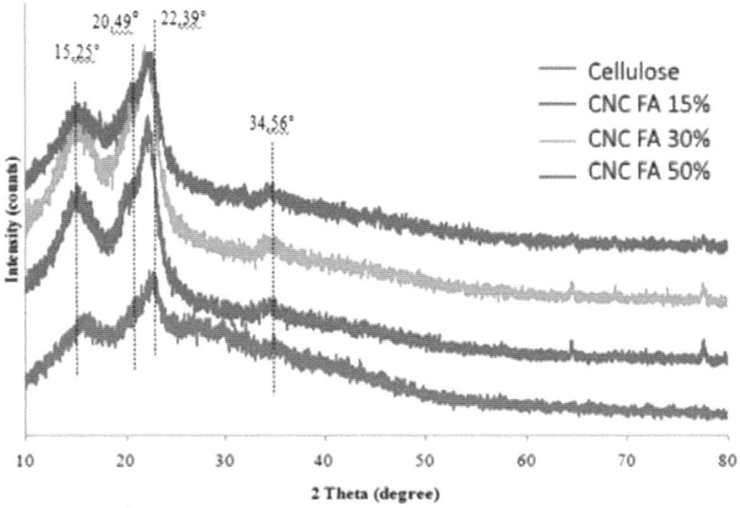

Figure 2. XRD-patern of cellulose and nano crystalline cellulose with different FA concentrations.

4. Conclusion

Nano crystalline cellulose was prepared from sugarcane bagasse with acid hydrolysis. The three level various FA concentrations has investigated. Functional group of the cellulose and nano crystalline cellulose shown the similarity parttern. It was indicated that no new bonds formed during acid hydrolysis. The crystalline index increase with increasing acid concentrations. But, for size of crystallite nano cellulose decrease when increasing of acid concentration. The higher crystalline index and the smaller the size of crystal of nano crystalline cellulose will be effect for good properties (tensile strength) of nano crystalline cellulose as reinforcement in polymer biocomposites.

Acknowledgment

The authors would like to thank The Research Institutes and Community Service University of Syiah Kuala and Directorate of Research and Community Service and Directorate General of Research and Development, Ministry of Research, Technology and Higher Education who have funded this research in accordance with the Research Contract No. 105/SP2H/LT/DPRM/IV/2017, April 03, 2017.

References

[1] Aprilia N A S, Davoudpour Y, Zulqarnain W, Abdul Khalil H P S, Hazwan C H, Hossain C I, Dungani R, Fezree H M, Zaidon A and Mohammad Haafiz M K 2016 *Bioresour.* p3875-3889
[2] Owalabi A W T, Arniza G, Daud W W and Alkharkhi A F M 2016 *Bioresour.* **11** 3013
[3] Karim Md. Z, Ali Md. E, Chowdury Z Z and Hamid S B A 2016 *Bioresour.* **11** 3840
[4] Du H, Liu C, Mu X, Gong W, Lv D, Hong Y, Si C and Li B 2016 *Cellulose* **1** 20
[5] Nuruddin M, Chowdhury A, Haque S A, Rahman M, Farhad S F, Jahan M S and Quaiyyum A 2011 *Cellulose Chem. Technol.* **45** 347
[6] Apprilia N A S, Ambarita A C, Karmila, Armando M A and Guswara F Y 2017 *Oriental J. Phys. Sci.* **2** 1
[7] Kumar A, Negi Y S, Choudhary V and Bhardwaj N K 2014 *J. Mater. Phys. Chem.* **2** 1
[8] Saele K, Yingkamhaeng N, Nimchua T and Sukyai P *The 26th Annual Meeting of the Thai Society for Biotechnology and International Conference.* 26-29 November 2014 Thailand
[9] Saurabh C K, Mustapha A, Mohd. Masri A, Owolabi A F, Syakir M I, Dungani R, Paridah M T,

Jawaid M and Abdul Khalil H P S 2016 *J. Nanomater.* **1**
[10] Lee S-Y, Mohan D J, Kang I-A, Doh G-H, Lee S and Han O 2009 *Fiber Polym.* **10** 77
[11] He W, Jiang S, Zhang Q and Pan M 2013 *Bioresour.* **8** 5678
[12] Hamid S B A, Chowdhury Z Z, Karim Md Z and Ali Md E 2016 *Bioresour.* **11** 3840
[13] Zhou Y M, Fin S Y, Zheng L M and Zhan H Y 2012 *eXPRESS Polym. Lett.* **6** 744

Effectiveness of inorganic membrane mixture of natural zeolite and portland white cement in purifying of peat water based on turbidity parameter

Elfiana, A Fuadi and S Diana
Chemical Engineering Department, Politeknik Negeri Lhokseumawe, Aceh

E-mail: elfiana@pnl.ac.id

Abstract. Peat water is water surface that brownish red colour caused by the contained constituents. Solving the peat watercolor problem requires special attention considering the quantity of peat water and suitable to be used to meet the daily needs. This study aims to know the inorganic membrane capability of mix nature zeolite and white Portland cement to purifying the peat water based on turbidity parameter. The study was conducted by varying the composition of nature zeolite (Za) and white Portland cement (Sp) in the ratio of Za:Sp is (25%:75%; 50%:50%; 75%:25%) with zeolite condition activated using HCl 2M and nonactivated zeolite treatments. The result of the characteristic test on membrane morphology using SEM (Scanning Electron Microscope) showed that the pore surface size of the membrane is 2 μm that could classified in microfiltration membrane an organic type. The characteristic test showed also resulted in the density of 0.77 to 0.86 gr/cm^3, porosity 26.22% to 35.93%, and permeability 2736.19 to 8428.15. While the water retention capacity is in range of 30.64% to 46.46%, The result of inorganic membrane application on peat water showed turbidity of peat water decreased 94.17%, from 10.3 NTU to 0.6 NTU.

1. Introduction

Peat water is the surface water found in peatlands. Visually, reddish-brown peat water, acidic and smelly. The Geology Resource Centre review from the Department of Energy and Mineral Resources reported that until 2006 peatland resources in Indonesia covered 26 million hectares (ha) spread over Borneo Island (± 50%), Sumatera (± 40%) and the rest spread over Papua and other islands. And for this peatlands, Indonesia occupies the 4th largest position in the world after Canada, Russia and the United States. Based on these data, peat water in Indonesia can potentially be utilized as daily human water resources if it is managed properly, and it also needs to be considered an adequate peat water treatment technology to convert peat water into clean water.

Several methods of peat water treatment have been done in both laboratory and field scales, however, the result is not maximal yet. Several previous studies have suggested that laboratory conventional coagulation method using coagulant of clamshell and coral in peat water treatment area of Geuredong Pase is only able to set aside 5-58% iron concentration and does not give significant water color change either so that the processed peat water still looks colorful brownish yellow. Two Stage Coagulation method is able to reduce the peat organic compounds of Bangkinang area in Riau up to 88% using Alum coagulant at a dosage of 280-300 mg/L but not significant either to the decrease of iron concentration which is stable to organic. Peat water treatment using eggshells as adsorbent powders is able to eliminate 95% of organic compounds peat water at a pH of 4.01 [1]. Low degree of acidity is not recommended for clean water. Therefore, the advanced process of peat water treatment using economical and reliable microfiltration membrane technology needs to be done.

Utilization of natural zeolite as a raw material for inorganic membrane production is considering. Aceh is one place whereas the plains contain natural zeolite could be found. In general, this natural zeolite has not been utilized optimally, especially in its utilization for membrane making. As it has been noticed that natural zeolites have the ability to perform a series of functions in various chemical processes such as catalysts, adsorbents and ion exchangers. Through-membrane technology, zeolites can be used as a medium for absorption and separation simultaneously [2].

Currently, the utilization of natural zeolite as an inorganic membrane is of much concern to the researchers because of its uniform porosity structure, stable to heat, good mechanical strength, and resistance to extreme chemical environment. Natural zeolite is a mineral that is sediment in nature, which is a compound of aluminosilicate that forms a three-dimensional structure framework between AlO4 and Tetrahedral SiO4. Natural zeolite is a suitable material for inorganic membrane fabrication because it is not easy to inflate in water and easily form a suspension to coat the membrane as support [3]. Natural zeolites have different characteristics than conventional membranes made from commonly used compounds such as Al_2O_3 or ZrO_2 and others since natural zeolites form inter-particle active pores when sintered in a dry state. Zeolites have the ability to perform a series of functions in various chemical processes such as catalysts, adsorbents and ion exchangers. Through-membrane technology, zeolites can be used as a medium for absorption and separation simultaneously [2]. The advantages of using membranes as filter media are greater efficiency, simpler operation and smaller energy requirements [4]. According to Mourato (2002), one of the most commonly used membranes in a water treatment process is a microfilter membrane, suitable for holding suspensions and emulsions. In addition, the microfilter membrane price is cheaper and the required operating pressure smaller i.e. less than 2 bars.

Previous research has been done Akbar (2010) regarding the making of natural zeolite microfilter membrane with the addition of Portland cement. The study was conducted by varying the mixture of natural zeolite Lampung with white Portland cement to see the best mixture as a filter on the 50:50% mix ratio as Z_1S, 66.67:33.33% as Z_2S, and 75:25% as Z_3S. From the results of this study obtained membrane separation efficiency of Fe^{2+} ion is 96.4%, 99.4% and 99.2%; for Mg^{2+} ions were 69.4%, 86.2% and 73.7%; as well as for Mn^{2+} ion obtained maximum results, i.e., up to 100%. While the efficiency of organic compounds separation and watercolor has not been tested. Previous research reported natural zeolite from Blang Pidie area of Southwest Aceh Regency that was chemically activated using 0.05 M HCl and calculated 3500C for 2 hours has been tested to improve the performance of zeolites in the ability to absorb copper metal ions [2]. Buggraaf (1996) reports that an inorganic membrane needs to be tested for characteristics of the membrane pore diameter that is formed [5]. Baker (2004) concluded that most membrane materials are made of synthetic organic polymers, such as MF and UF membranes made from the same material but different formation conditions so that the pore size becomes different [6]. Scheibler (2015) shows the zeolite membrane (ZSM-5 / γ- Alumina) performs well in oil and water separation processes. The resulting zeolite membrane was tested for its characteristics based on a morphological test with SEM, structural test with XRD and analysis with EDX showed the material formed with Si / Al = 19 [7].

Therefore, the research was conducted to find out the inorganic membrane capability of natural zeolite mixture and white Portland cement in reducing the turbidity of peat water to meet the standard of clean water. Inorganic membrane synthesis is carried out in a variation on the ratio of natural zeolite mixture and white Portland cement from several compositions to the activated and non-activated zeolite state. Membrane characterization test was performed based on morphology test with SEM, structuring test with XRD, flux test, perm-selectivity, hardness, porosity, and density. Membrane performance can be known based on the efficiency of peat land TDS concentration decrease.

2. Research method

The used instruments include analytical scales, 80 mesh sieves, SEM (Scanning Electron Microscope), turbid meter, oven, beaker glass1000 ml and 500 ml, 250 ml measuring flask, mixer, slipper, eraser mortar, module mold, picnometer. Materials used include peat water, natural zeolite; white Portland cement, chloride acid, aquades, and agent foam.

2.1 Preparation activation of natural zeolite

The zeolite is crushed and crushed with mortar until it is 80 meshes in size. Then 500 gr of fine zeolite was introduced into the Erlenmeyer and added with 300 ml of 2M HCl, heated while stirring until boiling for 1 hour, then cooled, filtered and washed with aquades until pH filtrate 6-7. The residue is dried for 5 hours at 700^{0}C. Finally, the active zeolite is ready to use.

2.2. Inorganic membrane synthesis

2.2.1 Membrane module

The membrane module is made of stainless steel or PVC pipe, dimensions t xd = 5x5.5 cm (module 1, permeability test) and 25x6.5 cm (module 2, hollow fiber, at 3 cm). The membrane molds are best prepared according to the specifications of Module 1 and Module 2 according to the following Figure 1.

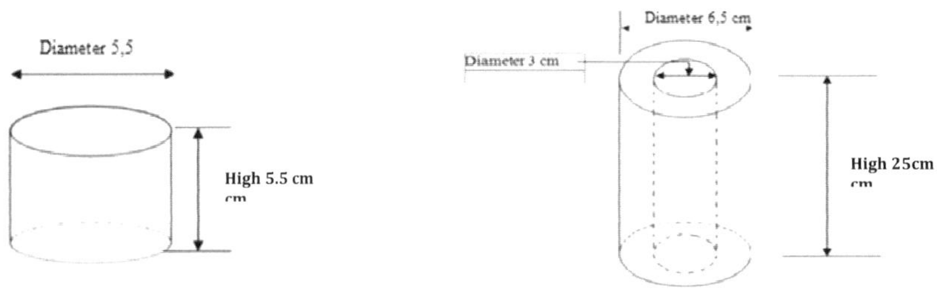

Figure 1. The module of inorganic membranes where (a) 5x5.5 cm as module 1; (b) 25x6.5 cm as module 2

2.2.2 Inorganic membrane synthesis of module 1 (permeability testing)

A total of 100 grams of natural zeolite mixture (Za) and white Portland cement (Sp) were prepared according to the composition $Za_{1.2.3}$:$Sp_{1.2.3}$ = 25%:75%; 50%:50%; and 75%:25%.Then each dough is put into a beaker glass and adds water little by little (ratio of zeolite:water = 1:10) to a paste. The paste dough is inserted in the module membrane mold 1 and dried at room temperature (29°C) for 3x24 hours. The resulting ZaSp inorganic membrane was rinsed with water to a neutral pH and continued by oven drying at 70^{0}C for 6 hours. Membrane characteristic test was performed on membrane density, membrane permeability, water storage capacity and morphology test with SEM.

2.2.3 Inorganic membrane synthesis module 2 (membrane performance testing)

A total of 500 grams of natural zeolite mixture (Za) and white Portland Cement (Sp) were prepared according to composition $Za_{1.2.3}$:$Sp_{1.2.3}$ = 125:375 gr; 250:250 gr, and 375:125 gr. Dough put in beaker glass and add water bit by bit (ratio of zeolite:water = 1:10) to form a paste. The paste dough is inserted in module 2, (hollow fiber) membrane mold, and dried at room temperature (29°C) for 3x24 hours, rinsed with water to neutral pH, dried oven at 70^{0}C for 6 hours. Insert the ZaSp hollow membrane into the housing membrane and drain the peat water at normal pressure, test the quality of permeate clearance based on turbidity parameter. Membrane performance was measured by the percentage of turbidity removal before and after filtration with inorganic membrane ZaSp.

2.3 Inorganic membrane characteristics

2.3.1 Permeability test

The ZaSp membrane permeability test can be carried out according to the following set of equipment.

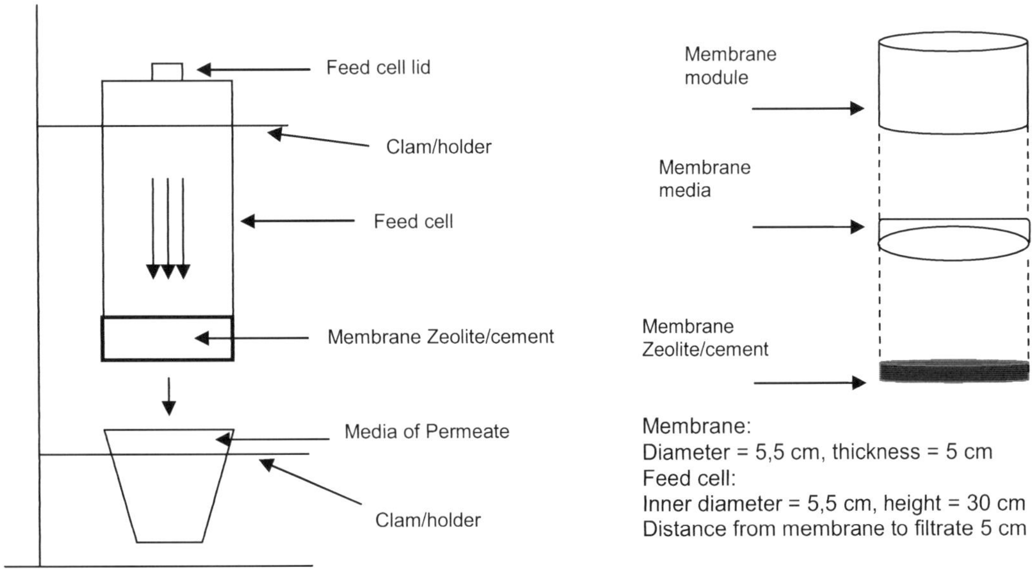

Figure 2. The module 1 of inorganic membrane ZaSp to permeability test.

The ZaSp membrane module 1 is inserted into a feed cell, passed water with gravity flow velocity into the feed cell at normal pressure. Calculate the time obtained to accommodate the Permeate of 500 ml; the flux will be obtained according to the following equation:

$$J = \frac{V}{A\,x\,t} \tag{1}$$

where: J = Fluks ($l/m^2.hr$)
V = volume permeate (ml)
A = Area of membrane (m^2)
t = time (hours)

2.3.2 Measurement of Membrane Density

Module 1 is the ZaSp inorganic membrane is retracted in the mold. If there is an excess volume or thickness of the membrane to the mold, then the erosion or smoothing using paper sandpapers. The membrane volume is calculated by measuring changes in membrane diameter and membrane thickness using the sliding term. ZaSp membranes are in a dry state prepared. Then weighed and recorded the result as a dry mass of membranes. Membrane density was calculated based on data of volume measurement and dry mass membrane using equation as follow:

$$Membrane\ Density = \frac{Mass\ Membrane\ (gr)}{Volume\ Membrane\ (cm^3)} \tag{2}$$

2.3.3 Morphological test of inorganic membranes

ZaSp membrane morphological tests include the pore diameter and membrane surface structure performed using SEM equipment (Scanning Electron Microscope)

2.3.4 Measurement of water storage capacity
Determination of water storage capacity in the membrane (degree of swollen), done by using data of measurement of the dry mass of membrane (m_d) and mass of water saturated membrane (m_s). The calculation of the %Ds water storage capacity (degree of swollen) uses the following equation:

$$\%Ds = \frac{m_s - m_d}{m_G} x\ 100\% \qquad (3)$$

2.4 ZaSp membrane performance testing procedures
Experiments testing the ability of the peat water treatment process using inorganic membrane hollow fibber ZaSp done by passing peat water into the housing inorganic membrane hollow at normal pressure. The turbidity of peat water before and after passing through the membrane is measured and calculated its turbidity removal efficiency at any given time. The prototype of peat water treatment with ZaSp inorganic membrane of hollow fiber type is shown in Figure 3.

Figure 3 illustrates the circulation of peat water fed into the filter cartridge 03μ - 01μ then forwarded to the membrane ZaSp. Retinted is returned to the feed tank while permeate is collected and analyzed for pH and turbidity parameters.

3. Results and discussion

3.1. Peat water characterization

Table 1. The characteristic result of Peat Water based on PERMENKES RI No.416/MENKES/PER/IX/1990

Parameters	Units	Level	Boundary Condition
pH	-	5,7	6,8-8,5
TDS	mg/L	250	1.500
Turbidity	NTU	4,55	25
Ion Iron	mg/L	16,18	1
Organic Compound ($KMnO_4$)	mg/L	395	10

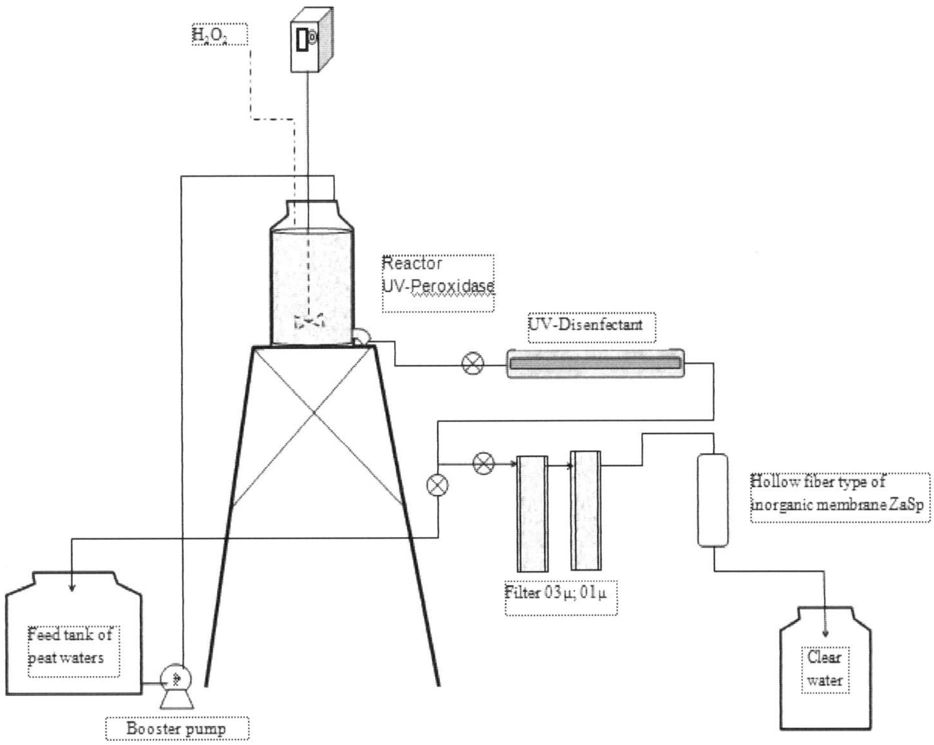

Figure 3. Prototype of peat waters treatment using hollow fiber type of inorganic membrane ZaSp.

3.2. The effect of zeolite weight on membrane permeability

Permeability or membrane flux is a measure of the velocity of a species across a broad union membrane and time with a pressure gradient as a driving force. Ceramic membrane flux is directly related to porosity, where the fine ceramic membrane is a high porosity membrane but does not decrease the mechanical strength of the membrane. Factors affecting permeability are the number and size of the pore, the interaction between the membrane and the feed solution, and the pressure from the outside. This measurement of the flow rate of the membrane is employed by the pressure of gravity based on the feed force which continues to decrease its pressure as it decreases the feed. Membrane permeability depends on the size of the pores on the membrane, so the larger the pores on the membrane then the resulting permeability will be large and vice versa on the small pore size will make the permeability on the membrane small or small.

In this research, the inorganic membrane production of zeolite mixture and white portland cement was prepared with several mixed compositions of natural Zeolite (Za) and white Portland cement (Sp) and given sample type Za_1Sp_1 (25%:75% = 125gr:375gr), Za_2Sp_2 (50%:50% = 250gr:250gr), and Za_3Sp_3 (75%;25% = 375gr:125gr). The results showed the largest permeability or flux membrane was 9577.32 kg/jam.m^2 in Za_3Sp_3 sample with Za activated. While the smallest permeability is 2736,19 kg/jam.m^2 obtained on sample Za_1Sp_1 with Za without activation. Permeability is influenced by pores on the membrane and sample size. The activated zeolite material has cleaner pores than impurities that make zeolite performance less than optimal. The larger the zeolite composition the greater the permeability of the membrane and the smaller the zeolite composition the smaller the permeability. The relationship of the mixture of zeolite and white portland cement to membrane permeability is shown graphically in Figure 4.

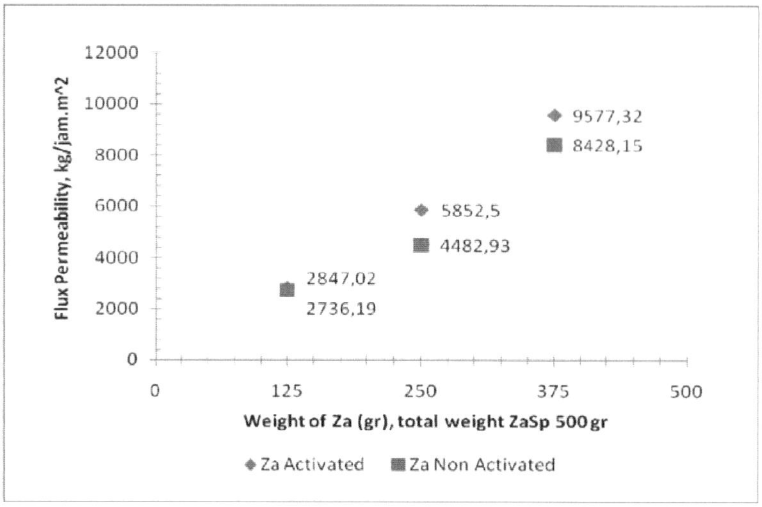

Figure 4. The effect of zeolite composition Za:Sp on flux permeability of membrane.

From Figure 4 it can be seen, the larger the zeolite composition the greater the membrane permeability flux obtained. Activated zeolite gives effect to better permeability value for each composition of zeolite. It can be concluded that zeolite activation improves the permeability performance of inorganic membrane of natural zeolite mixture and white Portland cement.

3.3. The effect of zeolite weight on membrane density

The results showed the greater the composition of the zeolite the smaller the membrane density and vice versa. These results are shown graphically in Figure 5.

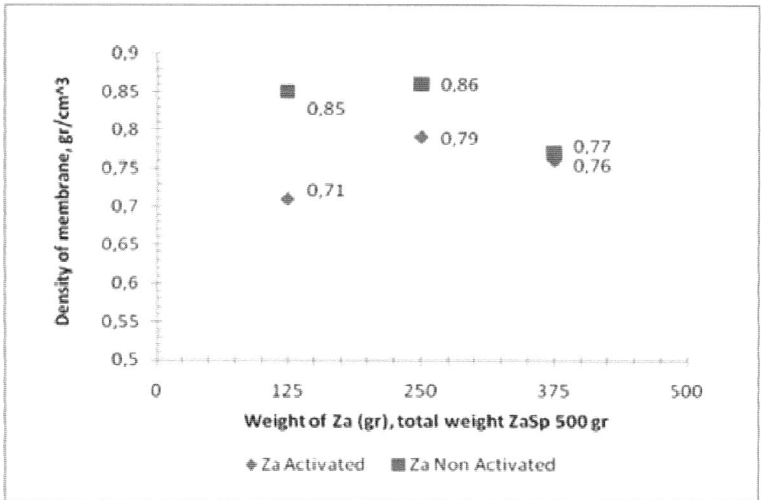

Figure 5. The effect of zeolite composition Za:Sp on membrane density.

The highest membrane density values were obtained in a sample of 250 gr of zeolite without activation (Za_2Sp_2) and the lowest density was in the 125 gr sample of activated zeolite (Za_1Sp_1). The magnitude of the membrane density values for zeolite conditions is not activated due to the release of impurities attached to the zeolite pore so that in the same membrane volume larger with zeolite without activation.

3.4. The effect of zeolite weight on membrane porosity

The aforementioned porosity is a void space occupying an object per unit volume. The formation of membrane pores is closely related to the particle size of the material, especially the zeolite. Tan et al. (2001) states that zeolites with large pores will produce a large absorption as well; so the smaller the size of the zeolite grains will result in greater absorption rate.

The membrane cavities explain that the amount of empty space occupied by the air or water of the membrane and the results of the inorganic membrane construction of the active zeolite mixture and the white portland cement has the largest cavities or porosity in the Za_3Sp_3 sample of activation and the lowest porosity present in the Za_1Sp_1 activation sample. The data are shown graphically in Figure 6, in which density and porosity are related to each other can explain the greater the density of an object, the smaller the pore or cavity.

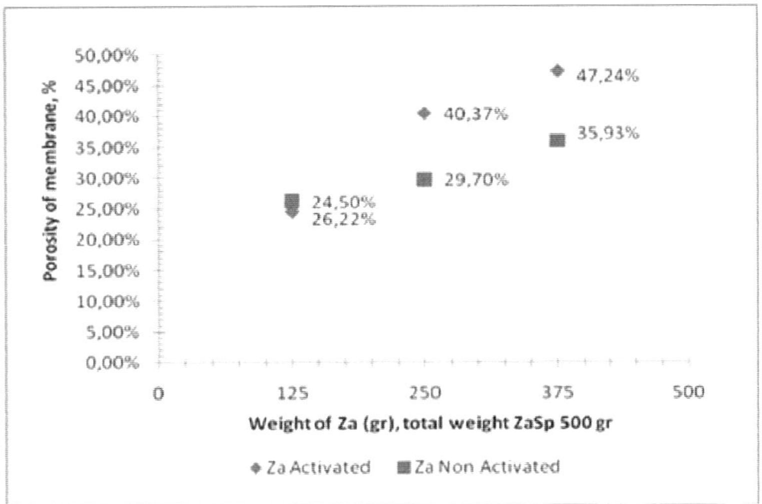

Figure 6. The effect of zeolite composition Za:Sp on Porosity of Membrane

3.5. The effect of zeolite weight on membrane capacity

The amount of water storage capacity in the membrane due to the number of cavities between the particles (void space) or porosity that is formed. At the time of membrane molding, the formation of fine bubbles due to zeolite particles that are filled with air by water and the addition of foam (foam) that makes the pores on the membrane more so that the membrane sample has large water storage cavities. The value of the membrane water storage capacity obtained is shown graphically in Figure 7. The highest membrane water storage capacity values obtained in the Za_3Sp_3 membrane sample with the activated zeolite and the lowest capacity were obtained in Za_1Sp_1 membrane samples with zeolite without activation.

From Figure 7, it can be seen that the water storage capacity is also associated with the resulting porosity, the larger the axis then the larger space or the cavity that can be occupied by water. Inactivated zeolites have the largest water storage capacity of zeolites without activation, with large cavities in the activation zeolite making uptake or storing water of great value.

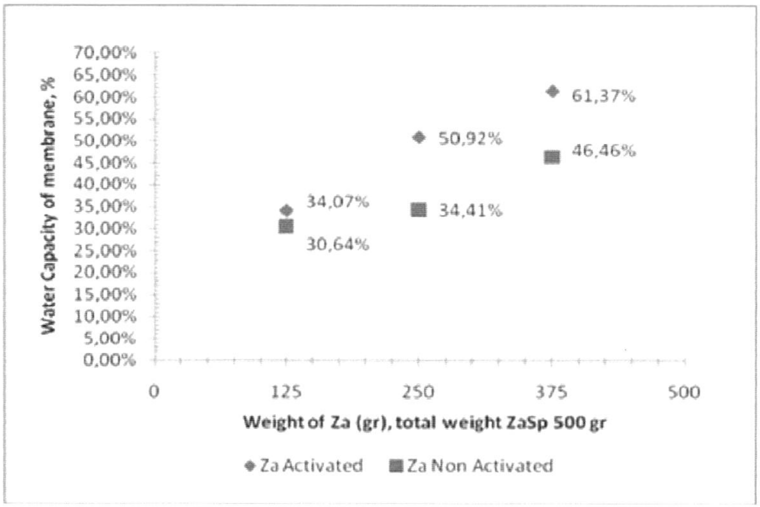

Figure 7. The effect of zeolite composition Za:Sp on Water Capacity of Membrane

3.6. Morphology of ZaSp inorganic membranes

Structure of the surface of a membrane or one way to know the morphology of a membrane is by SEM (Scanning Electron Microscope) test. SEM can also know the distribution of pores, pore geometry, pore size and porosity on the surface of the membrane, making it easier for us to know how membrane pore or pore membrane size. The results of membrane morphological testing with SEM test are shown in Figure 8.

From Figure 8 above it can be seen that the membrane surface has a size variation of 2μm-200μm. So it can be grouped that the resulting ZaSp inorganic membrane is a microfiltration membrane group with a pore size of 0.1μm -10μm. The smaller the pore size the smaller the permeability and the cleaner the water produced.

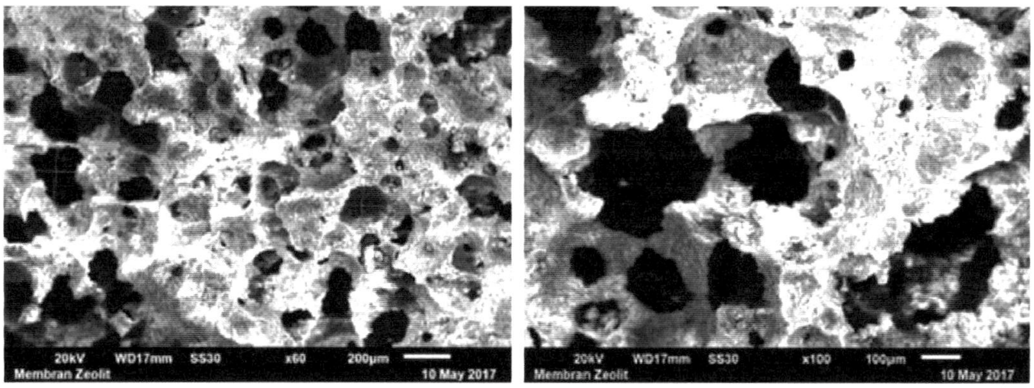

Figure 8. The Morphology of inorganic Membrane using SEM method

3.7. Performance of ZaSp inorganic membrane based on percentage reduction of turbidity of peat water

Based on the results of initial characteristics of peat water of Desa Ek Treun it is known that the turbidity level of peat water is 10.3 NTU. To find out the performance of ZaSp inorganic membrane in peat water processing is to make hollow type ZaSp membrane species as module 2 with 25 cm high dimension, outer diameter 6.5 cm and 3.5 cm inner diameter.

The experiment is carried out by draining the peat water at normal pressure to the inorganic membrane housing of ZaSp for a certain time, then the permeate is accommodated for its turbidity analysis while the retentate is returned to the feed tank. The results showed that the effectiveness of ZaSP inorganic membrane performance can decrease 95% turbidity of peat water from 10.3 NTU to 0.52 NTU on inorganic membrane Za_1Sp_1. The result of calculating the efficiency value of decreasing the turbidity with ZaSp inorganic membrane is shown graphically in Figure 9 below.

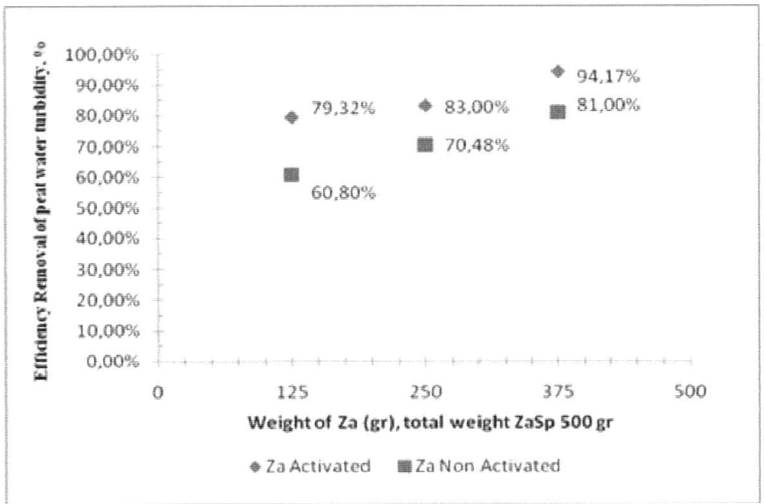

Figure 9. Performance inorganic membrane ZaSp based on efficiency removal of peat water turbidity before and after treatment

From Figure 9, it can be seen that the greater the composition of zeolites in the ZaSp membrane mixture, the efficiency of turbidity removal due to the smaller pore membrane the better the separation process of turbidity of peat water. The results showed the greatest turbidity decrease efficiency was 94.17% in Za_3Sp_3 with activation zeolite weight 375 gr.

4. Conclusion and recommendation

Based on the research that has been done can be concluded that activated natural zeolite can improve the inorganic membrane performance of the mixture of natural zeolite and white Portland cement. The greater the composition of the zeolite in the mixture, then the greater membrane permeability up to 9577.32 kg/hr.m^2. Performance of ZaSp inorganic membrane can be seen from the efficiency of turbidity reduction of peat water after treatment, with the largest value obtained is 94.17%. The results of the ZaSp membrane morphology test using SEM showed the inorganic membrane ZaSp grouped in the microfiltration membrane.

References

[1] M A Zulfikar and Henry S 2013 *Intl. J. Chem. Tech. Res.* **5** 1532-1540
[2] McLearly E E, Jansen J C and Kaptein F 2006 *Microporous Mesoporous Mater.* **90** 198-220
[3] Davis T M 2006 *Nat. Mater.* **5** 400-408
[4] Nandi B K, Uppalum R and Purkait M K 2008 *Appl. Clay Sci.* **42** 102-110
[5] Burggraaf A J 1996 *Membrane Sci. Technol.* **4** 21-34
[6] Guibert D, Rabie H, Cote P and Ben Aim R 2002 *Desalin.* **148** 395-400
[7] Scheibler J R, Santos E R F and Barbosa A S 2015 *Desalin. Water Treat.* **56** 3561-3567

Degradation of blending vulcanized natural rubber and nitril rubber (NR/NBR) by dimethyl ether through variation of elastomer ratio

A H Saputra[1], S Juneva[2], T I Sari[3] and A Cifriadi[4]
[1,2]Department of Chemical Engineering, Faculty of Engineering Universitas Indonesia Depok 16424 Indonesia
[3]Department of Chemical Engineering, Faculty of Engineering Sriwijaya University Kampus Indralaya, Indralaya 30662 Indonesia
[4]Indonesian Rubber Research Institute, Jl. Salak Bogor 16151, Indonesia

E-mail: sasep@che.ui.ac.id

Abstract. Dimethyl ether can cause degradation of the rubber material seal in some applications. In order to use of natural rubber in industry, research about a blending of natural rubber (NR) and nitrile rubber (NBR) to produce rubber to meet the standard seal material application were conducted. This study will observe the degradation mechanisms that occur in the blending natural rubber and nitrile rubber (NR/NBR) by dimethyl ether. Nitrile rubber types used in this study is medium quality nitrile rubber with 33% of acrylonitrile content (NBR33). The observed parameters are percent change in mass, mechanical properties and surface morphology. This study is limited to see the effect of variation vulcanized blending ratio (NR/NBR33) against to swelling. The increase of nitrile rubber (NBR33) ratio of blending rubber vulcanized can reduce the tensile strength and elongation. The best elastomer variation was obtained after comparing with the standard feasibility material of seal is rubber vulcanized blending (NR/NBR33) with ratio 40:60 NR: NBR.

1. Introduction

Dimethyl ether is a simple compound of ether which is able to substitute fossil energy and has meets the criteria as high energy [1]. Moreover, dimethyl ether is categorized as hydrocarbon compound which is interconnected with rubber seals and gaskets, so that makes it the cause of leakage [2]. Such interaction is referred to as polymer degradation. When the polymer is exposed to the liquid, the form of the degradation is swelling and dissolution [3]. The type of rubber widely used in order to overcome such problem is nitrile rubber (NBR) with its resistance to swelling degradation. In order to increase the consumption of natural rubber in automotive industry, especially in Indonesia, the research about natural rubber blending by utilizing its superior properties in the application of seal in machine parts are continuously performed.

The required physical properties are a rubber material compound which is resistant to swelling degradation and has good mechanical strength. Such properties are obtainable from natural rubber (NR) and nitrile rubber (NBR) materials. Natural rubber has good mechanical strength, while nitrile rubber is resistance to oil exposure. Both elastomers are able to be combined in order to obtain the best properties. However, the blending between natural rubber and nitrile rubber (NR/NBR) is not easy due to the significant level of polarity and solubility. Saputra and Johan (2016) has been conducted degradation characteristics of vulcanized natural rubber research by using minarex-B plasticizer [4]. Therefore, this study also can be developed to the vulcanized natural rubber research by replacing minarex-B with soybean oil as a plasticizer.

The objective of this study is to obtain the blending ratio (NR/NBR) used for rubber vulcanization. While, the type of NBR used is nitrile rubber containing 33% of acrylonitrile, Krynac 33.45F (NBR 33). The content of acrylonitrile determines the difficulty level of blending and oil resistance level.

Therefore, this study is to find out the leverage of blending ratio (NR/NBR) to the swelling degradation by dimethyl ether and its mechanical as well as morphological characteristics.

2. Materials and methods

The experiment is divided into: the synthesis of vulcanized natural rubber and nitrile rubber (NR/NBR26) as well as mechanical and morphological characterization.

Synthesis of vulcanized natural rubber and nitrile rubber has a purpose to produce the rubber product that will tested in this research. The synthesis of vulcanized rubber blending consists of 5 variations of NR/NBR ratio. They are (75: 25), (60: 30), (50: 50), (40: 60), (25: 75) of NR/NBR ratio. The study of elastomer mixture technique does a concern on consideration of typical behaviour from elastomer mixture and the combination of these properties in vulcanized process [5]. First step is preparation the equipment such as setting up the open mill, press moulding and rheometer. The material used in this synthesis process is natural rubber (NR), nitrile rubber (NBR33), anti-degradation, accelerator, activator, sulphur, carbon black and soybean oil. Natural rubber (NR) and nitrile rubber (NBR33) were masticated by the open mill equipment for 5-10 minutes. After the mastication process, the vulcanization material was mixed into the open mill until got the homogeneous mixture. Then, the rubber product was cooled for a day. A piece of the rubber product was used for determining time of vulcanization in rheometer and then put the rubber product in press moulding equipment. After that, the vulcanization process with vulcanization temperature and time that displayed in rheometer were conducted. The rubber product was cooled for more or less one hour. All procedure was repeated with different ratio natural rubber (NR) and Nitrile Rubber (NBR33).

The procedure of swelling of the blending vulcanization degradation test (NR/NBR33) is based on ISO 1817:2005. The first step of this test is to stick the rubber product into copper rod and it must be in perpendicular position between rod and the rubber product and then put them into pressure vessel and the PTFE insulate tape was installed in every channel of pressure vessel. Then the channel was closed tightly for avoiding the leakage. After that, the pressure vessel was cooled by putting it into plastic jar that filled with ice cubes and salt. The pressure vessel was connected with 50 kg gas cylinder that filled with dimethyl ether. The valve was opened slowly for 15 minutes and then was closed. The dimethyl ether was changed into liquid phase. For 7 days, the sample product was immersed in pressure vessel. The sample of rubber was put and dried for characterization steps. The procedure above was repeated with different ratio natural rubber (NR) and Nitrile Rubber (NBR33).

The other procedures are mechanical and morphology characterization. Mechanical characterization was conducted before and after swelling of blending vulcanization degradation test. There are 4 types of mechanical test. First the percentage of mass changing is measured by electronic mass balance, then maximum tensile and elongation strength were tested based on ISO 37:2005. The hardness characterization was tested with hardness testing machine (ISO 48:2010) and the compression set was tested based on ASTM D395 standard [6]. To observe the morphology of material, scanning electron microscopy (SEM) was used.

For calculate the percentage of mass changing is used the equation (1) as shown as below and m symbol is stated as sample rubber mass [7].

$$\%\boldsymbol{\Delta m} = \frac{m_{beforeimmerse} - m_{afterimmerse}}{m_{beforeimmerse}} X\ 100 \qquad (1)$$

Calculated data for tensile strength is maximum tensile force from blending vulcanized material (NR/NBR26) while retraction. The strength of tensile force can be calculated with the equation (2).

$$\boldsymbol{\sigma} = \frac{\boldsymbol{F_{maxsimum}}}{\boldsymbol{A}} \qquad (2)$$

The following variable is stated as σ = tensile strength (Mpa), F = Maximum Tensile force (Newton), and A = Area of tensile (m^2).

The value of blending vulcanized sample length (l) was measured before and after elongation. The equation (3) shows the percentage of maximum elongation

$$\% \Delta l = \frac{l_{beforeelongation} - l_{afterelongation}}{l_{beforeelongation}} X100\,\% \quad (3)$$

The calculated data for the Compression Set value is the value of the initial thickness and the final thickness of the rubber compound sample. In order to calculate compression set, the equation 4 is used as shown as below

$$C_B = \left[\frac{t_o - t_i}{t_o - t_n}\right] X100\,\% \quad (4)$$

The following variable is stated as C_B= compression set percentage, t_o = sample thickness before experiment, t_i = sample thickness after experiment and t_n= spacers thickness

3. Result and discussion

3.1 The result of physical characteristic change

The result is the physical characteristic change of the sample rubber; such as percentage of mass change, tensile strength change, elongation change, and hardness change. The percentage data of mass change is obtained from the mass calculation of the sample before and after immersion. The mass change value of each sample due to the DME immersion is displayed in Table 1.

The process of partial dissolution occurs with a decrease in rubber mass, while the swelling process is seen with the addition of rubber mass. The Table 2 is the value of the physical characteristic change of the sample rubber. The elongation value is in association with the tensile strength value which is the percentage change in accordance with the result of tensile strength test until. The elongation value of rubber samples decreases due to the immersion of dimethyl ether; such decrease tends to subside as the addition of nitrile rubber ratio into the rubber mixture. In addition to the elongation value, tensile strength value can also be obtained from tensiometer.

In Table 2, it shows that the tensile strength value of each sample changes. After immersion, the increase of nitrile rubber composition is able to reduce tensile strength. However, the tensile strength will increase after immersion in sample F and G of 75% and 100% of nitrile rubber content. Furthermore, the rubber sample is subjected to hardness test as also shown in Table 2. The hardness value is prone to increasing along with the addition of nitrile rubber composition (NBR33) within the rubber mixture. Nitrile rubber generally improves the physical properties of the mixed samples after dimethyl ether immersion.

3.2 Measurement of mass change

The presence of swelling and partial dissolution phenomena can be seen from the percentage mass change of the samples before and after dimethyl ether immersion. The mass change can be seen in Figure 1 with the positive value indicating mass addition is called swelling and negative value showing mass decrease is called partial dissolution.

Table 1. Percentage of mass changing of vulcanized rubber mixture (NR/NBR33) by DME.

Sample	Ratio (%)		Treatment	Mass Change, %
	NR	NBR		
A	100	0	After (direct)	10.94
			After (+ 3 hours)	-7.55
B	75	25	After (direct)	13.18
			After (+ 3 hours)	-5.69

C	60	40	After (direct)	6.71
			After (+ 3 hours)	-6.44
D	50	50	After (direct)	14.56
			After (+ 3 hours)	-4.50
E	40	60	After (direct)	11.52
			After (+ 3 hours)	-4.43
F	25	75	After (direct)	10.10
			After (+ 3 hours)	-4.27
G	0	100	After (direct)	11.10
			After (+ 3 hours)	-3,10

Table 2. Elongation, tensile strength and hardness of vulcanized rubber mixture (NR/NBR33).

Sample	Ratio, %		Treatment (Before/After immersion)	Elongation (%)	Tensile Strength, (Mpa)	Hardness, Shore A
	NR	NBR				
A	100	0	Before	590	16,7	60
			After	400	9,7	60
B	75	25	Before	560	16,3	65
			After	500	13,2	65
C	60	40	Before	570	16,5	67
			After	510	15,1	70
D	50	50	Before	550	16,3	67
			After	560	15,7	68
E	40	60	Before	540	15,8	67
			After	580	15,7	69
F	25	75	Before	580	17,3	67
			After	570	18,6	70
G	0	100	Before	520	15,9	64
			After	530	20,3	70

The result also shows that the swelling process occurs when DME fills the crack in the rubber sample which then extracts the soybean oil compound out of the rubber sample. The principle of this process is that similarity in terms of chemical structure will facilitate the occurrence of swelling and partial dissolution which is known as "like dissolves like" [3]. Soybean oil which contains most of the unsaturated fatty acid, when in large amount, is dissolved more easily into the DME which a polar compound is. It is also asserted that if more oil is used in the rubber vulcanizing formula, the oil extraction is greater [4]. Such process is unavoidable despite the presence of synthetic rubber (NBR33) within the rubber formula. Figure 1 informs that the presence of NBR makes the vulcanized rubber mixture is relatively resistant to soybean oil extraction. Further impact due to DME immersion is the occurrence of degradation based on physical property changes.

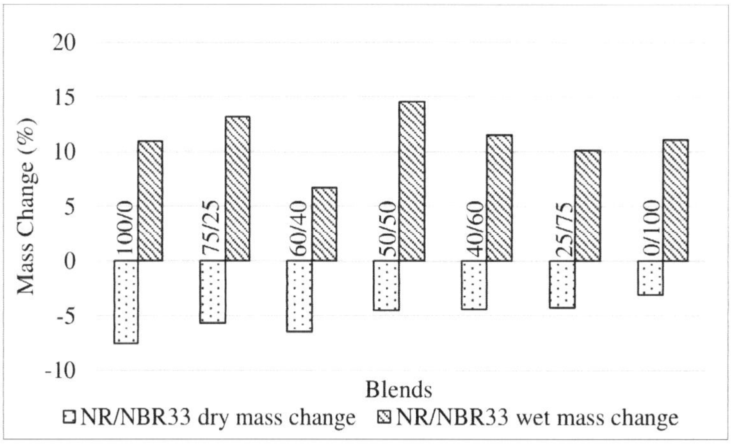

Figure 1. Percentage mass change of vulcanized rubber mixture (NR/NBR33).

3.3 Tensile properties change

The change of tensile strength value as a result of immersion is an indicator that the degradation mechanism due to the partial dissolution. Physical property degradation will occur if the rubber sample in the test is deteriorated. The change in tensile strength is shown in Figure 2. In the samples, natural rubber (NR) content is lower the tensile strength value. This is due to the fact that the absorption of DME makes the carbon black filler distribution become uneven within the vulcanized rubber. Carbon black distribution can be seen from SEM morphology test. Tensile strength value decrease approaching 50% is also predicted as a result of devulcanization, the breaking of cross linking bond of sulphur between natural rubber molecules. On the other hand, devulcanization phenomenon does not occur on natural rubber (NR) sample based on FTIR test; there is no change in the composition of the double bond number [9]. From Figure 2, it can be concluded that nitrile rubber prevents degradation as a result of swelling and partial dissolution phenomena by raising the value of tensile strength after DME immersion.

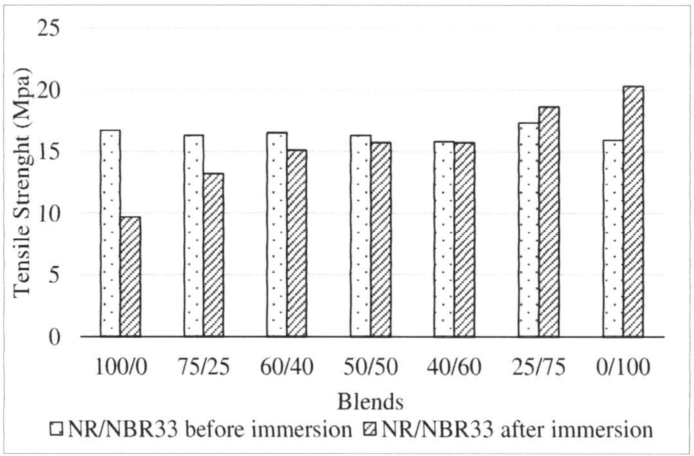

Figure 2. Tensile strength of vulcanized rubber mixture (NR/NBR33).

The elongation value will be associated with the tensile strength value. The elongation value is obtained after maximum tensile strength value is reached until eventually rubber sample breaks. The elongation change of the rubber sample after immersion can be seen in Figure 3. The elongation value in Figure 3 is similar to Figure 1 in terms of increased elongation value along with the increase of nitrile rubber ratio (NBR33) in vulcanized rubber mixture.

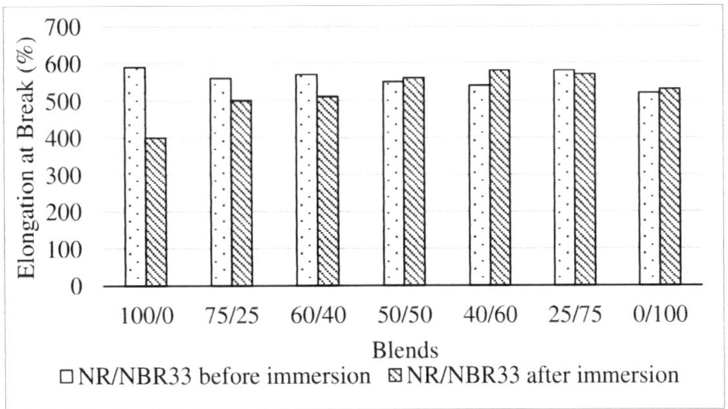

Figure 3. Elongation of vulcanized rubber mixture (NR/NBR33).

3.4 Hardness properties change

The presence of soybean oil is very influential on the hardness of rubber blends. Soy oil can act as a solvent or plasticizer. Plasticizer is an additive in rubber vulcanization that serves to increase the flexibility but reduces the hardness of a polymer [3]. Figure 4 shows a tendency to increase the value of violence with NBR33 contents increase. This was due to the blending of rubber to lose soybean oil after soaking. However, blends with higher NR is not affected by increased hardness values. It might be due to low compatibility between soybean oil and natural rubber. Figure 4 also informs about the hardness before immersion in NR / NBR mixtures is better if it is compared with pure natural rubber and pure nitrile rubber.

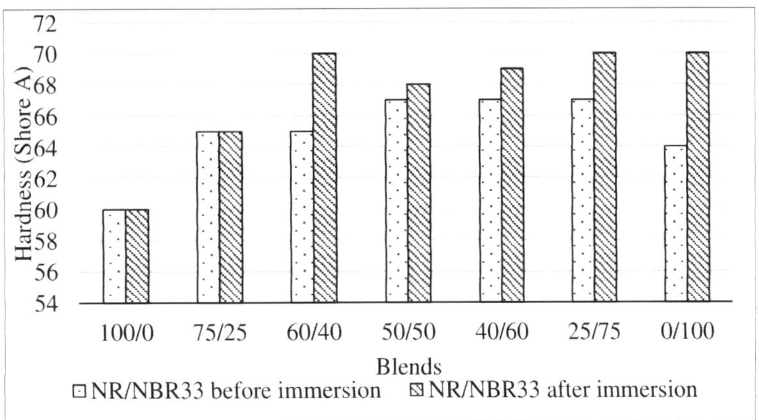

Figure 4. Hardness of vulcanized rubber mixture (NR/NBR33).

3.5 Compression set B properties change

Figure 5 informs that each blended sample has a similar compression set value before being immersed. This means that both NR and NBR resistance are relatively similar to plastic deformation. However, after the rubber blend is soaked, the compression set tends to rise with higher NBR33 to NR / NBR blends at 40/60 and drops back to pure NBR33. Figure 5 shows compression set NR / NBR33 blends at 40/60 and 50/50 tends to be lower. This was due to the lower compatibility of NR / NBR33 blends at those ratios than others. In addition, the lack of compatibility also affects the decreased tensile strength and elongation at break at the same blends before immersion that can be seen at Figure 2 and 3.

Thus, based on the physical properties, the increase of nitrile rubber ratio (NBR33) in rubber vulcanized mixture starting from 100% NR to 100% NBR33 can increase physical properties due to swelling and partial dissolution degradation which are % mass change, tensile strength, elongation and increase of hardness value. The best ratio of rubber vulcanized mixture (NR/NBR33) is 60: 40.

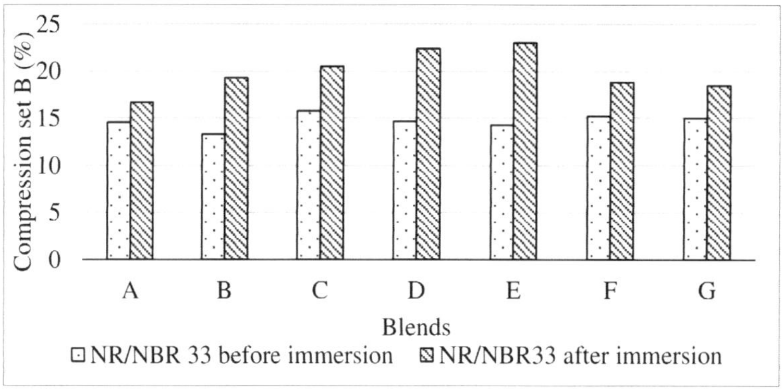

Figure 5. The compression set B change of NR/NBR33.

3.6 Morphology analytical (SEM)

Based on Table 3, it can be seen that pure NR before immersion has many white spots. The white spot shows uneven carbon blacks evenly distributed on the rubber molecule. This can facilitate the absorption of DME by filling an empty gap that is not filled with carbon black. While the SEM result of 50/50 NR/NBR33 before immersion shows the rougher surface due to low compatibility between NR and NBR33. Rough surfaces create an empty gap that can also facilitate DME absorption. On the other hand, pure NBR33 before immersion displays a smoother surface than others. NBR33 is very well tied to carbon black or plasticizer.

In compare to SEM result of after immersion condition, it shows failure distribution of carbon black on surface of pure NR, NR / NBR33 blends at 50/50 and pure NBR33 respectively after immersion by DME. According to pure NR, the surface of the visible rubber shrink was indicated with a rougher surface than pure NR before immersion. This was likely due to mass transfer of carbon black into the rubber due to absorption of DME. while picture 50/50 NR/NBR33 result does not look rough surface or white spot. The gap on the rubber surface was not visible, as if the rubber has a finer surface after immersion. It was possible that the bond between natural rubber and nitrile rubber prevents the presence of carbon black maldistribution. In pure NBR33 result, it was seen that nitrile rubber was not significantly altered by DME immersion. The pattern of the surface of nitrile rubber was similar to before immersion. The image proves that nitrile rubber was not easily affected by DME exposure.

Such phenomenon can be interpreted that based on SEM analysis carbon black is distributed more evenly on nitrile rubber (100% NBR33) than 100% NR before immersion, such that it has good physical properties and can prevent swelling and partial dissolution by dimethyl ether.

Table 3. SEM result of different ratio NR/NBR33 before and after immersion

Ratio (NR:NBR33)	Before Immersion	After Immersion
NR 100%	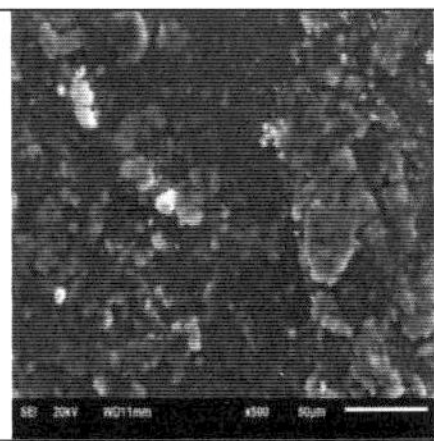	
NR/NBR33 50/50	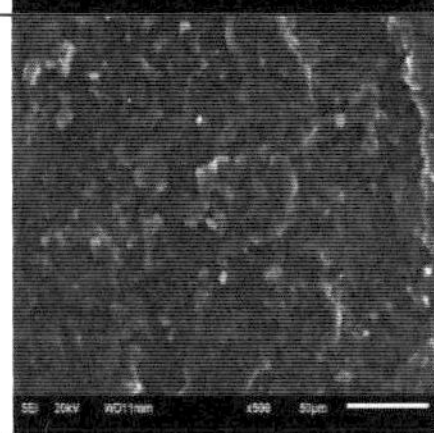	
NBR33 100%	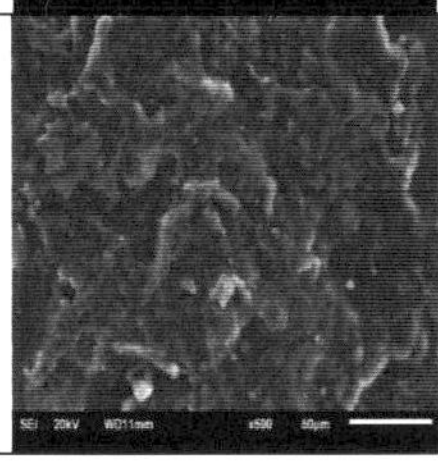	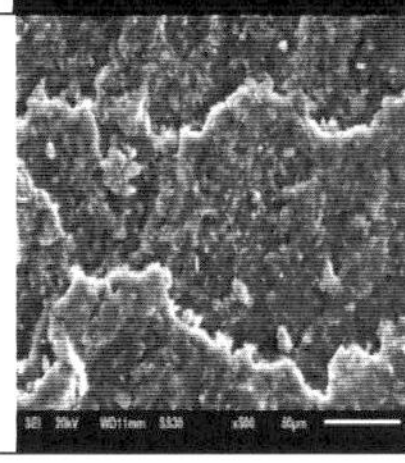

4. Conclusion

The increase of nitrile rubber ratio (NBR33) in rubber vulcanized mixture starting from 100% NR to 100% NBR33 can increase physical properties due to swelling and partial dissolution degradation which are % mass change, tensile strength, elongation and increase of hardness value. The best ratio of rubber vulcanized mixture (NR/NBR33) is 60: 40 NR NBR33 and it based on changes in physical properties that meet standard criteria. Based on SEM analysis, the distribution of carbon black is more evenly in 100% NBR33 than 100% NR before immersion.

Acknowledgement

The authors express a great gratitude to Ministry of Research, Technology and Education for the research financial support.

References

[1] Smelsberger T A, Borup, R L, and Greene H L 2005 *J. Power Sour.* **156** 497-515
[2] McKone T 2015 *California Dimethyl Ether Multimedia Evaluation* (United State: California)
[3] Callister W D and Rethwisch D G 2009 *Material Science and Engineering: An Introduction* ed 8 (United States: John Wiley and Sons, Inc.)
[4] Johan, Saputra A H, Indah S T, and Cifriadi A 2016 *Int. J. Technol.* **7** 616-624
[5] Tinker A J and Razak T A 1998 *Introduction – The Book and Rubber Blends* (London: Chapman & Hall) pp 1-7
[6] ASTM International 2008 *Standard Test Method for Rubber Property – Compression Set (ASTM D395)* (United State: ASTM International)
[7] Li G B and Zhou LB 2008 *J. Automob. Eng.* **222** 975-978

IOP Conf. Series: Materials Science and Engineering **345** (2018) 012036 doi:10.1088/1757-899X/345/1/012036

Coating hydroxiapatite on stainless steel 316 L by using sago starch as binder with dip-coating method

A Fadli , F Akbar, A Prabowo and P H Hidayah
Chemical Engineering Department, Faculty of Engineering, University of Riau

E-mail : fadliunri@yahoo.com

Abstract: Hydroxyapatite (HA) is a mineral form of naturally occurring apatite calcium with $Ca_{10}(PO_4)_6(OH)_2$ formula. One of the major innovations in the field of bone reconstruction is to apply HA as a surface coating on a mechanically strong implant metal and to improve the stability of bone implants thereby increasing the lifetime of the metal implants. Pure hydroxyapatite has poor mechanical properties so it is necessary to add sago starch as a binder to combine the strength and hardness of metal surfaces with bioactive properties of hydroxyapatite by Dip Coating method. Stainless steel 316L is the most commonly used alloy as an implant for bones and teeth due to its excellent corrosion and oxidation resistance and is easily formed. In this study, hydroxyapatite coatings used fixed variables as hydroxyapatite mass (10 grams), aquades mass (20 grams), dipping time (20 seconds), and calcination conditions (800°C, 1 hour). The variables are sago starch mass (1, 1.25, 1.5 gram) and stirring time (16, 20, 24 hours). The shear strength value is higher in the addition of 1.25, 10, 20, and again in the binder ratio of 1.5; 10; 20. The addition of stirring time causes a decrease in shear strength. The highest shear strength value obtained was 3.07 MPa. The layer attached to the substrate is a hydroxyapatite with a composition of 99.4% as evidenced by the results of XRD analysis.

1. Introduction

The number of accidentsemerge either light or heavy is relatively hugein every year. One of the most common cases of accidents are the fractures. This case can also occur due to osteoporosis. The high of fractures causes the high demand for bone implants. Hydroxyapatite has been known to be a good substitute for bone and dental implants in the health world due to its chemical and biological resemblance to human bone tissue. Stainless steel 316L is one of the most commonly used materials for implant. Stainless steel resistance to corrosion is one factor of stainless steel selection as bone and dental implants. In addition stainless steel has an affordable price compared to other metals such as titanium, so that affordable implants can be obtained using stainless steel as an implant[1]. One of the factors affecting the strength of adhesion on the hydroxyapatite coating is the binder, in which the binder serves to assist the deposition process that occurs on the surface of the substrate. Sago is a tropical plant that grows in Indonesia. Utilization of sago in the form of starch has been limited only as food. Sago can be used as a binder, thickener, and stabilizer. So in this study sago used as a binder in the process of hydroxyapatite coating on stainless steel 316L[2].The preparation of hydroxyapatite coatings can be done by various deposition methods such as thermal spraying, sputtering, EPD, and dip coating. Dip-coating method is a deposition method that has many advantages, such as low cost and simple process [3]. In addition, this method can also be used to coat the substrate with a complex form. Based on these advantages, the dip-coating method is chosen as a deposition method in hydroxyapatite coating on stainless steel 316L.

2. Method

The material used in this study were hydroxyapatite, Stainless steel 316L, Acetone and Aquadest. The main tool used in this research are the dip coating unit which serves as a hydroxyapatite coating device on the substrate. In addition, the tool used in this research is furnace (Nabertherm, Germany) which serves for sintering samples, stopwatch to measure the duration of dyeing, as well as sandpaper, glass chemical, scales, and magnetic stirrer used for the preparation of suspension and substrate. Preparation of a suspension solution hydroxyapatite is weighed as much as 10 grams and fed into beaker. Then added aquades as much as 20 grams, and sago that has been weighed as much as 1; 1.25 and 1.5 grams. The solution was stirred with a magnetic stirrer for 16, 20 and 24 hours. After that preparation of the substrate, stainless steel 316L cut with size 2x3 cm, then sanded by sandpaper. Stainless steel has been sterilized by soaking in acetone for 15 minutes, after it is rinsed using aquades. The substrate is then dried using an oven at 80°C for 10 minutes. Finally, the coating process,the sterilized substrate is immersed in the suspension for 20 second, and 1 dyeing. the hydroxyapatite coated substrate was then dried at 110 °C for 30 min, then sintered at 800 °C 1h.

3. Result and discussion

3.1 Result on visually analysis

Based on the results of the research, visual observations were madecoatings are generated on a variety of binder and time ratios stirring. Observations were made on the deposited layer on substrate. The observations is shown in Table 1.

Table 1. Observations visually

Sample Code	Stirring Time	Sago Ratio: HA: Affected	Results
S1	16 Hours	1.0: 10: 20	Layer Repositioned evenly No*cracking*
S2	16 Hours	1.25: 10: 20	Layer Repositioned evenly No*cracking*
S3	16 Hours	1.5: 10: 20	Layer uneven Cracking on deposition layer
S4	20 Hours	1.0: 10: 20	Layer Repositioned evenly No cracking
S5	20 Hours	1.25: 10: 20	Layer Repositioned evenly No cracking
S6	20 Hours	1.5: 10: 20	Cracking on deposition layer
S7	24 hours	1.0: 10: 20	Layer Repositioned evenly No cracking
S8	24 hours	1.25: 10: 20	Layer Repositioned evenly No cracking
S9	24 hours	1.5: 10: 20	Layer Repositioned evenly No cracking

Table 1 shown that almost all samples have a uniform layer and no cracking occurs. However, in the sample using a binder ratio of 1.5: 10: 20 with a stirring time of 16 hours (S3) and 20 hours (S6) layers are unevenly deposition. This is due to the high binder ratio, so most of the deposition layers are burned out during sintering.

Figure 1. Display surface samples (a) sample (S9) and (b) sample (S3).

Figure 1 (a) displays the sample with a uniform layer and is deposited on the entire surface of the substrate. Figure 1 (b) shows samples with uneven layers. This occurs because of cracking on the surface of the coating, thus causing the deposition layer to loose from the surface of the substrate and producing coatings with uneven layers

.

3.2 SEM analysis

3.2.1 Effect of binder ratio against hydroxyapatite coating

The binder ratio (sago) used in the manufacture of suspensions canaffect the thickness or surface state of the depositioned hydroxyapatite layer. Figure 3.2 is a cross-sectional view of stainless steel 316L which has been deposited by hydroxyapatite with 24 hours stirring time at various variations of binder ratio. Base on Figure 3.2, it shown that we can measure the thickness of the hydroxyapatite layer deposited on the surface of the substrate. In the binder ratio: HA: Aquadest 1:10:20 obtained layer with a thickness of 60 μm, while for the ratio of 1.25: 10: 20 and 1.5: 10: 20 obtained thickness of 58 and 35 μm. Hydroxyapatite coatings to be applied as implants should have a layer thickness of 50-200 μm, so that the coating is not lost when implanted and it still has good bond strength [4]. So the coating with the parameters is the coating produced in the ratio of 1:10:20 binder and 1.25: 10: 20 which is about 60 μm and 58 μm. While the binder ratio of 1.5: 10: 20 the resulting thickness is only about 35 μm.The binder used in this study is sago starch consisting of amylose and amylopectin. Amylose is a water soluble linear chain polymer, while amylopectin is a water-insoluble polymer. Amylose and amylopectin have a hydroxyl group (-OH) that will bind to the Ca^{2+} group on hydroxyapatite. In addition, amylopectin which is a branched polymer will form a network of polymers or matrices, and the hydroxyapatite particles will diffuse into the matrix with the help of stirring.

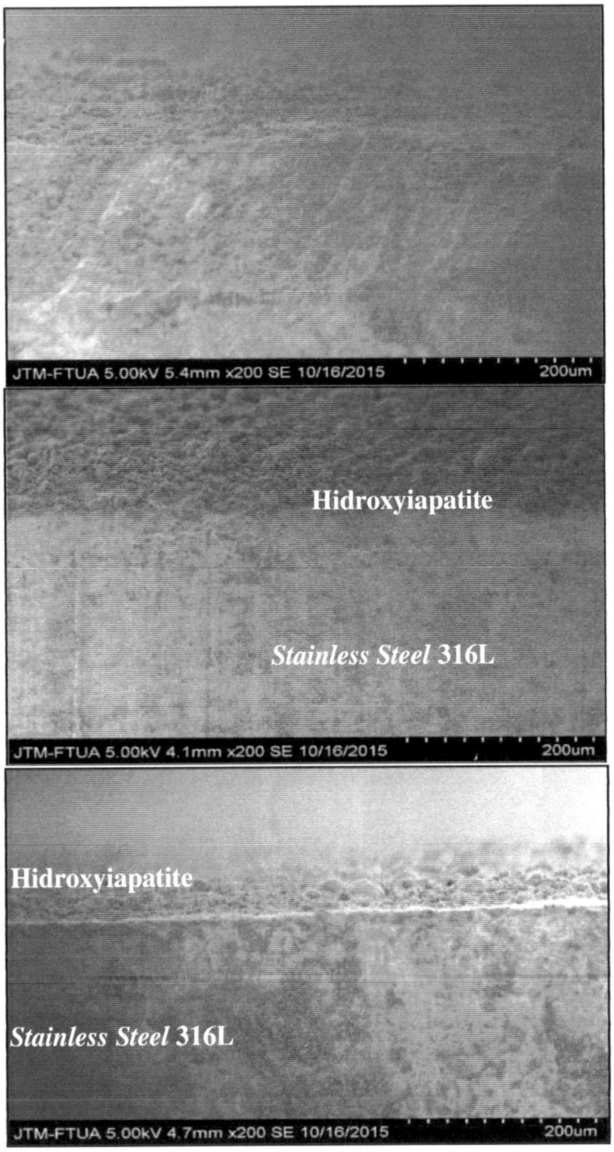

Figure 2. Visible stainless steel 316L strips with 24 hour stirring time and binder ratio: (a)1:10:10; (b) 1.25: 10: 20; (c) 1.5: 10: 20.

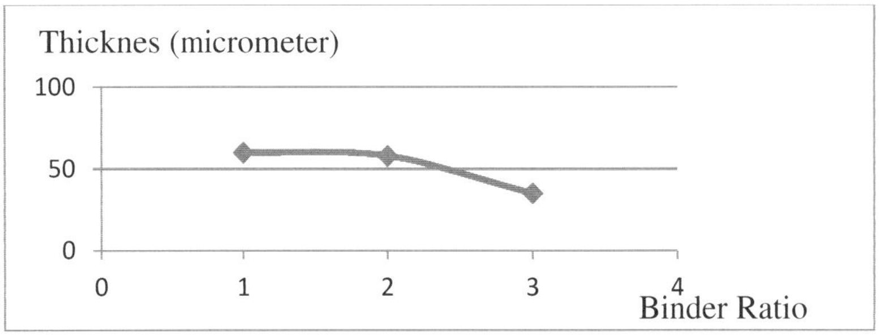

Figure 3. Graph of relation between coating thickness and binder ratio.

Figure 3 illustrates the relationship between the thickness of the coating and the mass of the binder.It can be seen that the resulting coating thickness decreases in the mass of the larger binder. The decrease of thickness due to the missing binder during the sintering process. The larger ratio of the binder, more hydroxyapatite particles bind to the sago starch, so the layers can run out during sintering will also increase. The layer obtained will be more thin with more binder used.

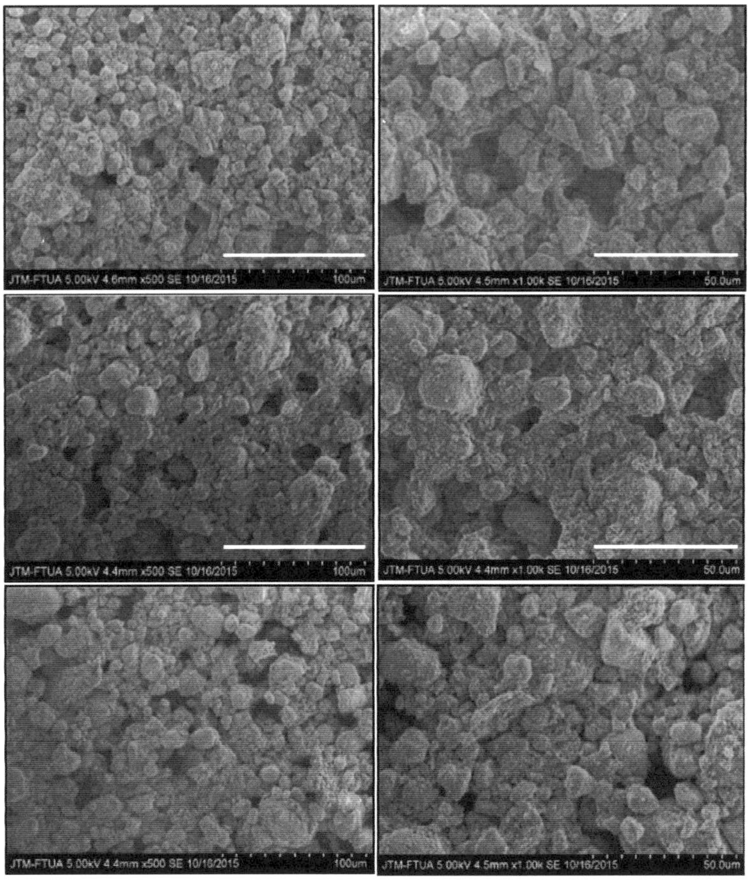

Figure 4. The hydroxyapatite surfaces repositioned with 24 hour stirring time and ratio: (a) and (b) 1:10:20; (c) and (d) 1.25: 10: 20; (e) and(f) 1.5: 10: 20 with magnification 500 and 1000 times.

The binder ratio in the suspension not only affected to the thickness of the resulting layer, but it also affected to the surface state of the layer. Figure 4 shown the the surface state of the deposited layer on a variety of binder ratios. Figure 4 (a) and (b) are coatings produced in a 1:10:20 binder ratio, it can be seen that there are some cavities on the surface of the layer. The cavity is more visible in larger binder ratios (Figure 4 (c) and (d)) and at the binder ratio of 1.5: 10: 20 (Figure 4 (e) and (f)) the cavity is seen more. The cavity on the surface of the coating is the result of a burning binder leaving. It can be seen that hydroxyapatite coats the entire surface of the substrate in the absence of cracking.Based on the results described previously, it can be seen that the binder ratio in the suspension is very influential on the coating produced, both thickness and surface conditions. The addition of the binder ratio in the suspension is inversely proportional to the thickness of the resulting layer, the larger the binder ratio used, the obtained layer will be thinner. In addition, the addition of the binder ratio can also cause more cavities to form on the surface of the layer.

3.2.2 Effect of stirring time on hydroxyapatite coatings

The hydroxyapatite coating process uses a dip coating method is performed by dipping the substrate into the suspension, where the physical properties of the suspension are one of the factors that may affected to the deposited layer. In this study coating with various stirring variations time to determine the effect on the thickness and surface conditions of the resulting hydroxyapatite coating. Figure 5 is a cross-sectional view of 316L stainless steel having been coated with hydroxyapatite using a binder: HA: aquadestion ratio of 1,25: 10: 20 at various stirring times, ie 16 hours, 20 hours and 24 hours.

Based on follow Figure 3.5 it is known that at the time of stirring 16 hours obtained a layer with a thickness of about 35 µm (Figure 5 (a)), and increased to 77 µm at 20 hours of stirring time (Figure 5 (b)). However, the thickness of the back layer decreased at 24 hours agitation time around 60 µm (Figure 5 (c)). From these results it is known that based on the coating thickness of the coating that qualifies as an implant is the coating obtained at the time of stirring 20 and 24 hours.

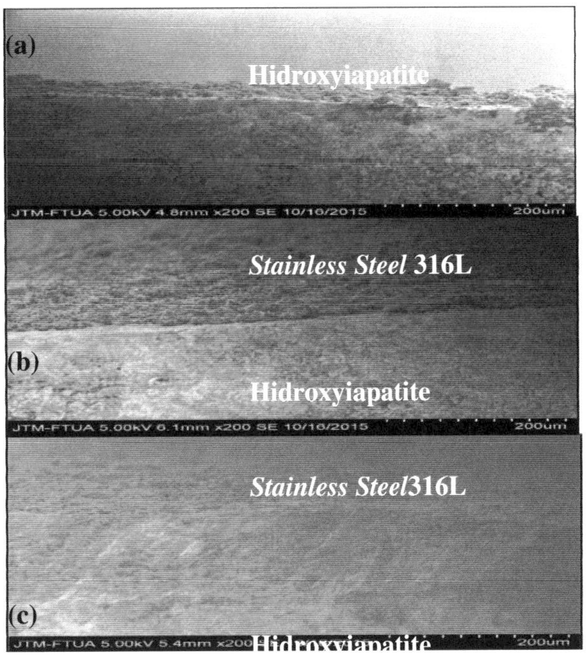

Figure 5. Transverse stainless steel looks repositioned with ratio binder 1:10:20 and stirring time, (a) 16 Hours, (b) 20 hours, (c) 24 hours.

3.3 Autograph analysis

Analysis using autograph is done to measure the strength value of the coating deposited on the surface of stainless steel. The shear strength of a layer shows how strong the layer is attached to its substrate.

3.3.1 Effect of binder ratio against hydroxyapatite Coating

The strength analysis is performed on the coatings produced at 24-hour stirring time with various variations of binder ratio. The graph of the relationship between shear strength and binder ratio can be seen in Figure 6.

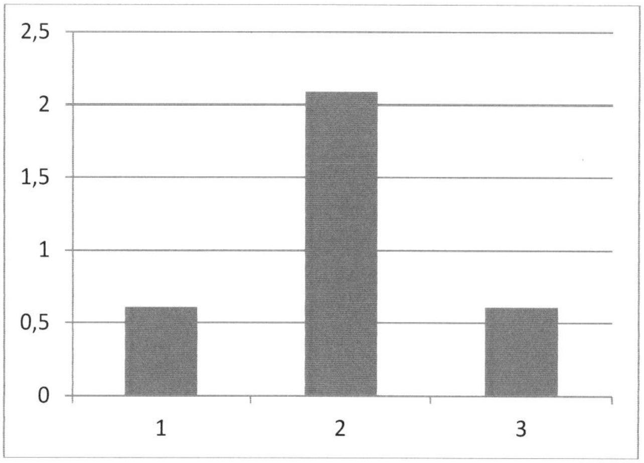

Figure 6. Graph relationships between shear strength and binder ratios used.

Based on Figure 6 can be seen that the ratio of 1:10:20 binder obtained shear strength value of 0.61 MPa, then at the ratio of 1.25: 10: 20 there was an increase in shear strength to 2.09 MPa, but at the ratio of 1, 5:10:20 shear strength value back down to 0.61 MPa. The addition of the binder ratio can increase bond strength to a certain point where the addition of the binder will cause a decrease in bond strength. This can happen because the bond strength is strongly influenced by the degree of homogeneity of the formed layer, the more homogeneous the layer is formed the bonding strength gets stronger [5]. In this study, the highest shear strength was obtained in the 1.25: 10: 20 binder ratio.

3.4 XRD analysis

XRD analysis is done to know the chemical compound in the sample and its composition. The results of the hydroxyapatite coating diffractogram with 24 hours stirring time and the 1.25: 10: 20 binder ratio can be seen in Figure 7.

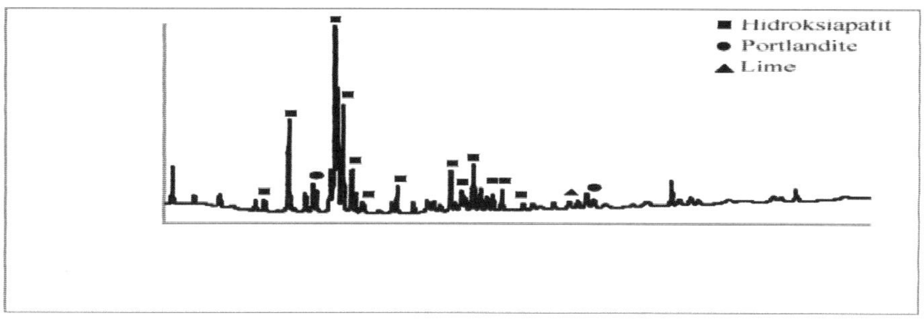

Figure 7 Diffractogram of hydroxyapatite coating results on stirring time 24 hours and binder ratio 1.25: 10: 20.

Figure 7 is a difractogram of the resulting coating. The hydroxyapatite peaks in the image have a hkl similar to the standard hydroxyapatite characterization pattern from JCPDS (002), 112 and 300 data with angle 2Θ 25.879°; 32,196° and 32,902°. The peak HA contained in Figure 7 with an angle of 2Θ is 25.927 °; 32.262° and 32.988°. From the results of the difractogram can also be known composition of compounds making up hydroxyapatite coating, which can be seen in Table 2.

Table 2. Composition of compounds available in hydroxyapatite coatings

Component	Composition (%)
Hidroxyapatite	99,4
Portlandite ($CaOH_2$)	0,5
Lime (CaO)	0,1

Based on the results of XRD analysis it can be concluded that this hydroxyapatite coating is qualified as an implant. The minimum permissible hydroxyapatite purity was> 95% [4]. and in this study the composition of the deposited hydroxyapatite was 99.4%

4. Conclusions

Hydroxyapatite coating on 316L stainless steel with sago starch binder has been successfully done by using dip dip coating method. The addition of stirring time causes a decrease in shear strength. The highest shear strength in this study was obtained at 16 hours of stirring time with a ratio of 1.25: 10: 20, ie 3.07 MPa.

References

[1] Hermawan H, Ramdan D, and Djuansyah J R P 2011. *Metals for Biomedical Applications*, http://www.intechopen.com/books/biomedical-engineering-from-theory-toapplications. Accessed 26 February 2017
[2] Belitz H D, Grosch W, and Schieberle P 2009 *Food Chemistry* 4th ed. (Berlin Heidelberg: Springer-Verlag)
[3] Yusoff M F M, Kadir M R A, Iqbal N, Hassan M A, and Hussain R Dipcoating of poly (e-caprolactone)/hydroxyapatite composite coating on Ti6Al4V for enhanced corrosion protection *Surf. Coat. Technol.* **245** 102-107
[4] Heimann R B 2002 Materials science of crystalline bioceramics: A review of basic properties and applications *Chiang Mai University J.* **1** (1) 23-46
[5] Tian T, Jiang D, Zhang J, and Lin Q 2007 Aqueous tape casting process for hydroxyapatite *J Eur. Ceram. Soc.* **27** 2671-2677

IOP Conf. Series: Materials Science and Engineering **345** (2018) 012037 doi:10.1088/1757-899X/345/1/012037

Green technology as a strategy in managing the black spots in Siak Highway, Indonesia

A Sandhyavitri[1], J Wira[1] and A Martin[2]

[1]Civil Engineering Department, University of Riau, Kampus Bina Widya, Pekanbaru, Indonesia

[2]Mechanical Engineering Department, University of Riau, Kampus Bina Widya, Pekanbaru, Indonesia

E-mail: ari.sandhyavitri@gmail.com

Abstract. It was identified that the total traffic accidents in the highway section of Siak, Indonesia within the period of 2011 to 2015 were 1,208 events (2 accidents per 3 days). This accidents figure were considered relatively high and it need to mitigate. The aim of this research are to; (i) analyze the location of Black Spot in the Siak highway, and (ii) drawn a strategy reducing the traffic accidents based on green technology. This study identified that the black spot area was located in the STA 44 + 050 (with a value of the weighted index was 86 and an accident severity rate was 6.21), these values were relatively high. The road horizontal alignment condition at this location was highlighted as a sub-standard high way, consists of low visibility, numerous turning pads, minimum road signs, and minimum road shoulders width. The technical strategy was then drawn as follow; conducting regular road rehabilitation and maintenance, equipping road markings and the street lights as well as road safety facilities based on the green technology such as solar cell traffic lights, solar cell street lights and deploying police statues in reducing traffic accidents within the black spot areas.

1. Introduction

It is acknowledged that the traffic accidents are a serious problem, hence it must be managed in a systematic procedure [1], [2]. Traffic accidents may involve several factors, such as driver/human errors, vehicle failures, road infrastructure deficiency, and environmental condition uncertainty such as heavy rain, rock fall and gush [2]. According to the Indonesia's Law No. 22 of 2009 concerning the traffic and transportation regulation in Indonesia, the traffic accidents are defined as unexpected and unintentional events which may occur on the road and have resulted in human casualties and/or property loss [3-5]. Based on the Article 226 of Indonesia Law No. 22 of 2009, in order to improve traffic safety, the prevention of accident programs should be put into actions [6], [7]. Thus, it is necessary to get involve the top level of decision maker commitments, stakeholder participations, community participations, law enforcements and global partnerships to wholeheartedly committing in the improvement of the existing traffic safety in Indonesia as well as reducing accident rates systematically [7].

Based on the traffic accident data released by the Directorate of the Traffic of the Riau Province Police, it was calculated that the number of traffic accidents in the Riau area has been relatively high. In 2010 as many as 1,719 accidents occurred, with 763 people died. 994 people were seriously and slightly injured 1,356 people. In October 2011, the Riau Police Traffic Directorate has recorded as many as 1,672 traffic accidents occurred in Riau. There were 674 victims died, 1,176 serious injuries and 1,361 slight injuries. This means that almost every single day, there were two people have died as a result of the traffic accidents in Riau. The number of the traffic accidents in Riau Province is at 10th the highest one at the Indonesia national level [3, 8, 9].

One of a region prone to accident in Riau was Siak district, as this district connecting Minas-Kandis sub-districts passed by the national highway of Sumatra Island [3, 9] (Figure 1).

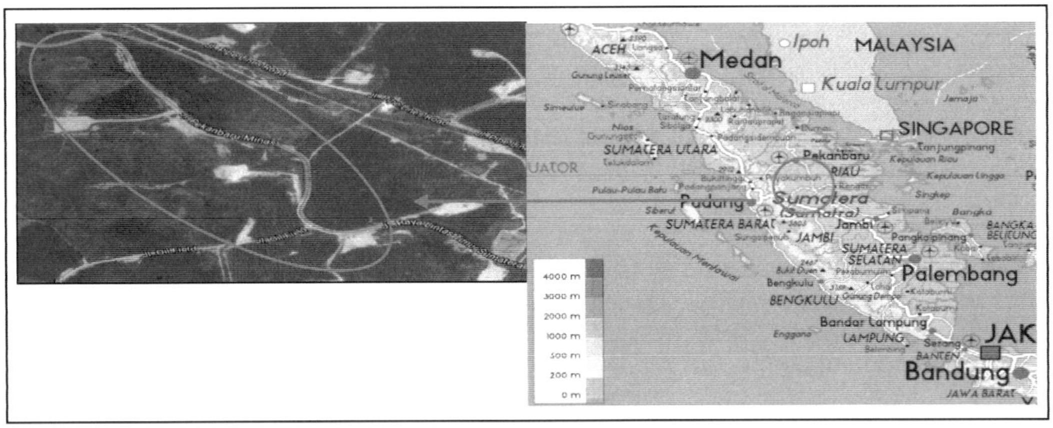

Figure 1. Research location at Siak, Riau, Indonesia.

It was identified that the total traffic volume from Minas sub-district = 2978 vehicle (45%), and from Kandis = 3640 vehicles (55%), with the total overall of 6618 vehicles [9] (Figure 2a). The dominant vehicles passing the highway were light vehicles (39%), and motorcycle (40%) (Figure 2b).

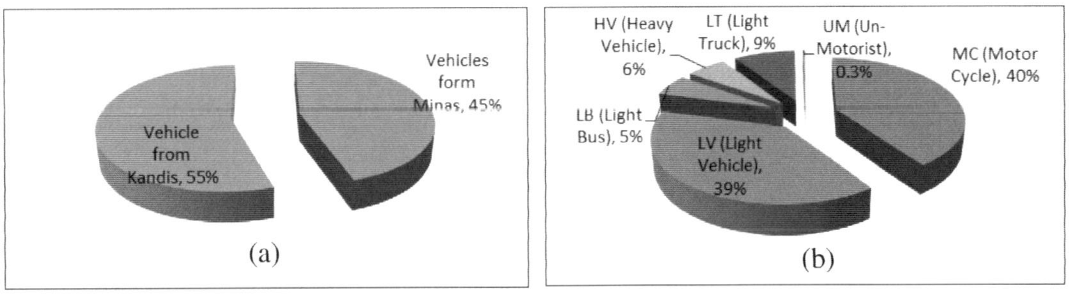

Figure 2. (a) Total traffic vehicles from Kandis and Minas, (b) Percentage of the vehicles source: Survey data, 2015.

The typical road segments may encompass a large number of sharp turnings (pads), limited visibility (no streetlights), and numerous patching holes. In anticipation of accidents on the road as the first step, it is a need to identify the accident-prone spots (black spots) (Figures 3a, b, and c)

(a) (b) (c)

Figure 3. (a) Sharp turnings (bends) without traffic signs and no streetlights, (b) the existing highway conditions with no road shoulder and no streetlights, (c) the existing patching holes. Source: Survey data, 2015.

Figures 3a, 3b and 3c shown the existing road turning and pavement conditions were need to repair. In order to repair the patch holes within the locations, the routine road maintenance programs can be conducted as regular bases.

The Figure 3 identified that there is no traffic signs and street lights on that locations. In the evening time, the locations were very dark and prone to traffic accident to occur.

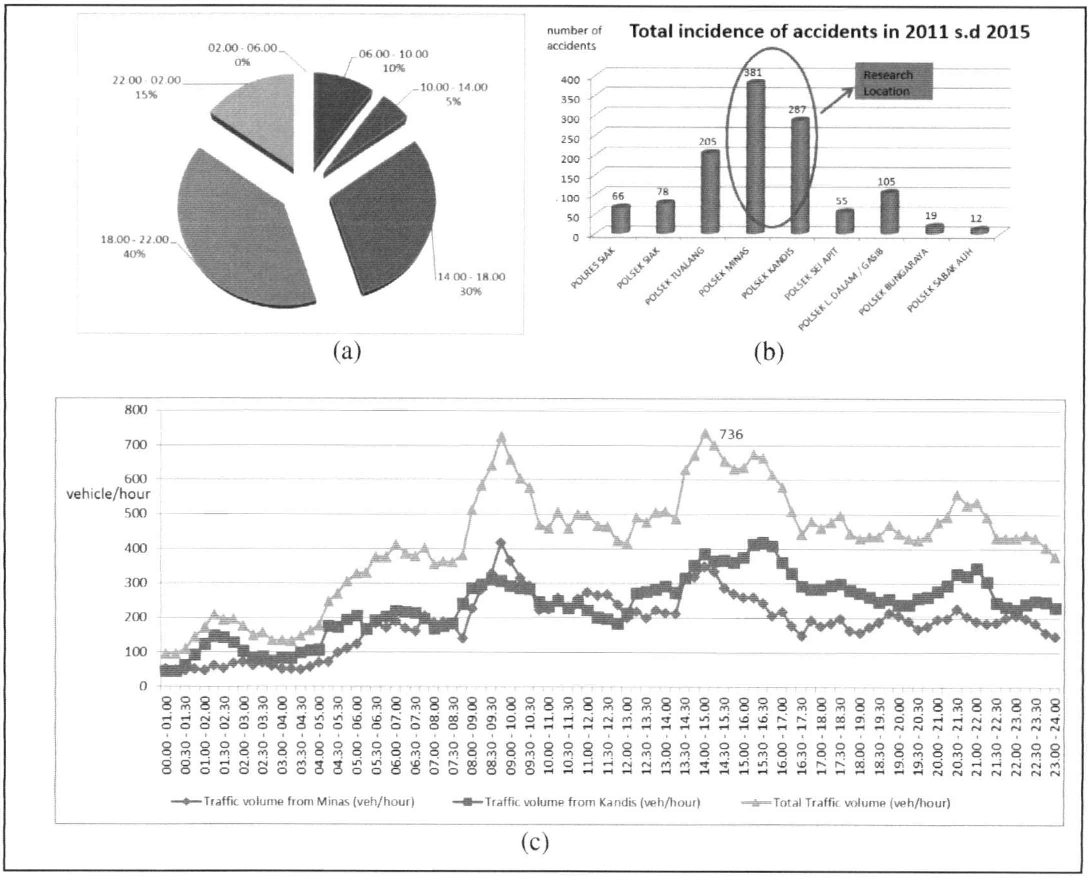

Figure 4. Traffic accident events in Siak, 2011-2015; (a) period of accident events, (b) the number of traffic accidents in Siak, (c) traffic volume (vehicle/hour). Source: Survey data, 2015.

Based on the period of accident events an evening time (18.00 - 22:00) (Figure 4a) was considered as the riskiest period of the traffic accident to occur in this location. This caused by the poor of street lights and as a consequence the road condition became very dark. Then at the evening time, some drivers have faced tiredness and physocological fatigue after long time driving during the day time period. The total traffic accidents in Siak within the five years period between 2011-2015 were calculated as 1,208 events [8, 9, 10] and 40% occurred in the evening time (18.00-22.00).

Figures 4b showed that 55% of traffic accidents in Siak was located at the Minas and Kandis Districts, contributing 381 and 287 accidents, respectively.

This analysis identified that the highest traffic volume within this highway was occurred during afternoon period of 14.00 – 15.00 with the number of traffic volumes of 736 vehicle/hour (Figure 4c). This traffic volume will be used to calculate a V/C ratio of this highway.

The highway capacity was calculated using PKJI 2014 formula [11]:

$$C = C_0 \times FC_W \times FC_{PA} \times FC_{HS} \tag{1}$$

C = Capacity
C_0 = Initial Capacity (3,000 passenger car/hour)
FC_W = Line width adjustment factors
FC_{PA} = Segregation of line direction adjustment factors
FC_{HS} = Obstacles aside and shoulder adjustment factors.

The highway capacity was calculated as follow;

C = 3,000 x 1.00 x 1.00 x 0.93 = 2.790 passenger car unit/hour

$$= \frac{736 \text{ pcu/hour}}{2{,}790 \text{ pcu/hour}}$$

The v/c ratio was calculated as v/c = 0.26

Hence, it was identified that v/c ratio was 0.26 (less than 0.8). It was determined that the traffic volume was relatively low compared to the existing highway capacity [11]. Thus, the vehicle speeds tend to be relatively high.

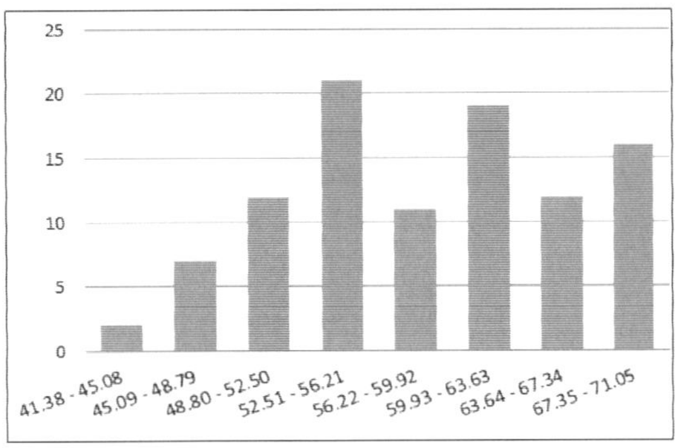

Figure 5. Private vehicle speed range in Siak, 2015.

Figure 5 shown that the average private vehicles travel speed passing this highway was 58 km/h. An approximately 16% of the vehicles travel speed was more than 67 km/h. Thus, the highway horizontal and vertical alignments, high pavement performance (Road Condition Index, RCI 8-10) and traffic signs required to be improved for accommodating the highway standard design for the designated speed limit at the average of 67 km/h.

The research objectives were to; (i) identify the location of black spot STA 43-45 in the Siak highway, Riau, Indonesia, (ii) analyze the accident occurrences, and (iii) develop an appropriate strategy to reduce the traffic accident events.

2. Methodology

A research methodology was developed into 5 stages; (i) identification and evaluation of the existing condition of the road alignment horizontal and vertical, hazardous conditions, encompassing; measuring the highway geometric conditions, horizontal curves, drainages, and road shoulders, pavement conditions, rutting, bleedings, potholes, cracks, and vertical curves, (ii) investigating traffic conditions, (iii) assessing the existing traffic signal conditions, road markings, and rail guards, and (iv) assessing black spots based on Weighted Severity Index (WSI) and Accident Rate (AR) approaches, then (v) this paper also give recommendations to manage the black spots.

The terminology of a black spot in a highway segment is defined as spot(s) that historically traffic accident has been regularly occurred [4], [5], [6], [12]. The black spot area in this highway was concentrated to 5 horizontal curves (bends) within 2 km road lengths (Figure 10). In order to identify the black spot areas, the calculated was performed using the Weighted Severity Index (WSI), as this WSI was commonly applied in many countries including India, Belgium, and Denmark [13, 14, 15]. The WSI was calculated as follow;

$$\mathrm{WSI} = (5 \times \mathrm{K}) + (3 \times \mathrm{GI}) + (1 \times \mathrm{MI}) \tag{2}$$

where, K = the number of persons killed,
GI = the number of injured persons,
MI = the number of non – serious injured persons.

The black spot areas were also identified using accident rate (AR). This was calculated using the following formula [15, 12,15]:

$$\mathrm{AR} = \frac{\mathrm{ACC} \times 1.000.000}{\mathrm{V} \times 365} \tag{3}$$

where; AR = Accident rate; ACC = annual accident cases, and V = traffic volume.

2.1. Calculation of the black spot based on the weighted severity index (WSI) and the accident rate (AR)

Based on the WSI and Accident Rate, it is identified that the black spot location as the following calculation; The highest WSI located at KM 44+050, with WSI = (5 x 4) + (3 x 17) + (1 x 15) = 86 (Table 1).

Table 1. Weighted severity index

Location (STA)	Victim conditions			WSI
	K	GI	MI	
43 + 250	1	11	9	47
43 + 600	2	4	16	38
44 + 050	4	17	15	86
44 + 400	1	9	4	36
44 + 950	2	2	1	17

where:

K = the number of persons killed,

GI = the number of injured persons,

MI = the number of non – serious injured persons.

The highest AR located at KM 44+050, with AR = 15 x 1,000,000/(6,618x365) = 6.21 (Table 2).

Table 2. Accident rate

Location (STA)	Accidents	Traffic Volume (vehicle)	Accident Rate (AR)
43 + 250	9	6618	3.73
43 + 600	7	6618	2.90
44 + 050	15	6618	6.21
44 + 400	11	6618	4.55
44 + 950	4	6618	1.66

Based on the results presented in the Table 1 and 2 above, it was identified that STA 44 + 050 and 43 + 250 were considered as the black spots as this highway segment yields the highest WSI compared to the other locations. Based on AR calculation, it was identified that STA 44 + 050 and 44 + 400 were considered as the black spots as this highway segment yields the highest AR. Hence, based on the results of WSI and AR, it was identified three black spots locations as follow; STA 44 + 050, 43 + 250, and 44 + 400.

This STA 44 + 050 was described as hilly road conditions, low visibilities, minimum traffic signs and markings, sharp bends, no road shoulders, no lightings, minimum guardrail, hence this STA was considered vulnerable to cause the higher road accidents at this highway segment (Figures 3 and 6).

Figure 6. Location of black spots.

2.2. The horizontal alignment

Based on the field study, it was identified that existing highway encompasses six turnings (one bend was classified as a full circle (STA 43 + 600), and five bends as a spiral-circle-spiral type (STA 43 + 250, 44 + 050, 44 + 400, and 44 + 950).

Table 3. Identified types of turnings, and minimum radius of the bends.

No	Bends	Type	Locate (STA)	Radius (m)	R min (m) (PKJI)	R min (m) (Arithmetic)	Information
1.	Bend 1	S-C-S	43+250	81,95	1000	112	Not Eligible
2.	Bend 2	FC	43+600	108,97	1000	112	Not Eligible
3.	Bend 3	S-C-S	43+850	165,34	1000	112	Qualify
4.	Bend 4	S-C-S	44+050	114,22	1000	112	Qualify
5.	Bend 5	S-C-S	44+350	213,11	1000	112	Qualify
6.	Bend 6	S-C-S	44+950	75,58	1000	112	Not Eligible

Table 3 and Figure 6 showed that three sharp bends which were a maximum speed less than 60 km/h, this is because of the radius of each bend radius is smaller than the minimum radius standards of SCS bends of 112 meters. The horizontal geometry of this highway segment less than 112 m, thus did not comply the highway bend specification standards [12, 16, 17, 18,19]. Then, it is need to improve the existing horizontal radius alignment of these three turnings (Figure 7a and 7b).

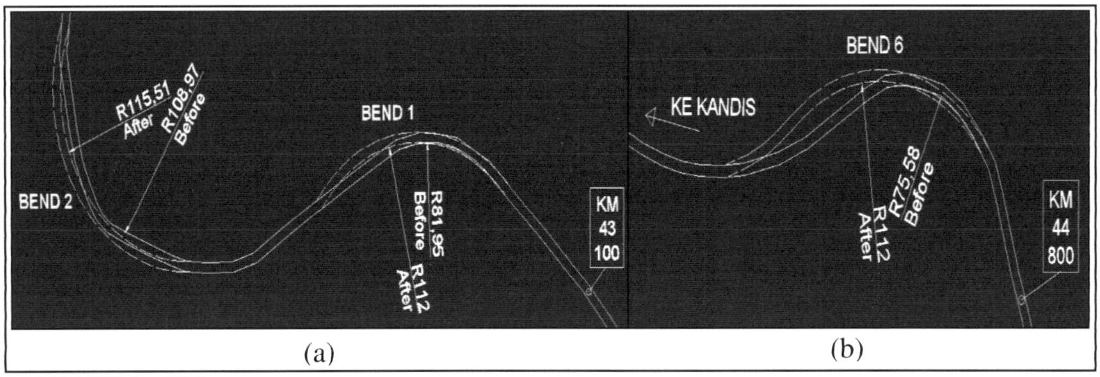

Figure 7. (a) A sharp turning of band 1 within the highway locations, (b) A sharp turning of band 6 (KM 44 + 950) within the highway locations.

Figure 8. The vertical bends within the existing highway locations at KM 44 + 95.

For example, it was calculated that the existing Bend 1 horizontal radius was 81.95 m, Bend 2 was 108.97 m, and Bend 6 was 75.58 m. These three turnings need to be extended by the following dimensions 30.05 m, 3.03 m, and 36.42 m respectively to fulfil the radius turning minimum standards of 112 m.

An environmentally friendly approach for mitigating this turning issues can be applied by an improvement of the existing horizontal radius alignment may reduce centrifugal force so that the risk of sliding and sliding vehicles may be declined and as a consequence the number of accidents will be reduced [11].

2.3. Managing the black spot at STA 44+050

The existing of black spot condition at the STA 44 + 050 showed that this segment did not have road shoulder, poor street lighting, inadequate traffic signs, no guardrails, and road delineators. The vertical road curve was considered high (>30%) (Figures 8 and 9). Hence, these conditions were urgently required to be managed systematically.

Figure 9. The situation at STA 44+050 (Before rehabilitation).

This paper highlighted the technical aspects of managing the black spot located in the STA 44 + 050. This may include the constructing of road shoulder to accommodate vehicles to pull over. There is a need to install traffic signs boards such as 'Reduce Speed' sign boards (blinking), road safety fences (guardrails), road markings and street lighting, and installing policeman statues equipped with flashlights in order to warn the traffic users to be aware of their safety as showed in Figure 10 [19, 20, 21,22].

Figure 10. A long climb segment (After rehabilitation).

In this study the overall traffic signs, street lighting and policeman statues equipped with flashlights will be proposed based on the green energy technology (Figure 10). Based on M. Richards, D. Carter, 2009 data the public lighting systems are considered as a major source of electricity consumption [23]. Approximately 3.19% of global electricity generation is used for lighting. This amount is greater than production of all hydro or nuclear plants and equals with the production from natural gas [23]. The depletion of fossil energy coupled with the climate change requires finding alternative energy production. Thus, alternative energies know a fast expansion [24].

Indonesia is ones of country that has a lot of solar energy resource and especially in Riau Province it can be seen in Figure 11, that the global horizontal irradiation of long-term average of annual sum is about 1600-1800 kWh/m^2 or about 4.5-5 kWh/m^2 for average daily sum [25]. The availability of solar energy resource in Riau Province is large enough for street lighting required.

The sun radiation in the form of shortwave radiation reaches the ground surface of earth may be estimated as a Global Horizontal Irradiance (GHI). The GHI in Indonesia was drawn as the following Figure 11 [26]. This GHI may indicate the potential of the solar photovoltaic (PV) technology may be applied in reducing electricity consumption in Indonesia.

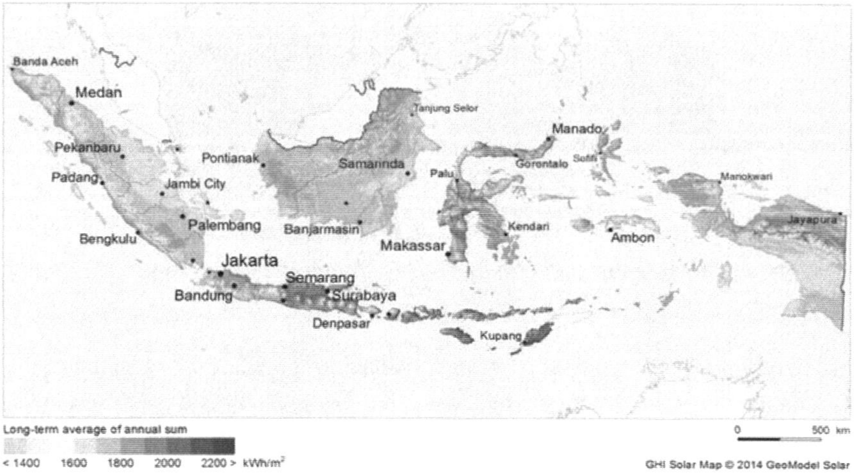

Figure 11. Global Horizontal Irradiation (GHI) in Indonesia [26].

For reducing the electricity consumption of street lighting, solar photovoltaic (PV) technology (Figure 11) is claimed as a solution for this part of electrical load because of its environmental advantages (e.g., cleaner, less emissions, and no fossil fuel) [26]. On the other hand, conventional street lightings with the solar photovoltaic usually utilize mercury lamps. The conventional street lighting utilizing mercury lamps consumed electrical power higher than 200 W per lamp in order to

IOP Conf. Series: Materials Science and Engineering **345** (2018) 012037 doi:10.1088/1757-899X/345/1/012037

meet the street lighting requirement standards. The optical efficacy of light emitting diode (LED) has exceeded 72 lm/W in 2006. This implies that energy can be saved about 75%, as compared to mercury lamps used in roadway lighting [27]. These lamps were proposed to be applied in this case study.

The diagram of Photovoltaic Block System can be seen in Figure 12. This Photovoltaic System for accommodating the road way lightings as well as traffic signs would be applied within the black spots areas in Siak, Indonesia (2017).

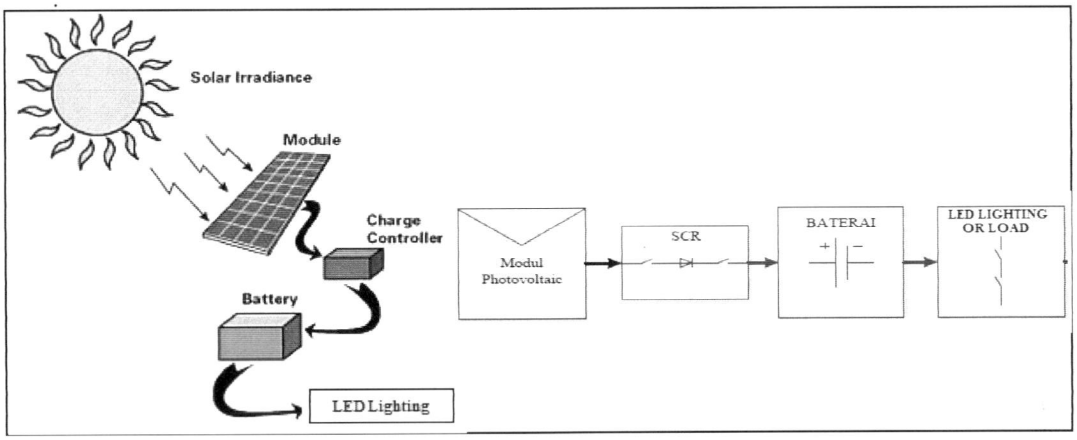

Figure 12. Schematic of photovoltaic system

3. Recommended strategy

There are 2 main strategies for reducing the number of accidents in the black spot area of KM STA 44 + 050:

a. Accomplishing the road safety (horizontal alignment improvements) by the extension of the road existing horizontal radius alignment within the black spot areas.

a. Installing the green road traffic signs, road markings, street lightings and the policeman statues equipped with flashlights based on solar energy technology as show in Figure 13.

Figure 13. Recommended strategy to reducing traffic accident on the black spots in Siak highway by using green energy technology.

4. Conclusions

It was identified that there were three horizontal alignments out of six turnings did not meet the specification technical radius standards (r bends <112 m). These turnings were in need to repair. This research identified the black spots within this highway located at 44 + 050 as a case study. The conditions of these black spots were described as poor street lightings, inadequate traffic signs, no guardrails, no road delineators, and limited road shoulders as well as a number of patching holes. Thus, it is necessary to develop strategies in managing these black spots as follow; (i) rehabilitation of horizontal alignment, road pavement, and widening the existing road shoulders, (ii) installing traffic signs, road markings, street lighting, and guardrail delineator based on the green technology approaches.

Acknowledgements

The author would like to thank the Department of Transportation of the Siak City and the Civil Engineering Department University of Riau for providing facilities and accident data for this paper.

References

[1] Abd K and Zaini 1998 The Traffic Accident Analyses in Riau Province (Analisa Kecelakaan Lalu lintas) Di Propinsi Riau (Studi Kasus Pada Ruas Jalan Rimbo Panjang – Bangkinang) *Unpublished Report* 7-12

[2] Anonymous 2011 *The National Traffic Safety Plan Rencana Umum Nasional Keselamatan (RUNK)* 2011 – 2035 Jakarta 21

[3] Anonymous 2013 Accidents Reports *Riau Police Unpublished Reports 2013* Pekanbaru Indonesia 2-9

[4] Homburger and Carter E C 1978 *Introduction to Transportation Engineering* (USA: Preston Publishing) p. 13-35

[5] Deacon J A, Zeeger C V and Deen R C 1975 Identification of hazardous rural highway locations *Transportation Research Record* **543** 16-33

[6] Karl-Olov H 2001 *Black Spot Manual*, General Directorate of Highways, Road Improvement and Traffic Safety Project Ankara 18-26

[7] Sekretariat Negara Republik Indonesia 2009 *Regulation of the Republic Indonesia No. 22 year 2009 (Undang-Undang Republik Indonesia Nomor 22 Tahun 2009), Tentang Lalu Lintas dan Angkutan Jalan* Jakarta **3-7** 35-45

[8] Indonesia Police of Republic Indonesia 2014 *Accident Data* Data Kecelakaan Jakarta 5-10

[9] Siak Police Report (Kepolisian Resort Siak) 2015 *Traffic Accidents Data Data Kecelakaan Lalu Lintas 2011-2015* Kabupaten Siak Indonesia 2-5

[10] Puskesmas Minas 2015 *The Traffic Accident Data Data Korban Kecelakaan Lalu Lintas 2011-2015* Minas Indonesia 2-4

[11] Direktorat Jenderal Bina Marga 2014 *Highway Capacity Manual (Panduan Kapasitas Jalan) Indonesia (PKJI, 2014)* Kementerian Pekerjaan Umum Jakarta

[12] Edisantoni 2012 Traffic Accident Characteristics and Road Safety Audit (Karakteristik Kecelakaan dan Audit Keselamatan Jalan) Ruas Jalan Kaharudin Nasution Pekanbaru Universitas Islam Riau *Unpublished Report* 22-32

[13] Vivek and Saini R 2015 *IJCEM* **2** 155-167

[14] Geurts K and Wets G 2012 *Black Spot Analysis Methods: Literature Review Verkeersonveiligheid Steunpunt Verkeersveiligheid Bij Stijgende Mobiliteit* p. 15-19

[15] Sandhyavitri A, Zamri, Wiyono S and Subiantoro 2017 Three strategies reducing accident rates at black spots and black sites road in Riau Province, Indonesia *Trans. Res. Procedia* **25** 2153–2166

[16] Nurkhotib 2010 *Road Safety Audit Audit Keselamatan Jalan* Study Kasus Jalan Wates Km 1-2,9 Bantul Yogyakarta
[17] Direktorat Jendral Bina Marga 1997 *A Safety Road Audit Manual (Pedoman Audit Keselamatan Jalan)* Departemen Pekerjaan Umum Jakarta
[18] Direktorat Jenderal Bina Marga 1997 *Highway Capacity Manual (Manual Kapasitas Jalan) Indonesia (MKJI) No. 036/T/BM/1997* Tata Cara perencanaan Geometrik Jalan Antar Kota Departemen Pekerjaan Umum Jakarta p. 44-45
[19] Peraturan Menteri Perhubungan RI 2014 Nomor 34 Tahun 2014 *Road Markings Tentang Marka Jalan* Indonesia
[20] Peraturan Menteri Perhubungan RI 2014 Nomor 49 Tahun 2014 *Traffic Signs Equipment Alat Pemberi Isyarat Lalu Lintas* Indonesia
[21] Peraturan Pemerintah RI 2011 Nomor 32 Tahun 2011 *Traffic Engineering and Management,Tentang Manajemen Dan Rekayasa, Analisis Dampak, Serta Manajemen Kebutuhan Lalu Llintas* Indonesia
[22] Peraturan Menteri Perhubungan Republik Indonesia Nomor 13 Tahun 2014 *Traffic Signs Rambu Lalu Lintas* Indonesia
[23] Richards M and Carter D 2009 *Lighting Res. Technol.* **41** 285
[24] Lagorse J J, Paire D and Miraoui A 2009 *Intl. Conf. Clean Electr. Power* p. 135-140
[25] Global Horizontal Irradiation (GHI) Solar Map 2014 GeoModel Solar http://solargis.com/products/maps-and-gis-data/free/download/indonesia 1
[26] Gang L 2014 *Elect. Power Energ. Syst.* **56** 168-174
[27] Wu M S, Huang H H, Huang B J, Tang C W and Cheng C W 2009 *Renewable Energ.* **34** 1934-1938

IOP Conf. Series: Materials Science and Engineering **345** (2018) 012038 doi:10.1088/1757-899X/345/1/012038

Performance of the electrical generator cell by the ferrous alloys of printed circuit board scrap and Iron Metal 1020

Y Sahan, S Sudarsono, E Silviana, Chairul and Wisrayetti
Department of Chemical Engineering, University of Riau, Pekanbaru 28293, Indonesia

E-mail: yusnisahan@lecturer.unri.ac.id

Abstract. Galvani cell is one of thealternative energy. This cell can be used as an electric resources. In this research, the generator cell was designed and builds to generate the electric. The generator cell consisted of the iron metal 1020 were used as anode, the ferrous alloys of printed circuit board scrapwas then used as chatode, and NaCl solution as an electrolyte. The aim of this research is to estimate the performance of this generator cell by using variation of NaCl concentration (i.e.1%, 3%, 5%, 7%, and 9%) with the electrodes pair (1 and 8 pairs). The performance of the cell was measured with a multi tester equipment and a LED bulb (5-watt 3Volt). The Results shown that the generator cell can produce the electric power of 3.679 Volt maximally by using NaCl 9% and 8 electrode pairs applied for this condition.

1. Introduction

Recently, some researchhave been done to create several things from the electronic good scraps, such as an alternative energy in order to overcome energy crisis issues in the future [1, 2, 3, 4, 5, 6 & 7]. One of them isto use electronic goods scraps such as jengkol transistors scraps as a power generator. By utilizing the power of the transistor jengkol 2N3055, the jengkol can produce a voltage of 0.3-0.6 Volt [8]. With the voltage of one transistor, then it arranged in parallel position as many as transistors to get the desired voltage and to increase the current by parallel the number of transistors to increase the electrical current.

In this research, development of the electrical generator cell has been designed. This generator cell can generate electric power based on the electrochemical reaction spontaneously of the cell. Basically, the generator cell consisted of two electrodes, the iron metal 1020 as an anode, the ferrous alloys of printed circuit board scraps as a cathode, and NaCl solution as an electrolyte solution.

Electrolyte compound is dissolved in water, then the electrolyte is dissociated into positively and negatively charged electric particles called ions (positive ions and negative ions). The amount of positive ion charge will equal to the amount of negative ion charge, so the charge of ions in the solution is neutral. These ionic functions to be as an electric current media. The solution transferred an electric current is called electrolyte [11, 12]. This solution gives a symptom of flashing a lamp or the incidence of a gas bubble in solution.

NaCI solution is solved in H_2O, it can be ionization as Na^+ and Cl^-ionics [9-12]. These ionics function as a electron transfer media from one electrode to another one. Because there is electron current between two electrodes, the reduction and oxidation reactions has been appeared spontaneously, as a result there is electric current produced of the generator cell [11-14]. The portable electrical generator cell was determined using variation concentration of NaC lwith the amount of electrode pair in the generator cell. The multitester equipment was used for estimating performance of the generator cell.

The objectives of this study are: (i) to study the influence of the variation concentration of NaCl solution to the performance of the generator cell (ii) study the reaction of the generator cell, iii) to determinethe efficiency of the generator cell.

2. Materials and methods

2.1. Metal material and supporting tools

Some material used such as sodium chloride p.a (Merck), aquadest DM, analytical balance, glasses, the multitester equipment, LED bulb (5 watt 3volt), the ferrous alloys of printed circuit board scrap and the iron metal 1020, and the generator cell with probes.

2.2. Design and built of the electrical generator cell

The following procedure were chosen the ferrous alloys, the iron metal is put in the generator cell as well. One unit of cell consisted of the ferrous alloys of printed circuit board scrap as a cathode (5 cm width x 10 cm length) and the iron metal 1020 as a anode (5 cm width x 10 cm length) (Figure 1). Then, a certain number of the unit of cell were put in the box, one unit of cell connected to other unit of cells in parallel position. Finally, NaCl solution filled into the generator cell, and this generator cell was connected to a LED bulb and the multitester equipment in order to estimate the electric power of the generator cell performance.

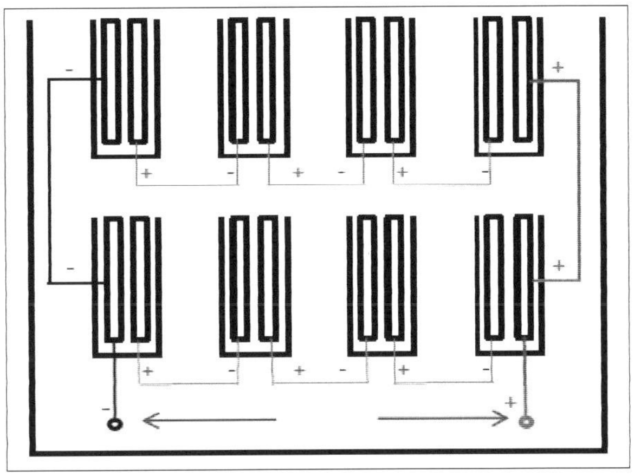

Figure 1. Circuit electron current of the generator cell which is consist of 8 electrode pairs.

The generator cell is consisted of several things, such as two Electrodes, the cell, the multitester equipment, terminal slot, LED bulb, Probe positive and Probe negative, also the electrolyte solution. The electrolyte solution is made from a solid mixture of NaCl with water (H_2O). Separator, between positive and negative plates, separator serves to separate or seal the positive and negative plates. The generator cell is a box containing liquid NaCl, plate positive and negative and separator.

Part of the generator cell consists of a box, it can be used as a house or container of its main components such as positive electrode (Fe_2O_3) and negative electrode (Fe) with electrolyte NaCl solution. The box lid above serves as the electrolyte filling hole cover, so the electrolyte solution is not easily spilled. The holeventilation serves to separate hydrogen gas from NaCl as well as a water evaporation channel. The metal plate is made of the ferrous alloys of printed circuit board scrap, while the negative plate is made using the iron metal 1020 (iron pure or Fe metal).

2.3. Determine performance of the electrical generator cell

Innitially, the ferrous alloys of printed circuit board scrap are put in the cell, then the iron metal is put in the generator cell as well. One unit of the cell consist of the ferrous alloys of printed circuit board scrap as a chatode and the iron metal 1020 as a anode (Figure 2). Then, a certain number of the unit of cell were put in the box, one unit of cell connected to other unit of cells in parallel position. Finally, NaCl solution filled into the generator cell, and this generator cell was connected to a LED bulb and the multitester equipment in order to estimate the electric power of the generator cell performance.

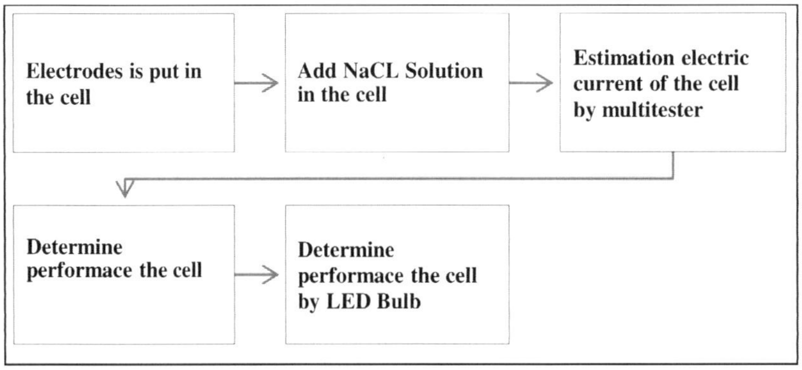

Figure 2. Diagram operation of the generator cell.

3. Results and discussion

3.1. Effect of variation concentration of NaCl solution to performance of the generator cell

This electrical generator cell is generated an electric power based on electrochemistry same as Galvani cell [12. 13, 14]. The effect of thevariation concentration of NaCl solution to the performance of the generator cell is presented in Table 1 and Figure 3.

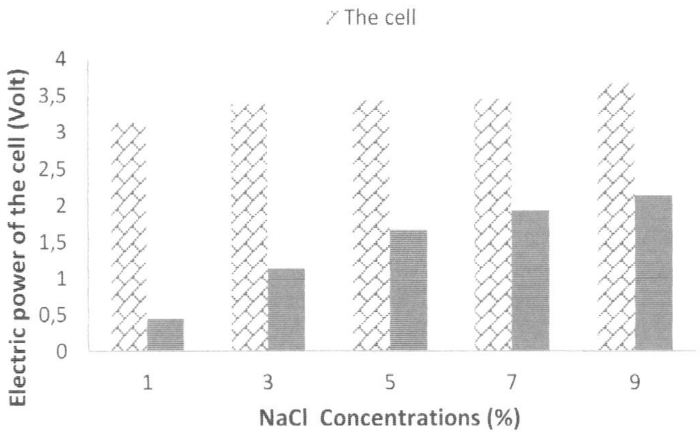

Figure 3. Influence of NaCl concentration variation to electric power of the cell.

Overall, the treatment of the variation NaCl concentration is affected to the performance of the generator cell. The highest of the NaCl concentration will be decreasing the electrical current significantly. There are a lot of wires used for connecting one unit of the cell to other unit of the cell in this generator cell. Another reason, exhoterm reaction is occured in this generator cell because of the differences enthalpi between ferro hydroxide (-925 kJ/mol) and NaCl (-411,153 kJ/mol) too high [10, 15].

Table 1. Performance of the generator cell over concentration variation of NaCl solution

NaCl Concentration (%)	Performace of the cell (Volt)	Performace of the cell by LED bulb (Volt)	Eficiency (%)
1	3.141	0.406	13.533
3	3.386	1.148	38.267
5	3.443	1.654	55.133
7	3.457	1.963	65.433
9	3.679	2.128	70.933

There were potential different of both electrodes, as can been seen on multitester equipment. This generator cell generated electric power around 3.679 Volt maximally under condition and it also unconnected with LED bulb and at 9 % NaCl solution applied. The generator cell produced an electric power 2.218 Volt under condition the generator cell connected by LED bulb at 9 % of NaCl solution. The cell was produced the lowest electrical power of 3.141 Volt at 1% of NaCl solution. It indicated tthat there was electrical current in the electricity generator cell.

The longest time LED bulb of 4 hours under condition at 9 % of NaCl solution and 8 pair of electrodes were applied. After 4 hours, LED bulb will suddenlyoff, because surface of both electrodes i.e, surface of ferrous metal 1020 was corrupted by reaction and the ferrous alloys of the printed circuit board scrap was covered by white solid residue, so there is no reaction and also there are no electron current among both of electrodes. In addition, from 1% until 3% of NaCl electrolyte solution the LED bulb was not on.

3.2. Study the reaction of the electrical generator cell

In the electrical generator cell designed and made electrochemical reactions occur. In the electrolyte solution used, each molecule of sodium chloride (NaCl) breaks into one positively charged sodium ion (Na^+) and a negatively charged chloride ion (Cl^-).

$$NaCl + H_2O \rightarrow Na^+ + Cl^- + H^+ + OH^-$$

Each ion Cl^- located near the iron metal (Ferrous 1020) or a pure iron plate will unite with one pure iron atom (Fe) into ferrous chloride compound ($FeCl_3$) while releasing three electrons.

$$Fe + 3\,Cl^- \rightarrow FeCl_3 + 3\,e-$$

The ion hydrogen from the decomposition of the NaCl solution in water will be drawn by a ferrous oxide plate (Fe_2O_3), receiving three electrons and uniting with one oxygen atom to form a water molecule (H_20).

$$H^+ + 3e + O_2 \rightarrow H_2O$$

In this process, there is an electron capture of Fe_2O_3 compound (so it becomes positive) and gives the electron on the pure Fe plate (so it becomes negative), this process results in an electric potential difference between the two poles of the electrode. The process occurs simultaneously, the chemical reaction is expressed as follows:

$$4\,Fe + 4\,H^{+} + \ 3e \ + 4\,O_2 \rightarrow 2\,Fe_2O_3 \ + 2H_2O$$

During the discharging process of $FeCl_3$, this will be reducing the reactivity of the NaCl electrolyte liquid. The resistance between the poles is very weak for practical use. In the chemical process during charging occurs after the battery is weak (unable to supply the current when the vehicle is started).

The battery condition can be restored to its original state by providing an electric current in the opposite direction of the current that occurs during discharging [14]. In this process, the water molecule decomposed and ions of the hydrogen pairs close to the negative plate unit with Cl^- ions on the negative plate to form hydrochloric acid molecules HCl. The free oxygen ion with each Fe atom on the positive plate formed Fe_2O_3. The chemical reactions that occur are;

$$2\,Fe + \ 8\,Cl^- + 4\ H_2O \rightarrow \ Fe_2O_3 + Na_2O + 8\ HCl$$

3.3. Determine efficiency of the generator cell

Efficiency of the electrical generator cell was calculatedusing this equation;

$$Unit\ eficiency = \frac{b}{a} X\ 100\%$$

where, b is electric power produced by the generator cell, and a is LED bulb's voltage.

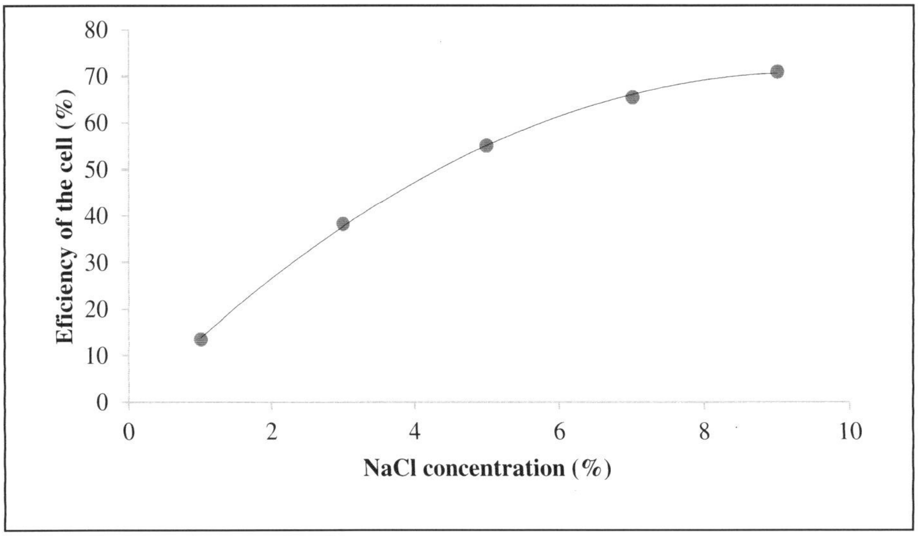

Figure 4. Efficiency of the generator cell by using LED bulb.

Overall, the increasing of NaCl content were appliedto the efficiency of the generator cell, it depicted in Figure 4. The most efficiency of the generator cell was 71 % under the system carried out by 9 % of NaCl solution. If the system operates with NaCl content over 9 %, it will be predicted that LED bulb still on. This trend is the same as the electrical power generator cell had been designed and built by Amri et. al (2015). They have been designed and built a portable electrochemical cell-based power cell which is consisted of three main components: magnesium AZ91 as an anode, MAB5A cell as a cathode, and NaCl solution as electrolyte [17]. Their cell produces an electrical power of 15.47 Volts using 10 %

of NaCl electrolyte solution. However, both electrodes are used to decay the surface rapidly due to their interaction with NaCl solution.

Corrosion is a natural process. The whole process will naturally move toward the lower energy [11, 12. 13, 14, 16, 17]. So iron and steel have a tendency to join elements or other chemicals to move to the lower energy. The principal disadvantage of many ferrous alloys is their susceptibility to corrosion [15, 16, 17]. For example, iron and steel will often bind to oxygen to form FeO or Fe_2O_3 iron oxide compounds called rust iron. Rust on iron and steel have the same chemical arrangement as iron ore FeO or Fe_2O_3 [20].

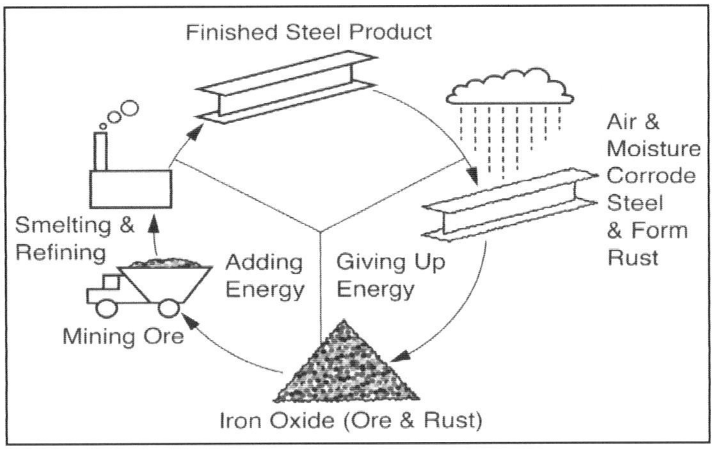

Figure 5. Corrosion Cycle on iron and steel (Adopted from Jenkins, 2005 [20])

Figure 5 illustrates the cycle of corrosion refining of iron and steel [20]. Metals and other alloys under high-energy conditions, they formed metal resistant to corrosion formed passive layer (usually oxide) on the surface. Stainless steel, aluminum alloys and titanium are metals with high-energy conditions. However, the passive layer form on its surface causes the relative resistance to corrosion. Especially in the case of stainless steel and aluminum alloys this layer is not immune to the entire natural environment and can be damaged in one or more special environments. Passive layer damage often progresses is very quickly, localized corrosion is caused by electrochemical activity part of the surface that remains passive.

4. Conclusions

The performance of the electrical generator cell is affected by variation concentration of NaCl solution. The performance of this generator cell was determinedusing the multitester equipment and Light Emitting Diode (LED) bulb. LED bulb was light on for 4 hours by using 9 % of NaCl solution and 8 electrode pairs were applied in this condition. This generator cell was produced the electric power of 3.141 – 3.679 Volt. The most efficiency of the electricity generator cell (71 %) is the system carried out by 9 % of NaCl electrolyte solution.

References

[1] Cheng Y L, Qin T W, Wang H M and Zhang Z 2008 Comparison of corrosion behaviours of AZ31, AZ31, AZ91, AM60and ZK60 magnesium alloys, *College of materials science and engineering* Hunan university Changsha 410082, China and Department of Chemistry Zhejiang University

[2] Cui L and Zhang L 2008 Metallurgical recovery of metals from electronic waste: A review *J. Hazard Mater.* **158** 228-256

[3] Susetyo F B, Qadri M, Maghfurah F and Yulianto S 2013 Analysis performance of metal iron

corrosion by rust remover X *Spektrum Ind.* **11** 117-242
[4] Kinoshitaa, Alkitaa S, khobayashib N, Niib S, Kawazumih F, and Takhashi K2003 Metal recovery from non-mounted printed wiring boards via hydrometal lurgical processing *Hydrometallurgy* 69 p73-79
[5] Oleszek S, Grabda M, Shibata E and Nakamura T 2013 Distribution of copper, silver and gold during thermal treatment with brominated flame retardants *Waste manage.* **13** 1835-1842
[6] Syed S 2012 Recovery of gold from secondary sources - A review *Hydrometallurgy* **115-116** 30-51
[7] Widmer R, Oswald-Krapf H, Sinha-Khetriwal D, Schnellman M and Boni H 2005 Global perspective on E-waste *Environ. Impact Asses.* **25** 436-458
[8] Riyanto 2013 *Electrochemical and it's application* 1st ed. (Yogyakarta: Graha Ilmu Press)
[9] Khopkar S M 1990 *Basic Analytical Chemistry Concept (*Jakarta: Indonesia University Press)
[10] Lide D R 2005 *CRC Handbook of Chemistry and Physics* 86th ed. (Boca Raton, FL: CRC Press)
[11] Keenan C W 1980 *Chemistry for University* 6th ed. (Jakarta: Erlangga Press)
[12] Petrucci R H 1985 *Basic chemistry, Principal and Modern Applied* 4th ed. (Jakarta: Erlangga Press)
[13] Oxtoby D W1999 *Chemistry Modern Principal* 4th ed. (Jakarta: Erlangga Press)
[14] Kaloko B S 2009 Acid battery modelling as energy alternative resources. *Skripsi*, Electrical Engineering Department of Jember University
[15] Callister Jr. WD 2007*Materials Science and Engineering, An Introduction* 7th ed. (New York: John Wiley & Sons, Inc.)
[16] Park Y Jand Fray D J 2008 Recovery of high purity precious metals from printed circuit boards *J. Hazard. Mater.* **164** 152-158
[17] Zhang Sand Forssberg E 1997 *J. Waste Manage. Resour. Recov.* **3** 157-167
[18] Tuncuk A Stazi V Akcil A Yacizi E Yand Deveci H 2012 Aqueos metal recovery techniques from e-scrap.
[19] Amri I Sahan Y Martina and Sari J M 2015 Determination performance of the electric generator portable based on NaCl concentration variation *Proc. 1st.Intl. Conf. Oleo and Petrochemical Engineering 2015* Pekanbaru Indonesia
[20] Jenkins, James F, Drisko,and Richard W 2005 *Eur. J. Mineral.* **2** p 63-77

IOP Conf. Series: Materials Science and Engineering **345** (2018) 012039 doi:10.1088/1757-899X/345/1/012039

Production of palm frond based wood plastic composite by using twin screw extruder

Russita M and Bahruddin
Chemical Engineering Departement, Riau University
Kampus Binawidya Jl. HR. Subrantas KM 12,5 Simpang Baru, Panam, Pekanbaru, 28293, Indonesia

E-mail: russitam@yahoo.com; bahruddin@lecturer.unri.ac.id

Abstract. Wood plastic composite (WPC) is the blending product from wood as filler and polymer thermoplastic as matric. Palm frond waste is a material with selulose about 68%, so it has potential to be developed as raw material for WPC. The purpose of this research was to learn how to produce WPC based on palm frond use twin screw extruder. It used popropilen as matric. As for aditif, it used*Maleated Polypropilene* (MAPP) as *compatibilizer* and paraffin as *plasticizer*. The size of palm frond is 40 – 80 mesh. WPC is made from blending polipropylene, palm frond, MAPP and paraffin with dry mixing method in room temperature. Then, PP, Palm frond and additive from dry mixing is fed into twin screw extruder at 190°C and 60 rpm. It use palm frond/polypropylene 60/40, MAPP 5% w/w and paraffin 2% w/w. From the result, it shown that WPC based on palm frond met the standards forcommercial WPC. It has tensile strength up to 19.2 MPa, bending strength 43.6 MPa and water adsorption 0,32% w/w. So, WPC based on palm frond has prospective to be developed for commercial WPC.

1. Introduction

Wood plastic composite is composite product from blending wood and thermoplasstic. The development of WPC is driven by the need for solid wood materials that are durable but still environmentally friendly. So far, preserved wood using chomated copper arsenate (CCA) is widely used for a variety of exterior needs, such as docks, wood fence, playgrounds and others, especially in America and Europe. However, since 2003, CCA for exterior is prohibited by the world [1]. So that the alternative wood are required for durable and environmentally friendly. Wolcott and Englund, 1999 [2] finally get WPC as an alternative material to replace the use of wood that is durable and can be recycled so that more environmentally friendly.

WPC consists of thermoplastic polymer as matrix and wood fibers as fillers. Basically, all high cellulose-containing biomass can be used as a filler on WPC manufacturing. Wood is used as filler because it has the highest cellulose composition than other biomass. Palm frond has cellulose just like wood, about 68.3% [3]. Now, palm frond is waste without further processing. Waste of palm frond reaches 6.3 ton/ha/day [4]. Large amount and high cellulose provide large potential for palm frond waste to be developed as filler for WPC.

WPC production used pellet thermoplastic that has density 500 kg/m^3. Instrumentation for plastic production has problem about wood fiber, because wood fiber has low density. The next ploblem is temperature process.Plastic has high melting points, but wood fiber will be degraded althought it used the lowest melting point of plastic.

High water content in wood or other natural fibers is also a problem for plastic industry. Usually, instrumentation of plastic industryused material with 1 – 2 % water content. Althought some instruments be equipped with equipment, it can remove water vapor, it can not remove water vapor from wood fiber that has 5 – 7 % water content. The resin dryer for plastic is not suitable for wood fiber and it has risk of fire.

Some problems begin to be solved with the development of technology and additives. So, it produces WPC that has high resistance, durable and can be recycled. Currently, WPC is experiencing significant growth. Since its introduction in the american market in 1990s, WPC production has increased 300% in a decade [5].

Some of WPC research that has been done, such as Chen et al, 2006 [6] has made WPC from recycled wood waste and HDPE with soybean oil as plasticizer. Nzokou at al, 2006 [7] has made WPC from oak and PVC. Shi et al. 2008 [8] also made WPC from pine particle and PP with additive carbon nanofibers and maleated polypropylene (MAPP). Jarusombuti and Ayrilmils, 2011 [9] also made WPC from rubber wood powder and PP with MAPP as additive. Bahruddin et al. 2013 [10] also made WPC from palm rod particle and polypropylene. The results of these research show that the content and size of biomass particle affects of carachteristic of WPC.

WPC that have been commercially developed are generally based on pine wood and other types of fibrous one. Palm frond as material for WPC has not been developed because of relatively low quality of wood. Ussually, palm frond waste will be burned to produce ash for plant fertilizer or just stacked among palm tree, so it is not economically profitable. Based on compotition, palm frond waste has considerable potential for further processing, especially for WPC [3].

Some researchers are specifically discussing WPC using palm fiber such as Fadly et al, 2015 [11] that made from palm frond and polypropylene with MAPP as compatibilizer. Parsaulian et al. 2015 [12] made WPC from palm frond with pretreatment oxalic acid solvent and polypropilene.

For this research, it will study the production process of WPC based on palm frond and polypropylene. For additive, it uses MAPP as compatibilizer and paraffin as plasticizer.

2. Research methodology

2.1 Material

Material for this reseach are palm frond as filler from local plantations; Polypropylene (PP) HE 2.0 TF produced by PT Chandra Asri Petrochemical Tbk as matrix; MAPP produced by Aldrich Chemistry, USA as compatibilizer; and paraffin wax as plasticizer. It used pretreatment for Palm frond particle use oxalid acid as solvent.

2.2 Filler preparation

The palm frond cleaned from leaf, crushed using sawmill to produce particle. Particle of palm frond macerated by 0.01 M oxalic acid in autoclave at 120 ^{o}C for 15 minutes. Particle from maserate process strained and dried at sunlight for about 24 hours. Particle palm frond is dried again with oven at 105 ^{o}C for 2 hours for maximum drying, so water content no more than 5% w/w. The dry palm frond is crushed again and stir so it get palm frond powder size 40 – 80 mesh.

2.3 WPC blending

WPC blending needed four steps, dry mixing, blending at twin screw extruder, pelleting and pressing. The first step, palm frond powder and polypropilene ratio 60/40 mixed with MAPP 5% w/w and paraffin 2% w/w with dry mixing method at room temperature. The second step, the result from dry mixing wasfed on twin screw extruder at 190 ^{o}C and 60 rpm motor speed. The result of this process was a long cylindrical WPC. WPC ferom extruder need cooling water because hot WPC has low tensile strenght. The next step, WPC from cooling water was cut into pallet shape using pelleting unit. Water content of WPC from cooling water caused low tensile strength of WPC. So, it needed to be dried at 60 ^{o}C for one night. Pallet WPC from this process can be molded or pressed to become final product.

The last step for WPC production was pressing process. Pressing process for WPC used hot press at 190 ^{o}C for 5 – 10 min with 5 bar pressure. WPC from hot press can be cut for tenting. For this process, it used weather test, QUV for 108 hours.

3. Results and discussion

3.1 Mechanical properties

WPC producedin this reseach is a mixture of palm frond and polypropylene as matrix of palm frond as filler and polypropylene as matrix ratio 60/40 with MAPP 5% w/w as compatibilizer and paraffin 2% w/w as plasticizer.

Technically, all types of wood can be used as raw material for WPC, but wood from hardwood trees are better than softwood trees. Palm frond is one of softwood, so the fiber from palm frond is random fiber.

60% w/w palm frond particle gave 3x volume than 40% w/w polypropylen. This made problem for mixing process. Polypropylene as matrix has too small volume compared to palm frond as filler. So, some of the palm frond particle can not be mixed perfectly with polypropylene. This caused mixed compatibility to be weak.

The problem of mixing plastics and wood are the consistency of wood that has low density into the small pore of plastics at extrussion process. Another problem is plastic's melting point is too high for wood fiber. Wood fiber will be degraded at high temperature. Ussually, the water content of wood particle is high, more than 1 – 2 %.

Wood particle should be dried before use for WPC, and water content will evaporate at high temperature and put pressure on. Bubbles will also provide an empty space that will reduce mechanichal strength. Wood particle could be dried in various ways such as rotary drum drier. Water content may also remove when extruding with venting. WPC is thermoplastic composit from wood particle or fiber from other sources. Usually, USA used wood material, europe used natural fiber such as jute, hemp and kenaf, and Asia used rice husks and bamboo fiber. Wood mixing can be expressed in low aspect ratio such as wood flour also high ratio such as wood fiber. Althought wood fiber have problem in process, but it better tensile propertise.

MAPP as compatibilizer and paraffing as plasticizer help to overcome mixed deficiency of polypropylene and palm frond. Paraffin serves as lubricant that facilitates the mixing process. MAPP works to modify prolypropylene surfaces to provides radical surfaces. So, palm frond particle can be more easily mixed with popypropylene. So, the mixture becomes more compatible.

WPC with MAPP will provides better physical properties than WPC without MAPP. MAPP for WPC increased the WPC mechanichal properties. This is confirmedthat the increasing ofthe compatibilizer strength of the interface bonds with chemically modified. One of compatibilizer is MAPP [13]. However, excessive use of MAPP will reduce the mechanichal properties of WPC [14]. This is due to the lack interactions of OH from MAPP with palm frond particle and more likely to interact with PP [15]. This is consistent that excessive additive may decrease the machanical properties of the composite [16].

WPC production used twin screw extruder for blending process at 190°C and 60 rpm rotor speed. Melting point of polypropylene is about 170 °C, so it needed high temperature. However, at 150°C, palm frond particle as filler will be burn. This causedthe WPC blend from this process to be black. Fe_2O_3 as color for WPC also not work because the burning black color of palm front particle defeats the yellow color of FeO_3. So, it is prefer to make WPC without using dyes. Although the palm frond particle is burnt, but the result of this process is WPC with excelent properties, exceeding the minimum standart of WPC properties for commercial. Table 1 is shown the a comparison of results test from WPC generated to standart of commercial WPC based on Kronotex, 2017 [17].

Table 1. Comparison of WPC test result with commercial WPC

Test	WPC SPS/PP	WPC Standards
Tensile Strength	19,2 Mpa	≥ 15 MPa
Bending Strength	43,6	≥ 30 MPa
Water absorption	0,32 %	≤ 2%

3.2 Weather resistance

Weather testing use QUV for WPC at 60°C for 8 hours and condensate at 50°C for 4 hours. Testing have done for 108 hours or 9 cycle.

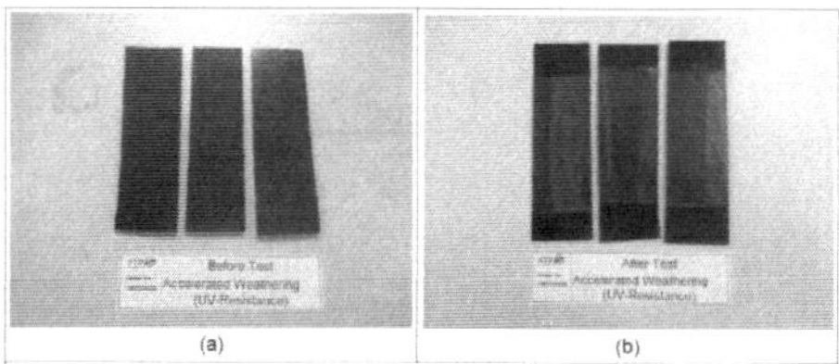

Figure 1. WPC Sample, (a) before, and (b) after weather test (UV-Resistance).

Figure 1 shown the color changed of WPC sample. Irradiance at 340 nm is 0,76 W/m^2. Changed color of WPC shown that mechanical properties of WPC have been decrease for weathering test 108 hours and 9 cicle.

The fast deterioration of weather is loss of wood. For outdoor aplication, the change of color and texture will increase. It made tensile strength of wood decrease. Usually, the cause of wood damage is weather (sun light and rain), dust, temperature and air humidity, and organism (fungi or microba). Sunlight with UV is dominant factor for depolimerization process of lignin and rain is perfect process [18].

Fotodegradation for sintetic polymer of polyolefin such as polypropylene get stimulation from polymer-oxigen complex, expecially because of catalyst, hydroperoxide, carbonyle, and double bond of polymer formation[19; 20]. Polymer degradation of fotooxidation will decrease mechanical propertise of WPC such as tensile strength and bending strength.

The same thing happened to the wood that will be degraded because of fotochemical at outdoor. The research shown that the weather damage to wood is a process at surface wood, involve photoinduced that damage lignin into product that can react with water (loss lignin) [20]. This causedfunctional group generation of chomophores sach as carbonyl, carboxylic acid, quinone and radical hydroperoxide.

Johnson at al,1999 [21] found that if WPC was used for outdoors, it would be open to UV radiation, air humidity and microorganism. Simonsen, 1996 [22] found that WPC or other biofillers with thermoplastics can not withstand the effects of outdoor exposure. decline in properties is particularly noticeable in stiffness and the influence of summer and cold on the composite mechanical properties. Some examples studied were cracked and bent.

The influence of tropical weather in Indonesia on wood-plastic composites has been done [23]. After removal of discoloration on the composite surface directly exposed to UV and decreased the mechanical properties of the composite.

4. Conclusion

From the result of the reseach, it can be concluded that WPC production based on palm frond waste with matrix polypropylene has potential to be developed as commercial WPC, but it need MAAP as compatibilizer and paraffin as palsticizer to get excelent WPC.

References

[1] Lebow S 2004 Alternative to chromated copper arsente (CCA) for residential contruction. *Proc. Environ. Impact Presentative Treated Wood Conference Orlando*, 8–10 February 2004, Wisconsin

[2] Wolcott M P and Adcock T 1999 New Advaces in wood fiber-polymer formulation *Proc. Wood-Plastic Conf. Plastics Technol. Magazine Polym. Proc. Comm.* 107-114

[3] Fauzi I, Bahrudddin, and Irdoni H S 2016 *JOM FTEKNIK* **3** Oktober 2016

[4] Badan Pusat Statistik 2017 *Luas Tanaman Perkebunan Menurut Jenis Tanaman*

[5] Eckert C 2000 Opportunities for natural fibers in plastic composite *Proc. Progr. Woodfibre-Plastic Composite*, Toronto Canada
[6] Chen H C, T Y Chen and C H Hsu 2006 *Holz als Roh-und Werkstoff* **64** 17
[7] Nzokou P, J Freed and D P Kamdem 2006 *Holz als Roh-und Werkstoff* **64** 90
[8] Shi J, J Zhang C U, Pittman Jr, H Toghiani and Y A Xue 2008 Holz als Roh-und Werkstoff **66** 313
[9] Jarusombuti S and Ayrilmils N 2011 *Eur. J. Wood Prod.* **69** 375
[10] Bahruddin R, Efrizal, Helwani Z, Zulfansyah and Utami S P 2013 Effect of blending process condition on morphology and properties of palm oil trunk based wood plastic composite *Proc. Innov. Polym. Sci. Technol. (IPST)-Asian Polymer Assotiation (APA) Conf.*, Yogyakarta
[11] Fadly N A T, Irdoni HS and Bahruddin 2015 *JOM Bidang Teknik dan Sains*
[12] Parsaulian A S, Bahruddin and Helwani Z 2015 *JOM FTEKNIK* **2** Februari 2015
[13] Li X L, Tabil G and Panigrahi S 2007 J. Polym Environ. **15** 25
[14] Sombatsompop N, Yotinwattanakumtorn C and Thongpin C 2005 *J. Appl. Polym. Sci.* **97** 475
[15] Harper D P 2003 *A thermodynamic, Spectroscopic, and Mechanical Characterization of the Wood-Polypropylene Interphase* Washington State University Dept. Civil Environ. Eng.
[16] Schneider M H and K I Brebner 1985 *Wood Sci. Technol.* **19** 67-73
[17] Kronotex 2017 Wood plastic composite's parameter. www.kronotex.gr/wp-content/uploads/technical-data.doc. Accessed 9 Maret 2017
[18] Sudiyani Y Y, Imamura Y, Doi S and Yamamuchi S 2003 J. Wood Sci. **49** 86-92
[19] Stark N M and L M Matuana 2002 *ANTEC 2002* 2209-2213
[20] Matuana L M, Kamdem D P and Zhang J 2001 *J. Appl. Polym. Sci.* **80** 19943-1950
[21] Johnson D A, Urich J L, Rowell R M, Jacobson T and Caufield D F 1999 Weathering characteristics of fiber-polymer composites *Proc. Fifth Intl. Conf. Woodfiber-Plastic Comp.* Forest Prod. Soc., Madison WI
[22] Simonsen J 1996 *The Mechanical Properties of Wood Fiber Plastic Composite: Theoritical vs Experimental.* Forest Product Society, Madison p. 134-143
[23] Sulaeman R 2003 *Deteriorasi Komposit Serbuk Kayu Plastik Polipropilena Daur Ulang oleh Cuaca dan Rayap.* Tesis Program Pascasarjana Institut Pertanian Bogor.

Synthesis of hydroxyapatite nanoparticles from egg shells by sol-gel method

Y Azis, M Adrian, C D Alfarisi, Khairat and R M Sri
Department of Chemical Engineering, Riau University, Pekanbaru 28293, Indonesia

E-mail: yelmida.azis@gmail.com

Abstract. Hydroxyapatite, [$Ca_{10}(PO_4)_6(OH)_2$, (HAp)] is widely used in medical fields especially as a bone and teeth substitute. Hydroxyapatite nanoparticles have been succesfully synthesized from egg shells as a source of calcium by using sol-gel method. The egg shells were calcined, hydrated (slaking) and undergone carbonation to form Precipitated Calcium Carbonate (PCC).Then the PCC was added $(NH_4)_2HPO_4$ to form HAp with variation the mole ratio Ca and P (1.57 ; 1.67 and 1.77), aging time (24, 48, and 72 hr) and under basic condition pH (9, 10 and 11). The formation of hydroxyapatite biomaterial was characterized using XRD, FTIR, SEM-EDX. The XRD patterns showed that the products were hydroxyapatite crystals. The best result was obtained at 24 hr aging time, pH 9 with hexagonal structure of hydroxyapatite. Particle size of HAp was 35-54 nm and the morphology of hydroxyapatite observed using SEM, it showed that the uniformity crystal of hydroxyapatite.

1. Introduction

Hydroxyapatite, $Ca_{10}(PO_4)_6(OH)_2$, is a compound with hexagonal structure and white solids. Hydroxyapatite is known as one of the ingredients with bioactive, biocompatible and osteoconductive properties that chemically resemble for bone and also for tooth mineral components [1, 2]. Therefore, hydroxyapatite is widely used in the field of biomedical as filler and coating on bone and dental implants. Hydroxyapatite is also widely used as a catalyst and adsorben because their structure is porous and also has heat resistance [2, 4].

Synthetic hydroxyapatite can be prepared using basic ingredients of calcium from chemicals such as Ca $(NO_3)_2$ [4], Ca $(OH)_2$ [5], $CaCO_3$ or from natural-made materials such as limestone [6] and bioanorganic materials such as bone [7,8], cockle shells[2] shellfish [9,13], coral or egg shell [5, 6, 14, 16]. In this study the synthesis of hydroxyapatite was done using the eggshell as base material. The selection of egg shell as a source of calcium is due to abundant availability of egg shells in Indonesia.

In the previous work, we have synthesized hydroxyapatite from the cockle shells' PCC using hydrothermalmethod [2]. Precipitated Calcium Carbonate (PCC) is a calcium carbonate compound ($CaCO_3$) that can be processed from the material through a series of chemical reactions. PCC particles' is homogeneous in the same size with micro-scale particle [10,11]. The common methods used to produce synthetic nanoparticle HAp including the precipitation [18], hydrothermal [2,16], mechanochemical [17] and sol-gel [8,12]. Sol-gel method offers some advantage in the molecular-level of the calcium mixing andphosphorusprecursors,itiscapable to improve the results of the chemical homogeneity HAp to a significant extent, rather than conventional techniques [8].

Agrawal et al. [8] used the sol–gel process in Hap synthesized with $Ca(NO_3)_2.4H_2O$ as calcium source and P_2O_5 as phosphate sources. Ca/P ratio of 1.67 with the aging time variation of 10-15 hr. Anuar et al.

[3] performed hydroxyapatite preparation using sol gel method of $Ca(NO_3)_2.4H_2O$ as a source of calcium and P_2O_5 as a phosphate source with a Ca/P mole ratio of 1.67. some variations used are made with the speedof 100-500 rpm and temperature of 600-800 °C. In the present work, the synthesis of a nanoparticle HAp powder via sol-gel method was prepared using the egg shells' PCC. To the best our knowledge,there are no reports aboutthe synthesis HAp from the egg shells' PCC using sol-gel method.

2. Materials and methods

2.1 Materials preparation

The egg shells was collected from Pekanbaru, Riau; diammonium hydrogen phosphate $(NH_4)_2HPO_4$), Merck); 2M HNO_3; ammonium hydroxide (NH_4OH) 65% (Merck); CO_2 and aquadest. The egg shells samples were washed and cleaned, then sun dried for two days followed by drying oven at 110°C for 2 hr. Dried and cleaned egg shells were then crushed and grounded with blender into egg shells powder form.

Figure 1. Chicken egg shells.

2.2 The formation of precipitated calcium carbonate (PCC)

Azis et al. [2] reported about the procedure for PCC formation from eggshells using the carbonation modification method. The PCC results are then used as the raw material for the synthesis of hydroxyapatite using the sol-gel method. PCC has a low solubility in water, so in this study PCC was dissolved using 0.3 M HNO_3.The formation of PCCcan be represented by the following reactions [10, 11]:

Calcination:	$2CaCO_3$ + Heat	→	$2CaO + 2CO_2$
Hydration:	$CaO + 2HNO_3$	→	$2Ca(NO_3)_2 + H_2O$
	$Ca(NO_3)_2 + 2NH_4O_3$	→	$Ca(OH)_2 + 2NH_4NO_3$
Precipitation	$Ca(OH)_2 + CO_2$	→	$CaCO_3 + H_2O$

2.3 Synthesis of hydroxyapatite

Vijayalaksmi and Rajeswari [19] procedures were used for synthesis of hydroxyapatite by sol-gel method. The egg shells' PCC and $(NH_4)_2HPO_4$ were mixed by varying aging time (24, 48, 72 h), mol ratio of Ca and P (1.57; 1.67 and 1.77) and pH (9, 10, 11). PCC was dissolved in 0.3M HNO_3, while (NH_4) $_2HPO_4$ was dissolved with aquadest. Furthermore, PCC solution and $(NH_4)_2HPO_4$ solution were mixed in beaker glass. The solution was stirred for 3 h at a stirring speed of 300 rpm until the gel formed. Then, the aging process was done according to the adjusted variable.After the aging process, the formed gel is dried in an oven at 80 °C for 24 hr. The obtained solids were cleanly washed by distilled water to separate the hydroxyapatite from the rest of the reactants and dried out for 3 h in the oven. Finally, the obtained solidswerecalcinated for 1 hour at 500 °C

The chemical reaction is shown as follow :

$$10Ca(NO_3)_2.4H_2O + 6(NH_4)_2HPO_4 + 8\ NH_4OH \rightarrow Ca_{10}(PO_4)_6(OH)_2 + 20NH_4NO_3 + 20\ H_2O$$

2.4 PCC and HAp characterization

The mineralogy of PCC and HAp were characterized using X-ray diffraction (X'Pert Powder DY 3688) with Cu Kα radiation. The surface morphology was probed using scanning electron microscopy(SEM) linked to energy dispersive X-ray micro analysis (EDX) (JEOL JED 2300).The Fourier Transform Infrared Spectroscopy (FTIR, Perkin Elmer Spectrometer Frontier) was used to analysis bonding structure inthe samples. The surface area of the hydroxyapatite were characterized using BET measurements, Surface Area Analyzer (SAA, Quantachrome NovaWin2).

3. Results and discussion

The synthesis of Precipitated Calcium Carbonate (PCC) was acquired through the modification of the carbonation method. The comparison of XRD pattern among of obtained PCCand the standard calcium carbonate (ICDD – 01-083-1762 (rhombohedral calcite)) are shown in Figure 2 (a-b). The characteristic peak with the highest intensity of the obtained PCC (Figure 2a) matches with the pattern of calcium carbonate standard and the pattern also clearly shows the PCC is the only phase.

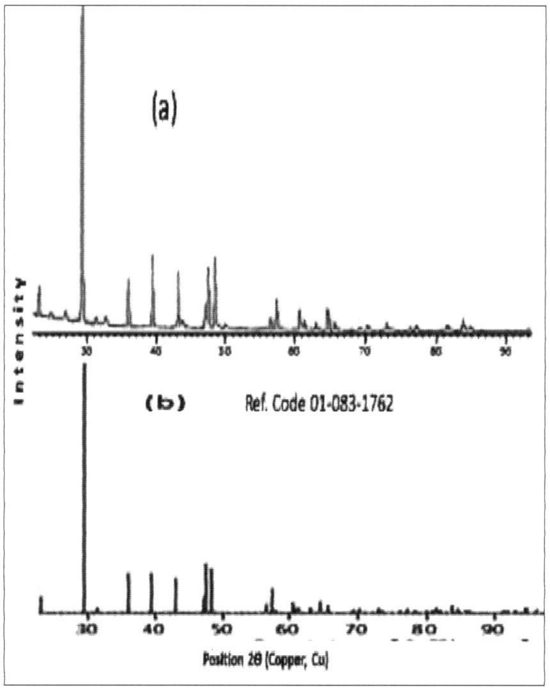

Figure 2. XRD diffractograms of product: (a) PCC of egg shells, (b) calcite, Calcium Carbonate standard (ICDD 01-083-1762).

At the synthesis stage of HAp from PCC egg shell by sol-gel method, a very similar pattern of FTIR spectra was obtained for all synthesized compounds and show PO_4^{-3} groups at 1025-1029 cm^{-1} and no ammonia or NH_3 group uptake at 3500 cm^{-1} wave numbers. The result of characterization with FTIR spectroscopyat varied mole rasio Ca/P (1.57; 1.67; 1.77), pH 9 with aging time 24 h rare depicted in Figure 3.

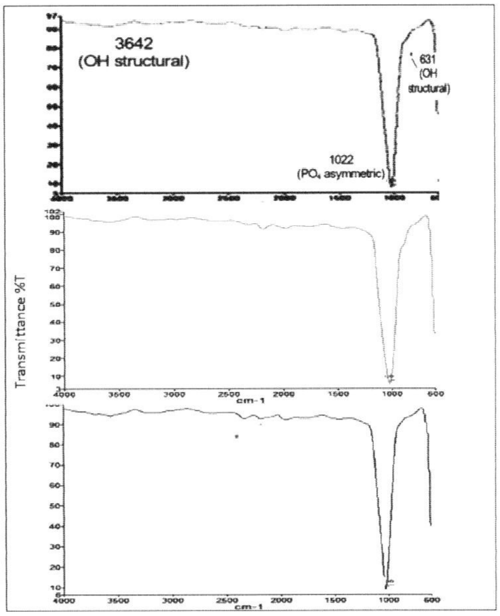

Figure 3. FTIR spectrum of hydroxyapatite at varied mole rasio Ca/P (1.57; 1.67; 1.77), pH 9, aging time 24 hr.

X-ray diffraction analysis of synthetic hydroxyapatite showed a spectral pattern similar to the standard hydroxyapatite spectrum of ICDD 01-089- at 2θ value: 25.8°, 31.9°, 32.2°, 32.9°, 33.96°, 39.74°. Figure 4. shows some hydroxyapatite diffractogram of the synthesis result at 24 hr aging time, compared to ICDD standard

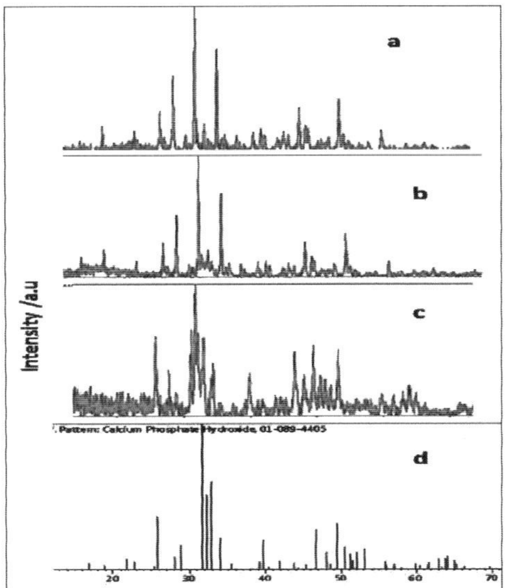

Figure 4. XRD diffractograms for the products of synthesis at varied mole rasio Ca/P at aging time 24 h (a)1.57 (b)1.67, (c) 1.77 (d) standard of hydroxyapatite (ICDD01-089-4405).

Hydroxyapatite made with the ratio of reactant of Ca / P 1.77 aging time of 24 h (Figure 4c) produces X-ray diffraction patterns very similar to the ICDD standard (Fig.4d) and there are no other crystalline phases are detected other than phase. Analysis using SEM-EDX was performed to find out the hydroxyapatite morphology of the synthesis. The results of characterization shown that the hydroxyapatite morphology in the form of agglomerates, as shown in Figure 5.

Figure 5. SEM image of hydroxyapatite compound of egg shells' PCC at varied mol rasio Ca/P at aging time 24 h (a)1.77 (b)1.67

From the surface area analysis with BET, the hydroxyapatite surface area is about 8.968 m^2 / g, at 24 hr aging time and the mole ratio of Ca / P is 1.77. The mean grain size of the crystallites of HA particles was obtained using the Scherrer's equation [20],with calculated crystal size of 35-54 nm.

4. Conclusion

Hydroxyapatite nanoparticle was successfully synthesized from egg shell via formation Precipitated Calcium Carbonate (PCC) followed with sol-gel process. The results from the analysis of compound proved that hydroxyapatite synthesis of egg shells through the formation of PCC produced the excellent result. Synthesis HAp at 24 h aging time and mole ratio of Ca/P is 1.77 achieved hydroxyapatite with hexagonal crystal structures. Furthermore, the X-ray diffraction patterns and characterization with FTIR are matched with standard HAp. The dried Hap particles powder was very pure with the specific surface area of 8.968 m^2/g

Acknowledgements

The authors would like to acknowledge the Kemenristekdikti and Universitas Riau for supporting this research through Dana Penelitian Desentralisasi and Dana DIPA UNRI.

References

[1] Nayak A K 2010 *International J. of Chem. Tech. Res.* **2** 903-907
[2] Azis Y, Jamarun N, Arief S and Nur H 2015 *Oriental J. Chem.* **31** 1099-1105
[3] Anuar A, Salimi M N A, Zulkal M, Daud M and Yee Y F 2013 Characterizations of hydroxyapatite (HAp) nanoparticles produced by sol-gel method *Intl. Conf. Adv. Mater. Eng. Technol.* Bandung, Indonesia p. 3587-3590
[4] Choi A K, Nissan B 2016 *J S M Dent. Surg.* 2016 **1** 1002
[5] Gergely G, We´ber F, Luka´cs I, T´oth A and Horva´th Z E 2009 *Ceram. Intl.* **36** 803-806
[6] Wu S H, Hsu H C, Hsu S K, Chang Y C and Ho W F 2016 *J. Asian Ceram. Soc.* **4** 85-90
[7] Stadelman W J 2000 *Encyclopedia of Food Science and Technology* 2nd ed, ed F J Francis (New York: John Wiley and Sons) p. 593-599
[8] Agrawal K, Singh G, Puri D and Prakash S 2011 *J. Miner. Mater. Charact. Eng.* **10** 727-734
[9] Sebastian A, Kumar D A and Divya S 2014 *Intl. J. Eng. Sci. Manage.* ISSN: 2277-5528

[10] Kontrec J, Ukrainczyk M, Babic-Ivancic V and Kralj D 2011 Synthesis of calcium carbonate by semicontinuous carbonation method in the presence of dextrans *Croat. Chem. Acta* **84** (1) 25-32
[11] Kemperl J and Macek J 2009 *Intl. J. Miner. Proc.* **93** 84-88
[12] Sari T K, Jamarun N, Arief S, Azharman Z and Asril A 2015 *Oriental J. Chem.* **30** 1799-1804
[13] Venkatesan J and Kim S K 2010 *J. Marine Drugs* **3** 4761-4772
[14] Denymol P V and Joseph R 2014 *Intl. J. Sci. Technol.* ISSN: 2321-919X
[15] Cimdina L G and Borodajenko N 2012 *Research of Calcium Phosphates Using Fourier Transform Infrared Spectroscopy.* InTech, Institute of General Chemical Engineering, Riga Technical University, Latvia.
[16] Hui P, Meena S L, Singh G, Agarawal R D and Prakash S 2010 *J. Miner. Mater. Charact. Eng.* **9** 683-692
[17] Yeong K C B, Wang J and Ng S C 2001 *Biomater.* **22** 2705-2712
[18] Eslami H, Solati-Hashjin M, Tahriri M 2008 *Iranian J. Pharmaceutical Sci. SPring* **4** 127-134
[19] Vijayalakshmi U and Rajeswari S 2006 *Trends Biomater. Artif.* Organs **19** pp 57-62
[20] Cullity B D and Stock S R 2013 Elements of Xray diffraction 3rd ed. (New-Jersey: Prentice Hall)

IOP Conf. Series: Materials Science and Engineering **345** (2018) 012041 doi:10.1088/1757-899X/345/1/012041

Characteristic of betel nuts activated carbon and its application to Jumputan wastewater treatment

L Cundari, K F Sari and L Anggraini
Chemical Engineering Department, Faculty of Engineering, Universitas Sriwijaya, Palembang 30139, Indonesia

E-mail: liacundari@ft.unsri.ac.id

Abstract. Wastewater from Jumputan production contains synthetic dye which is harmful to the environment. The contaminant can be reduced by adsorption process using activated carbon. The activated carbon was prepared from betel nuts with carbonization temperature of 500°C and 0.5 M HCl as an activator. Batch mode experiments were conducted to study the effect of various factors, such as the size particle of adsorbent, the dosage of adsorbent, and the contact time on Jumputan's dye adsorption. The volume of treated solution was 200 mL. This solution agitated using a Jar Test at 150 rpm. The objectives of this work were to analyze the characteristic of the betel nuts, to analyze the characteristic of the activated carbon and to determine adsorbent's ability to dye adsorption. Betel nuts compositions were analyzed with proximate analysis method. The adsorbents were carried out by SEM-EDS analysis. The dye adsorptions were analyzed with a portable spectrophotometer. The result shows betel nuts contains 60.86% carbohydrate, 32.56% water, 2.17% fat, 3.35% protein, and 1.06% ash. The major component of the activated carbon is carbon (C) of 86.27%, and the rest is Oxygen (9.18%) and Aurum (4.55%). The best condition is the adsorbent that has a particle size of 250 pm (60 mesh), the dosage of 20 grams, and the contact time of 15 minutes with dye removal of 76.4%.

1. Introduction

In general, water pollution is described as harmful substances or unexpected materials that found in the water in sufficient quantities[1]. These reduced the dissolved oxygen levels for aquatic organisms, and so water quality. Jumputan wastewater is an example of water pollution. Jumputan wastewater contains dyes of organic compounds such as napthol, indigosol, procion, erionyl, auramin, and rodhamin.

Industrial wastewater can easily be known for its color. Pollutant contamination varies both types and quantities. The most widely used dye is the napthol and the indigosol substance. Naphtol is called Azo Dyes because when combined with salt diazo a new color arises. These substances are soluble in water, contains cyclic nuclei and aniline acids compounds. Naphtol invented in 1911. It is manufactured by dyestuff factories in Europe, Japan, and China. This substance is applied in batik traditional Palembang such as Songket, Jumputan, Blongsong. Indigosol is also called Leuco Sodium Esters, if the color oxidized it becomes an insoluble solution and gives the real color. The process used sodium nitrite and acids. The dye wastewater in the textile industry, as well as Jumputan, produce from dyeing and washing process.

One method to overcome water pollution is adsorption. Activated carbon is the most widely used adsorbent. The raw materials of the activated carbon were rice husk ash [2], coalite [3], banana stem [4], bark peel [5], betel nuts [6-9], and others. The previously activated carbon that utilized in Jumputan wastewater treatment were coalite [3], the shell of kemiri [10], coffee powder [11], and betel nuts [6-9]. In general, factors that affect adsorption process are the surface area (particle size), the type of adsorbent, the adsorbate molecular structure, the concentration of adsorbate, temperature, pH, stirring rate, and contact time.

The primary component in Jumputan wastewater is a synthetic color that is utilized. The study of dye removal from Jumputan wastewater by using betel nuts activated carbon is prior being done. Considered factors such as particle size, weight of adsorbent, and contact time were varied. The aims of this study were to analyze the characteristic of betel nuts and the activated carbon, and to investigate the adsorbent's ability to adsorb dye contaminant on Jumputan wastewater in batch mode.

2. Research methodology

The furnace, oven, analytical balance, mortar, beaker glass, measuring cup, funnel, desiccator, and pH meter were used in this work. Materials were betel nuts, HCl 0.5 M, Jumputan wastewater, and aquadest.

The activated carbon was produced from betel nuts with carbonization temperature of 500°C and 0.5 M HCl as an activator [6]. Betel nuts compositions were analyzed with proximate analysis method. The adsorbents were carried out by Scanning Electron Microscope (SEM) - Energy Dispersive Spectroscopy (EDS) analysis. The dye adsorptions were analyzed with a portable spectrophotometer. Variation in particle size, dosage of adsorbent, and contact time were applied to investigate dye adsorption. The volume of Jumputan wastewater was 200 mL. This solution was agitated using a Jar Test at 150 rpm. The size of the adsorbent was 1 mm (18 mesh) and 250 pm (60 mesh). The dosage of the adsorbent was 5, 10, 15, 20, and 25 grams. The contact time was 5, 10, 15, 20, and 25 minutes.

3. Result and discussion

Betel nuts is an ornamental plant that isn't utilized yet. It has a hard texture that can be used as a raw material to produce activated carbon. The characteristics of betel nuts by using proximate analysis method are presented in Table 1.

Table 1. Proximate analysis of betel nuts[a]

Composition	Percentage (%)
Water	32.56
Fat	2.17
Protein	3.35
Ash	1.06
Carbohydrate	60.86

[a] Chemical Analysis Laboratory, Faculty of Agriculture, Universitas Sriwijaya, 2017

Carbohydrate's content on betel nuts is a prime source of carbon. It produces to become activated carbon with carbonization and chemical activation process based preliminary research procedure [6]. The dried betel nuts were carbonized in an electrical furnace at temperature of 500°C and then activated with 0.5 M HCl [6]. Betel nuts that become activated carbon are 22.6%wt from its original. Highly water content on betel nuts effects the quality and quantity of the activated carbon produced. This value is lower than coconut shell activated carbon which was 30-35%wt [12]. However, betel nuts is still a promising raw material consider high ability to adsorb COD, BOD, TSS, and dye in Jumputan wastewater [6-9].

Betel nuts activated carbon was characterized by Scanning Electron Microscope (SEM) - Energy Dispersive Spectroscopy (EDS) analysis. It presents topography (surface texture), morphology, porosity, and composition of the blank adsorbent. SEM micrographs of betel nuts activated carbon are shown in Figure 1. The adsorbent exhibits a caves-like, porous, uneven, and rough surface morphology [13]. The shortest pore diameter and area are 1.564 pm and 1.92 pm^2 respectively. The longest pore diameter and area are 1.787 pm and 2.508 pm^2 respectively. Figure 2 describes the composition of activated carbon. The major component is carbon (86.27%), and the rest are oxygen (9.18%) and aurum (4.55%). In the SEM-EDS analysis, the sample should be a conductive material, so aurum is added to cover the activated carbon. The carbons tie up impurities in the Jumputan wastewater, so the dye, BOD, COD, TSS, and Crom can be removed. This similar result also showed by Cundari L., et. Al. [6-9]. Although activated carbon was the most effective sorbent, other low-cost sorbents, such as betel nuts activated carbon, could be used for color removal to substitute for high- cost adsorbent [14].

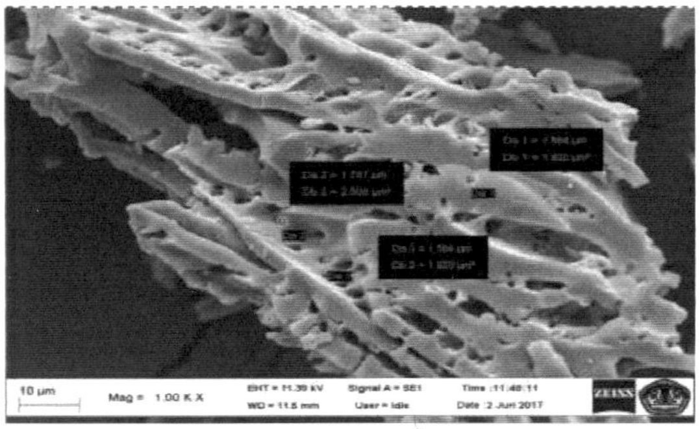

Figure 1. Scanning Electron Microscope of betel nuts activated carbon.

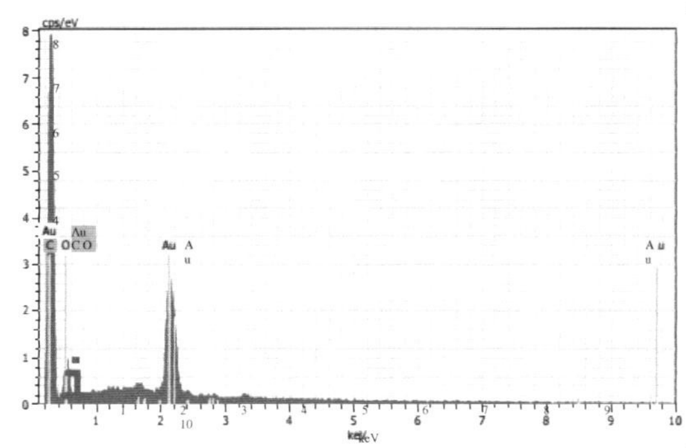

Figure 2. Composition of betel nuts activated carbon

Another analysis that applies to the adsorbent is water content and iodine adsorption. The water content of activated carbon is equal to 1.76% and 0.05% for the size of 1 mm (18 mesh) and 250 pm (60 mesh) respectively. The water content of the adsorbent has fulfilled the standard of activated carbon quality according to SII 0258-88 which the maximum value is 4.4% for activated granulated carbon and maximum 15% for powder activated carbon. The adsorption capacity of activated carbon to Iodine for is 476.95 mg/g and 593.94 mg/g for 1 mm (18 mesh) and 250 pm (60 mesh) respectively. The result of adsorption analysis of Iodine that obtains from this research has not fulfilled the standard of active carbon quality according to SII 0258-88 which is at least 750 mg/g.

Betel nuts activated carbon is applied on dye adsorption from Jumputan wastewater. The experiment conducts in batch mode. The initial dye concentration of the Jumputan wastewater is 1293 ppm. The analysis result of the standard solution is seen in Table 2 and Figure 3.

Table 2. The analysis result of standard solution[b]

Absorbance	Concentration (ppm)
1.229	1102.20
0.521	551.10
0.390	440.88
0.246	330.66
0.188	220.44
0.158	110.22

[b] Laboratory Analysis of Polytechnic Sriwijaya, 2017

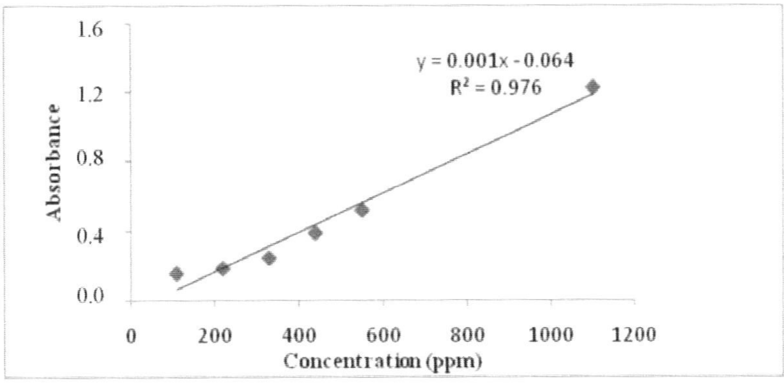

Figure 3. Standard calibration curve

Table 3. Analysis data of Jumputan's dye adsorption onto betel nuts activated carbon[c]

Dosage of adsorbent (gr)	Contact Time (minute)	Absorbance		Concentration		% Adsorption	
		1 mm	250 pm	1 mm	250 pm	1 mm	250 pm
5	5	0.729	0.553	793	616.5	38.7	52.3
	10	0.489	0.505	552.5	569	57.3	56.0
	15	0.463	0.426	526.5	490	59.3	62.1
	20	0.285	0.282	348.5	346	73.0	73.2
	25	0.304	0.242	368	306	71.5	76.3
10	5	0.374	0.404	438	467.5	66.1	63.8
	10	0.357	0.335	421	398.5	67.4	69.2
	15	0.286	0.301	350	365	72.9	71.8
	20	0.247	0.280	311	344	75.9	73.4
	25	0.275	0.278	339	341.5	73.8	73.6
15	5	0.371	0.315	434.5	379	66.4	70.7
	10	0.316	0.283	379.5	346.5	70.6	73.2
	15	0.286	0.270	349.5	333.5	73.0	74.2
	20	0.280	0.247	343.5	310.5	73.4	76.0
	25	0.264	0.232	327.5	295.5	74.7	77.1
20	5	0.454	0.297	517.5	360.5	60.0	72.1
	10	0.346	0.268	410	331.5	68.3	74.4
	15	0.281	0.241	345	305	73.3	76.4
	20	0.274	0.234	337.5	298	73.9	77.0
	25	0.256	0.219	319.5	282.5	75.3	78.2
25	5	0.506	0.258	569.5	322	56.0	75.1
	10	0.351	0.240	415	303.5	67.9	76.5
	15	0.270	0.235	334	298.5	74.2	76.9
	20	0.258	0.219	321.5	283	75.1	78.1
	25	0.241	0.214	305	277.5	76.4	78.5

[b] Laboratory Analysis of Polytechnic Sriwijaya, 2017

The linear regression that presented in Figure 3 is $y = 0.0011x - 0.0064$, where y as absorbance value and x as dye concentration (ppm). By using that equation, the absorbance data from spectrophotometer analysis is used to determine the dye concentration of Jumputan wastewater. The absorbance, dye concentration, and adsorption percentage data are shown in Table 3. The percentage of dye adsorption is determined by equation (1).

$$\% \, Adsorption = \frac{C_o - C_t}{C_o} \, x \, 100 \, \% \tag{1}$$

Where C_o and C_t are the dye concentration of Jumputan wastewater at initial and time t. The average of dye adsorption is 68.6% for 1 mm adsorbent and 72.2% for 250 pm adsorbent. The percentage maximum of adsorption is 76.4 (1 mm adsorbent) and 78.5 (250 pm) at 25 grams of adsorbent dose and 25 minutes of contact time. The minimum is 38.7 and 52.3 for size particle of 1 mm and 250 pm at 5 grams adsorbent dose and 5 minutes contact time, respectively. The dosage of adsorbent and contact time effect the adsorption of dye. The greatest of adsorbent dose and the longest of contact time indicate maximum adsorption.

Figure 4 and 5 show the effect of contact time to dye adsorption process. Betel nuts activated carbon dosage of 5, 10, 15, 20, and 25 grams is added to each 200 mL of Jumputan wastewater. Experiments are conducted at a temperature of 298 K by variation of contact time to test the effect of contact time on the adsorption process. The results (Figure 4 and 5) indicate that the adsorption of dye onto adsorbent is very rapid in the first 15 min hence the adsorbed amount (q_t) reached from 43.65% to 71.6% and 50.94% to 74.2% for 1 mm and 250 pm respectively. Then the adsorption of dye increases gradually during the following 20 min until reached equilibrium at about 25 min. The results show that the removal of dye onto betel nuts activated carbon depends on contact time [15]. The dye adsorption increase as the contact time increased. Time is required to encounter the boundary layer, diffuse to surface of adsorbent and diffuse to the porous structure of adsorbent [15]. The equilibrium conditions are reached within 20-25 minutes. At this time, the increase of contact time resulted only in about 3% more to reach the maximum of dye removal. Thus 15 minutes of contact time is chosen as the optimum contact time. Both at 1 mm and 250 pm of particle size of adsorbent show similar trend. This also is seen from the highest correlation factor (R^2) as 0.876 for 1 mm and 0.978 for and 250 pm, respectively.

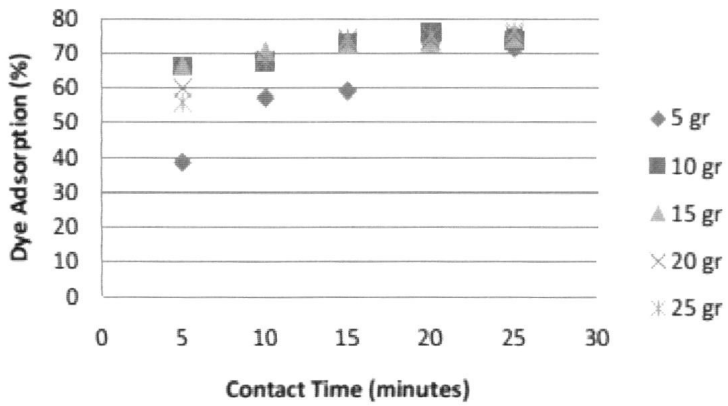

Figure 4. Effect of contact time to dye adsorption percentage at 1 mm adsorbent

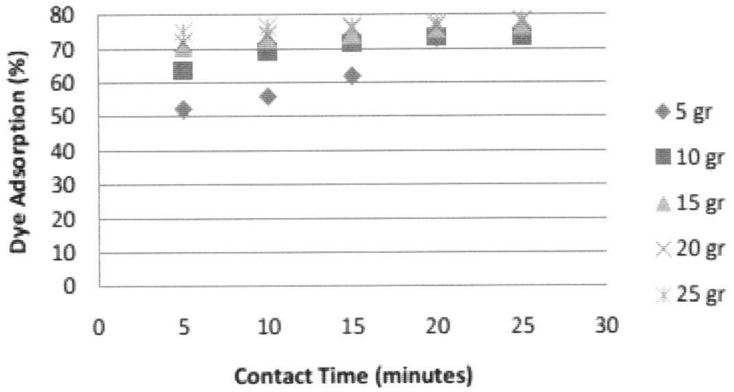

Figure 5. Effect of contact time to dye adsorption percentage with 250 pm adsorbent

The influence of dosage adsorbent to dye adsorption is presented in Figure 6 and 7. The results in Figure 6 and 7 indicate that the dye adsorption increase as the adsorbent dosage increased. At 1 mm size of adsorbent, the optimum of adsorbent dose is 25 grams that reached the most at 76.4%. At 250 pm particle size of adsorbent, the optimum shows at 20 grams of dosage adsorbent that reached 78.2%. This value is seen from the highest correlation factor (R^2) as 0.940 for 1 mm adsorbent's size and 0.946 for 250 pm adsorbent's size, respectively. The dye adsorption is not enhanced effectively by increasing the amount of the adsorbent (from 5 to 25 grams). It may be due to increasing the overlapping and/or aggregation of adsorbent sites at high dose [16].

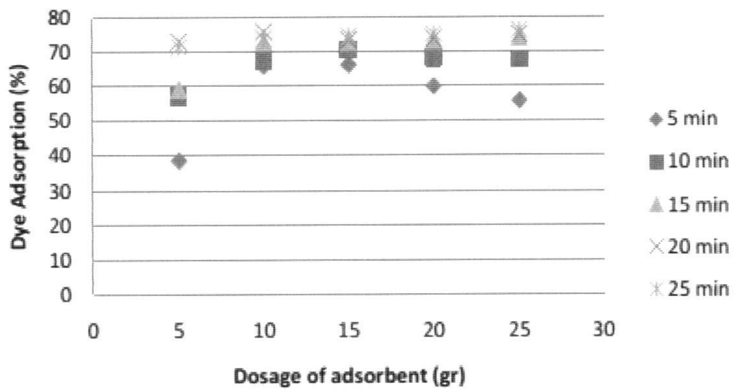

Figure 6. Effect of dosage adsorbent to dye adsorption percentage with 1 mm adsorbent

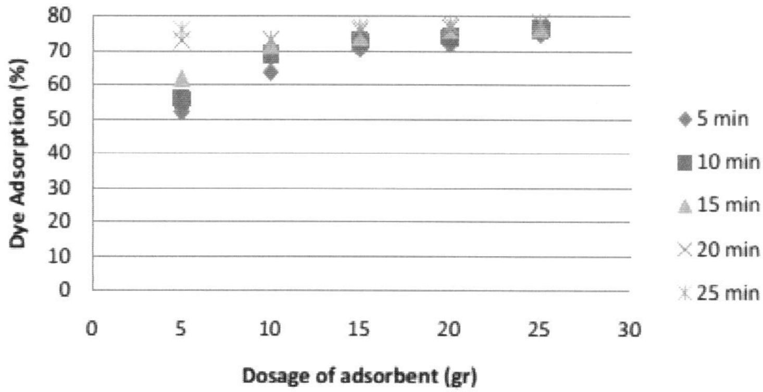

Figure 7. Effect of dosage adsorbent to dye adsorption percentage with 1 mm adsorbent

Based on Table 3 and Figure 4-7, a particle at 250 pm shows a better result than 1 mm. It's seen from the highest value of R^2. The best condition of Jumputan's dye treatment for batch mode is 15 minutes of contact time and 20 grams of adsorbent dosage with dye adsorption 76.4%.

4. Conclusion

The result shows betel nuts contains 60.86% carbohydrate, 32.56% water, 2.17% fat, 3.35% protein, and 1.06% ash. The major component of the activated carbon is carbon (C) at 86.27%, the rest is oxygen at 9.18% and aurum at 4.55%. The best variation is adsorbent that the size of 250 pm (60 mesh), the dosage of 20 grams, and the contact time of 15 minutes with dye removal of 76.4%.

Acknowledgments
Authors wishing to acknowledge Universitas Sriwijaya for financial support under SATEKS scheme (No. 1013/UN9.3.1/PP/2017).

References

[1] Said N I 2010 *Method of removing mercury in industrial wastewater* (Indonesia: Agency for Assessment and Application of Technology, BPPT)
[2] Cahyonugroho O H 2007 *J. Chem. Eng.* **1** 59-64
[3] Ramdja A F, Kurniawan A and Ahmad S 2008 *J. Teknik Kimia* 15 1-7
[4] Adeolu A T, Okareh O T, Dada A O 2016 *American J. Environ. Protect.* **4** 7-20
[5] Angela N S 2015 Synthesis of activated carbon from leather of salak fruit by activation of H_3PO_4 as adsorbent of blue methylene substance *Proc. National Seminar on Teknik Kimia Kejuangan* UPN Veteran Yogyakarta Indonesia
[6] Cundari L, Putra P P and Febrina Y 2015 Jumputan wastewater treatment by using betel nuts activated carbon *Proc. AVoER VII* Universitas Sriwijaya Indonesia
[7] Cundari L, Kemit A S and Usman B R 2016 Effect of bed height and sampling time to reduction of BOD level on Jumputan wastewater *Proc. AVoER VII* Universitas Sriwijaya Indonesia p. 190-197
[8] Cundari L, Melsi M F and Fiat C 2016 *J. Teknik Kimia* **22** 19-26
[9] Cundari L, Kemit A S and Usman B R 2017 Adsorption of jumputan liquid waste by betel nuts activated carbon in a continuous fixed-bed adsorber *Proc. Sriwijaya International Conference on Engineering, Science & Technology* vol 101 MATEC Web of Conferences France
[10] Surest A H, Kasih J A F and Wisanti A 2008 *J. Teknik Kimia* **15** 17-22
[11] Fernianti D 2013 *Berkala Teknik* **3** 563-72

[12] Budi E 2011 *Study offormation and application of coconut shell charcoal as fuel* (Indonesia: Universitas Negeri Jakarta)
[13] Hameed B H, Mahmoud D K and Ahmad A L 2007 *J. Hazard. Mater.* **158** 65-72
[14] Velmurugan P, Rathina K V and Dhinakaran G 2011 *Int. J. Env. Sci.* **1** 1492-1504
[15] Elmorsi T M, Mohamed Z H, Shopak W and Ismaiel A M 2014 *J. Environ. Protect.* **5** 1667-81
[16] Manzoor Q, Nadeem R, Iqbal M, Saeed R and Ansari T M 2013 *Bioresour. Technol.* **132** 446-452

IOP Conf. Series: Materials Science and Engineering **345** (2018) 012042 doi:10.1088/1757-899X/345/1/012042

Immobilization of *Candida rugosa* lipase by adsorption-crosslinking onto corn husk

A Nuraliyah[1,2], A Wijanarko[2] and H Hermansyah[2]
[1] Department of Chemical Engineering, Faculty of Engineering, Universitas Fajar, Jl.Prof Abdurahman Basalam No. 101, South Sulawesi, Indonesia
[2] Department of Chemical Engineering, Faculty of Engineering, Universitas Indonesia, Kampus UI, Depok 16424

E-mail: andi.nuraliyah@yahoo.co.id

Abstract. Corn husk is one of the agricutural waste that has not been used optimally. corn husk waste allows to be used as immobilized support for biocatalyst because it is easy to obtain, available abundant, renewable and easy to decompose. This research was conducted in two phases, namely the adsorption of enzyme immobilization on the support, followed by cross-linking between the enzyme and support through the addition of glutaraldehyde. The optimum conditions for cross-linked adsorption immobilization using support of corn husk were achieved at concentrations of 0,75 mg / ml at 4 hour reaction time. The biggest unit activity value is obtained at 2,37 U / g support through 0.5% glutaraldehyde addition.

1. Introduction

Corn crop (Zea mays ssp.mays) is grain crops that have the largest production capacity every year compared to other grain crops [1]. The composition of the corn crop morphology consists of roots, stems, leaves, flowers and fruit. The fruit consists of a corn cob, seeds and leaves wrapping (skin). Corn husk is a part of the plant that protects the corn kernels which is light green when young, and dries up when the old tree. Corn husk as agricultural waste is generally used conventionally as a source of raw materials for crafts and fodder, while on an industrial scale, among others, have been used as raw material for bioethanol industry [2] and the paper industry [3].

The main composition of the corn husk are composed of cellulose and lignin [3]. Corn husk can be used as a raw material for enzyme immobilization support because its availability is abundant, are renewable and biodegradable [4]. Some criteria of a good support is inert, physically strong, stable, can be regenerated, and able to increase and decrease the activity of enzymes inhibitors [5].

Utilization of agricultural waste biomass as a biocatalyst immobilization support has been carried out which utilizes coconut fiber to immobilize α-amylase [6,7]. Through method of adsorption-crosslinking, an agricultural waste such as of corncobs, banana stem, rice husk, coconut fibers, and stems salak were reported as the support in immobilization lipase from acinotebacter [8]. Utilization of corn husk was studied to be used as a support on proteinase and amylase enzyme immobilization by adsorption [4]. In 2013, continue research on core utilization of corn stalks as support for the immobilization amylase, lipase and proteinase enzymes [9].

Enzyme immobilization process is highly dependent on the type of support and immobilization method used [10]. Adsorption method is a method of enzyme immobilization as it is the most simple, easy and economical. A disadvantage of adsorption immobilization method is the weak interaction between the enzyme and the support that results in low activity and stability [11]. The use of glutaraldehyde as a cross linker is one way to improve the interaction between the enzyme and support. This process begins with enzyme adsorption on its support, followed by cross-linking between enzymes with the addition of glutaraldehyde. Immobilization by adsorption-cross-linking has been made using 0.01% glutaraldehyde as a cross-linker activating agent however it was not successful [6,8]. In this research, effect of different concentration of glutaraldehyde was investigated.

The use of support materials from waste biomass in the immobilization process is largely determined by the preparation of the biomass. The main purpose of the pretreatment is to increase the surface area, decrystallize cellulose, hemicellulose and eliminate lignin [12][13]. Pretreatment of the support of corn husk is important in physics and chemistry to facilitate the process of immobilization.

The purpose of this research is to investigate the abilities of corn husk as the support for lipase obtain from commercial *Candida rugosa*.

2. Material

Commercial C. rugosa lipase was purchased from Sigma- Aldrich (Singapore). Agricultural waste biomass corn husk was obtained from people's a plantation in Ciomas-Bogor, West Java. NaCl, HCl and other chemicals standard uses analytic of merck.

3. Methods

3.1. Pretreatment of the corn husk

A modified pre-treatment method using NaCl and HCl from previous studies was used [8]. Corn husk was cut to a length of 10 - 20mm, washed with clean water, then soaked in 0,8% (w /v) NaCl solution for 30 minutes. It was subsequently rinsed with distilled water 3 times. It was put into the autoclave for sterilization and dried at 65°C for 24 hours. Corn husk was then crushed and screened to obtain a particle size of 0.1 - 0.5 mm. As much as 10 grams of corn husk that has been filtered and then was soaked in 2N HCl solution, and triple washing with 0.1 M phosphate buffer (pH 7.0). Final was obtained by drying at 60 °C for 24 hours.

3.2. Immobilization adsorption-crosslinking

The process of immobilization was done using adsorption-crossliking method [8, 12]. One gram support was suspended in 100 ml solutions (0.5 mg/ml) prepared in a phosphate buffer 20 mm (pH 7.5), stirred (350 rpm) for two hours at room temperature, then 0.5% glutaraldehyde was added. Variations crosslinking reaction time used were 2, 4 and 6 hours. Support that has been immobilized separated from the filtration. Lipase activity was tested in the supernatant.

3.3. Measurement of hydrolytic activity for Candida rugosa lipase

The activity of lipase was done by a reaction call hydrolysis [14]. The substrate was prepared by mixing 50 ml of olive oil with 150 ml of polyvinyl alcohol solution (2%, w/v) to obtain emulsion. The reaction mixture consists of phosphate buffer 5.0 ml (0.025 M, pH 7) and 5 ml of the emulsion above and a certain amount of free *C. rugosa* lipase was incubated for 15 min at 37 oC in a vessel. The reaction was terminated by adding 15 ml of 95 wt.% ethanol. The liberated fatty acid was titrated with 0.05M sodium hydroxide solution in the presence of phenolphthalein indicator. In addition, a blank experiment for comparison without adding *C. rugosa* lipase was also carried out. One unit (U) of CRL activity was defined as the amount of CRL which produced 1 mol of free fatty acids per min under the assay conditions.

4. Result and discussion

4.1. Effect of pretreatment on corn husk

Biomass used in this study is a waste of agricultural produce. Require a special pretreatment to clean the debris from the source of the waste attached to biomass, as well as to facilitate the absorption of lipase *C.rugosa* at the time of immobilization. Pretreatment early done in physics by the process of grinding to get smaller size of particles (0,1 and 0,5 mm). This process aims to get the surface area of particles greater. The corn husk were soaked in NaCl for 30 minutes. This stage serves as a impurities solvent still attached to agricultural waste biomass when washing. Further pretreatment using HCl solution. The use of solution HCl at the time of treatment is too destructive to structures lignocellulose that is a component authors biomass. That accessibility of adsorption enzyme when immobilization increased. Treatment using hydrochloric acid (HCl) serves to damage the lignocellulosic structure followed by the removal of hemicellulose thereby increasing the porosity and the ability of biomass digestion enzymatically. [12]. Initial treatment of lignocellulose with acid at ambient temperature is done to improve anaerobic digestibility [15].

The result on the influence of the surface structure before and after treatment using scanning electron microscopy (SEM) can be seen in figure 1 and 2.

Figure 1 was the SEM test results before pretreatment. Looks larger pore structure, contrary to the Figur 2, which look having a smaller pore structure. Effect of HCl addition at the time of treatment could have damaged the rice straw structure thus enables increased accessibility of the enzyme to the support at the time of immobilization (Figure 1). The opposite occurs before the pretreatment, the larger pore size so that the enzymes are adsorbed more easily dislodged (Figure 2).

Figure 1. SEM corn husk before treatment.

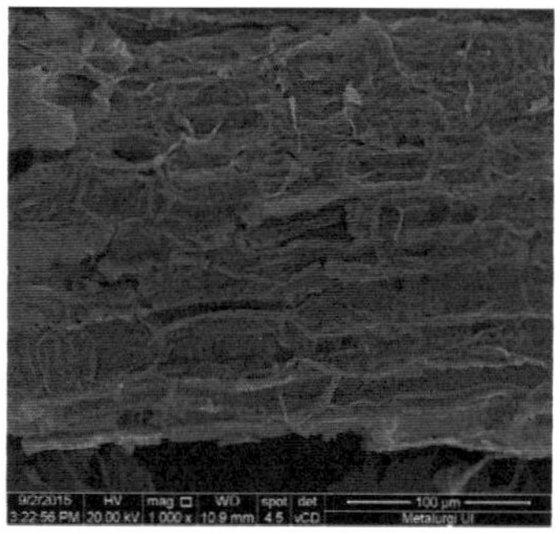

Figure 2. SEM corn husk after treatment.

4.2. *Influence of unit activity on reaction time and concentration of Candida rugosa lipase*

Candida rugosa activity for 2 hour reaction time is shown in Figure 3. Activity lipase was the results of immobilization adsorption was seen to rise fit with the increase in time reaction. After the crosslinking process through the addition of glutaraldehyde precisely causes a decrease in the value of C. rugosa lipase activity for all lipase concentration variations (Figure 3). This is probably caused by the reaction conditions that are inconsistent with glutaraldehyde concentration so that the crosslinking process is not been successful [11]. The biggest activity value is the optimum condition because it shows the number of enzymes adsorbed on the support of corn husk is increasing [8]

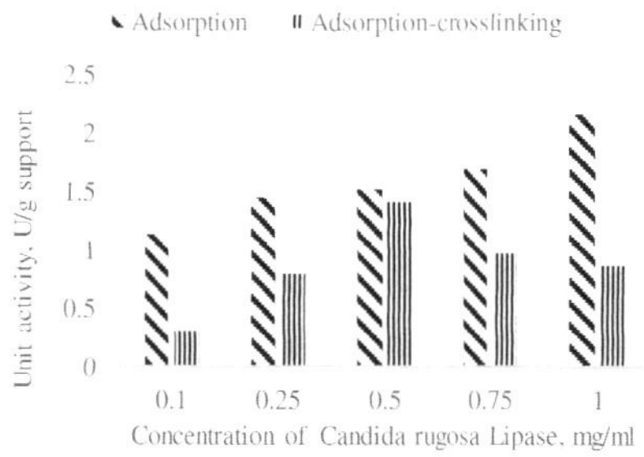

Figure 3. Unit Activity for reaction time 2 hours.

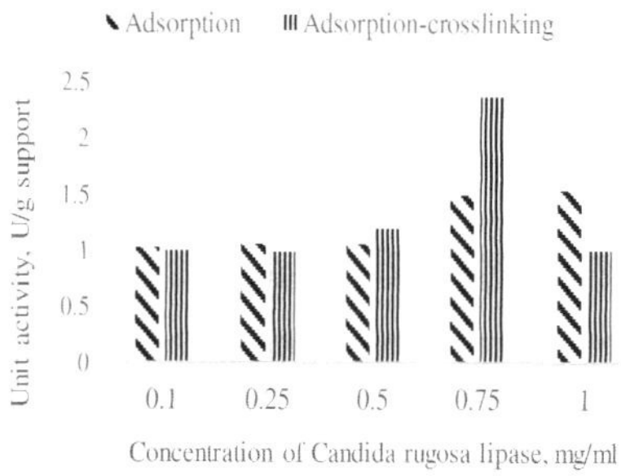

Figure 4. Unit Activity for reaction time 4 hours.

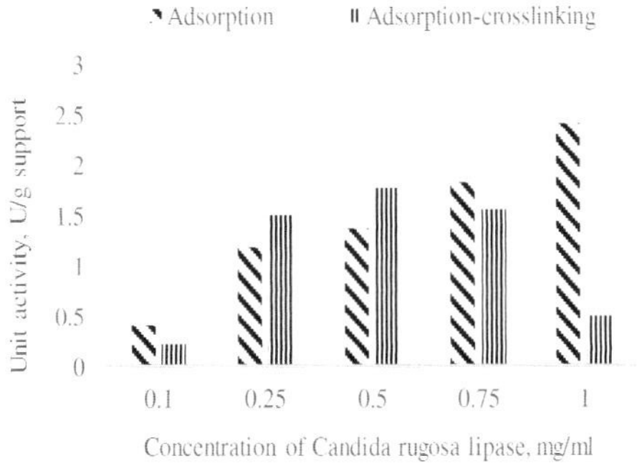

Figure 5. Unit activity for reaction time 6 hours

For reactions which lasted 4 hours, the addition of glutaraldehyde led to an increase in value of the activity of 1.2 and 2.37 U / g support respectively at lipase concentrations of 0.5 and 0.75 mg / ml (figure 4). When 6 hours reaction there was increased activity at smaller cockroach concentrations of 0.25 and 0.5 mg / ml with an activity value of 1.77 and 1.56 U / g support (figure 5).

5. Conclusions

The optimum activity for adsorption-crosslinking immobilization using the corn skin support for 2.4 and 6 hours was obtained with an activity value of 1.42; 2.37 and 1.77 U / g. This result was obtained from lipase concentration of 0.5 mg / ml for reaction time of 2 and 6 hours. whereas for 4 hours is obtained from lipase concentrations of 0.75 mg / ml.

Acknowledgments

This research was funded by the Ministry of Research and Technology of the Republic of Indonesia through Research Grants of the Doctoral dissertation in 2017.

References

[1] Malti, Ghosh, Kaushik, Ramasamy, Rajkumar, Vidyasagar 2011 *Intl. J. Bioresour. Stress Manage.* **2** (3) 250-256
[2] Oyeleke S B and Jibrin N M 2009 *African J. Microbiol. Res.* **3** (4) 147-152
[3] Fagbemigun T K, Fagbemi O D, Otitoju O, Mgbachiuzor E and Igwe C C 2014 *Intl. J. AgriSci.* **4** (4) 209-213
[4] Lv J S, Shen T T and Nie L 2012 *Adv. Mater. Res.* **550-553** 1477-1481
[5] Varalakshmi V and Siva rama prasas D 2013 *JECET* **2** (3) 840-853.
[6] Dey, Gargi, Nagpal, Varima, and Rintu B 2002 *Appl. Biochem. Biotechnol.* **102** (1-6) 303-313.
[7] Samanta S, Jana M, Maity C, Kar S, Mohapatra P K D, Pati B R and Mondal K C 2009 *Biocat. Biotransform.* **27** (2) 131 135
[8] Ittrat, Pairat, Chacho, Thidarat, Pholprayoon, Jitchon, Suttiwarayanon, Navadon, Charoenpanich, Jittima 2014 *Biocat. Agric. Biotechnol.* **3** (3) 77-82
[9] Lv J S, Liu X Y, Xu J X, Deng Y F, Wu Z, Wang Y M and Xu H 2013 *Indust. Crops Prod.* **50** 787-796.
[10] Jin-Shun Lv J S, Shen T T and Nie L 2012 *Adv. Mater. Res.* **550-553** 1477-1481
[10] Elnashar, Magdy M M 2010 *J. Biomater. Nanobiotechnol.* **1** (01) 61.
[11] Wang, Wenqin, et al. 2011 *Appl. Biochem. Biotechnol.* **164** 561–572.
[12] Nee' Nigam P S, Gupta N and Anthwal A 2009 DOI 10.1007/978-1-4020-9942-7 2, *Springer Science Business Media* B.V. 2009
[13] Agbor V B, Cicek N, Sparling R, Berlin A, David B and Levin B D 2011 *Biotechnol. Adv.* **29** 675–685

[14] Yu W H, Fang M, Tong D S, Shaoc P, Xua T N and Zhou C H 2013 *Biochem. Eng. J.* **70** 97– 105
[15] Hendriks A T W M and Zeeman G 2009 Pretreatments to enhance the digestibility of lignocellulosic biomass *Bioresour. Technol.* **100** (1) 10-18

IOP Conf. Series: Materials Science and Engineering **345** (2018) 012043 doi:10.1088/1757-899X/345/1/012043

Trichoderma Reesei single cell protein production from rice straw pulp in solid state fermentation

M Zaki and S D Said
Bioprocess Engineering Laboratory, Chemical Engineering Department, Syiah Kuala University, Darussalam-Banda Aceh, Indonesia 23111

E-mail: syahiddin@che.unsyiah.ac.id

Abstract. The dependency on fish meal as a major protein source for animal feed can lead toit priceinstability in line with the increasing in meat production and consumption in Indonesia. In order todeal with this problem, an effort to produce an alternative protein sources production is needed. This scenario is possible due to the abundantavailability of agricultural residues such as rice straw whichcould be utilized as substrate for production of single cell proteins as an alternative proteinsource. This work investigated the potential utilization of rice straw pulp and urea mixture as substrate for the production of local *Trichoderma reesei* single cell protein in solid state fermentation system. Some parameters have been analyzed to evaluate the effect of ratio of rice straw pulp to urea on mixed single cell protein biomass (mixed SCP biomass) composition, such as total crude protein (analyzed by kjedhal method) and lignin content (TAPPI method).The results showed that crude protein content in mixed SCP biomassincreases with the increasing in fermentation time, otherwise it decreases with the increasing insubstrate carbon to nitrogen (C/N) ratio. Residual lignin content in mixed SCP biomass decreases from 7% to 0.63% during fermentationproceeded of 21 days. The highest crude protein content in mixed SCP biomasswas obtained at substrate C/N ratio 20:1 of 25%.

1. Introduction

The consumption of meat in Indonesia continues to increase from year to year along with the increasing in population and people's income. This trend will affect the increasing demand on animal feed which in turn will disrupt the market stability of feed prices. The increase in feed prices is due to the higher price of raw materials of fish meal which has limited availability. One way to reduce the dependency on fish meal is to find out alternative sources such as single cell protein (SCP). Single cell proteins can be produced from various organic substrates by various microorganisms in fermentation process. One of the promising organic substrate to be utilized is agricultural residues such as rice straw. The utilization of rice straw has attracted much attention among researchers since it is available abundantly and as low-cost carbon source for SCP production [1-3].

In utilizing lignocellulosic biomass as substrate, it is required to select an appropriate microorganisms that have the ability to consume lignocellulosic biomass carbon source for its growth. Among of the microorganisms having these capabilities are fungal families such as *Trichoderma reesei.* According to Ugalde et al and Schuster, et al [3,4], *T. reesei* is a fungi which has good capability on digesting lignocellulosic material, as it can produce cellulolytic and hemicellulolytic enzymes. However, studies on the performance of *T. reesei* in utilizing rice straw pulp as substrate for its growth is still lacking . In this study, the potential of local *T. reesei* to produce single cell protein (SCP) with rice straw pulp as substrate in solid state fermentation and the capability of *T. reesei* in degrading the lignin compounds contained within the substrate were examined.

2. Materials and methods

2.1. Rice straw liquid hot water pretreatment.

Rice straw which was used in this study was obtained from a local rice field in Aceh Besar District , Indonesia. Rice straw was collected and then put into a plastic bag and stored in a cool place. It was then crushed using a blender to increase surface area about 60 mesh. Pretreatment was conducted in an autoclave by using liquid hot water at 121 °C for 60 minutes and sequently rinsed by distilled water and dried in oven at 50 °C for 2 days.

2.2. Microorganism

The fungus used was *Trichoderma reesei* which obtained from Microbiology Laboratorium, School of Life Sciences and Technology, Bandung Institute of Technology, Indonesia. It was subcultured periodically on potato dextrose agar (PDA) plates.

2.3. Inoculum

Aerial spores of *T. reesei* were removed from 7 days old potato dextrose agar plates by scraping with a spatula and suspending in sterile water. This suspension of spores was used to inoculate 200 ml medium in a 500 ml erlenmeyer flask. The flask was incubated for 3 days in an orbital shaker at 150 rpm and 30 °C. The resulting biomass was used as an inoculum for the medium in solid state fermentation. Media for inoculum consists of 40 g/l of molasses and 7 g/l of ammonium sulfate.

2.4. Solid state fermentation (SSF)

Single cell protein production experiment were conducted using 500 ml erlenmeyer flask. 100 grams of sterilized rice straw pulp were put into the flask and added with urea solution according to C/N ratio variation, and then added 100 ml of the inoculum were sequently added in. The cultivation was carried out for 21 days at temperature of 30 °C.

2.5. Measurement of crude protein content

The total amount of SCP in fermented product was estimated as crude protein contents by Kjeldahl method [5]. Subsequently 0.51 grams of samples (for incubation time of 4, 7, 14 and 21 days), was put into 100 ml of kjedahl flask. Added 2 grams of selenium mixture and 25 ml of concentrated H_2SO_4. Heated over an electric heater until boiling and the solution becomes clear greenish (about 2 hours). Allow to cool and then diluted and put into a 100 ml measuring flask, right up to the line mark. Pipetted 5 ml of solution and put in to the distiller, added 5 ml of 30% sodium hydroxide and several drops of phenofthalein indicator. Distilled for approximately 10 minutes, distillate dropping into the flask contained 10 ml of 2% boric acid solution and indicator. Titrated with 0.01 N HCl solution.
Calculation:

$$crudeproteincontent\ (\%) = \frac{(V_1 - V_2) x N x 0{,}014\ x fk x fp}{w} x 100\% \qquad (1)$$

where :
w : weight of sample (gram)
V_1 : volume of HCl used for sample (ml)
V_2 : volume of HCl used for blank test (ml)
N : normality of HCl (N)
fk : conversion Factor (6.25)
fp : dilution factor

2.6. Lignin content analysis

Analysis of lignin content using kappa number test [6]. This Standard procedure is adopted from TAPPI Useful Method 246-modified micro Kappa number of standard methods for Kappa number (T-236). After the moisture content of the pulp samples was determined, 0.5 gram (dry weight) pulp was taken. Samples were soaked with 10 ml of distilled water in a beaker glass for 10 minutes. Then, the mixture is transferred into the rotary blender with 40 ml of distilled water. After a minute, the mixture was transferred into the 250 ml beaker glass and rinsed with 20 ml distilled water before stirred by a magnetic stirrer. A mixture of 10 ml of 0.02 M $KMnO_4$ and 2.0 M H_2SO_4 solution that has been prepared beforewas inserted into the beaker. After 10 minutes, 2 ml of 1.0 M KI added to the beaker followed by titration with a solution of 0.1 M sodium thiosulfate solution and a few drops of 0.2% starch.Total volume of 0.1 M sodium thiosulfate solution used was recorded. Control test without the dissolved pulp sample was performed with the same procedure. Kappa number was calculated by the following equations:

$$p = \frac{(b-a)N_1}{0,1} \quad (2)$$

$$\log K = \log \frac{p}{w} + 0.00093\,(p - 50) \quad (3)$$

where

p : volume of potassium permanganate actually consumed by the test specimen, ml
b : volume of sodium thiosulfate consumed in black test, ml
a : volume of sodium thiosulfate consumed by pulp sample test, ml
N_1 : normality of sodium thiosulfate
w : weight of sample (pulp), gram
K : Kappa number

The obtained kappa number is converted to lignin content using the following equation:

$$\%\ \text{lignin} = 0.147 \times \text{Kappa Number} \quad (4)$$

3. Results and discussion

3.1 Degradation of lignin

The main obstacles on utilization of agricultural waste such as rice straw as animal feed are related to low nutritional content and digestibility level. Lignin chemically binds to carbohydrates and physically acts as a barrier to the process of reshuffling cell walls by microorganisms in the digestive process. According to Bhargav, et al., [7], lignin is a biopolymer with complex phenylpropanoid structure which can deteriorate digestion ability. In this experiment, liquid hot waterpretreatment was used for eliminating lignin content in rice straw substrate.

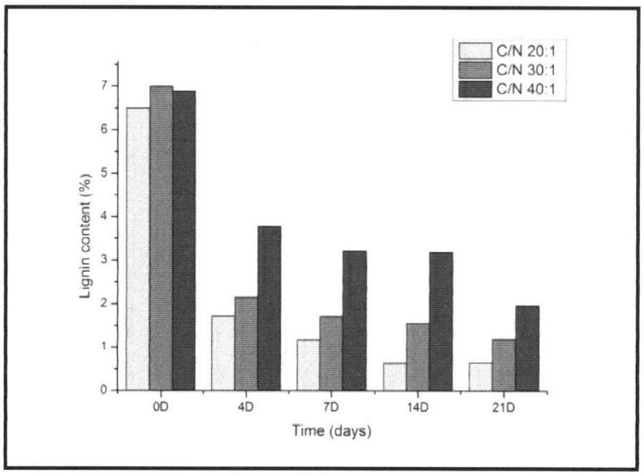

Figure 1. The lignin content of mixed SCP biomass.

But, unfortunately the pulp result from LHW pretreatment still has lignin content of 7% (dry basis). Therefore to fulfill the poultry feed standard [8] which is lignin content in the feed not allowed, the efforts should be made to eliminate the remaining lignin content in rice straw substrate using *T. reesei*. Schuster, et al., [4] reported that *Trichoderma reesei* is a fungus possesing ability to digesting lignocellulosic material very well, as of it can produce cellulolytic and hemicelulolytic enzymes, the same result reported by Helal [9].In this study the ability of local *T. Reesei* in removal of the remaining lignin content in rice straw pulp was evaluated. In Figure 1shows the lignin content in mixed SCP biomass against fermentation time. The minimum lignin content in mixed SCP biomass (0.63%) was observed at 14 days. From the Figure 1, it can be seen that the lignin content is influenced by the C/N ratio of the substrate. The lower C/N ratio (20: 1), the greater decrease of lignin level. At 14 days, the lignin content in mixed SCP biomass was 0.63%. This evidence is due to higher fungal cellular growth on the substrate with a lower C/N ratio. The increase infungal cells growth on substrate would affected the increasing of lignin disruption level.

3.2 Effect of initial C/N ratio on crude protein production

Crude protein is a single cell protein which extracted from mixed single cell biomass (mixed SCP biomass). Mixed SCP biomass is a fermentation product containing SCP and undigested pulp. The profile of crude protein production at different initial carbon to nitrogen rasio are shown in Figure 2. The increase of crude protein content in all mixed SCP biomass begin after 4 days incubation and reached the maximum after 14 days. It can be seen from the Figure 2, that the crude protein increase with decreasing initial C/N ratio. This trend is due to different amount of nitrogen supplied in substrate [10,11]. The substrate initial C/N ratio of 20:1 had more nitrogen content, so that the cell of *T. reesei* could utilize more carbon source and therefore would produce more single cell biomass than other C/N ratio combinations. The maximum crude protein in substrate with C/N ratio 20:1 was 25% (based on ratio of crude protein/mixed SCP biomass) after 14 days of fermentation.

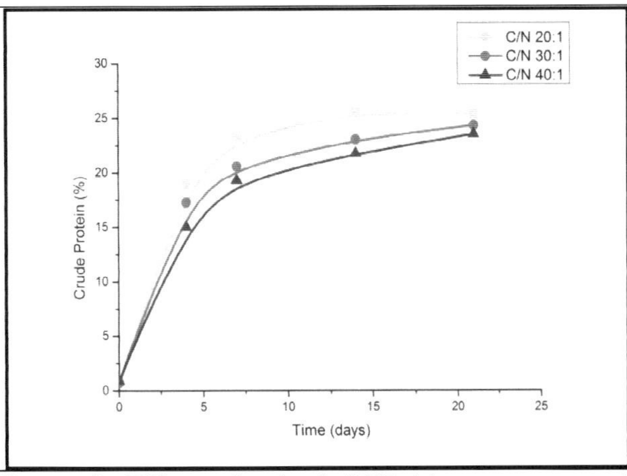

Figure 2. Effect of mixed SCP biomass on protein production by *T. reesei* at temperature of 30 °C and relative humidity of 95%

4. Conclusion

This study shows that it is possible to grow a locally *Trichoderma reesei* on mixed urea-rice straw pulp substrate in solid state fermentation to produce single cell protein. *T. Reesei* proves to have ability to reduce lignin content in substrate from 7% (dry basis of biomass) to 0.63%. The maximun of total crude protein produced on substrate with C/N ratio of 20:1 was 25%.

Acknowledgment

The authors would like to thank the Ministry of Technology, Research and Higher Education of Republic of Indonesia for research funding under Syiah Kuala University Research Competitive Grant 2017.

References

[1] Anupama and Ravindra P 2000 Value-added food : Single cell protein *Biotechnol. Adv.* **18** 459-479

[2] Samadi S, Mohammadi M and Najafpour G D 2016 Production of single cell protein from sugarcane bagasse by *Saccharomyces cerevisiae* in tray bioreactor *Intl. J. Eng.* **29** (8) 1029-1036

[3] Ugalde U O and Castrillo J I 2002 Single cell proteins from fungi and yeasts *Appl. Mycol. Biotechnol.* **2** (C) 123-149

[4] Schuster, Andre, Schmoll and Monika 2010 Biology and biotechnology of *Trichoderma Appl. Microbiol. Biotechnol.* **87** (3) 787-799

[5] Standar Nasional Indonesia 1992 *Analisa Bahan Pangan* SNI 01-2892-1992 Badan Standarisasi Nasional

[6] Standar Nasional Indonesia 2008 *Cara Uji Bilangan Kappa* SNI 0494:2008 Badan Standarisasi Nasional

[7] Bhargav S, Panda B P, Ali M and Javed S 2008 Solid state fermentation: An overview *Chem. Biochem. Eng.* Q 22 (**1**) 49-70

[8] Standar Nasional Indonesia 2006 *Pakan Ayam Ras Petelur Dara* SNI 01-3928-2006 Badan Standarisasi Nasional

[9] Helal G A 2005 Bioconversion of straw into improved fodder: Mycoprotein production and cellulolytic activity of rice straw decomposing fungi *Mycobiol. J.* **33** Issue 2 90-96

[10] Khanifar J, Ghoorchian H, Ahmadi A R and Hajihosaini R 2011 Comparison of essential

IOP Conf. Series: Materials Science and Engineering **345** (2018) 012043 doi:10.1088/1757-899X/345/1/012043

and non essential amino acids in the single cell protein (SCP) of white rot fungi from wheat straw *African J. Agric. Res.* **6** (17) 3994-3999

[11] Sarnklong C, Coneja J W, Pellikaan W and Hendriks W H 2010 Utilization of rice straw and different treatments to improve its feed value for ruminants: A review *Asian-Attustralasian J. Anim. Sci.* **23** (5) 680-692

IOP Conf. Series: Materials Science and Engineering **345** (2018) 012044 doi:10.1088/1757-899X/345/1/012044

Planning of Eka Hospital Pekanbaru wastewater recycling facility

A Jecky, D Andrio and A Sasmita
Environmental Engineering Program, Faculty of Engineering
Riau University

E-mail: jecky.asmura@gmail.com

Abstract. The Ministry of Public Works No. 06 2011 required the large scale of water to conserve the water resource, Eka Hospital Pekanbaru have to improve the sewage treatment plant through the wastewater recycling. The effluent from the plant can be used to landscape gardening and non-potable activities. The wastewater recycling design was done by analyzing the existing condition of thesewage treatment plant, determine the effluent quality standards for wastewater recycling, selected of alternative technology and processing, design the treatment unit and analyze the economic aspects. The design of recycling facility by using of combination cartridge filters processing, ultrafiltration membranes, and desinfection by chlorination. The wastewater recycling capacity approximately of 75 m^3/day or 75% of the STP effluent. The estimated costs for installation of wastewater recycling and operation and maintenance per month are Rp 111,708,000 and Rp 2,498,000 respectively.

1. Introduction

Water is a vital needed for human life, so the availability of water becomes the most attention in various activities. Water have to follow the quality aspects, quantity, and continuity is needed for meet human needs that continue to increase, so it takes conservation efforts of water resources. Conservation of water resources is done through the protection and conservation of water resources, water preservation, and quality management and control of water pollution. Every person or business entity utilizing water resources is encouraged to conduct water preservation through water saving movement and controlling groundwater use by stakeholders by recycling liquid waste into raw water [1].

The recycling of liquid waste into clean water needs to be carried out by large groups of water users such as hotels, hospitals, commercial and industrial areas to maintain the sustainability of water use and efforts to minimize waste disposed into the environment [2]. Liquid waste recycling results can be utilized for agricultural irrigation or landscaping, industrial uses, groundwater replenishment, clean water supplies, and for non-potable uses such as flushing and fire water [3].

Eka Hospital as one of the hospitals in Pekanbaru City. The use of groundwater should be regulated and supervised so the existensing of the underground water as water resources still support and can anticipate the demands of sustainable development [4]. Planning of a liquid waste recycling system can maintain the availability of groundwater and as a form of responsibility for the utilization of water resources by stakeholders [1].

Eka Hospital Pekabaru has conducted liquid waste treatment through Sewage Treatment Plant (STP) so that the waste discharged into the environment fill in the quality standard specified in PERMEN LH No.5 year 2014 [5] on Quality Standard of Waste Water for Business and/or Activity of Health Service Facility. Waste water treatment at Eka Hospital is conducted in Sewage Treatment

Plant (STP) physically, chemically, and biologically with processing capacity of 235 m^3/day. Liquid waste generated from various hospital activities is channeled into the chamber inlet, whereas liquid waste from kitchen, cafes, restaurants and sources that produce oil and fat is flowed to the grease trap. The liquid waste is then channeled into the gravitational equalization basin through the overflow channel.

Wastewater treatment is also equipped with biosystem bios tub or biocontrol tub it basically water quality control boiler using biological indicator. Based on the data of the effluent quality test results on effluent STP Eka Hospital shown that the processed water are fill in the quality standard, so it has the potential to be developed and utilized as a source of recycled water

2. Methodology

2.1. Research sites
The research has been done on EKA Hospital in Pekanbaru City, Riau Province.

2.2. Types and data sources
Secondary data obtained from the Eka Hospital Pekanbaru was used as the planning location to know the existing condition of the planning area. The data was used to determine the quality of liquid waste. Secondary data are also obtained from manufacturers of processing units used as a reference in conducting economic analysis. The required data can be seen in Table 1.

Table 1. Types and sources of data

Data Description	Data Type	Data source
Location characteristic	Secondary data	Company data, survey identification, literature
Clean water needs	Secondary data	Company data, survey identification, literature
Capacity and dimension of Existing WWTP	Secondary data	Company data, survey identification, literature
Debit and characteristic liquid waste	Secondary data	Company data, survey identification
Recycle needs	Secondary data	interview, survey identification
Monetary Funds	Secondary data	Treatment unit factory

2.3. Data collection technique
The data was used as source of planning data. Processing and data analysis based on technical aspects to determine the form of processing, it can be applied in Eka Hospital Pekanbaru. Selection of processing form with technical aspect analysis consisted of:
1. Existing Sewage Treatment Plant (STP) condition analysis
2. Determination of water quality targets from recycling plants
3. Selection of Recycling Installation processing unit

The recycling installation has been done by analyzing STP condition. The analysis has been done by knowing the efficiency of pollutant removal by STP, and base on the efforts was used to optimize STP performance. Based on the results of the analysis, technical and non-technical advice is given to improve the condition and quality of the effluent STP in order to support the planned recycling plant planning.

The STP effluent condition determined the type of advanced processing unit and the type of pre-treatment required. The analysis has been done by examining the results test of each sample parameter. Recycled water have to established the standard quality. Recycled water quality requirements are adjusted with reference to Government Regulation No.82 of 2001 [6] on Water Quality Management and Water Pollution Control and United States Environmental Protection Agency (U.S.EPA).

The standard on PP 82/2001 is used as a consideration of the effluent quality standards of the Recycle Installation in accordance with the water quality standards in Indonesia, while the USEPA standard is chosen because Indonesia does not have specific quality standards for recycled water use, and some countries that have not has a quality standard on water recycling has been pushed to this standard [3].

Table 2. Effluent standard of recycle water by U.S. EPA

Parameter	Landscape and Parks	Non-Potable Needs
pH	6-9	6-9
BOD	$\leq$ 30 mg/l	$\leq$ 10 mg/l
TSS	$\leq$ 30 mg/l	-
Turbidity	-	$\leq$2 NTU
Coliform/100 ml	0	0
Residu Cl2	$\leq$ 1 mg/l	$\leq$ 1 mg/l

3. Discussion

3.1. Recycle unit design

The targeting of recycled water quality targets is based on the results of effluent quality measurements and used as parameters for planning the recycling of wastewater, as recycling plants are planned by utilizing the output from STP. The test results are compared with Government Regulation No.82 of 2001 [6] on the standard of clean water quality class II. The STP effluent test results are also compared with the recycled water quality standard by U.S.EPA for landscape irrigation and non-potable activities because Indonesia does not have a quality standard for recycled water. The U.S.EPA standard is selected because Indonesia does not have a specific quality standard for recycled water use, and some countries that do not have a quality standard on water recycling have proceeded to this standard [7].

The comparison of recycled water quality standards based on PP No.82 of 2001 [6] and U.S. EPA Guidelines for landscape and non-potable needs as a consideration of the effluent standard of recycled water can be seen in Table 3.

Table 3. Recycle water standard

Parameter	Value	Efluent STP quality	PP 82/2001 Class 2 [6]	USEPA Landscape and parks [8]	USEPA Non Potable needs [8]
pH	-	7,44	6-9	6-9	6-9
BOD	mg/l	70,2	3	30	10
COD	mg/l	118	25	td	td
TSS	mg/l	42	50	30	td
Residu Cl2	mg/l	-	td	1	1
Oil and Fat	mg/l	1	1 x 10-3	td	td
MBAS	mg/l	0,474	0,2 x 10-3	td	td

The installation planning is equipped with pre-treatment using a filter cartridge to extend the membrane oflife, and the disinfection process in an attempt to optimize STP conditions for reducing the total number of coliforms. Re-planning of disinfection process has been done by placing disinfection process after treatment with ultrafiltration membrane, and disinfecting was done by adding chlorine and injection is shown in reservoir. Liquid waste processing and recycling plants is shown in Figure 1.

3.2. Ultrafiltration membrane design
The liquid waste to be recycled is pumped from the effluent tub to the ultrafiltration unit. The ultrafiltraation membrane is a micron scale filter that can filter solids and turbidities up to a size of 0.005 - 0.2 μm [8]. Pretreatment (pre-treatment) is necessary to protect the membrane from damage and blockage. Micro screening or pre-filtration will exclude larger size particles of 0.1-0.5 mm [9].

3.3. Disinfection process design
The main purpose of the disinfection process is to kill pathogenic bacteria that pass from wastewater treatment units. In this planning chlorination is used with the use of chlorine as a disinfectant. Chlorination process is done post-chlorination, that is after the water undergoes the processing.

3.4 Container tub design (reservoir)
Reservoirs are planned to accommodate recycled water before being used for garden irrigation or non-potable activities. Recycled water was used for the use of irrigation parks with an area of 1890 m^2, and non-potable needs include vehicle wash water, washbasin carts, backwash tube filtration at STP, and fountains with a discharge of ± 75 m^3.

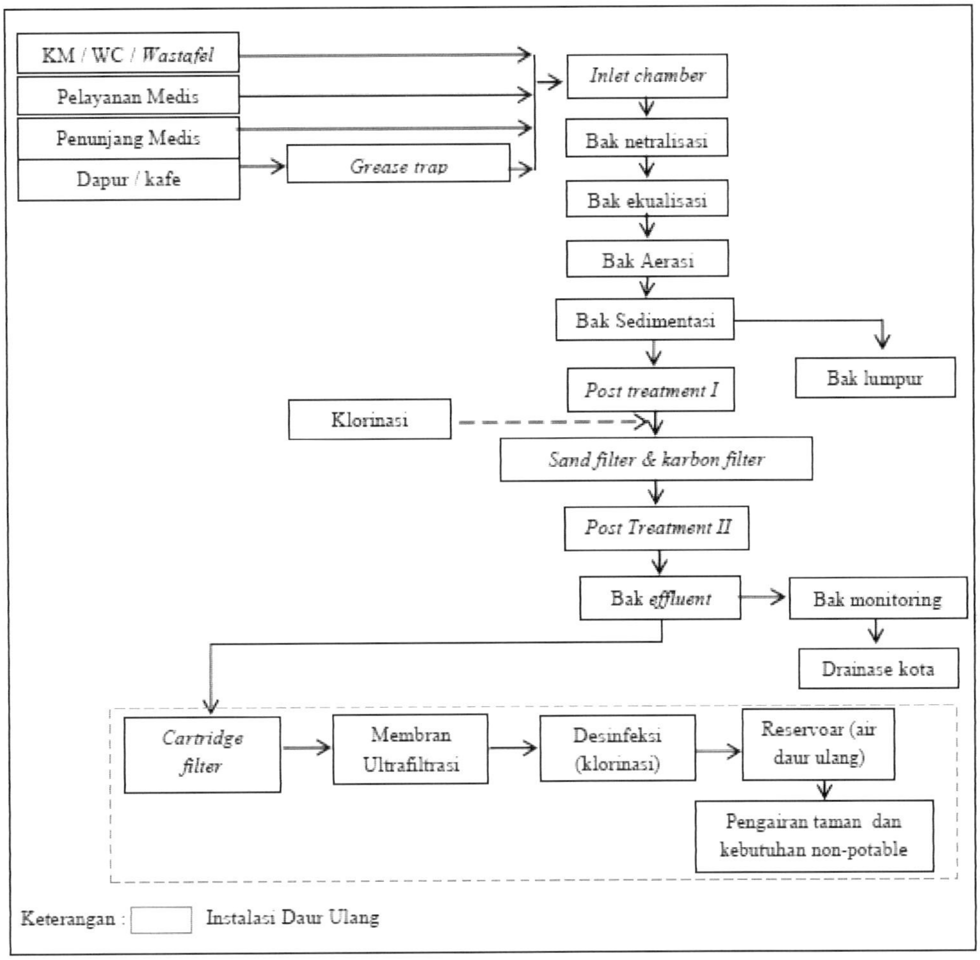

Figure 1. *Sewage treatment plant* (STP) and recycle unit plan.

3.5. Budget plan

The Budget plan is undertaken for predicting the cost for initial investment and operating costs as well as maintenance of waste water recycling installations. Economic analysis is also done by calculating the costs required for the operational and maintenance costs of the planned recycling plant. The required operating costs are calculated on the basis of electricity consumption and the required chemicals. The initial investment cost required for the construction of this recycling plant is Rp 111,708,000. Initial investment costs can be cheaper or higher than the budget design offered, as this depends on the conditions in the field. Operational and maintenance costs incurred are Rp 2.498.000, - per month. Operational and maintenance costs can be reduced, if the unit replacement period is longer than expected.

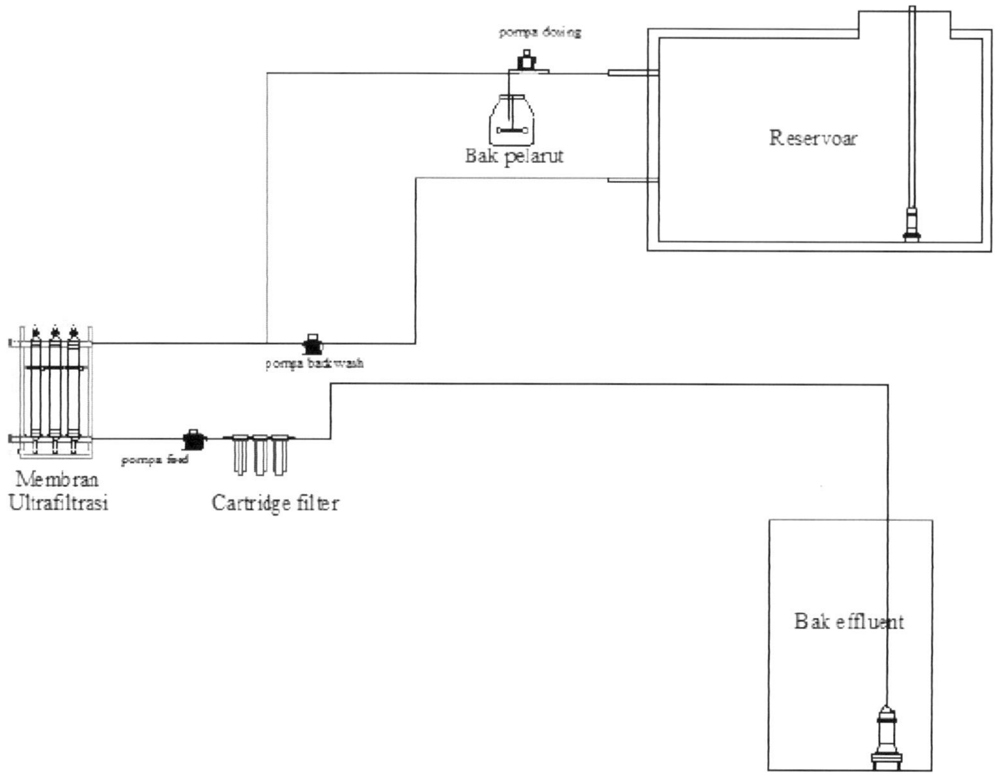

Figure 2. Recycle unit facility.

4. Conclusion

Based on the results, it can be concluded that the process of recycling liquid waste was planned by utilizing membrane technology. The planned recycling installation scheme is processing with filter cartridge, ultrafiltration membrane, disinfection process with chlorination method, and reservoir with production capacity of 75 m^3/day. Liquid waste recycling process required 3 units of filter cartridge, 6 units of ultrafiltration membrane, disinfecting process with $CaO(Cl)_2$ injecting using dosing pump, and 1 reservoir unit to hold recycled water before used for garden irrigation and non-potable requirement.

Estimated cost of initial investment for the construction of waste water recycling installation amount to Rp 111,708,000,- covering the cost of procurement of materials, equipment, and labor wages. The estimated operational and maintenance costs required per month amounted to Rp 2.498.000,- which includes electricity costs, chemical purchase costs, and membrane replacement and other maintenance costs.

References

[1] Peraturan Presiden Republik Indonesia 2011 *Kebijakan Nasional Pengelolaan Sumber Daya Air* No. **33** (Jakarta: Sekretaris Kabinet RI)

[2] Peraturan Menteri Pekerjaan Umum 2011 *Pedoman Penggunaan Sumber Daya Air.* Berita Negara No. **250** (Jakarta: Kementerian Pekerjaan Umum RI)

[3] Said N I 2006 Daur ulang air limbah (water recycle) ditinjau dari aspek teknologi, lingkungan dan ekonomi *Jurnal Air Indonesia* **2** (2)

[4] Peraturan Daerah Pekanbaru 2004 *Pengelolaan Air Bawah Tanah* No. **04**
[5] Peraturan Menteri Lingkungan Hidup *2014 Baku Mutu Air Limbah* No. **5** (Jakarta: Kementerian Lingkungan Hidup Republik Indonesia)
[6] Peraturan Pemerintah Republik Indonesia 2001 *Pengelolaan Kualitas Air dan Pengendalian Pencemaran Air* No. **82** (Jakarta: Sekretariat Negara)
[7] Suprihatin 2009 *Makalah Workshop Air Daur Ulang dalam Perspektif Hukum Islam* (Jakarta: Majelis Ulama Indonesia)
[8] Asano T, Burton F L, Leverenz H L, Tsuchihashi R and Tchobanoglous G 2007 *Water Reuse: Issues, Technologies, and Applications* (New York: McGraw-Hill Book)
[9] Howe K J 2012 *Principal of Water Treatment* 1st ed. (New Jersey: John Wiley & Sons, Inc.)

Effect of direct electric current on contaminants removal from the peat water with continuous system

I Amri[1], A Azis[2] and Drastinawati[1]
[1]Chemical Engineering Department, Faculty of Engineering, University of Riau
[2]Mechanical Engineering Department, Faculty of Engineering, University of Riau

E-mail : idral_amri@eng.unri.ac.id

Abstract. This research was analysed the essentially of treat peat water using an electric current. Initially, the characterization of peat water was determined including of three parameters they are pH, colour, and conductivity solution exhibited values that exceeded the water standard limit. There are two factors influencing the electric coagulation such as electric current and voltage that were observed in the continous study. The results obtained indicated that the majority of the electric current were very effective for removing TDS, and pH. The research variable for the voltage from 23,5 to 42,5 volt and the electric current from 2,2 to 4,1. The optimum electric current and voltage was found around 1,5 Ampere and 25 volt, it was exhibited at 4 L/minute. In unit study, continous electric reactor showed that the optimal reduction on the 20 minutes treatment were found pH = 7, 256 ppm. It was meet to the minimum standard government permition.

1. Introduction

Electrochemically is one of the promising methods for treating hard to-treat waste water streams. Electrochemical processes (electrolysis and electrocoagulation) have been successfully demonstrated for removing pollutants in various industrial wastewaters [1-3]. Generally, removal mechanisms reported in the electrolysis process are including the oxidation, reduction and decomposition proceses, whereas the mechanisms in the electrocoagulation process included the coagulation, adsorption, precipitation and flotation [4-6]. Electrocoagulation utilizes iron anode to produce iron hydroxide flocks by reaction at the anode followed by electrolysis.

The electrocoagulation is simple and efficient method for the treatment of many water and wastewater. In the recent years, many investigations have been done, it focused on the using of the electrocoagulation owing to increase the effluent waste water in the environmental restrictions [7-9]. Electrocoagulation has been applied successfully to treat potable water, food and protein waste water [10-12] yeast waste water [13], urban wastewater [14], restaurant wastewater [8,15] tar sand and oil shale waste water [16]. The aims of this work is to study the possibility of using aluminium electrode for removal of peat water from simulated waste water by electric reactor method.

IOP Conf. Series: Materials Science and Engineering **345** (2018) 012045 doi:10.1088/1757-899X/345/1/012045

2. Experimental

The samples were collected from a peat water river and stored in a water drum. The untreated samples have high contaminants concentration according to pH and conductivity meter measurements. The physico- chemical properties of the samples before and after ER treatment are presented in Table 1. pH and Conductivity Water was measured before and after treatment to see the trend of percent removal of contaminants of the peat water. In this study, we made a variation of voltage, ten ER cells were constructed, reactor was made for 10 aluminum plates; 10 anodes and 20 cathodes. Cell A made of aluminum anodes (10 plates) and inert cathodes (10 plates). Cell Cathode (C) was similar to Cell Anode (A), but inert plates were used as anodes and aluminum plates as cathodes. Inert electrode material used in this study was commercially available dimensionally stable anode (DSA) electrodes. Aluminum alloy 1050 (99.5% Al) was used in the experiments. All electrode plates were 2 m long and 1inch and 1,5 inch diameter. Distance between electrode plates was 0,6 cm. ER treatment was done in continous cells where electrodes were arranged into monopolar configuration. Water volume during the treatment was 10,000 ml.

Table 1. The condition of peat water before and after treatment.

No	Time	TDS	Cond.	Res	PH	Voltage	Current	Flow Rate lt/mnt
	Minute	*ppm*	*Micro Siemen*	*Ohm*		*Volt*	*Ampere*	
1	5	323,671	522,097	1915	5,8	23,5	2,2	0
	10	320,804	521,986	1916	6	23,5	2,2	0
	15	317,037	518,318	1929	6,1	23,5	2,2	0
2	5	321,098	519,018	1927	6,2	32	3,4	0
	10	318,563	515,197	1941	6,2	32	3,4	0
	15	316,901	512,863	1950	6,3	32	3,4	0
3	5	315,436	515,85	1939	6,2	45,5	4,1	0
	10	311,732	509,721	1962	6,4	45,5	4,1	0
	15	305,645	502,396	1990	6,5	45,5	4,1	0

Samples were flowing inside the pipe during the ER treatment wih flowrate 4L/minute. After the treatment, samples were entering the separator to separate the flocculation using settling gravity. Settling time after flocculation was 20 min. The samples were determined the pH and conductivity meter before and after treatment.

2.1. Chemical analysis

The content of Peat water was determined the pH and conducitivity meter from the samples. Dissolved aluminum samples were taken after the electrochemical treatment before slow settling stage. Conductivity was determined using a Conductivity Meter, and pH Meter by depoinovasi electronics.

2.2. The variable of the research

The research variables were conducted to optimized the unit. The variables covered; electric current, voltage and running times of the reaction.

3. Results and discussion

3.1. Peat water characterization

The surface of water originating from the peat land of Riau was taken from Kampar river along the Kampar River on several occasions as the main sample. The characterization of peat water was carried out at the sampling point (in-situ measurement) using conductivity and pH meter. Three parameters were successfully determined, including pH, voltage, current, and conductivity were measured at the sampling point. Acidic pH of the peat water was predicted due to the composition of the surrounding peat soil itself which had been formed by decaying material possessing humic substances [3]. Besides that, humic substances also lead to high the organic content as humic substances are comprised of numerous oxygen containing functional group and fractions (humic acid, fulvic acids and humin) with different molecular weights. Moreover, composition of peat soil may also have an impact on the aluminium ion concentration of peat water [5].
All the three parameters were indicated comply to the standard limit. These parameters were pH, TDS and alum that showed values of 6,9, 140 ppm, and 25,439 grams respectively while the standard limit of these parameters are 6.5 - 9.0, 1500 ppm accordingly.

3.2. Effect of electric voltage for contaminant removal

Electric voltage is adjusted to desired the removal level by increasing the voltage Influence of initial voltage for reduction of TDS and pH, Al ion were investigated. Figure 1 to Figure 4 below, displayed the percentage removal of TDS, and alum ion against pH respectively.

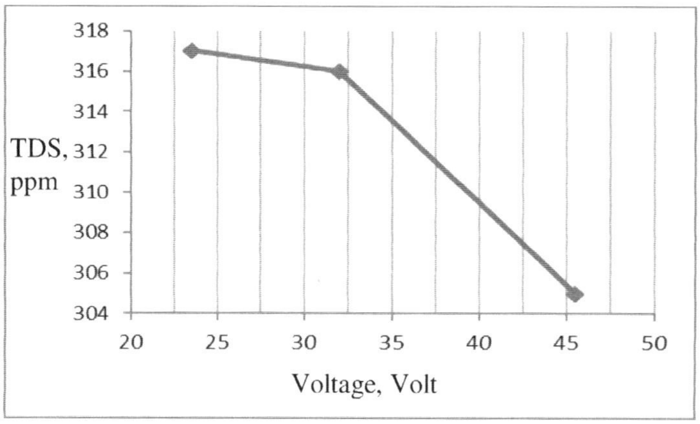

Figure 1. Effect of voltage to TDS removal.

Based on the results of research, with increasing of the voltage will remove the TDS solution, within 20 minutes first and stable until the chatode losses. The better removal, it showing for 45 volt to decrease the TDS solution from 317 to 305 ppm.

3.3. Effect of electrict current to TDS removal

The results of the research, giving the graph is as below;

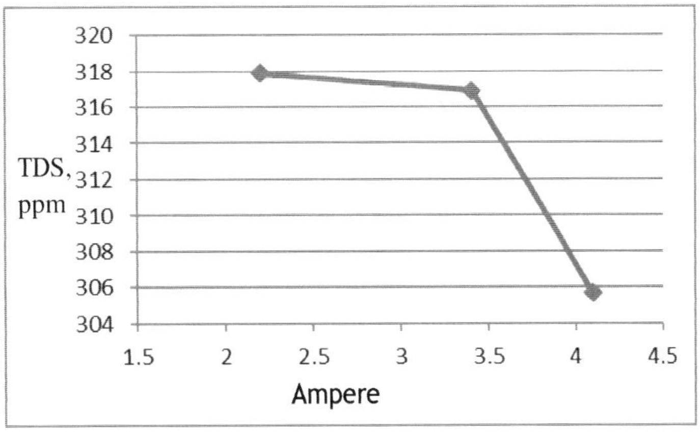

Figure 2. Effect of current on TDS removal.

Based on the graph, it can be explained that with the increasing of the electric current, will remove that TDS level from 310 ppm to 305 ppm with using 4,2 electric current. The following graph explained the correlation between voltage to ampere.

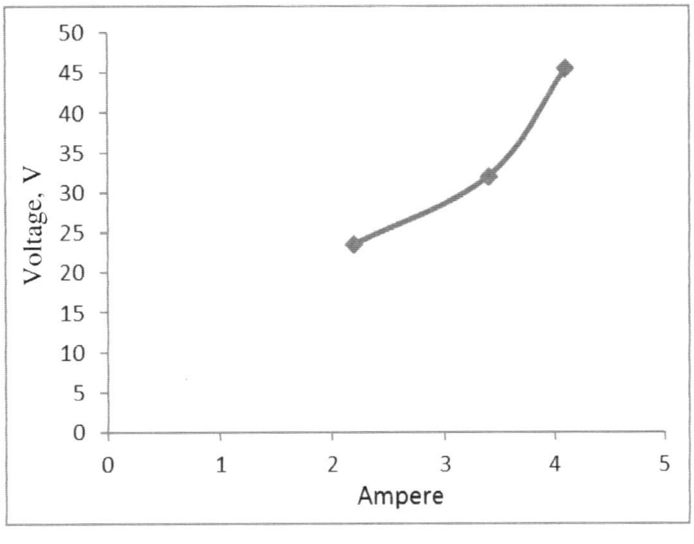

Figure 3. Correlation between electric current and voltage.

When voltage increase, the current increased as well, it is show the results that more current will get more voltage and will increase the removal level.

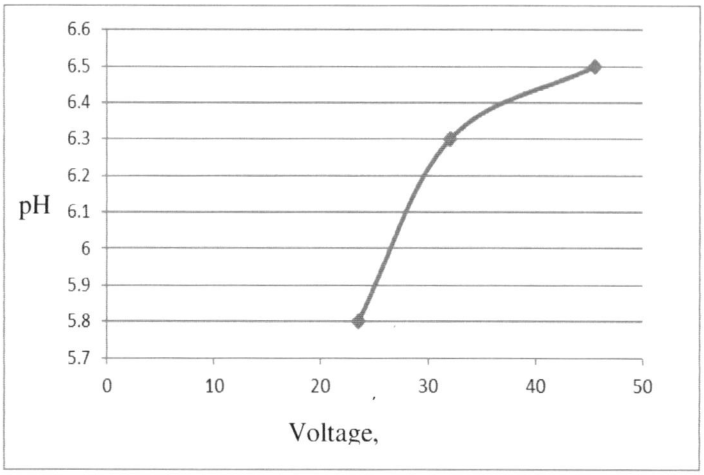

Figure 4. Effect of voltage to pH neutralization.

Based on the results, it shows that the trend when voltage increase will increase the pH to the limit pH as comply to the govermet regulation.

3.4. Life time of the cathode

According to the results for the 220 voltage and 1,8 Ampere, the contaminant has reduced from 322 ppm to 140 ppm during 6 hours more and pH have increased from 5,9 to 6,9 for 20 minutes first and stable until 85 minutes and increased more until 6,9 after 6 hours running.

Table 2. The sacrifice metal lossis.

No	Quantyty/pcs	Length/cm	weight/ gr		Weight losses (gr)
			before	after 42 hours	
1	1	185	175,13	149,691	25,439
2	5	185			127,465

The Al loss for each reactor, show the result from the 175,13 gram to 149,69 gram after 42 hours operation, so that mean, the life time of Al for continous operation could be calculted by weighting Al before processes and after processed, in this research gained that weight loss 25,439 gram during the experiment mentioned.

4. Conclusions

This study analyzed the effect of electric reactor construction on contaminants removal. Ten cells of reactor were studied: one only had aluminum electrodes, one had aluminum anodes and inert cathodes, and one had inert anodes and aluminum cathodes. The main parameters measured were final pH, TDS and times of the water produced. The following conclusions can be made from the results:

Aluminum dissolved from the Cells A was constructed from aluminum anodes and inert cathodes followed theoretical values calculated according to Faraday's law.

Aluminum dissolved from the cathodes did not follow electric charge per volume linearly. Low initial pH (pH 5,8) increased the dissolving reaction rate, probably because the hydroxyl ions produced were consumed by the acidic water.

The efficiency of aluminum in TDS removal was similar in all, cell constructions in the pH range tested. According to these results, the efficiency of electric reactor cannot be enhanced significantly by using either aluminum anodes or cathodes or both.

The results of this research indicate that reactions between TDS and aluminum occur in the bulk water and not on the electrode surface. It seems that destabilization mechanisms of TDS are similar compared to chemical coagulation. According to the results, optimum TDS matter can be obtained simultaneously with minimum residual metal concentrations when surface water is treated with electric reactor. The potential applications of water after the treatment could be potable water or industrial fresh water.

Acknowledgements

We would like to thank our team research member, the dprm ristek dikti for financial support for this research. The authors would like to thank all students.

References

[1] Chen G 2004 *Sep. Purif. Technol.* **38** 11-41
[2] Emamjomeh M M and Sivakumar M 2009 *J. Environ. Manage.* **90** 1663-1679
[3] Liu H, Zhao X and Qu J 2005 *Electrochemistry for The Environment* (New York: Springer) p. 245- 262.
[4] Sarkar M S K A, Evans G M and Donne S W 2010 *Miner. Eng.* **23** 1058-1065
[5] Mollah M Y A, Schennach R, Parga J R and Cocke D L 2001 *J. Hazard. Mater.* **84** 29-41
[6] Chen G and Hung Y 2007 *Advanced Physicochemical Treatment Technologies* (Humana Press) p. 57-106

IOP Conf. Series: Materials Science and Engineering **345** (2018) 012046 doi:10.1088/1757-899X/345/1/012046

Perspectives of high power ultrasound in food preservation

Evelyn[1,2] and F V M Silva[1]
[1]Department of Chemical and Materials Engineering, University of Auckland Private Bag 92019, Auckland 1142, New Zealand
[2]Department of Chemical Engineering, University of Riau, Pekanbaru 28293, Indonesia

E-mail: evelyn@eng.unri.ac.id

Abstract. High Power ultrasound can be used to alter physicochemical properties and improve the quality of foods during processing due to a number of mechanical, chemical, and biochemical effects arising from acoustic cavitation. Cavitation creates pressure waves that inactivate microbes and de-agglomerate bacterial clusters or release ascospores from fungal asci. Bacterial and heat resistant fungal spores' inactivation is a great challenge in food preservation due to their ability to survive after conventional food processing, causing food-borne diseases or spoilage. In this work, a showcase of application of high power ultrasound combined with heat or thermosonication, to inactivate bacterial spores *i.e. Bacillus cereus* spores in beef slurry and fungal spores i.e. *Neosartorya fischeri* ascospores in apple juice was presented and compared with thermal processing. Faster inactivation was achieved at higher TS (24 KHz, 0.33 W/g or W/mL) temperatures. Around 2 log inactivation was obtained for *B. cereus* spores after1 min (70 °C) and *N. fischeri* ascospores after 30 min (75 °C). Thermal treatments caused <1 log in *B. Cereus* after 2 min (70 °C) and no inactivation in *N. Fischeri* ascospores after 30 min (80 °C). In conclusion, temperature plays a significant role for TS spore inactivation and TS was more effective than thermal treatment alone. The mould spores were more resistant than the bacterial spores.

1. Introduction

Ultrasound technology utilizes the energy of sound waves that travel through a medium (solid, a liquid, or a gas) at a specific speed or velocity. This technology has a number of mechanical, chemical, and biochemical effects, which are used to modify the physicochemical properties and improve the quality of various food systems during processing [1]. A wide variety of application of ultrasound technology in industrial food processing has been known including drying, extraction, crystallization, freezing, emulsification, degassing, meat tenderization, defoaming, oxidation processes, and microbial or enzyme inactivation. High power ultrasound with a typical sound intensity between 10 and 1000 W/cm^2 and frequency between 20 and 100 kHz are commonly used for the inactivation of microorganisms. Microbial cell death is believed to arise from acoustic cavitation, which refers to the creation, growth, and collapse of microgas bubbles due to regions of pressure change [2]. Microbial killing involves thinning of the cell membrane, localized heating, and production of free radicals which induces adverse chemical changes in the DNA or protein denaturation [3-5].

Some microbial species can produce spores, which are resistant structures able to survive under environmental stresses such as nutrients deprivation. The outgrowth of these contaminating spores in foods can cause food spoilage, and if the microbe is pathogen, food borne illnesses and outbreaks. Among the main bacterial and fungal spore-formers of concern for the food industry are *Bacillus cereus* bacteria and *Neosartorya fischeri* mould. *B. cereus* is pathogenic bacterium commonly found in meat, rice, cereals and spices, and food intoxification can cause diarrhea or emesis depending on the type of toxin produced [6]. Psychrotrophic strains of *B. cereus* are frequently found in low-acid chilled foods [7, 8] due to their ability to grow at low temperatures (T< 8 °C) [9]. *N. fischeri* mould is typical microbes associated with spoilage of high-acid and acidified foods (pH<4.6). Thus, incidents or spoilage with these species have been registered in pasteurised fruit products such as juices, purees, jellies, jams, and canned fruits [10, 11].

The objective of the present study was, therefore to: (i) study TS inactivation of *B. cereus* spores in beef slurry at different temperatures; (ii) study TS inactivation of *N. fischeri* ascospores in apple juice at different temperatures; and (iii) compare TS and thermal inactivation of *B. cereus* spores in beef slurry and *N. fischeri* ascospores in apple juice.

2. Materials and methods

2.1. Microorganisms and sporulation

Psychrotrophic *Bacillus cereus* NZRM984 (= ATCC 11778) from the New Zealand Reference Culture Collection and *Neosartorya fischeri* JCM 1740 (= ATCC 1020) from the Japan Collection of Microorganisms were selected for this study. For sporulation of *B. cereus*, fresh cells from initial cultures were inoculated on tryptic soy agar (TSA; Difco, USA) supplemented with 0.05 g/L $MnSO_4 \cdot 4H_2O$ and incubated aerobically for 14 days at 37 °C. With respect to sporulation of *N. fischeri*, spores were obtained after growth for four weeks at 30 °C on malt extract agar (MEA). The sporulation was confirmed using phase-contrast microscopy (Motic microscope BA410 Series).

2.2. Food inoculation and preparation

The following foods were chosen since they are prone to contamination by the microorganisms studied: beef slurry (pH 6.5) for *B. cereus*; and apple juice (pH 3.7, 10.6 °Brix) for *N. fischeri*. Each microbial spore was inoculated in each food in TS experiments to yield a final concentration of approximately ~10^6–10^7 cfu/g or cfu/mL prior to processing.

2.3. Spore enumeration

The *B. cereus* spore concentration in beef slurry before and after processing was determined by spread plating into nutrient agar (NA) supplemented with 0.1% soluble starch and incubation at 37 °C for 24-48 h. Prior to plating, samples were placed in 110 × 230 mm sterilized stomacher bags (Inter science, France) and decimal diluted ten times with 0.1% (w/v) buffered peptone water. These dilutions were considered in the calculation of the final spore concentration. Samples were then homogenized in the stomacher (Masticator Stomacher, IUL Instruments, Germany) for 2 min prior to plating onto NA plates. With respect to *N. fischeri*, the mould ascospore concentration in apple juice before and after TS processing was determined by spread plating onto MEA and followed by incubation at 30 °C for 3 to 5 days.

2.4. Thermosonication

A Hielscher UP200S ultrasonic processor (Hielscher-Ultrasonic GmbH, Germany) with a sonotrode tip diameter of 3 mm was used for all TS experiments. The processor has a frequency of 24 kHz and was operated at 100% amplitude (460 W/cm^2, 0.33 W/g or W/mL). For each temperature (70, and 50 °C for *B. cereus*; 75 and 65 °C for *N. fischeri*), a 200 mL round bottom-flask containing the food sample (100 g or 100 mL) was placed in a thermostatic water bath inside a biosafety cabinet (class II, type A2, Esco Micro Pte. Ltd., Singapore) to bring the sample temperature to the designated temperature. Inoculation of spores was carried out aseptically when the sample reached the designated temperature, and then the ultrasound was turned on. The temperature of the food sample during processing was monitored and a thermostatic water bath was used to keep it at the desired value (±1 °C). Ultrasonic treatments were carried out for up to 70 min, depending on the microbes and TS temperature. Food samples were removed from the water bath at pre-specified intervals, and spore survivor counts were immediately performed.

2.5. Thermal processing

Regarding the thermal treatments, transparent packed food samples containing the inoculated spores were processed in a thermostatic water bath for up to 150 min with the temperature set to the processing temperature: 70 °C for *B. cereus*, and 80 °C for *N. fischeri*. The pouches were maintained at the desired temperature during the thermal treatments, taken out at different time intervals, and kept in an ice water bath until microbial enumeration. At least two independent survival experiments were carried out for each TS and thermal treatment temperature with duplicate samples processed for each time.

2.6. Statistical analysis

At least two survival experiments were run for each temperature, subjected to the same or different processing method, and average values ± standard deviation were calculated. Then, significant differences in the microbial log reductions from different temperatures under the same method and similar temperature under different methods were investigated by performing a one-way analysis of variance (ANOVA) followed by Tukey's tests, with a confidence level of 95% ($p < 0.05$).

3. Results and discussion

3.1. Effect of TS temperature on the inactivation of B. cereus spores in beef slurry

The effect of increasing temperature from 60 to70 °C in a 0.33 W/g TS process on the log survivors of *B. cereus* spores in beef slurry is illustrated in Figure 1. A higher spore inactivation was obtained with a higher temperature. For example, for a 2 min process, >5.0 log was obtained at 70 °C compared to 1.7 log at 60 °C ($p < 0.05$). The higher spore inactivation at the higher temperature (i.e. around 70 °C) for the survival curves are in agreement with previous study [12] with *Bacillus subtilis* spores in milk. The results suggest benefit of TS method for the inactivation of this bacterial spores.

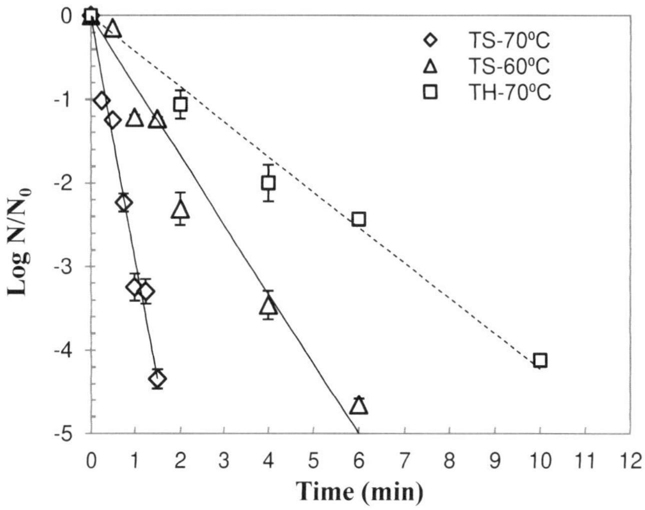

Figure 1. Effect of different temperatures on the thermosonication (TS, 24 kHz, 0.33 W/g) inactivation of *B. cereus* spores inbeef slurry and its comparison with 70 °C thermal treatment (TH) (data from Evelyn and Silva, 2015 [13]).

3.2. Effect of TS temperature on the inactivation of N. fischeri ascospores in apple juice

The log survivors of *N. fischeri* ascospores in apple juice at 65 and 75 °C after 0.33 W/mL TS treatments are shown in Figure 2. Activation shoulders were observed for all TS temperatures tested, with a maximum at 10 min for 75 °C and 30 min for 65 °C, followed by approximately linear spore inactivation. Increase in the spore number at the beginning of the treatments could be due to the role of sonication in de-agglomerating fungal clusters and producing suspension of free ascospores [14-16]. Lower spore activation and faster inactivation were obtained as the TS temperature was increased from 65 to 75 °C, indicating significant role of temperature on the inactivation of these ascopores for up to 75 °C. The 75 °Cwas the maximum temperature supported by the ultrasound equipment. Overall while 30 min at 75 °C achieved ≃ 2 log reductions, > 80 min at 65 °C were required to obtain the same spore inactivation. Although more log reductions at 75 °C can be obtained at longer times (>30 min), the long processing times and activation shoulders for this type of microorganism and food make TS should be re-evaluated for commercial application.

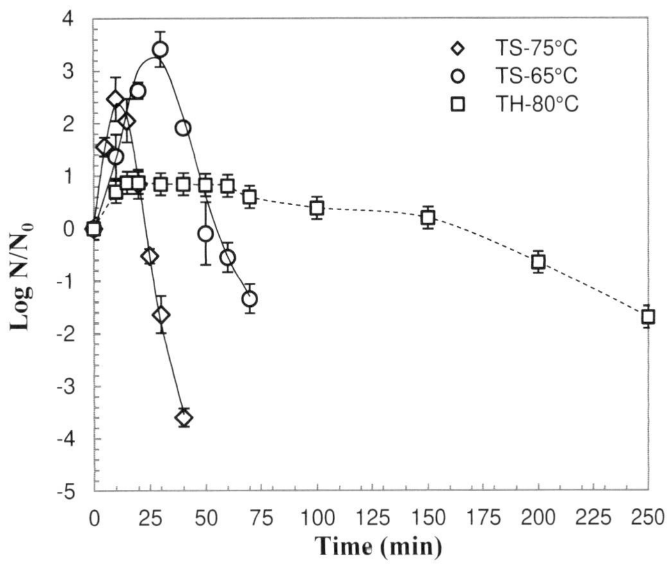

Figure 2. Effect of different temperatures on the thermosonication (TS, 24 kHz, 0.33 W/mL) inactivation of *N. fischeri* spores in apple juice and its comparison with 80 °C thermal treatment (TH) (TS data from Evelyn et al. 2016 [15]).

3.3. TS versus thermal inactivation of B. cereus and N. fischeri spores

Figure 1 and Figure 2 also show a comparison of the log survivors of *B. cereus* spores in beef slurry by 0.33 W/g TS-70 °C vs. 70 °C thermal process, and the log survivors of *N. fischeri* ascospores in apple juice by 0.33 W/mL TS-75 °C vs 80 °C thermal process, respectively. The TS inactivation for both microbes was much faster than thermal inactivation alone. Regarding *B. cereus* spores, >5.0 log reduction for TS vs. 0.9 log reduction for thermal alone were obtained after a 2 min process at 70 °C (Figure 1, $p< 0.05$). In fact, similar log reduction (0.9 log, $p> 0.05$) were obtained after 2 min for the 70 °C thermal process and 50 °C TS process, thus demonstrating that TS treatment requires 20 °C lower temperature than a thermal process alone to obtain the same level of lethality. With respect to *N. fischeri* ascospores, in order to compare TS with thermal results, 80 °C was used for the thermal process since no inactivation of the ascospores by 75 °C thermal treatment was observed (not shown). Similar spore inactivation (≈ 1.7 log, $p> 0.05$)was obtained for the 75 °C TS process (after 30 min) and 80 °C thermal process (after 250 min), indicating the remarkable advantage of ultrasound. The benefit of TS vs. thermal for spore inactivation suggesting less impact of heat in the food quality was also documented by other authors in past studies [12, 13, 17].

4. Conclusions

Increasing TS temperature for up to 70 or 75 °C caused an increase in the inactivation of *B. cereus* spores in beef slurry and *N. fischeri* ascospores in apple juice. 0.33 W/g TS–70 °C for a short time period (2 min) was a better method than 70 °C thermal for the inactivation of *B. cereus* spores in beef slurry, and 0.33 W/mL TS–75 °C was also a better process than 80 °C thermal treatment for *N. fischeri* ascopore inactivation in apple juice, confirming the benefit of ultrasound technology. A reduction of >5.0 log can be obtained for psychrotrophic *B. Cereus* spores after 2 min of TS–70 °C, however further research must be carried out for other bacterial spores. Activation shoulders and long treatment times for *N. fischeri* ascospore inactivation by TS–75 °C make TS not yet feasible for processing high acidic foods containing this heat resistant spores. The TS process might be applicable at longer treatment times (≥30 min) and higher temperatures (75 °C), thus further research must be conducted to reduce the activation shoulders and design an ultrasound probe that can withstand higher temperatures. The results from this study show that TS might be a better option over thermal treatment for the preservation of food products prone to contamination of these microbial spores, however caution must be taken for each microbial spores.

References

[1] Rastogi N K 2011 Opportunities and challenges in application of ultrasound in food processing *Crit. Rev. Food Sci. Nutr.* **51** 705-722
[2] Chen 2012 *Handbook on Application of Ultrasound: Sonochemistry for Sustainability*, ed S S Chen and A Mudhoo (Boca Raton, US: CRC Press) p 373-406
[3] Butz P and Tauscher B 2002 Emerging technologies: Chemical aspects *Food Res. Intl.* **35** 279-284
[4] Piyasena P, Mohareb E and McKellar R 2004 Inactivation of microbes using ultrasound: A review *Intl. J. Food Microbiol.* **87** 207-216
[5] Riesz P and Kondo T1992 Free radical formation induced by ultrasound and its biological implications *Free Radical Bio. Med.* **13** 247-270
[6] Dierick K, Van Coillie E, Swiecicka I, Meyfroidt G, Devlieger H, Meulemans A, Hoedemaekers G, Fourie L, Hendrickx M and Mahillon J 2005 Fatal family outbreak of *Bacillus cereus*-associated food poisoning *J. Clin. Microbiol.* **43** 4277-4279
[7] Carlin F, Girardin H, Peck M W, Stringer S C, Barker G C, Martinez A, Fernandez A, Fernandez P, Waites W M and Movahedi S 2000a Research on factors allowing a risk assessment of spore-forming pathogenic bacteria in cooked chilled foods containing vegetables: A FAIR collaborative project *Intl. J. Food Microbiol.* **60** 117-135
[8] Carlin F, Guinebretiere M H, Choma C, Pasqualini R, Braconnier A and Nguyen-the C 2000b Spore-forming bacteria in commercial cooked, pasteurised and chilled vegetable purees *Food Microbiol.* **17** 153-165
[9] Choma C, Guinebretiere M, Carlin F, Schmitt P, Velge P and Granum P 2000 Prevalence, characterization and growth of *Bacillus cereus* in commercial cooked chilled foods containing vegetables *J. Appl. Microbiol.* **88** 617-625
[10] Pitt J I and Hocking A D 1997 *Fungi and Food Spoilage* vol 2 (London, UK: Blackie Academic and Professional).
[11] Silva F V M, Gibbs P A, Nunez H, Almonacid S, and Simpson R 2014 *Encyclopedia of Food Microbiology* vol 2, ed C A Batt and M L Tortorello (Amsterdam: Elsevier) p 577-595
[12] Garcia M L, Burgos J, Sanz B and Ordonez J 1989 Effect of heat and ultrasonic waves on the survival of two strains of *Bacillus subtilis J. Appl. Microbiol.* **67** 619-628
[13] Evelyn and Silva F V M 2015 Thermosonication versus thermal processing of skim milk and beef slurry: Modeling the inactivation kinetics of psychrotrophic *Bacillus cereus* spores *Food Res. Intl.* **67** 67-74

[14] King A D, Michener H D and Ito K A 1969 Control of *Byssochlamys* and related heat-resistant fungi in grape products *Appl. Microbiol.* **18** 166-173
[15] Evelyn, Kim H J and Silva F V M 2016 Modeling the inactivation of *Neosartorya fischeri* ascospores in apple juice by high pressure, power ultrasound, and thermal processing *Food Control* **59** 530-537
[16] Michener H D, and King AD1974 Preparation of free heat-resistant ascospores from *Byssochlamys* asci *Appl. Microbiol.* **27** 671-673
[17] López-Malo A, Palou E, Jiménez-Fernández M, Alzamora S M and Guerrero S 2005 Multifactorial fungal inactivation combining thermosonication and antimicrobials *J. Food Eng.* **67** 87-93

IOP Conf. Series: Materials Science and Engineering **345** (2018) 012047 doi:10.1088/1757-899X/345/1/012047

Adsorption kinetics of ion of Pb^{2+} using Tricalcium Phosphate particles

A Fadli , S R Yenti, F Akbar, Maihendra and F Mawarni
Department of Chemical Engineering, Faculty of Engineering, Universitas Riau

E-mail: fadliunri@yahoo.com

Abstract. One of the heavy metals that can pollute water is Pb^{2+}. The concentration of ion Pb^{2+} can be removed using the adsorption method. The purpose of this research is to determine the adsorption kinetics model of ions Pb^{2+} using tricalcium phosphate (TCP) particles with variation of the temperature and adsorbent dosage. Five hundred mililiter Pb^{2+} solution with of 3 mg/L were added 0,5 gr, 1 gr and 1,5 gr of TCP in a glass beaker and stirred with rate of 300 rpm at a temperature of 30 °C, 40 °C and 50 °C. Pb^{2+} concentration in solution was analyzed by AAS (Atomic Adsorption Spectroscopy). The results showed that the rate of adsorption increased with the increasing of the temperature and adsorbent dosage. Minimum constant value of adsorption kinetic was 1,720 g/mg.min obtained at temperature of 30 °C and adsorbent dosageof 0,5 gr. The maximum value of adsorption kinetic constant was 9,755 g/mg.min obtained at temperature of 50 °C and adsorbent dosage of 1,5 gr. The appropriate model for adsorption kinetics followed the pseudo second order.

1. Introduction

Lead metal pollution becomes an environment problem. Lead metal can cause human health effects such as allergens, encephalopathy, anemia and carcinogens [1]. Currently, several method were used for removing the heavy metals from aqueous solution, such as ion exchange, carbon adsorption andneutralization. Adsorption have been considered high efficiency and low cost method, it can be used to remove the heavy metals concentration in solution[2]. Tricalcium phosphate (TCP) can be used to reduce the lead metals. TCP as adsorbent have also been considered amongst ideal low cost with smaller particle (nano and micrometer) so the surface area is larger, the highest stability and low density [3-4].

Yunizar (2013) has been studied the application of lead adsorption in water using tricalcium phosphate particle by determining the equilibrium of adsorption. The optimal TCP adsorption capacity at operating temperature of 30 °C with concentration 15 mg/L is 6,7880 mg/g with value of ΔH equal -8,8 kcal/mol.K and exothermic [5]. Kinetic study of lead in aqueous solution using tricalcium phosphate particle obtained the value of TCP optimal rate adsorption occured at concentration 14,619 mg/L with stirred 300 rpm is 3,129 g/mg.min [6].

This paper reported the effects of temperature and adsorbent dosage on the adsorption of lead using TCP. In addition, the kinetics of adsorption model with two orders kinetics adsorption of pseudo-first-order and pseudo-second-order are also reported to determine a suitable kinetic model for metal adsoption.

2. Materials and methods

The materials used in this study are Pb $(NO_3)_2$, aquadest, and tricalcium phosphate $Ca_3(PO_4)_2$ as adsorbent. Adsorption process was carried out by addingfive hundred mililiter Pb^{2+} solution with 3 mg/L into 0,5 gr, 1 g and 1,5 gr of TCP in a glass beaker and stirred with rate of 300 rpm at a temperature of 30 °C, 40 °C and 50 °C. The solution was taken at intervals of 1 minute for 5 minuted as much as 10 mL, then separation by centrifuged. Pb^{2+} concentration in solution was analyzed using Atomic Adsorption Spectroscopy (AAS).

3. Results and discussion

3.1 Effect of time and temperature on Pb^{2+}concentration

The Pb^{2+} metal solution was prepared by dissolving Pb $(NO_3)_2$ into the aquadest to obtain Pb^{2+} at the concentration of 3 ppm. The solution was added tricalcium phosphate as adsorbent. In Adsorption process the concentration of lead metal in solution will be decrease each time, itshown in Figure 1.

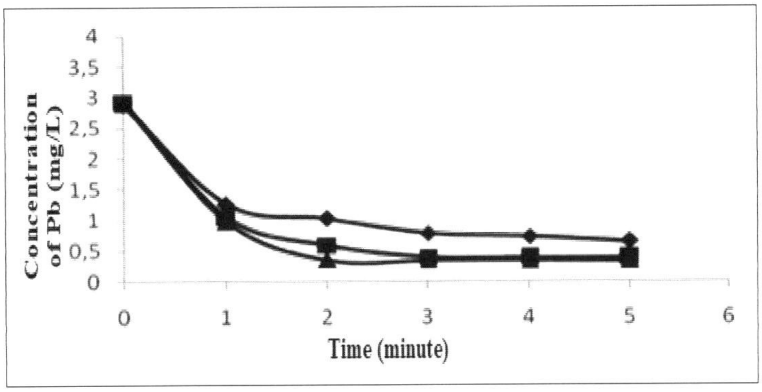

Figure 1. Relation of un-adsorbed Pb initial concentration (Co) and time t (min) at 30 °C.

In five minute, Pb concentration was constant, so this time can be taken as equilibrium time. The smallest Pb concentration in solution at 5 minute with process condition at 30 °C, 0,5 gr adsorbent dosage and initial concentration (Co) 2,872 mg/L is 0,653 mg/L. Its due to the addition of adsorbent, PO_4^{3-} adsorbent active site is increase, and the amount of lead metals was adsorbed is high. In the most cases, the increasing of the adsorbent amount causes the increasing of the adsorbent capacity. Due to the surface contact area of adsorben is increases with the more adsorbent dosage [7]. It also happen that the increasing of the activation time will increase adsorption ability, until become constant or decrease [8].

3.2 Adsorption kinetics

3.2.1. Pseudo first order

The lead (Pb) concentration againts time for pseudo first order equation with adsorbent dosage and temperatures variation are shown ini Figure 2. The data shown adsorbent dosage for variation oftemperature i.e. 30°C, 40 °C, 50 °C. The first order model follows Equation 1 [9]:

$$\ln (q_e - q_t) = \ln q_e - k_1 t \quad (1)$$

where q_e is adsorption equilibrium capacity (mg/g), q_t is amount of adsorbate adsorbed at time t (min) (mg/g), and k_1 is adsorption rate constant.

Based on Figure 2, we get the result of q_e and R^2 values for first order. Table 1 shown the k and %SSE for first order kinetics. The concentration of lead metals was adsorbed in variation of adsorbent dosage are 0,5 gr, 1 gr, 1,5 gr at 30 °C for this study indicates a considerable correction range from

18,747% to 53,987%,at 40°C the range from 6,88% to 81,492%, and at 50 °C the range from 14,652% to 29,403%.

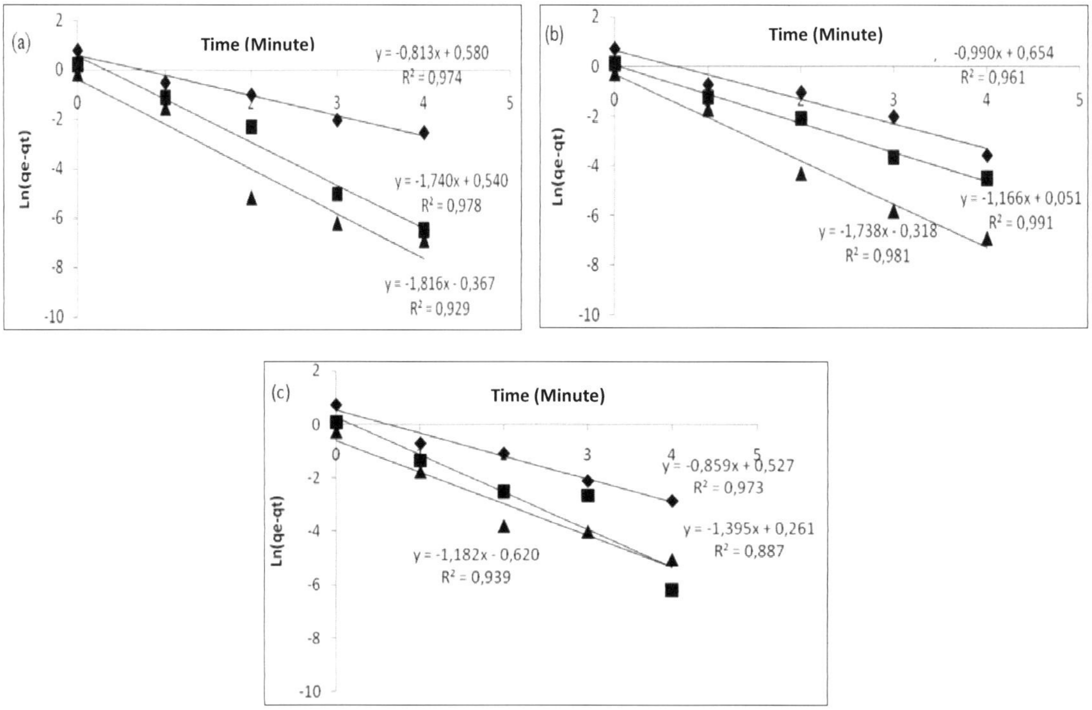

Figure 2. Adsorption process follows pseudo first order kinetics model (a) 30 °C gr, (b) 40 °C, (c) 50 °C.

Table 1. Parameter of kinetics adsorption pseudo first order model with temperature variation

Temperature	30 °C			40 °C			50 °C		
TCP Dosage (gr)	k_1 (1/min)	R^2	%SSE	k_1 (1/min)	R^2	%SSE	k_1 (1/min)	R^2	%SSE
0,5	-0,813	0,974	19,646	-0,993	0,961	6,88	-0,859	0,973	19,646
1	-1,740	0,978	53,987	-1,166	0,991	15,304	-1,395	0,887	14,652
1,5	-1,816	0,929	18,747	-1,395	0,981	81,492	-1,182	0,939	29,403

3.2.2 Pseudo second order

The lead (Pb) Concentration againts time for pseudo second order equation with adsorbent dosage and temperatures variation are shown ini Figure 3. The data shown for temperature variation 30 °C, 40 °C, 50 °C and 1,5 gr adsorbent dosage. The second order model follows equation [10]:

$$\frac{t}{q_t} = \frac{1}{k_2 {q_e}^2} + \frac{1}{q_e} t \quad (2)$$

Based on Figure 3, we get the result of q_e and R^2 values and comparison between experimental data for second order model. Where the value of q_e is 0,762 mg/g and R^2 0,997. Table 2 shown the k and %SSE for second order kinetics. The concentration of lead metals was adsorbed in temperatures variation 30°C, 40°C, and 50°C and 0,5 gr adsorbent for this study indicates a considerable correction range from 3,23% to 3,578%, at 1 gr adsorbent the range from 2,897% to 3,28%, and at 1,5 gr adsorbent the range between 2,47% and 2,636%. The small value of correction indicates that the equation of second order can be used.

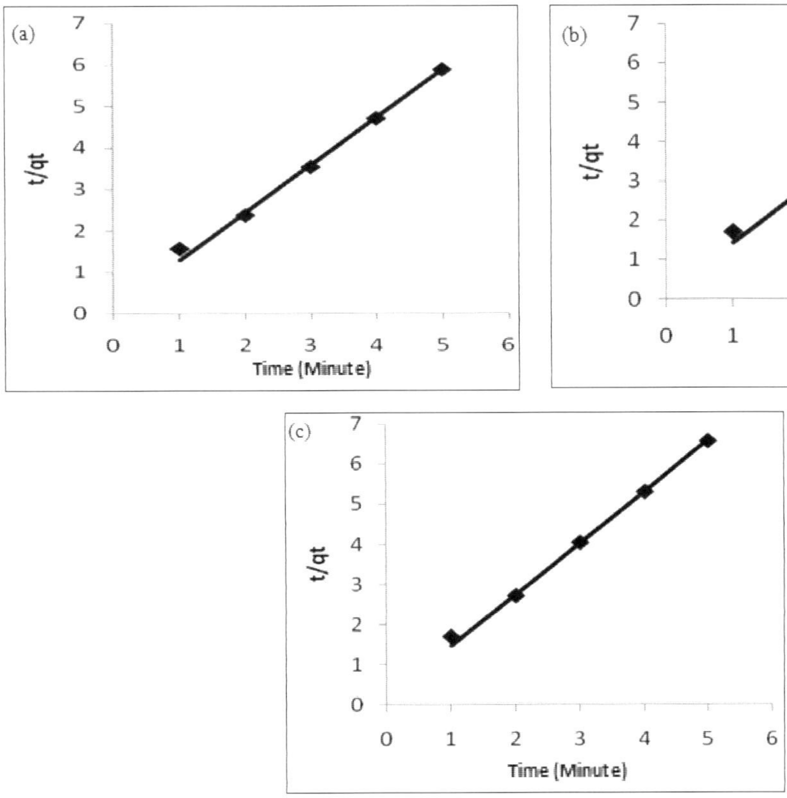

Figure 3. Adsorption process follows pseudo second order for 1,5 gr adsorbent at (a) 30 °C, (b) 40 °C and (c) 50 °C.

Table 2. Parameter of kinetics adsorption pseudo second order model with temperature variation

Temperature	30°C			40°C			50°C		
TCP Dosage (gr)	k_2 (g/mg.min)	R^2	%SSE	k_2 (g/mg.min)	R^2	%SSE	k_2 (g/mg.min)	R^2	%SSE
0,5	1,720	0,991	3,578	1,982	0,992	3,281	2,022	0,992	3,230
1	4,117	0,994	3,217	4,330	0,994	3,280	4,996	0,995	2,897
1,5	8,479	0,995	2,636	9,6	0,996	2,605	9,755	0,997	2,470

From Table 2 we find a suitable kinetics model for adsorption of lead metal ions using TCP is pseudo second order. The highest linear regression (R^2) parameter is 0,997 with the lowest %SSE value is 2,470. The kinetics adsorption constant of 9,755 g/mg at 50 °C and adsorbent dosage 1,5 gr. The value of adsorption velocity constant increases with the increasing of the temperature. Base on the theory that the adsorption rate increases with the highest temperature, where at high temperatures the activation of adsorbent surface side activates and give more kinetic energy of metals ion [11].

3.3 Effect of adsorbent dosage on constant value (k) adsorption rate

Kinetics adsorption follows pseudo second order model to seen the effects of adsorbent dosage for adsorption of lead metal ions rate constant value. Data of adsorption lead metals ions is calculated using Equation (2).

Based on Figure 4, we get result of q_e and R^2 values. The comparison between experimental data with second order model is increases with the addition of adsorbent. Where the highest of $R^2$of 0,997 at temperature 50 °Cand stirred at 300 rpm. From the figure we find R^2, and value of k and %SSE shown in Table 3.

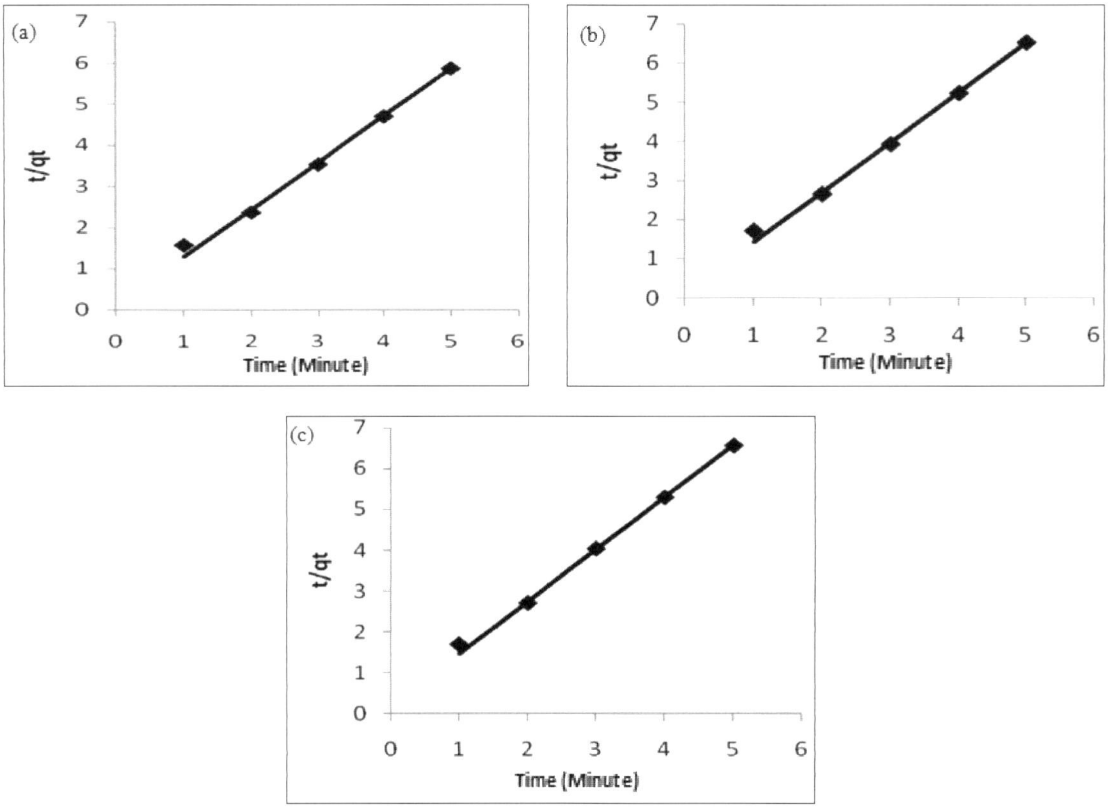

Figure 4. Adsorption process follows pseudo second order at 50 °C for adsorbent dosage (a) 0,5 gr, (b) 1 gr dan (c) 1,5 gr.

Table 3. Parameter of kinetics adsorption pseudo second order model with adsorbent dosage variation

TCP Dosage	0,5 gr			1 gr			1,5 gr		
Temperature (°C)	k_2 (g/mg.min)	R^2	%SSE	k_2 (g/mg.min)	R^2	%SSE	k_2 (g/mg.min)	R^2	%SSE
30	1,720	0,991	3,578	4,177	0,994	3,217	8,479	0,995	2,636
40	1,982	0,992	3,281	4,330	0,994	3,280	9,6	0,996	2,605
50	2,022	0,992	3,23	4,996	0,995	2,897	9,755	0,997	2,47

From Table 3 indicate that the increase of constant (k) adsorption rate was increased with addition of adsorbent dosage. The increase of constant (k) shown in Figure 5.

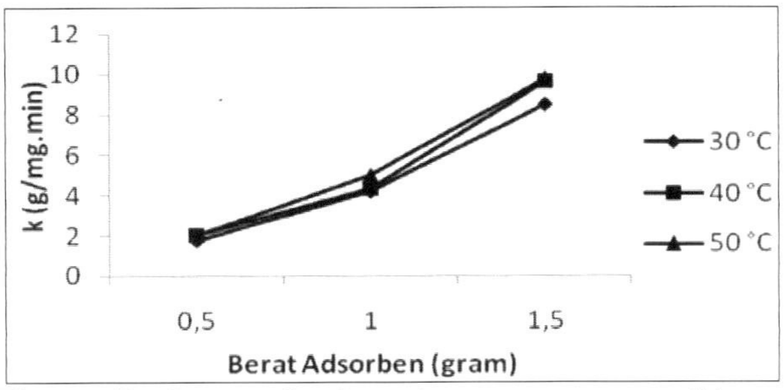

Figure 5. Curve of adsorption rate constant (k) to adsorbent dosage.

The addition of adsorbent dosage will increased the k value, caused the amount of adsorbent will activated the adsorbent active surface area, so the adsorption process progressively faster and increase the rate of adsorption (k) [12]. In this research, the results obtained for pseudo second order model with linear regression value $R^2$is 0,997, %SSE 2,47 % and kinetics constant (k) 9,755 g/mg.min. The optimal condition were obtained at 50 °C with 1,5gr adsorbent dosage.

4. Conclusion

The adsorption process of lead metal ions in aqueous solution using tricalcium phosphate as adsorbent are influenced by temperature and adsorbent dosage. The higher temperature and amount of adsorbent dosage can causes the adsorption constant become larger. The highest lead metal ions adsorption constant (k) using tricalcium phosphate shown at optimal condition of 50 °C and adsorbant dosage 1,5 gr is 9,775 g/mg.min. The kinetics adsorption models to adsorb lead metal ions using tricalcium phosphate followed pseudo second order with k 9,775 g/mg.min.

References

[1] Mahurpawar 2015 Effect of heavy metals on human health *Int. J. Res. Granthaalayah* **2** 350-0530

[2] Yuchi N, Tadashi H, Jun Y, Yoko Y and Toshiro O 2015 Adsorptiom and removal of strontium in aqueous solution by synthetic hydroxyapatite *J. Radioanal Nucl. Chem.* **307** 1279-1285

[3] Davod G, Iman M and Seyed A M 2017 Sorption thermodynamic and kinetic studies of lead removal from aqueous solutions by nano tricalcium phosphate *Bull. de la Société Royale des sciences de Liège* **86** 96 – 112

[4] Krestou A, Xenidis A and Panias S 2003 Mechanism of aqueous uranium (VI) uptake by hydroxyapatite *J. Miner. Eng.* **17** 373-381

[5] Yunizar F 2013 *Penjerapan ion timbal (Pb^{2+}) terlarut dalam air menggunakan partikel tricalcium phosphate sebagai adsorben* Skripsi Teknik Kimia Universitas Riau

[6] Fitriani I 2015 *Kinetika adsorpsi pada penjerapan ion timbal (Pb^{2+}) terlarut dalam air menggunakan partikel tricalcium phospate* Skripsi Teknik Kimia Universitas Riau

[7] Kumar P S, Vincent C, Kirthika K and Satish K 2010 Kinetics and equilibrium studies of Pb^{2+} ion removal from aqueous solutions by use of nano-silversol-coated *J. Chem. Eng.* **27** (02) 339-346

[8] Kaushik N, Thomas C and McWiliam I G 2008 *Adsorption Study of Cupper (II)* (New York: Jhon Wiley & Son Inc.)

[9] Hui K S, Chao C Y H and Kot S C 2005 Removal of mixed heavy metal ions in wastewater by Zeolite 4A and residual products from recycled coal fly ash *J. Hazard. Mater.* **B137** 89-101

[10] Michael H J and Ayebaemi I S 2005 Effects of temperature on the sorption of Pb^{2+} and Cd^{2+} from aqueous solution by caladium bicolor (Wild Cocoyam) biomass *J. Biotech.* **8** (02)

[11] Otero M, Rozada F, Calvo L F, Garcia A I and Mordn A 2003 Kinetic and equilibrium modeling of the methylene blue removal from solution by adsorbent materials produced from sewage sludges *J. Biochem. Eng.* **15** 59–68
[12] Kumar P S and Kirthika K 2009 Equilibrium and kinetic study of adsorption of nickel from aqueous solution onto bael three leaf powder *J. Eng. Sci. Technol.* **4** 351-363

IOP Conf. Series: Materials Science and Engineering **345** (2018) 012048 doi:10.1088/1757-899X/345/1/012048

Assesment of $PM_{2.5}$ emission from corn stover burning determining in chamber combustion

Hafidawati[1,2], P Lestari[1] and A Sofyan[2]
[1]Environmental Engineering, Faculty of Engineering, University of Riau HR Subrantas, Pekanbaru, Riau, Indonesia
[2,3]Environmental Engineering, Faculty of Civil and Engineering, ITB Ganesha 10, Bandung, Indonesia

E-mail: afi.waldi@gmail.com

Abstract. Chamber measurement were conducted to determine Particulate Matter ($PM_{2.5}$) emission from open burning of corn straw at Garut District, West Java. The of this study is to estimate the concentration of $PM_{2.5}$ for two types of corn (corncobs and cornstover) for five varieties (Bisma, P29, NK, Bisma, NW). Corn residues were collected and then burned in the chamber combustion. The chamber was designed to simulate the burning in the field, which was observed in the field experiment that meteorological condition was calm wind. The samples were collected using a minivol air sampler. The assessment results of $PM_{2.5}$ concentrations (mg/m^3) from open burning experiment in the chamber for five varieties of corn cobs (Bisma, P29, NK, Bisi, NW) was 9.187; 2.843; 7.409; 3.781; 1.895 respectively. Concentration for corn stover burn was 2.060; 5.283; 4.048; 5.306 and 5.697 respectively. Fluctuations in the value of concentration among these varieties reflect variations in combustion conditions (combustion efficiency) and other parameters including water content, biomass conditions and the meteorological conditions. The combustion efficiency (MCE) of the combustion chamber simulation of corncobs ia lower than the MCE of corn stover, that the concentration $PM_{2.5}$ more emitted from the burning of corn stover. The results of this study presented provide useful information for the development of local emission factors for $PM_{2.5}$ from open burning of corn stover in Indonesia.

1. Introduction

Field burning of corn straw was a common practice in Garut district, West Java. About 58% farmers burn corn straw after harvest for eliminate corn residue from the field. Corn straw burning was contributed more than 80% of the total biomass open burning emissions [1]. One of the major source of aerosol emissions emits type of particulate matter (PM) [2]. Many research projects have been initiated to assess the intensity of air pollution and to identify the major contributors. The purpose of this study was to estimating and analyzing $PM_{2.5}$ emissions from burning straw during harvest season in West Java Province. In this study, laboratory simulation experiments were conducted to characterize concentration of $PM_{2.5}$. Burning experiments were carried out using the burning chamber of the Open Burning Test Facilities (OBTF) type[3]. Corn straw burning experiments were conducted in the chamber to simulate open burning in order to determine concentration of $PM_{2.5}$ through real time equipment measurement. The chamber was designed to simulate open burning in the field, which was observed in the field experiments that meteorological condition was calm wind.

2. Experimental methods

2.1 Source of experimental material

Corn straw were chosen as representatives of agricultural crop residue. The corn straw is taken from a rural field in the Garut districts. Burning straw conducted into five varieties, they were collected from five villages (Baba kansaung, Batu karut, Cisalam, Bojong and Dangdeur) in the districts Banyuresmi, Garut. It is located at 6°56'49"-7 °45'00" South latitude and 107°25'8"-108°7'30" East Longitude. District Banyuresmi is located in the West Java Province it is about 66 km from Bandung. This district as one of the centers of corn production in West Java with corn production 6.3 tons/year (Figure 1).

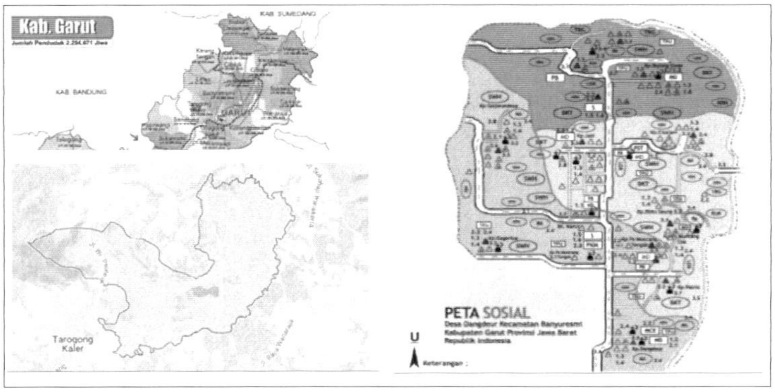

LS1 = Babakan Saung Village
LS2, LS3, LS5 = Cisalam Village
LS4 = Batu Karut Village
LS6 = Bojong Village
LS7 = DangdeurVilllage

Figure 1. A map showing the location of study area.

2.2 Experimental process

The Experimental has been performed in the combustion chamber system consisted of stainless-steel. The design of chamber is presented in Figure 2. Three sides of combustion zone are closed with heat durable steel sheet and one side is opened for the air circulation. The straw was burned to measure the weight before and after burning. During combustion, time was recorded and emission was measured using the same equipments and the same method. The experimental of $PM_{2.5}$ was collected a samplers minivol (Anderson 214 series) on the teflon filter (whatman 47 mm). The emission measurement equipments were installed in front of opened side chamber at ground level for 1 m height. Meteorological condition (temperature, pressure and humidity) and ambient air quality (concentration of CO and CO_2) were also measured.

2.3. Analysis methods

Particulte Matter quartz fiber filters of 47 mm in diameter (whatman international Ltd. UK) were used for Particulate Matter sampling. Before and after sampling filter were condition for 24 hours at desicator and weighted at five times after and before sampling using scale Boeco Germany with accuracy up to 0,00001 g and deviation of ±0.00005 g. This value (m) is used in the calculation of gravimetric analysis to determine the concentration of $PM_{2.5}$ in aerosol burning corn straw. The net mass was obtained by subtracting the average of pre-sampling weights from the average of the post sampling weight. All samples were analyzed for particulate matter ($PM_{2.5}$) mass methods gravimetry to be analyzed at Air Quality Laboratory Bandung Institute of Technology. For the QA/QC purpose during transport between field and laboratory locations, the filters were carefully placed in the filter holders and used for sample collection. After sampling the filters were taken out of the holder and placed in a petri dish. Blank filter was treated in the same way and used for the correction of the calculation results of concentration.[12]

Figure 2. Design chamber combustion.

2.4 Computing method

In order to estimate the concentration of $PM_{2.5}$ of corn straw, burning experiments were conducted in the simulate open burning chamber. Calculation of burned corn straw was done by[19]

Particulate weight (M) were calculated using Equation 1

$$M_p = M_2 - M_1 \tag{1}$$

Where M_P is particulate weight (g), M_1 is filter's weight before sampling (g) and M_2 is filter's after sampling (g). The particulate weight (Mp) obtained is corrected by the weight of the blank filter.

Actual Volume (Vact) was determined using Equation 2.

$$V_{act} = Q\ .\ t \tag{2}$$

where Vact is actual volume (m^3), Q is flowrate (lpm) and t is sampling duration (minutes)

For conversion actual volume (Vact) to standard volume (V std) was calculate using Equation 3.

$$\frac{P_{std} \times V_{std}}{T_{std}} = \frac{P_{act} \times V_{act}}{T_{act}} \tag{3}$$

where *Pstd* is standar pressure at 1 atm, *Vstd* is volume at standard condition (P=1 atm and T 25 °C) (Nm^3), *Tstd* is temperature at standard condition (25 °C), *Pact* is actual pressure (atm), *Vact* is volume aktual (m^3) and *Tact* is actual temperature (°C).

The concentrations of $PM_{2.5}$ emission, from the smoke emision Wet calculated on equation 4 and net burning smoke concentration Wet calculated on Equation 4:

$$C = \frac{M_p}{V_{std}} \times 10^6 \tag{4}$$

Where, C is concentration of particulate mass (mg / Nm^3), Mp is particulate weight (g) and Vstd is calculated volume in standard state (Nm^3), was done by Equation.5.

$$C\ PM2.5\ (net) = C\ PM_{2.5}\ (background) - C\ PM_{2.5}\ (sampling) \quad (5)$$

3. Results and discussion

3.1. $PM_{2.5}$ concentration

In this study conducted an investigation into the estimation of $PM_{2.5}$ concentration and chamber combustion characteristics of corn straw to five corn varieties (Bisma, P29, NK, BISI 18 and NW). Combustion characteristics is shown in Table 1.

Table 1. Characteristic of corn straw burning on chamber combustion

Parameter	Bisma (cobs)	Bisma\ (stover)	P29 (cobs)	P29 (stover)	NW (cobs)	NW (stover)	NK33 (cobs)	NK33 (stover)
Temp (°C)	30.7	34.9	34.7	29.2	36.6	30.2	32	31.3
P (mbar)	933.9	935	940	938.7	941.8	939.2	936	930.1
CO (ppm)	17.8	17.5	21.1	12.37	17.58	14.75	9.26	13.6
CO_2 (ppm)	414.0	409.6	391	390.7	345.2	439.5	340.9	391.3
Waste weight (kg)	4	5	5	3	5	5	4	5
Ash Weight (kg)	0.32	0.5	0.21	0.25	0.21	0.6	0.3	0.9
Efficiency (%)	63.2	79.8	77.8	78.3	84.6	79.5	85.5	75.9

Varying concentration of $PM_{2.5}$ was observed in the study for different varieties corn straw. Concentration $PM_{2.5}$ from corn straw combustion in units of ug/m^3. Summary value $PM_{2.5}$ concentrations for corn residue (cobs and stover) for the five types of varieties is shown in Table 2.

Table 2. Concentration $PM_{2.5}$ of corn residue burning (mg/m^3)

	Corn cobs Avarage	Corn stovers Avarage
Bisma	9.187	2.060
P29	2.843	5.283
NK	7.409	4.048
Bisi 18	3.781	5.306
NW	1.895	5.697

Figure 3 illustrates the comparison value of $PM_{2.5}$ concentrations for both types of corn residue (corn cobs and stover) for the five types of varieties. From the figure 3, this asses result show that average concentration of $PM_{2.5}$ from the chamber combustion of five varieties of corncobs and cornstover was around 284.3 – 918.7 μg/m3 and 404.8 – 602.0 μg/m3 respectively. Level $PM_{2.5}$ concentration from Bisma varieties of corn cobs had the highest than the other relative low. Variation of this value is influenced by the efficiency of the combustion simulation and the characteristic of the straw [17]. The combustion efficiency of corn corb is lower than the combustion efficiency of stover that the concentration $PM_{2.5}$ more emitted from the burning of corn stover.

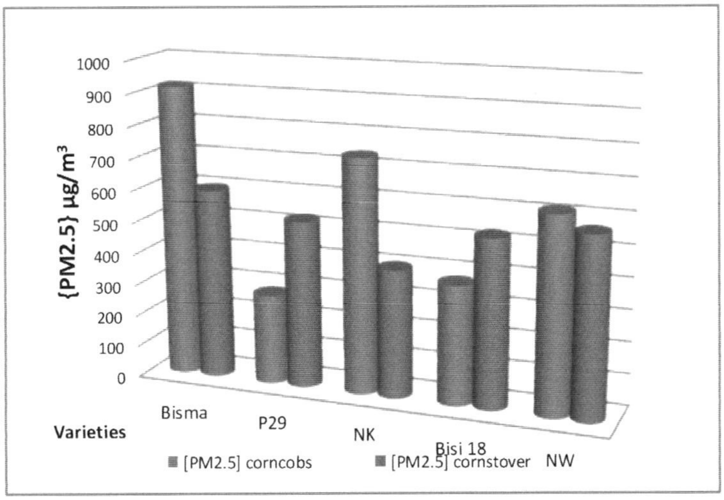

Figure 3. Level concentration of $PM_{2.5}$.

3.2. Analysis of $PM_{2.5}$ emission

Particulate matter 2.5 micrometer emitted as a smoldering combustion product. The emission for $PM_{2.5}$ observed in this study are slightly lower to those compiled by CAO (2008) [4] and Ni (2015). Concentration of $PM_{2.5}$ is highly variable and dependent on the fuel type (corn varieties) and burning behavior (combustion efficiency). While $PM_{2.5}$ has been established previously as a product of smoldering combustion [27]. Concentation of $PM_{2.5}$ decreases with increasing MCE with a linear fit of $[PM_{2.5}] = -6.0997\ (MCE) + 85.828$ with an R_2 value of 0.604. The inverse dependence of $PM_{2.5}$ on MCE in our study is likely due to different PM $_{2.5}$ content in the corn straw. For example, in our study, the straw type sampled (Bisma varieties) burned with the lowest MCE the concentration of $PM_{2.5}$ was highest. Figure 4 shows the instantaneous excess concentration of $PM_{2.5}$ as a function of combustion efficiency (MCE) for a single fire.

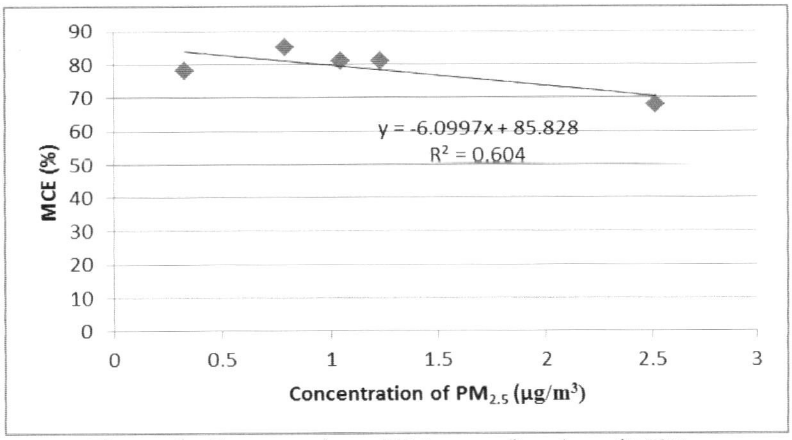

Figure 4. Concentration of $PM_{2.5}$ as a function of MCE.

From this figure it is obvious that $PM_{2.5}$ is undirectly correlated to combustien efficiency indicating that $PM_{2.5}$ is a smoldering combustion product [27]. A possible explanation for the loop trajectory of $PM_{2.5}$ as a function of combustion efficiency observed in Figure 1 is that at the beginning of flaming combustion, and completely, then as smoldering combustion diminishes the higher $PM_{2.5}$ concenstration of cornburning.

Comparison with field measurements of concentration for conrcob and corn stroves burning [3]. Result of comparison were presented in Table 3.

Table 3. Comparison of chamber burning with field measurement

	Chamber buning		Field burning	
$PM_{2.5}$ Concentration	Corn cobs average	Corn stovers average	Corn cobs average	Corn stovers average
Bisma	9.187	2.060	6.48	5.23
P29	2.843	5.283	2.85	1.05
NK	7.409	4.048	6.42	0.33
Bisi 18	3.781	5.306	2.93	1.23
NW	1.895	5.697	1.67	0.79

There has been concentration previously measured in the field for $PM_{2.5}$, for five varieties. We determined the average concentration $PM_{2.5}$ for all varieties of corn cobs and corn stover as well as the averages for all corn straw collected from Banyuresmi, disticts Garut. The average corn burning MCE for these two data sets presented the emission are in general higher for our laboratory data with chamber combustion than field burning.

4. Conclusion

- There are different concentration between five varieties of corn straw.
- This asses result show that average concentration of $PM_{2.5}$ from the chamber combustion of five varieties of corncobs and cornstover was around 284.3 – 918.7 μg/m^3 and 404.8 – 602.0 μg/m^3 respectively.
- Variations of this value are influenced by the combustion efficiency of the combustion simulation and the corn residue charactersistic.
- Emision $PM_{2.5}$ from straw burning on chamber are higher than field burning.
- The results of this study presented provide useful information for the development of local emission factors for $PM_{2,5}$ from open burning of corn straw

References

[1] Permadi D A 2013. Assesment of biomass open burning emissions Indonesia and potential climate forcing impact *Atmos. Environ.* **28** 250-258
[2] Andreae and Merlet P 2001. Emissions of trace gases and aerosols from biomass burning *Global Biogeochem. Cycles* **15** 955-966
[3] Bice K 2009 *Woodrow Wilson School of Public and International Affairs* (New Jersey: Princenton University)
[4] Cao G L, Zhang X Y, Wang, Y Q and Zhang F C 2008 Estimation of emission from field burning of crop strawa in China *Chinese Sci. Bull.* 53 784-790
[5] Christian T J, Kleiss B, Yokelson R J, Holzinger R, Crutzen P J, Hao W M, Saharjo B H and Ward D E 2003 *Geophys. Res.* **108** 4719-4732.
[6] Crutzen P J and Andreae M O 1990. *Chem. Biogeochem. Cycles Sci.* **250** 1669-1678
[7] Dhammapala R, Claiborn C, Simpson C and Jimenez J 2007 *Atmos. Environ.* **41** 1512-1520
[8] Environmental Protection Agency 2008 *Direct Emissions from Stationary Combustion Sources* (United States: Environmental Protection Agency)
[9] Garivait S and Coauthors 2013 *Monitoring and Assessment of Biomass Open Burning in Agricultural Area/Lands in Thailand* (Thailand: Pollution Control Department) p. 60
[10] Hayashi K, Ono K, Kajiura M, Sudo S, Yonemuro S, Fushimi A, Saitoh K, Fujitani Y and Tanabe K 2014 *Atmos. Environ.* **95** 36-44

[11] UNFCCC 2010 *Indonesia Second National Communication Under the United Nation Framework Convention on Climate Change* Jakarta
[12] Kanokkanjana, K, Cheewaphongphan, P and Garivait, S. B;ack Carbon Emission Paddy Field Open Burning in Thailand, 2nd International Confrence Environmental Science And Technology. IPCBEE Vol.6 2011.
[13] Kim Oanh N T, Thuy L B, Tipayarom D, Manandhar D R, Pongkiatkul P and Simpson C D 2011 *Atmos. Environ.* **456** 483-502
[14] Krejcie, Robert V, Daryle W and Morgan 1970 *Educ. Psychol. Meas.* **30** 607-610
[15] Lee Y C, Lam Y F, Khlmann G, Wenig M O and Chan K I 2013 *Atmos. Environ.* **80** 478-467
[16] Liousse C, Penner J E, Chuang C and Walton J J 1996 *J Geophys. Res.* **101** 19411–19432
[17] Munthe K K J 2011 *Swedish Environ. Res. Inst.*
[18] Ni H, Han Y, Cao J, Huang R and Li H 2015 *Atmos. Environ.* 1-4
[19] Seinfield, J.H. Atmospheric Chemistry and Physics of Air Pollution, New York : John Willey and Sons,
[20] Sanshis E, Ferrer M, Calvet S, Coscola C, Yusa C and Cambra-Lopez M 2014 *Atmos. Environ.* **98** 25-31
[21] Sillapapiromsuk S, Chantara S, Tengjaroenkul U and Prasitwattanaseree 2013 *Chemosphere*
[22] Singh N, Mittal S K, Agarwal R, Awasthi A and Gupta P K 2010 *Intl. J. Environ. Anal. Chem.* **90** 829–843
[23] Standard Practice for Proximate Analysis of Coal and Coke Active Standard ASTM D3172
[24] Tiparayom D and Kim Oanh N T 2007 *J. Sci. Asia* **33** 339-345
[25] Tumuluru J S, Kremer T, Wright C T and Boardman R D 2012 Proximate and ultimate compositional changes in corn stover during torrrefaction using thermogravimetric analyzer and microwaves the Ohio State University Columbus USA
[26] Watson J, Chow, Chen and Cample 2011 *Total Environ.* **109** 2384-2396
[27] Yokelson R J, Kart T, Artaxo P, Blake D R, Christian T J, Groffith D W T, Guenther A and Hao W 2008 *Atmos. Chem. Phys.* **7** 5175-5196.
[28] Zhang Y, Obrist D, Zielinska B and Gertler A 2013 *Atmos. Environ.* **72** 27-35

IOP Conf. Series: Materials Science and Engineering **345** (2018) 012049 doi:10.1088/1757-899X/345/1/012049

Synthesis of biodegradable plastic from tapioca with N-Isopropylacrylamid and chitosan using glycerol as plasticizer

Syaubari, S Safwani and M Riza
Chemical Engineering Department, Engineering Faculty of Syiah Kuala University - Darussalam, Banda Aceh

E-mail: medyan_riza@yahoo.com

Abstract. One of natural polymers that can be used as raw material in the manufacture of biodegradable plastic is tapioca and chitosan. The addition of other compounds such as glycerol as plasticizer is to improve the characteristics of the plastic that already produced. N-Isopropylacrylamid (NIPAm) is an organic compound that can be synthesized into a polymer or polymer grafting which also biodegradable too. This research aims tostudy the synthesis of biodegradable plastics from tapioca with the addition of chitosan, NIPAm, poly(NIPAm) and analyze the characteristics of biodegradable plastics that already produced. This research was done in three stages, there are (1) polymerization NIPAm, (2) the grafting of chitosan-poly NIPAm and (3) the synthesis of biodegradable plastics from starch mixture with variation of addition chitosan, NIPAm, poly(NIPAm), chitosan-graft-poly(NIPAm) and also variations of glycerol as plasticizer. The results of this research is a thin sheet of plastic which is will get analyzed for the characteristics of functional groups, mechanical, morphological and its biodegradability. FTIR spectra showed the grafting process with the new group formation of CO single-bond at 850 cm^{-1}. Plastic with the addition of NIPAm and 1 ml glycerol has the highest tensile strength value about 31.1 MPa. Plastic with poly(NIPAm) and 4 ml glycerol produces the highest elongation value about 153.72%. Plastic with Chitosan-graft-poly(NIPAm) with 1 ml glycerol has the longest biodegradation because of the small mass-loss for six weeks which is about 6.6%.

1. Introduction

The using of plastics as food packaging has increased along with the increase of population and human needs. This is because plastic has many advantages. Some plastic advantages are include having good mechanical properties, a good barrier against water and air, cheaper price, light weight material compared to the other materials, sheet-formed so that it can be made into bags, and the ease of the process and its application. However, plastic also has a disadvantage because it is hard to degradable in a long term, so it will cause environmental pollution and environmental damage. Therefore, it needs a package that can be degradable naturally by microorganism or biodegradable. Some natural biopolymer which could be synthesized into a plastic sheet are starch, protein, cellulose, chitosan, etc.

Tapioca is a cassava starch which has been widely used as raw material of biodegradable plastic and showed a promising result [1]. To improve the characteristic of biodegradable plastic needed some other materials such as plasticizers. Plasticizers is an organic material which has a low molecular weight which is added to weaken the inflexible character of the polymer, increased the flexible and extensibility of the polymer. Plasticizer is dissolved in each polymer chains so it will facilitate the polymer molecule movements and makes the glass transition temperature lower, crystalized temperature or melting temperature from polymer. The purposed of the addition of plasticizer is to producea film that is more flexible and not brittle [2].

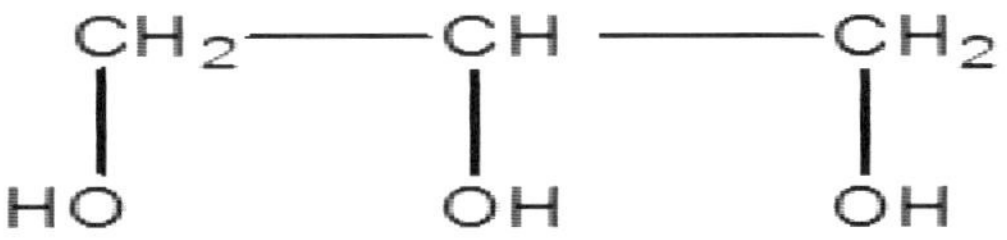

Figure 1. Structure of the glycerol molecule.

N-Isopropylacrylamid (NIPAm) is one of acrylamide's monomer which is white, odorless, crystalline solid that is highly soluble in water and easily to react through amide reaction or double bond. These compounds can form long polymer chains are called poly(N-isopropylacrylamid) or poly(NIPAm). Poly(NIPAm) is a thermosensitive material that can form a hydrogel. Applications poly(NIPAm) has been widely applied in various fields, such as in wastewater treatment, thickeners, papermaking, iron ore processing or to form other compounds. The main weakness of hydrogel homopolymerpoly (NIPAm) is having a relatively low mechanical properties and fragile, so the development of applications is limited [3]. Considering that the last few years of intensive research is being done to modify poly(NIPAm) [4]. McDowall et al. (1984) suggested that chemical modification through graft copolymerization (grafting) with various synthetic monomers are known to improve the properties such as the ability to absorb water, elasticity, ion exchange capability, thermal resistance, and resistance to microbial attack.

The purpose of this research is to study the synthesis of biodegradable plastics from the main raw material starch and chitosan were added to the treatment variations of NIPAm, poly(NIPAm) and grafting of chitosan-poly(NIPAm) using glycerol as a plasticizer and see its effect on the functional groups are formed, mechanical properties, morphology, and biodegradability of plastic produced.

2. Materials and methods

The materials that is needed in this study are tapioca, chitosan, 2% acetic acid, glycerol, ethanol, monomer NIPAm, NaOH, azobisisobutyronitrile initiator (AIBN), mercaptopropionic acid initiator (MPA), N activator, N-Dicyclohexylcarbomide (DCC) and distilled water.

2.1 Nipam polymerization into poly (Nipam)

Polymerization NIPAm made by dissolving 10 grams NIPAm in 20 ml of water. Then added 0.1 grams of AIBN and 10 ml of MPA and stir until dissolved. The mixture precipitated in ethanol at 60 °C for 20 hours. The precipitate obtained was dried in an oven for 12 hours at a temperature of 80 °C.

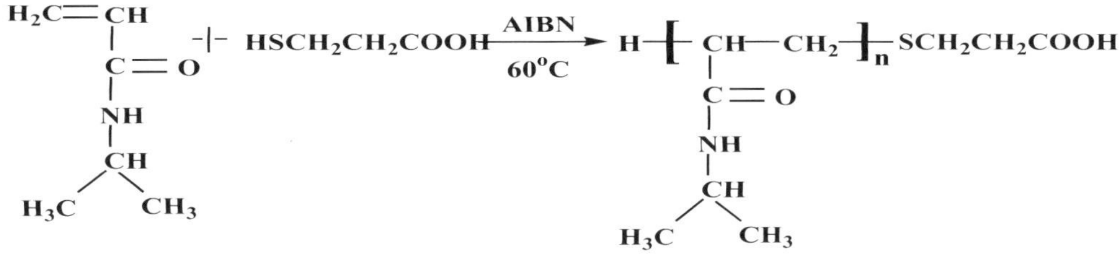

Figure 2. The polymerization reactions NIPAm into poly(NIPAm).

2.2 Grafting Chitosan and Poly (NIPAm)

Synthesis grafting of poly(NIPAm) with chitosan to form khitosan- graft- poly(NIPAm) by mixing a solution of chitosan and poly solution (NIPAm) was then added NaOH to pH 6.5. After that DCC was added step by step for 24 hours. Then the solution is precipitated in ethanol for 24 hours, then filtered the sediment. The precipitate that formed was washed with Tris HCl and distilled water and then dried at 80 °C for 5 hours.

2.3 Synthesis of biodegradable plastic

Synthesis of biodegradable plastics made by blending method that is mean to mix the material with variation 1) tapioca and chitosan, 2) tapioca, chitosan and NIPAm, 3) tapioca, chitosan and poly (NIPAm) and 4) tapioca and chitosan- graft- poly(NIPAm). The mixture was heated with stirring using hot plate and magnetic stirrer until form a paste (gelatinization) on temperature 70 ° C. Then each solution of starch paste is added at a concentration of glycerol as plasticizer 1ml;2 ml;3 ml and 4 ml. After that, the casting procedure in hot conditions with pouring material on the surface of polyethylene evenly and dried at room temperature for 24 hours.

Table 1. Study design synthesis of biodegradable plastics from tapioca and variations NIPAm treatment with the addition of glycerol as plasticizer

Treatment variation of NIPAm on plastic material (A)	**Gliserol (B)**			
	1 ml	2 ml	3 ml	4 ml
1. Tapioca+ chitosan	A1B1	A1B2	A1B3	A1B4
2. Tapioca + chitosan + NIPAm	A2B1	A2B2	A2B3	A2B4
3. Tapioca + chitosan + polyNIPAm	A3B1	A3B2	A3B3	A3B4
4. Tapioca+ chitosan-*graft*-polyNIPAm	A4BI	A4B2	A4B3	A4B4

2.4 Analysis of materials

The biodegradable plastic that was formed will be to be analyzed to identify the compound that contained in that plastic using a Fourier Transform Infra-Red (FTIR) brand Shimadzu in Instrument Analysis Laboratory, Department of Chemical Engineering Unsyiah and analysis of mechanical properties such as tensile strength and elongation using autograph shimadzu based on ASTM D638 in Instrumentation Laboratory, Department of Physics Unsyiah. Morphological analysis was conducted to study the surface structure of biodegradable plastics using Scanning Electron Microscope (SEM) in the Corrosion Laboratory, Department of Mechanical Engineering Unsyiah. Biodegradability analysis quantitatively by measuring the percentage of plastic weight loss after six weeks.

IOP Conf. Series: Materials Science and Engineering **345** (2018) 012049 doi:10.1088/1757-899X/345/1/012049

3. Results and discussion

3.1 Characteristics of functional groups

FTIR spectra at all biodegradable plastics showed an absorption in the range 3500-3300 cm^{-1} with the presence of -NH on chitosan which is reinforced by -OH groups and C-H on tapioca. The addition of NIPAm will presence the absorption of C = O at 1760 cm^{-1} and C = C at 1690 cm^{-1}. The presence of polymerization NIPAm into poly (NIPAm) is shown with the release of the C = C group into clusters C-C on absorption area at 1000-500 cm^{-1}. For the presence of characteristics of grafting is shown by the functional absorption of new single bond of C-O in the region 1000-750 cm^{-1} (Figure 3).

Table 2. Analysis of functional groups for biodegradable plastics from tapioca with treatment variations NIPAm and glycerol as plasticizer

Composition of plastic material	Functional group	Spectrum number (cm^{-1})	Compounds
1. Tapioca+chitosan	N-H	3500	Chitosan
	C-H	2850	Tapioca
	O-H	3600	Tapioca
2. Tapioca+chitosan+NIPAm	N-H	3500	Chitosan
	C-H	2850	Tapioca
	O-H	3600	Tapioca
	C=O	1760	NIPAm
	C=C	1690	NIPAm
3. Tapioca+chitosan+poly(NIPAm)	N-H	3500	Chitosan
	C-H	2850	Tapioca
	O-H	3600	Tapioca
	C=O	1760	NIPAm
	C-C	1200	Poly(NIPAm)
4. Tapioca+chitosan-*graft*-poly(NIPAm)	N-H	3500	Chitosan
	C-H	2850	Tapioca
	O-H	3600	Tapioca
	C=O	1760	NIPAm
	C-C	1200	Poly(NIPAm)
	C-O	850	*Grafting* chitosan-NIPAm

The amount of yield that was formed during the grafting process is influenced by a variety of variables including monomer concentration, activator concentration, reaction time and reaction temperature [5]. In this grafting process, the yield of chitosan- graft -poly (NIPAm) was obtained for 12.76%.

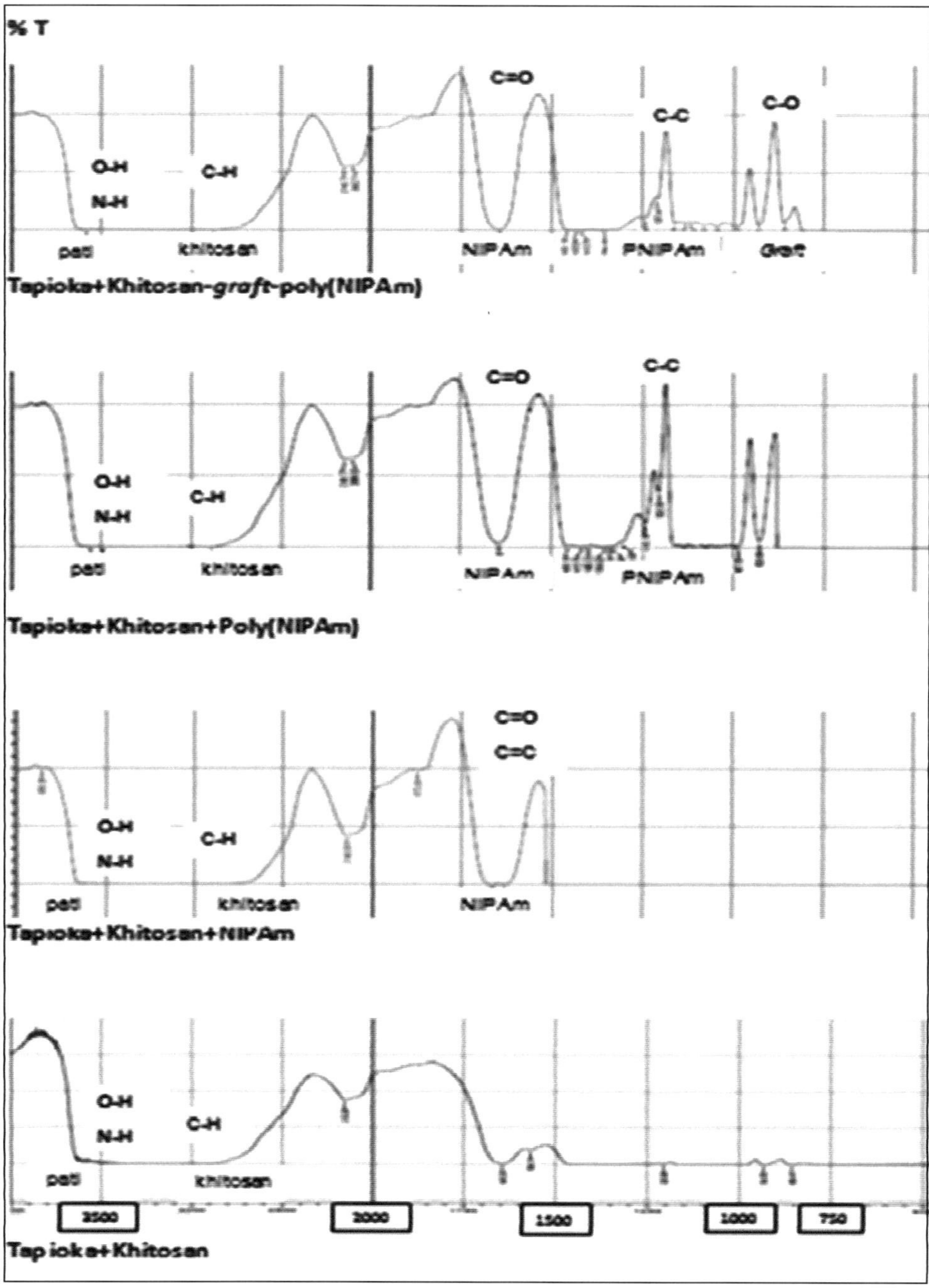

Figure 3. FTIR PB spectra with treatment variations of PNIPAm and the addition of glycerol as plasticizer.

Grafting occurs through the hydroxyl group at C6. Hydroxyl groups at C2 and C3 are hydroxyl groups attached to the secondary carbon atom, for the hydroxyl group at C6 are attached to the primary carbon atom. Reactivity and acidity of primary and secondary hydroxyl group is different. Besides, the modified polysaccharide can produce macromolecular structured products such as gel or hydrogel, polymer resins, membranes or composite materials that can be applied as a separator material in separation technology [6].

3.2 Mechanical characteristics

Tensile strength is the maximum tensile force that can be held by the film during the measurement takes place. And then the elongation percent is a change in the maximum length of the film before it is broken.

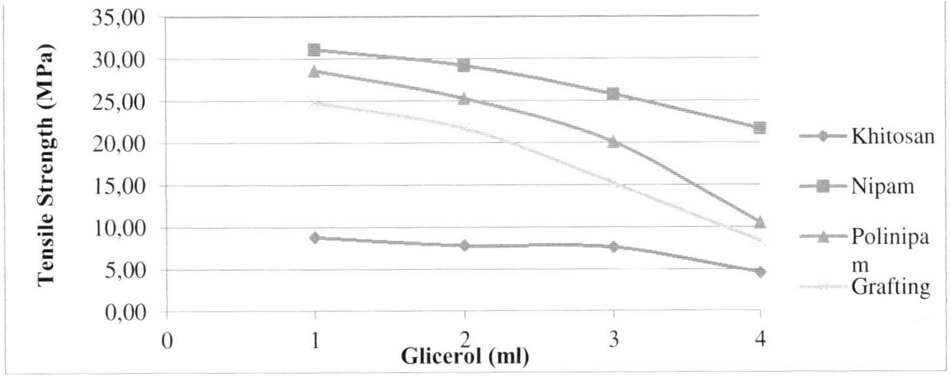

Figure 4. The effect of glycerol concentration and treatment variations NIPAm against tensile strength.

Figure 4 shows the effect of glycerol concentration and the addition of NIPAm treatment variations against the biodegradable plastic tensile strength value. In the picture, we can note that the higher the concentration of glycerol, the value of the tensile strength decreased. However, the higher concentration of glycerol will provide an increasing elongation values as shown in Figure 5.

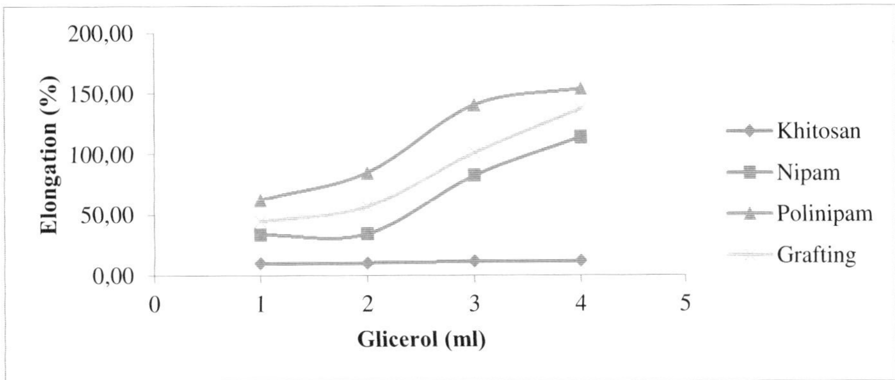

Figure 5. The effect of glycerol concentration and treatment variations NIPAm against elongation.

Research by using chitosan with low molecular weight which is dissolved in 1% acetic acid at a concentration of 3% by weight of chitosan / volume of solution, found that the more the concentration of glycerol were added to the chitosan solution will produce the film with the smaller tensile strength [7].

The addition of plasticizer in the polymer solution will form hydrogen bonds between molecules that cause the interaction between polymer molecules become less so facilitate the movement of polymer molecules. This is case that leads reduced tensile strength of plastic with the addition of plasticizers.

Plastics with the addition NIPAm are hard and stiff because they contain little polymer compound, thus requiring a large force to broke it and less elastic. From the graphic it can be seen that the plastic with the addition of NIPAm has the highest tensile strength value, but has a lower elongation value from plastics with the addition of poly (NIPAm) and plastics with the addition of chitosan-graft-poly (NIPAm).

Grafting will produce a branching structure that lowers the crystallinity. Irregularities in the structure of the polymer chain, for example branching, will inhibit the chain to be able to interact with each other, and therefore the crystallinity be limited [8]. Polymer which has a higher crystallinity generally harder, stronger, more turbid, more resistant to solvents, and its density is higher than amorphous polymers [9]. Therefore, plastics with the addition of poly(NIPAm) has a tensile strength greater than the plastic with the addition of chitosan-graft-poly(NIPAm) because plastic with poly(NIPAm) will formed a more crystalline structure. While plastics with the addition of chitosan-graft-poly (NIPAm) will formed a more amorphous structure.

3.3 Biodegradability

The simplest quantitative methods to characterize the occurrence of biodegradation of a polymer is to determine the mass loss of the polymer material [10]. Mass loss is determined by weighing the mass of the polymer before and after biodegradation process for a certain time. The greater the mass loss plastic means the faster degradable plastic.

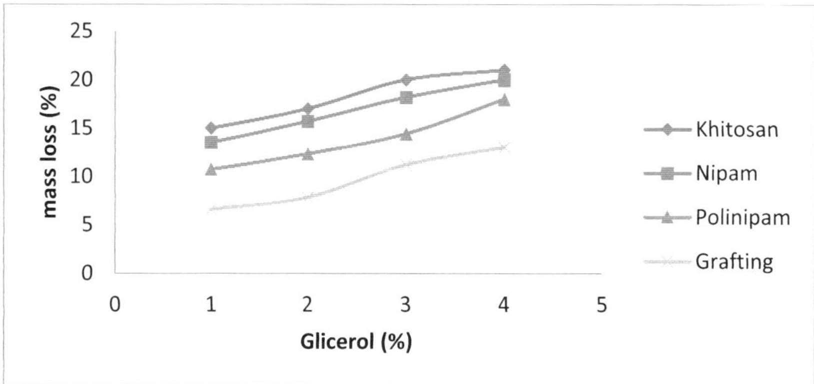

Figure 6. The effect of glycerol concentration and treatment variations of NIPAm against of weight loss from biodegradable plastic from tapioca for 6 weeks.

Biodegradable plastic mass loss increased with increasing concentrations of glycerol. It shows that the high glycerol cause more plastic hydrophilic because of glycerol and starch having an OH group which can initiate the hydrolysis reaction, so that the polymer into small pieces [11] Plastics with the addition of chitosan-graft-poly(NIPAm) has the smallest mass loss. This suggests that the grafting of chitosan and poly(NIPAm) can improve the biodegradability of plastic as a branching structure that forms require more enzymes to decide. [12] also added that polymer degradation is influenced by the composition, polymer properties such as molecular weight, melting temperature, crystallinity and the transition temperature, the more hydrophilic, the faster degradable polymer.

4. Conclusions

From the research results above, it can be concluded that:

1. The process of grafting of chitosan and poly (NIPAm) indicated by the uptake of new functional groups on a single C-O at 1000-750 cm $^{-1}$ with a yield of 12.27%.
2. The higher the concentration of glycerol, then the value of the tensile strength (tensile strength) plastic lower, but the elongation value will be higher.
3. Within six weeks, plastics with high glycerol content will have a greater mass loss so more quickly degraded.
4. Plastics with the addition NIPAm is hard so it has the highest tensile strength values but it has a low elongation value and rapid degradation.
5. Plastics with the addition of poly (NIPAm) has the highest elongation value and the longer degradation of plastic with the addition NIPAm
6. Plastics with the addition of chitosan-graft-poly (NIPAm) has the lower tensile strength and elongation compared to plastic with theaddition of poly (NIPAm) which is the longest to degraded
7. In general, the addition NIPAm, poly (NIPAm) and chitosan-graft-poly (NIPAm) in the synthesis of biodegradable plastics from tapioca can improve the mechanical characteristics and biodegradability of plastic instead of plastic produced by the addition of chitosan itself.

References

[1] Indrasti S, Nasti, Yuliani S and Permatasari C 1999 Aplikasi edible coating dari pektin pod coklat pada penyimpanan tomat suhu dingin dan suhu kamar *J. Tek. Ind. Pert.* **9** (1) 20-30
[2] Erizal 2010 Sintesis dan karakterisasi hidrogel superabsorben poliakrilamida (PAAM) berikatan silang – karagenan hasil iradiasi gamma *Indo. J. Chem.* **10** (1) 12-19
[3] Khan G M A, Shaheruzzaman M D, Rahman M H, Razzaque S M 2009 Surface modification of okra bast fiber and its physico-chemical characteristics *Fiber Polym.* **10** 65-70
[4] Crini G 2005 Recent development in polysaccharide-based materials used as adsorbent in wastewater treatment *Prog. Polym. Sci.* **30** 38-70
[5] Butler B L, Vergano P J, Testin R F, Bunn J M and Wiles J L 1996 Mechanical and barrier properties of edible chitosan films as affected by composition and storage *J. Food Sci.* **61** (5) 953-956
[6] Stevens and Malcolm P 2001 *Polymer Chemistry: An Introduction* (Oxford University Press, Inc.)
[7] Owen S, Masaoka M, Kawamura R and Sakota N 1995 *Degradable Polymers Recycling, and Plastics Waste Management*, ed A C Albertsson and S J Huang (New York: Marcel Dekker Inc.) p. 81-85